Long-wave Optics

Long-wave Optics

The Science and Technology of Infrared and Near-millimetre Waves

GEORGE W. CHANTRY

National Physical Laboratory, Teddington, Middlesex, United Kingdom

Volume 1: Principles

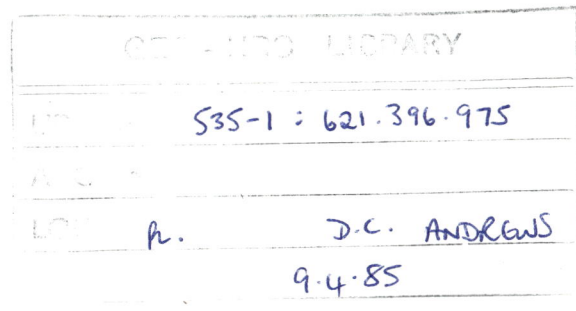
1984

ACADEMIC PRESS

(Harcourt Brace Jovanovich, Publishers)

London   Orlando   San Diego   San Francisco   New York
Toronto   Montreal   Sydney   Tokyo   São Paulo

ACADEMIC PRESS INC. (LONDON) LTD
24/28 Oval Road,
London NW1

United States Edition published by
ACADEMIC PRESS INC.
(Harcourt Brace Jovanovich, Inc.)
Orlando, Florida 32887

British Library Cataloguing in Publication Data
Chantry, G. W.
  Long-wave optics.
  Vol. 1: Principles
  1. Infra-red rays
  I. Title
  535'.012   QC547

  ISBN 0-12-168101-7

Typeset by Macmillan India Ltd., Bangalore
and printed in Great Britain by
Thomson Litho Ltd., East Kilbride, Scotland

To my wife Diana and my children Richard, Catherine and Paul whose vigorous protests at being deprived of my time and company did much to expedite the completion of this book.

Preface

Infrared science and technology is far from being a new field of endeavour, indeed it has a venerable history stretching back almost to the middle of the last century. What is new is the integration of the infrared into the general area of electromagnetic physics and engineering and the deployment of infrared devices into a bewildering and rapidly increasing number of technical areas. As some token of this, one could say with fair accuracy that nearly all the significant work in the field has been done in the last twenty years or so and that most of the books devoted to the art, written more than ten years or so ago, are quite out of date. This state of affairs is true, of course, of technology as a whole and some areas, such as semi-conductor device fabrication, have had even shorter periods before obsolescence set in, but the rate of change in infrared technology is faster than most.

This rapid rate of advance has been fired undoubtedly by the evident fact that infrared devices provide the best solutions for a huge range of technically important problems, but this "market need" would have gone unrequited were it not for the "technology push" provided by the development of some radically new components. Principal amongst these were infrared lasers and infrared detectors with sensitivities approaching the fundamental limits. The lasers gave the infrared engineer coherent radiation and opened up to him all the possibilities which had been enjoyed for so long by his microwave colleagues. The detectors made possible new applications of incoherent radiation and the interaction between these two areas, by a "knock-on" effect, led to the opening up of quite novel areas, for example ultra-high resolution spectroscopy and high-definition infrared imaging. Infrared devices are now used almost across the board: examples would include, analysis, pollution monitoring, imaging, movement sensing, espionage, medicine, machining, materials fabrication, thermonuclear research, telecommunications and warfare. In this latter, heat-seeking missiles have repeatedly proven that they are the most deadly and effective air-to-air weapons available.

The subject is now clearly far too large to be covered in adequate depth in a single volume and such a treatment would anyway be beyond the reach of a single author no matter how diligent. What is attempted here is an overview which hopefully covers all the important topics in sufficient depth to enable the interested reader to continue his or her reading in the learned literature. To this end the book includes a general bibliography and also a particular bibliography which includes over 1500 references.

In writing the book I have received help and information from a huge number of people spread all over the world. To list them all would be an impossible task. All I can do is to mention those who have contributed individual points. I therefore thank Mohammed Afsar, John Baker, Andre Bellemans, Tom Blaney, David Buckingham, Ken Button, James Calderwood, the late John Chamberlain, Alan Costley, Mike Cudby, Myron Evans, John Fleming, Werner Frank, Alistair Gebbie, Ludwig Genzel, Bob Gott, Jose Goulon, Armand Hadni, John Harries, Dean Hodges, Harry Jones, Udo Kaatze, Fritz Kneubuhl, David Knight, Derek Martin, Jim McConnell, Elisabeth Nicol, Takeshi Oka, Terry Parker, Roger Partridge, Carl Pidgeon, Reinhard Pottel, Ernest Putley, Jean Louis Rivail, Joseph Sattler, George Simonis, Des Smith, the late Robin Smith, Tony Stradling, Wally Stone, Mary Tobin, Brian Walker, Harry Willis and David Whiffen. Especially though I would like to thank Dr James R. Birch of the National Physical Laboratory and Professor George R. Wilkinson of King's College London who read the manuscript critically in its early drafts and made countless suggestions for improvements and for the removal of obscurities.

December 1983   G. W. CHANTRY

Contents

Contents of Volume 2

Applications

Chapter 1
General Introduction and Mathematical Background

1.1 Historical introduction

Although Newton had discovered the composite nature of white light by 1704 and had thereby erected, virtually single-handed, the entire basis of spectroscopy, it was not realised for almost another century that the solar spectrum contained radiation lying at longer and shorter wavelengths. The discovery of the infrared was made by Sir William Herschel in 1800. He was investigating the heating effect of the various wavelengths in a dispersed spectrum, using sensitive thermometers as monitors, and found that the greatest heating effect occurred just beyond the red end of the spectrum. Herschel went on to investigate the properties of this invisible radiation and showed that it obeyed the laws of reflection and refraction which had been established for the visible radiation. His son Sir John Herschel was, in 1840, the first to discover absorption bands in the infrared. He soaked a piece of black paper in alcohol and placed the paper in the infrared region of a dispersed solar spectrum. The alcohol was found to evaporate more rapidly from some regions of the paper than from others and as the paper dried, a crude visualisation of the variation of infrared power with wavelength was obtained. These variations are now known to be due to water vapour and carbon dioxide in the Earth's atmosphere which selectively absorb the solar infrared radiation. Although he could hardly have realised it, it is not too fanciful to ascribe the discovery of the near-infrared atmospheric "windows" to Sir John Herschel.

In the period following 1830 steady advances occurred in the field of infrared physics. This was because the development of new, much more sensitive detectors facilitated experimental work and because the growth of theoretical understanding helped to guide the experimentalists and provided a basis for the interpretation of their observations. The discovery of the thermoelectric effect by Seebeck in 1826 was soon followed in 1830 by Nobili's invention of the radiation thermocouple. The sensitive element in a thermocouple is the union of two dissimilar metals—usually antimony and bismuth—and when this junction is at a different temperature to that of the free ends, a potential difference is set up which can be detected by a Wheatstone bridge or other means. The potential difference to be measured is rather small but can be readily increased—provided the radiation beam is wide enough—by using several thermocouples in series. The junctions are then clustered together—but insulated from one another—to make a larger but much more sensitive element, a thermopile. The first thermopile was made in 1833 by Melloni. A difficulty with thermopiles is that the metal wires should ideally be poor conductors of heat and good conductors of electricity and these twin requirements are not readily met. Nevertheless the thermopile was the dominant detector for nearly fifty years, until the invention of the bolometer by Langley in 1880 presented it with a serious challenge. The bolometer, just like the thermocouple, depends upon physical changes brought about by the heating effect of infrared radiation. In the case of the bolometer the change is in the resistance of the sensitive element. Langley's bolometers were much more sensitive than the thermopiles then available and with their boost he was able to use the rather inefficient gratings, which were the best that could be made at that time, to study the solar spectrum out to 18 μm. Since Langley's time, first the bolometer, then improved thermopiles jostled for pre-eminence but eventually, with the introduction of cryogenically cooled bolometers in the last decade, the question was finally settled in favour of the bolometer as the ultimate in sensitivity. Thermopiles are, however, still commonly used in commercial instruments because of their simplicity, ruggedness and room temperature operation but they are progressively encountering stiff competition from other room-temperature thermal detectors such as the Golay cell and the pyroelectric detector.

The use of photodetectors in the infrared was very slow to develop. Becquerel showed in 1843 that very near infrared (< 1 μm) radiation could cause phosphorescent and photographic effects and Abney in 1880 produced photographic plates that could be used out to 0·98 μm. With the development of emulsions specially sensitised by cyanine dyes, infrared photography was extended in wavelength to 1·3 μm by Merrill in 1918 but there is little likelihood of any significant further extension. The first solid-state photodetector (based on thallium sulphide and introduced by Case in 1920) was likewise limited to the very near infrared ($\lambda < 1·1$ μm), but with the rapid modern development of semiconductor technology, solid-state photodetectors are now available out to the limits of the far infrared. Some highlights in the development of photodetectors were the introduction of the Ag–O–Cs photocathode by Koller in 1929, the development of the lead sulphide, lead selenide and lead telluride photoconductive cells in Germany during the Second World War, the introduction of the gold-doped germanium extrinsic photodetector by Lasser, Cholet and Wurst in 1958 and the introduction of the variable band-gap photodetectors based on mercury cadmium telluride by Lawson in 1958.

In the early days of infrared physics, the device which finally produced an analogue record of the infrared intensity was a galvanometer fed by the electrical signals from a detector/Wheatstone bridge combination. This system is very prone to trouble from stray DC currents and there are fundamental limitations on the sensitivity, which prevent the realisation of the ultimate Johnson noise limit. A major advance came in the 1930s when Lehrer in Germany and Firestone in America introduced "chopping" of the radiation and detection at the modulation frequency. When this technique was supplemented in the early 1940s by the invention of tuned electronic amplifiers great improvements in the signal-to-noise ratio of the resulting spectra were obtained. In the 1960s, especially with the help of solid-state electronics, came the introduction of lock-in or phase-sensitive amplifier/demodulator units which gave still further improvements in quality.

The development of infrared technology has always been dependent upon the availability of suitable materials. The most ubiquitous of these has always been rock-salt, i.e. crystalline sodium chloride, which occurs naturally as fairly large crystals but which can nowadays be made synthetically as crystals of almost arbitrarily large size. The infrared transmitting qualities of rock-salt were apparently known to Leslie in 1804 who remarked that it was not opaque like glass in the mid infrared. By 1859 Muller had extrapolated the dispersion laws for rock-salt to $1.9\,\mu m$ and in 1868 this was extended still further, out to $7\,\mu m$, by Desains and Curie. In 1880 Ketteler published some accurate values for the variation of refractive index of rock-salt with wavelength. In the year 1889, Ångstrom used a rock-salt prism to obtain infrared spectra of simple gases and by 1894 Rubens had studied the optical properties of rock-salt out to $18\,\mu m$. Since that time many other alkali halides have come into use as infrared materials and exotics such as KRS-5 and KRS-6 (mixed thallium bromo-halides) have been introduced as high refractive index materials for the mid infrared. Nevertheless the range of infrared materials is still not sufficient to meet all requirements.

The growth of infrared theory began in the late eighteenth century essentially with Prevost's Theory of Exchanges. According to this suggestion, the energy radiated by a body is determined solely by the condition of that body and is not in any way determined by the condition of its surroundings. At the time this was considered a revolutionary theory for it seemed obvious to common sense that cold bodies radiated cold just as hot bodies radiated heat. In 1833 Richie showed experimentally that good absorbers of radiation were also good emitters and in 1859 this observation was elevated to a law of physics (Kirchhoff's law) by Kirchhoff and also independently by Steward. The theory advanced significantly in the period from 1879 to 1884 when, following the experiments of Tyndall, first Stefan and then Boltzmann showed that the total power radiated from a black body was proportional to the fourth power of the temperature. This Stefan–Boltzmann law was finally derived from a microscopic theory in the first decade of the present century by Planck. Early infrared determinations of the spectra of gases and vapours showed unmistakeable evidence that the absorption bands had discrete fine structure. By the early years of this century many infrared absorption bands had been resolved. Examples which might be quoted were the resolution of the HCl fundamental by Imes in 1919 and the observation of pure rotational fine structure in the far infrared by Czerny in 1925. Attempts to explain this discrete structure proved crucial to the development of quantum theory.

Herschel senior had apparently assumed that the infrared rays were of the same physical essence as the visible rays and this assumption was supported by Melloni, Ampere and Herschel junior. The matter was not, however, generally accepted and the wave nature of infrared rays was not finally settled until 1847 when Fizeau and Foucault showed that interference effects could be demonstrated in the infrared. Similar results were obtained by Knoblauch a year later. In 1862 came one of the intellectual triumphs of the nineteenth century— Maxwell's theory of the electromagnetic field. One prediction of this theory was that the dielectric and optical parameters of a medium were simply related. This prediction is not readily confirmed at visible wavelengths because of the great disparity in frequency between the visible and the electrical regions of the spectrum. However, as the frequency disparity is reduced, i.e. as one goes to longer and longer infrared wavelengths, the optical behaviour of materials comes more and more to be treatable by relativele simple approximations of Maxwell's field equations. In particular, the restrahlen—or residual rays— discovered by Rubens and his colleagues are readily explained. From these experiments and those of Hertz in 1887 it became clear that electromagnetic waves were available with wavelengths ranging from the indefinitely large down to the vanishingly small.

The use of infrared spectroscopy as a tool for diagnosing molecular structure stems from the work of Abney and Festings in the 1880s on the near infrared absorption spectra of organic compounds. Julius followed this up in 1892 by showing that all compounds containing a methyl group showed an absorption band at $3.45\,\mu m$. Coblentz in the first years of this century carried out his classic experiments culminating in his publishing a catalogue of the infrared spectra of numerous compounds out to $15\,\mu m$. During this period infrared science was dividing into two separate disciplines, infrared spectroscopy and infrared physics, the former, principally of interest to chemists, was concerned with the question of what molecular properties governed the positions and intensities of absorption bands and the latter, principally of interest to the physicists, was concerned with the generation, transmission and detection of electromagnetic radiation having wavelengths between 0.7 and $5000\,\mu m$. This division exists to this day but fortunately there is a good dialogue between the two sets of practitioners and the constant cross fertilisation ensures the steady advance of infrared science.

The serious study of infrared dispersion is very much a twentieth-century interest, but the theoretical tools which are used have their roots in the nineteenth century work of H. A. Lorentz (1853–1928). Contributions came also from Sellmeier (1871), Helmholtz (1875), Ketteler (1885) and Drude (1890s). Dispersion has always been of interest to practical spectroscopists since it is the variations of refractive index with wavelength that on the one hand make prism spectroscopy possible, and on the other make the construction of achromatic lenses difficult. The possibilities of dispersion studies as a discipline in their own right was, however, only realised in the late 1920s with the work of H. A. Kramers. Today analysis of complex refraction and reflection by means of the powerful Kramers–Kronig dispersion formulae is becoming more and more an important part of infrared science.

Early work in the application of infrared measurements to astrophysics, associated with the names of Draper, Fizeau, Lamansky and Abney, proved that the near infrared structure of the solar spectrum which Herschel had observed was due to absorption in the earth's atmosphere by water vapour and carbon dioxide. This work established the existence of near-infrared windows and hinted at the existence of a further window in the 10 μm region. Langley and Abbot did some measurements at moderate resolution on atmospheric water vapour and showed that the method could be used to determine the amount of water vapour along the absorbing path. As early as 1862, Rutherford reported absorption regions in the spectrum of Jupiter which did not correspond with those of the Earth's atmosphere. These features are now known to be due to methane and ammonia. Spectroscopy on stars and planets with usable resolution was not practicable before 1930, but in the preceding period, observers using the known laws of black-body emission, determined the temperatures of the planets Mars, Jupiter, Saturn and Uranus (Pettit and Nicholson) and many measurements were made of the photosphere temperatures of the brighter stars (Huggins, 1868; Coblentz 1922). In recent years with the development of ultrasensitive detectors and interferometric methods (Connes), ground-based measurements of stars and planets have given us a great deal of understanding of the molecular composition of the atmospheres of the planets and the cool stars. The introduction of satellite laboratories and the use of deep-space probes promise still more.

The development of the infrared spectrometer as an integrated automatically operating machine began in Germany in the 1920s. The German workers at the Ludwigshafen Laboratory of the BASF company had, by the late 1930s, developed automatic scanning instruments which featured modulated radiation, electronic amplification, varying slit width and presentation of the resulting spectra as percentage absorption versus linear wavelength on a chart recorder. Infrared technology was given an enormous boost by the 1939–1945 war because infrared devices were needed for night vision and range-finding purposes on the battlefield and sophisticated spectrometers were needed in the factories to help the war effort. After the war, British and American industry was not slow to incorporate the German technical innovations together with their own wartime advances into a wide range of commercial infrared spectrophotometers. In the immediate post-war period, the prism monochromator dominated the market, but by the middle 1960s grating instruments were becoming more and more commonplace and even the most inexpensive instruments tended to feature grating monochromators. Infrared technology has always benefited in this way from technical innovation developed for the adjoining regions. Thus the early development of infrared spectroscopy was greatly helped by the availability of sources such as the Nernst glower, the globar and the medium-pressure mercury-in-quartz lamps, all of which were originally introduced as commercial contenders in the field of electric illumination. In more recent times infrared practitioners have drawn heavily on the concepts and on the technology which were developed during the war years, 1939–1949, for microwave radar systems. The infrared has consequently tended to be the region where radically different techniques, introduced from outside, have jostled together and by a process of cross fertilisation produced new technical advances. Very recently this process has started to work the other way round and novel concepts introduced to deal with difficulties peculiar to the infrared have found applications in other spectral regions. An excellent example is provided by Fourier transform spectrometry (FTS) which was originally introduced to deal with the energy shortage problems endemic in working in the far-infrared but which nowadays is not only used in the mid and near infrared but is also finding uses in the visible and even the ultraviolet regions.

Fourier transform spectrometry is better in energy limited situations because of fundamental limitations to the operation of conventional spectrometers. These stem, as first shown by Jacquinot and Fellgett, from the poor use of radiant flux and of observing time in a dispersive spectrometer. Radiant flux can be transferred much more efficiently, normalised to some such criterion as resolving power, in an instrument (e.g. an interferometer) which has full cylindrical symmetry than in one (e.g. a grating spectrometer) which has merely two-fold symmetry. Interferometric spectrometers also operate in a multiplex mode which makes for efficient use of the observing time. The drawback to interferometric instruments is that they do not give the spectrum directly, rather they give its Fourier transform and a numerical or analogue inverse Fourier transformation is required to derive the desired spectrum. All of this was known to Michelson who in the early years of this century carried out monumental investigations using the interferometer which now bears his name but even a genius of his standing could not advance beyond the analysis of simple line spectra because the technology required to analyse the interferograms produced from broad-band sources simply did not exist. This technology, the high-speed digital computer, was developed rapidly in the 1950s and interferometric spectroscopy in the hands of the Connes, of Strong Vanasse, Gebbie, Genzel and Richards made rapid headway. Without this advance, it is difficult to see how much more could have been done in the far infrared than had already been achieved by Rubens. Commercial far-infrared interferometers began to appear in the 1960s but for these instruments where the element of fierce competition from conventional instruments is absent, only access to a general purpose computer is necessary and the basic hardware can be relatively cheap. However in the 1970s several manufacturers began to produce interferometric instruments for the mid- and near-infrared regions and they felt that their instruments should be able to compete in ease of operation with conventional instruments and should therefore feature their own built-in and dedicated computer which would not only do the Fourier transformation but would also control the interferometer and do all the data housekeeping. Not surprisingly such instruments are rather expensive. They differ in one major respect from the far-infrared versions in that they operate in a rapid scan mode using an air-bearing mirror movement. This device was originally developed for use on interplanetary probes in the days before sophisticated guidance was available. The probe simply crashed onto the planet being probed and it was necessary to get the spectrum of its atmosphere in a hurry! Since those days all the planetary missions have featured infrared instrumentation and there is little doubt that the enormous amount of manpower and money which has been devoted to the space programme has had several useful spin-offs in the field of infrared technology. Back on Earth, the increasing realisation in the 1970s of the threats to our atmosphere due to various forms of pollution led to the development of numerous infrared spectroscopic techniques for monitoring air quality. Hadamard transform and correlation spectrometers have been used together with a wide range of laser techniques for remote sensing.

The manufacturers of conventional dispersive instruments have not been slow to acknowledge the great benefits which accrue to interferometric spectrometers from the production of the final data in digital form and from an instrument which has a built-in computer ready and able to process the data further. The newest top-of-the-range dispersive spectrometers now feature built-in dedicated mini-computers and associated software which can not only process the output data but also control all the operations of the instrument. The costs of computer hardware are constantly falling and now that the manufacturers have to a large extent recovered the costs of producing the necessary software, it is becoming increasingly rare to find any commercial machine of reasonable quality which does not feature some built-in computery.

The explosive development of lasers has had profound consequences for infrared science and technology. The extremely powerful Neodymium/Yttrium Aluminium Garnet (or YAG) laser, which operates near 1·08 μm has proved invaluable for non-linear optics experiments and has found widespread use in industry, medicine and warfare. The double-heterostructure solid-state lasers working in the 1–3 μm region are prime candidates for the sources in optical communication systems. The concurrent development of extremely transparent glasses ($< 20\,\mathrm{dB\,km^{-1}}$) and of the associated optical-fibre technology to go with them has led to a revolution in long-distance communications and to the situation where the optical systems working in the near infrared have been able to overwhelm the rival microwave trunk waveguide systems. In the mid infrared near 10 µm, the powerful, flexible and easy to operate CO_2 laser discovered in the middle 1960s has been used in an extraordinarily wide array of spectroscopic experiments and technological applications. Lasers were initially fixed-frequency devices and as such could only make a limited impact on conventional infrared spectroscopy. However, in recent years, many devices which produce tunable coherent radiation have been invented and some of these have been developed beyond the laboratory *tour de force* stage to the point where they can be used in earnest in fairly routine spectroscopy. It is much too early to forecast the demise of incoherent infrared spectroscopy based on the traditional black-body source, but it is nevertheless clear that diode lasers, colour centre lasers, parametric oscillators, difference frequency generators etc. are here to stay and equally clear that they will find increasing use especially for the more difficult type of infrared problem. Examples would be found in the detection of ultra-low concentration (parts per thousand million) pollutants in the atmosphere, in the study of specimens in rather inaccessible milieu (for example in high-pressure cells), and especially in infrared spectroscopy at the limits of resolution. The non-tunable, but potentially very high powered, primary lasers are now starting to make a significant come-back in molecular spectroscopy through the development of the novel techniques of non-linear spectroscopy, Stark spectroscopy, Zeeman spectroscopy and microwave-infrared double-resonance spectroscopy. The availability of these coherent sources combined with the advances in microfabrication techniques which have come from the semiconductor electronics industry has led to the extension of heterodyne detection techniques into the infrared. When operating at adequate local oscillator power, a heterodyne detector can approach the fundamental quantum limits and with such detectors established in infrared and millimetre-wave practice, the old barriers which isolated the infrared from its neighbouring regions, the visible and the microwave, are finally being removed.

The notion of converting infrared images into visual images, for the convenience of the human eye, has a long history. Infrared photography at wavelengths shorter than 1 µm has been used for aesthetic purposes, for discovering via aerial camera work, the sites of ancient habitations, for monitoring crop health and for numerous military applications. The conversion via the photoelectric effect is the basis of several night-vision devices. However, the production of visible analogues of longer wavelength infrared images has only been a practical proposition in recent years with the development of sensitive detectors and reliable scanning techniques. Once this was achieved though, many extremely useful applications arose. This is mainly because the ambient thermal radiation from everyday objects lies mostly in the

10 μm atmospheric window. Thermal imagers have been used in industry to locate both "hot spots" on working machinery and cold spots where the insulation is inadequate, in civil engineering to reveal the effects of the discharge of warm water from power stations, in the fire service to locate unconscious people lying in smoke-filled rooms and in meteorology to gain some understanding of the heat exchanges in the atmosphere. These imagers have also made an impact in medicine where their ability to detect very small differences in skin temperature has opened up new diagnostic techniques which may prove valuable for the early detection of breast cancer and for the diagnosis of several other diseases.

Infrared and millimetre-wave science and technology has had a long and fascinating history. The reader interested in probing rather more deeply will find further details in the historical references given in the general list at the end but for the particular case of the history of far-infrared science, the excellent set of five articles published in the Journal of the Optical Society of America, Volume 67, No. 7, July 1977, pp. 857–894, can be thoroughly recommended. These give a rather nice cameo of how the topic, as a whole, has developed and in particular show how much we owe to the giants of the past.

1.2 Definitions and units

1.2.1 *The Système International*

Spectroscopists traditionally use the cgs system of units with occasional forays into the esu or emu systems on those rare occasions when electrical quantities are involved. This tradition will gradually disappear in the future since there has now been international agreement to adopt the MKSA system in the form of the Système International (SI) throughout science and technology [1]. One immediate consequence of this decision is that students at schools and colleges are being taught only the SI. Clearly since future generations of spectroscopists will use the SI only, it is necessary to consider the impact this will have on both spectroscopic theory and practice.

Spectroscopic theory will only be affected in those areas where electrical quantities enter, since non-electrical equations are independent of any particular system of units. Thus the equation

$$\mathscr{E} = h\nu \tag{1.2.1}$$

does not presume any system of units. In the cgs one would have $h = 6.626176 \times 10^{-27}\,\mathrm{erg\,s^{-1}}$ and $\mathscr{E}$ would therefore be in ergs: in the SI one would have $h = 6.626176 \times 10^{-34}\,\mathrm{J\,s^{-1}}$ and $\mathscr{E}$ would be in Joules. In both cases the frequency would have the same magnitude but would be called by different names in the two systems. Thus in the cgs one would say "cycles per second" and in the SI one would say "hertz".

The reason why exceptions have to be made for the electrical quantities is so that these can be expressed in the familiar practical units. Thus the SI units for potential difference, current and resistance are the volt, the ampère and the ohm. In order that this reconciliation can be brought about the basic equations have to be modified. Thus if one takes Coulomb's law,

$$F = \frac{q_1 q_2}{\varepsilon_r r_{12}^2}, \qquad (1.2.2)$$

where F is the force between two charges q_1 and q_2 separated by a distance r_{12} in a medium of relative permittivity ε_r, then in the esu, by definition, if $q_1 = q_2 = 1\,\text{esu}, r_{12} = 1\,\text{cm}$ and $\varepsilon_r = 1$, the force is 1 dyne. One therefore seeks an SI equation which would produce a force of 10^{-5} Newton between two charges of 3.3356×10^{-10} (i.e. $1/10c$) coulombs separated by a distance of 10^{-2} metre in a vacuum. Clearly a suitable multiplying factor will produce the desired result and Coulomb's law is therefore rewritten as

$$F = \frac{q_1 q_2}{4\pi\varepsilon_0 \varepsilon_r r_{12}^2}. \qquad (1.2.3)$$

The factor 4π is introduced because the SI is an extension of the "rationalised" MKS system. The remainder of the multiplying factor, ε_0, is usually called the "permittivity of free space" and has the value

$$\varepsilon_0 = 10^7/4\pi c^2 = 8.541878 \times 10^{-12}\,\text{F m}^{-1} \qquad (1.2.3a)$$

where c is the velocity of light namely $2.99792458 \times 10^8\,\text{m s}^{-1}$. Students are now taught Coulomb's law in the form of (1.2.3) rather than in the older form (1.2.2). The appearance of electrical quantities, charge, dipole moment, field strength, etc., in spectroscopic equations, can always be traced back to relations of the form (1.2.3). It follows, therefore, that an equation written in SI form can be converted to the corresponding cgs/esu by setting $\varepsilon_0 = (4\pi)^{-1}$.

For completion it should be mentioned that to make magnetic equations compatible with the SI it is necessary to introduce another quantity, μ_0, which is known as the "permeability of free space". This has the value

$$\mu_0 = 4\pi/10^7 = 1.256637 \times 10^{-6}\,\text{H m}^{-1}. \qquad (1.2.4)$$

It arises, just as did ε_0, from the need to make equations, originally introduced with one set of units in mind, still numerically correct within the new system. Magnetic effects were formerly discussed within the framework of the cgs/emu system in which the unit of charge was 10 coulombs, the unit of magnetic field strength was the Oersted and the unit of magnetic induction or magnetic flux density was the gauss. To effect the transition to the SI one considers a suitable equation and then sees how the new units must be related to the old in order that the equation still be true. An excellent starting example is provided by the classical equation for cyclotron resonance:

$$\nu_{\text{cr}} = eB/2\pi m. \qquad (1.2.5)$$

In the SI, e will be ten times larger and m will be 10^3 times smaller so, B will have to be 10^4 times smaller to compensate. The SI unit for magnetic induction, the tesla, is thus numerically equal to 10^4 gauss, a result which is usually remembered in the form

$$1\,\text{T} = 10\,\text{kG}. \tag{1.2.6}$$

The next step is to go to the fundamental definition of the magnetic field H produced by a current i flowing in a wire. If the wire can be considered to be essentially infinitely long, then the field in free space at a distance r from the wire is given by

$$H = \frac{i}{(4\pi)r}, \tag{1.2.7}$$

where the (4π) only applies for the SI case. One then has that a wire carrying a current of 10 A will produce, by definition, a field of 1 Oe at a distance of 1 cm. The SI current will be ten times smaller and the distance one hundred times greater; therefore the SI unit for magnetic field strength (not so far named) will be related to the Oersted by

$$1\,\text{A}\,\text{m}^{-1} = 4\pi \times 10^{-3}\,\text{Oe}. \tag{1.2.8}$$

In all systems of units, the flux density and the field strength are related by

$$B = \mu H, \tag{1.2.9}$$

where μ is the permeability. In the cgs/emu system, μ is unity for empty space and in fact essentially so for all non-magnetic media, but to reconcile equations (1.2.6) and (1.2.8), it is necessary for μ to have a different value in the SI. In fact, a field of 1 Oe in free space will be equivalent to a flux density of 1 gauss and hence to 10^{-4} T. One then has

$$B = 4\pi \times 10^{-3} \times 10^{-4}H = \mu_0 H, \tag{1.2.10}$$

the result given previously as (1.2.4).

The powers of ten which appear in the definitions of ε_0 and μ_0 reflect essentially the differences, g/kg, cm/m, between the cgs and the MKS systems but the remainder, the factor of c, reflects the constant of proportionality between the esu and the emu systems which has just this numerical value and which also has the dimension of a velocity. The SI arises essentially from a merger of the esu and the emu together with a switch of base units, so not surprisingly c appears in the adjusting constant ε_0. The constant c is not, however, just accidentally equal to the velocity of propagation of electromagnetic waves in free space. Maxwell's theory of the electromagnetic field which essentially takes equations such as (1.2.3) and (1.2.7) as its starting point shows that waves will propagate in this field with a velocity

$$c = (\varepsilon_0 \mu_0)^{-1/2}. \tag{1.2.11}$$

Normally in the electromagnetic field E and H are perpendicular to one another and to the direction of propagation (TEM waves) but this is not necessarily always so. However unless E and H are parallel there will always be a flow of energy and it can be shown that the magnitude and direction of this flow are given by the Poynting vector

$$\mathbf{S} = \mathbf{E} \times \mathbf{H}. \tag{1.2.12}$$

It will be seen at once that since $\mathbf{E}$ has dimensions of $V\,m^{-1}$ and $\mathbf{H}$ has dimensions of $A\,m^{-1}$, $\mathbf{S}$ has dimensions of $W\,m^{-2}$, i.e. it is an intensity.

If one goes further and asks how much energy is propagating in the magnetic wave and how much in the electric wave, one finds these quantities to be equal, from which it follows that

$$\sqrt{\varepsilon_0}\,|\mathbf{E}| = \sqrt{\mu_0}\,|\mathbf{H}|. \tag{1.2.13}$$

This equation can be interpreted as

$$|\mathbf{E}| = \sqrt{\left(\frac{\mu_0}{\varepsilon_0}\right)}\,|\mathbf{H}| \tag{1.2.14}$$

which is completely analogous with Ohm's law. The quantity $\sqrt{(\mu_0/\varepsilon_0)}$ is therefore formally an impedance:

$$\sqrt{(\mu_0/\varepsilon_0)} = 376 \cdot 7\,\Omega. \tag{1.2.15}$$

It is usually called the impedance of free space. E and H are free-space quantities and both will alter as the field enters a material medium of permittivity ε and permeability μ. Inside the medium it makes more sense to consider just the electric displacement

$$\mathbf{D} = \varepsilon\mathbf{E} = \varepsilon_r\varepsilon_0\,\mathbf{E} \tag{1.2.16a}$$

and the magnetic induction

$$\mathbf{B} = \mu\mathbf{H} = \mu_r\mu_0\mathbf{H}. \tag{1.2.16b}$$

Here ε_r and μ_r are the relative permittivity and permeability respectively. ε_r will always differ significantly from unity but, as mentioned above, μ_r can be taken to be unity for all save magnetised media. So far it has been assumed that the waves propagate without loss. When this is not the case ε_r (and in principle μ_r) will become complex. Usually, however, one need only consider the components of complex ε_r and can still assume μ_r to be unity. The nature of the complex permittivity ε_r will be examined in more detail in Chapter 3.

1.2.2 *The Système International: symbols for quantities*

The various international bodies concerned with developing and promulgating the SI have taken the further step of issuing a unified list of symbols for physical quantities [1]. Unfortunately the suggested symbols have been adopted under the influence of various pressure groups and there is very little evidence of any concerted effort to avoid illogicality and awkward duplications. As an example it shall be mentioned that the SI has adopted the suggestion of the joint commissions for Spectroscopy (Rome, 1952) that the symbol for wavenumber be the Greek letter sigma. This decision is unacceptable since σ apart from being the accepted symbol for Stefan's constant is already used to mean several other things in spectroscopy—thus the symbols for conductivity and wavenumber would become the same and this would make a mathematical discussion of the dispersion of conductivity with wavenumber rather confusing! Throughout this work, $\tilde{v}$ will be used for wavenumber—the symbolism being deliberate—that is to emphasise the close connection with frequency v. However the use of σ for wavenumber will still be encountered in the literature, especially in the French literature. This duplication and others of a similar kind are very unfortunate, especially since they have provided grist for the mills of those who are implacably opposed to the extension of the SI into their own particular territory. It is devoutly to be hoped that the approved symbol list will be suitably modified in the not too distant future, but since there is no acceptable internationally promulgated set of symbols, at the moment, it may be useful to suggest a list which does appear to find fairly widespread support in the spectroscopic community. This is given in Table 1.1.

It is somewhat unsatisfactory to have to use the same symbols, B, S and ϕ with merely a subscript to indicate the dimensional difference. However despite this difference the quantities are so conceptually alike and the custom is so well established that it does not seem worthwhile to challenge it now and to propose a more uniform system. There are other duplications in common use which it will probably also be difficult to eliminate. Thus R is used both for resistance and power reflectance as well as being the agreed symbol for the molar gas constant; τ is used for transmittance and as a symbol to indicate a parametric temporal quantity; ρ is used for resistivity and reflectivity, ε is used for emissivity and also for dielectric permittivity and α is used for polarizability and also for absorption coefficient. One will probably just have to accept these confusing duplications for the time being, but spectroscopists, who are in a unique position, because of the multidisciplinarian nature of their art, should constantly press for a rationalisation of scientific notation.

With this clarification of symbols, one can develop a clear mathematical exposition of the theory. The various radiant quantities are related by simple integral and differential equations. Thus one has

$$I = \int_0^\infty S_v \, dv = \int_0^\infty S_\lambda \, d\lambda; \quad P = \int_0^\infty \phi_v \, dv = \int_0^\infty \phi_\lambda \, d\lambda,$$

$$\mathscr{I} = \int_0^\infty B_v \, dv = \int_0^\infty B_\lambda \, d\lambda; \quad \mathscr{E} = \iiint \mathscr{I} \, dA \, d\Omega \, dt \text{ etc.} \quad (1.2.17)$$

where A stands for source area, Ω for solid angle and t for time. Analogously one has:

$$S_\lambda = -(c/\lambda^2)S_v; \quad \phi_\lambda = -(c/\lambda^2)\phi_v \text{ etc.} \qquad (1.2.18)$$

Theoretical physicists usually work in terms of the angular frequency, that is radians per second, whose symbol is ω. They do this to avoid the continual factors of 2π which appear in the arguments of the trignometric functions when time frequency, v, is the variable. Planck's law then takes on the form $\mathscr{E} = \hbar\omega$ where $\hbar = h/2\pi = 1{\cdot}054589 \times 10^{-34}$ J^{-1} Hz^{-1}: with this convention understood, the extended SI forms a good basis for spectroscopic theory.

One notable exception, in practice, is the choice of unit to describe dipole moment. Almost universally this is given in terms of the non-SI unit the Debye. One Debye unit (D) is the dipole moment of two charges, one positive and the

TABLE 1.1
Some important quantities used in infrared physics

Quantity	Symbol	Unit	Dimensions
Charge	Q	coulomb	$M^{1/2} L^{3/2} T^{-1} \varepsilon_0^{1/2}$
Current	i	ampere	$M^{1/2} L^{3/2} T^{-2} \varepsilon_0^{1/2}$
Potential	V	volt	$M^{1/2} L^{1/2} T^{-1} \varepsilon_0^{-1/2}$
Resistance	R	ohm	$L^{-1} T \varepsilon_0^{-1}$
Impedance	Z	ohm	$L^{-1} T \varepsilon_0^{-1}$
Resistivity	ρ	metre ohm	$T \varepsilon_0^{-1}$
Conductivity	σ	siemen per metre	$T^{-1} \varepsilon_0$
Electric field strength	E	volt per metre	$M^{1/2} L^{-1/2} T^{-1} \varepsilon_0^{-1/2}$
Magnetic field strength	H	ampere per metre	$M^{1/2} L^{1/2} T^{-2} \varepsilon_0^{1/2}$
Magnetic flux density	B	tesla	$M^{1/2} L^{-3/2} \varepsilon_0^{-1/2}$
Capacitance	C	farad	$L \varepsilon_0$
Inductance	L	henry	$L^{-1} T^2 \varepsilon_0^{-1}$
Energy	$\mathscr{E}$	joule	$M L^2 T^{-2}$
Dipole moment	μ	metre coulomb	$M^{1/2} L^{5/2} T^{-1} \varepsilon_0^{1/2}$
Power	P	watt	$M L^2 T^{-3}$
Intensity	I	watt per square metre	$M T^{-3}$
Circular frequency	ω	radian per second	T^{-1}
Radiant frequency	v	hertz	T^{-1}
Wavenumber	$\tilde{v}$	reciprocal metre	M^{-1}
Wavelength	λ	metre	M
Source brightness	$\mathscr{I}$	W m^{-2} sr^{-1}	$M T^{-3}$
Source spectral/ brightness	B_v	W m^{-2} sr^{-1} Hz^{-1}	$M T^{-2}$
	B_λ	W m^{-2} sr^{-1} m^{-1}	$M L^{-1} T^{-3}$
Spectral intensity	S_v	W m^{-2} Hz^{-1}	$M T^{-2}$
	S_λ	W m^{-2} m^{-1}	$M L^{-1} T^{-3}$
Spectral radiant power	ϕ_v	watt per hertz	$M L^2 T^{-2}$
	ϕ_λ	watt per metre	$M L T^{-3}$

other negative, each of magnitude 10^{-10} esu separated by a distance of one ångstrom unit (10^{-8} cm). The reason for this choice is that the dipole moments of most polar molecules turn out to be of the order of one Debye. It is therefore a particularly suitable unit for molecular physics. The best thing to do in this situation is to retain the Debye and to have available a conversion factor to transform to the proper SI unit. One finds that one Debye unit equals $3\cdot3356 \times 10^{-30}$ C m. Some more examples of practical resistance to the implementation of the full SI will unfortunately emerge in the next section.

1.2.3 *The use of the SI in practical spectroscopy*

The Système International has raised far more controversy in spectroscopic practice. One of the basic precepts of the SI is that multiples and sub-multiples of a basic unit are generated by multiplying or dividing by 10^3. These multiples and submultiples are designated by internationally agreed prefixes thus:

$$\begin{array}{ll} \text{atto (a)} = 10^{-18} & \text{exa (E)} = 10^{18} \\ \text{femto (f)} = 10^{-15} & \text{peta (P)} = 10^{15} \\ \text{pico (p)} = 10^{-12} & \text{tera (T)} = 10^{12} \\ \text{nano (n)} = 10^{-9} & \text{giga (G)} = 10^{9} \\ \text{micro (}\mu\text{)} = 10^{-6} & \text{mega (M)} = 10^{6} \\ \text{milli (m)} = 10^{-3} & \text{kilo (K)} = 10^{3} \end{array}$$

Within the SI there is therefore no room for the centimetre and its inverse the cm^{-1} and since most spectroscopists quote "frequencies" in cm^{-1} confrontation is inevitable. The SI unit for wavenumber is m^{-1}, but few spectroscopists are prepared to adopt it since the benefits are not obvious and the magnitudes of the wavenumbers are thereby increased by a factor of 10^2. There is a deeper reason which, though less logical, is very much more powerful. The wavenumber in cm^{-1} is so deeply entrenched and so familiar to spectroscopists that to many of them there has ceased to be any difference between the quantity and the unit used to measure it. Thus it is commonplace to hear spectroscopists refer to a band having a "frequency" of so many wavenumbers! This is the basic reason why the attempt to introduce a name for the cm^{-1}— the kayser—failed†. There was just no need for a name, spectroscopists already had one!

The confrontation situation puts authors into a dilemma: if they use the SI their spectroscopic colleagues will disapprove and if they do not, students will have difficulty in following the arguments. Not surprisingly there has been an outcry from spectroscopists and they, in concert with other groups equally threatened, have been able to mount an effective lobby which has succeeded in

---

†Oddly enough the millikayser (i.e. 10^{-3} cm^{-1}) is still used occasionally, especially by people engaged on high-resolution spectroscopy by means of tunable lasers.

gaining some concessions from the CGPM (Conférence Générale des Poids et Mesures) and the CIPM (Comité International des Poids et Mésures) the two bodies principally concerned with promulgating the SI. The latest position (P. Vigoureux, private communication) is that the committee for units of the CIPM has stated that "there is no objection whatever to the use of the cm^{-1} in spectroscopy". However this statement has all the hallmarks of a political compromise and really it avoids the issue. The point is that if one *calculates* a wave number within the SI it will come out in "metres to the minus one". Likewise if one substitutes a wave number into an SI equation one will have to quote it in m^{-1}. The refusal of the CIPM in 1974, under pressure from several distinguished elder statesmen, to grant a special name and symbol to the m^{-1} will in the long term seem retrograde. In the meantime some form of compromise is necessary. One can simply do a mental juggling act and remember that

$$100\,\text{m}^{-1} = 1\,\text{cm}^{-1} \qquad (1.2.19)$$

and use this to translate back and forth. A reasonable alternative would be to avoid wave number altogether except as a descriptive term. It would then only appear in the text and could be expressed without ambiguity—one might say, for example, "carbon hydrogen stretching vibrations occur in the 3000 cm^{-1} region". Another helpful practice is to use the expressions "waves per metre" or "waves per cm" to make one's meaning completely clear. In mathematical expressions the best approach is to use the frequency exclusively. In fact there is quite a case for sidestepping the whole dispute and using the time frequency both for the descriptive term and for the unit.

If one asks spectroscopists why they cherish a particular unit, one will get several answers most of which boil down to the unit being familiar. More substantial reasons are:

1. The unit is the one which has been extensively used in the past and to abandon it would lead to a break in scientific continuity.
2. The unit should be such that the magnitudes of the quantities commonly measured in terms of it shall be small numbers but not too small.
3. The experimental uncertainties in the measurements shall be in digits close to the decimal point.

The cm^{-1} certainly meets requirements (1) and (3) for there is a vast literature in terms of it and the large majority of spectrometers have absolute precision of wave number measurement lying between 1 and 0·1 cm^{-1}. The cm^{-1} does not satisfy reason (2) very well since the lowest wave numbers that most infrared spectroscopists measure are already in double figures. If now one considers the frequency—and particularly the terahertz—as unit it will be seen that the three reasons can be met just as well. Frequency units would promptly make for ease of communication between infrared and microwave spectroscopists: the infrared region would extend from 0·3 to 400 terahertz; the absolute accuracy of measurement would commonly be about 0·03 THz. In this work, therefore, the terahertz will be the preferred frequency unit but wave-numbers (in cm^{-1}) will be quoted whenever this could prove helpful. The conversion between the two can be made to an accuracy acceptable in the majority of cases by putting 1 cm^{-1} equal to 30 GHz from which 33·33 cm^{-1} is equivalent to 1 THz.

Other alternatives which should be mentioned are the use of wavelength and, despite the non-cooperation of the CIPM, to go ahead with the introduction of a name for the inverse metre. Wavelength is to be deprecated even though the micrometre (μm) is a proper SI unit. The reason for this is that wavelength is a medium-dependent quantity (via the refractive index) whereas frequency is not. It is argued that spectroscopists commonly *measure* wavelength and, whilst this is true, it can equally well be argued that this is an intermediate quantity which is to be used (via the speed of light) to arrive at the fundamental quantity—frequency. Accurate line positions in the far infrared are determined by calibrating the instrument against the pure rotational spectrum of carbon monoxide. The positions of the lines in this spectrum are known from extrapolations of microwave frequency data and therefore it is perfectly arguable that far-infrared spectroscopists are doing frequency measurements. In recent years it has become possible to make absolute measurements of time frequencies in the mid and near infrared. Thus the old arguments in favour of wavelength are now collapsing and, since the velocity of electromagnetic radiation in vacuo is known [2] to extreme high accuracy, viz.

$$c = 2·99792458[8] \times 10^8 \, \text{m s}^{-1} \qquad (1.2.20)$$

there seems little reason to retain wavelength any longer. The notion of introducing a unit for the inverse metre is more attractive. Of course there is already a unit which is formally the m^{-1}, namely the dioptre, but this is used in so specialised a context that it would be best to have a new name. Since the kayser has been rejected for the cm^{-1} perhaps one could have the Kayser for the m^{-1}. The milliKayser (or alternatively "waves per mm") would then be a convenient unit—formally the mm^{-1}—for describing the positions of infrared bands. In terms of the milliKayser the infrared would stretch from 1 to 1000 mK.

Apart from wave number, the other spectroscopic quantity which is in very common use and which involves the cm, is the absorption coefficient α. This can be defined via the differential equation

$$\frac{dI}{dl} = -\alpha I \qquad (1.2.21)$$

where I is the intensity of a beam propagating through a lossy medium and l is the distance into that medium. By integration it follows that

$$I = I_0 \exp(-\alpha l) \qquad (1.2.22)$$

and

$$\alpha = \frac{1}{l} \ln (I_0/I).$$ (1.2.23)

In virtually all work, l is measured in cm and, since the logarithms are often natural or Napierian logarithms, authors quote as the unit for α the neper cm^{-1}. Again there is confrontation with the SI system. The confrontation is virtually complete since the only other unit used is that based on decimal logarithms and thus still involves the cm. At present there does not seem to be any acceptable compromise so absorption coefficients will continue to be quoted in cm^{-1} for some time ahead. It follows from the definition of α that $1 \, cm^{-1}$ equals $100 \, m^{-1}$ and this conversion can be used if it is desired to substitute SI-only units into equations.

The word "neper" has a slightly different meaning to electrical engineers. They use the neper as a measure of attenuation of voltage—one thus has an attenuation coefficient α_v which is defined in a way analogous to (1.2.23), namely

$$\alpha_v = \frac{1}{l} \ln (V_0/V).$$ (1.2.24)

Since intensity is proportional to the square of voltage it follows that

$$\alpha = 2\alpha_v.$$ (1.2.25)

Some authors give a subscript to α, i.e. they write α_p, to stress that it is a *power* attenuation coefficient. Electrical engineers themselves avoid the difficulty by using a different unit for power attenuation—the decibel per metre. This is defined by the equation

$$\alpha_p [dB \, m^{-1}] = l^{-1} \times 10 \lg (I_0/I)$$ (1.2.26)

from which it follows that

$$\alpha_p [dB \, m^{-1}] = 4{\cdot}343 \, \alpha_p [neper \, m^{-1}] = 434{\cdot}3 \, \alpha_p [neper \, cm^{-1}].$$ (1.2.27)

The different practices reflect the different types of detector which are available to electrical engineers and to infrared spectroscopists. The latter are, at the moment, restricted mostly to the use of power detectors because devices capable of following oscillations at frequencies in excess of 10^{11} Hz are only just starting to become available. Infrared spectroscopists therefore prefer to use a power attenuation coefficient α_p. To avoid confusion α (or α_p) will be referred to as an *absorption* coefficient, thus stressing its optical nature, and on the rare occasions that α_v occurs it will be referred to as an *attenuation* coefficient.

The units for the ordinate axis of emission spectra present no problems *vis-à-vis* the SI system but there is still an unresolved difference of opinion between spectroscopists. This arises because the "intensity at a given frequency" can only have a meaning in a differential coefficient sense. The quantity that one can measure is power. From this one can abstract, by imagining detectors with smaller and smaller windows, the concept of intensity, i.e. power per unit area. The next stage of abstraction is to conceive of spectral intensity, i.e. power per unit area per unit spectral interval. Unfortunately there is a well entrenched custom of using wavelength as the spectral interval instead of the more natural frequency. This custom leads to the ridiculous situation that plots of emission spectra cannot be meaningfully drawn at either abscissa extremity and furthermore to the unsatisfactory situation that the graphs in the regions where they can be drawn differ in shape, peak position etc. from the corresponding graphs in terms of frequency. It is high time that wavelength was abandoned as a working unit and therefore the quantity $[W] [Hz]^{-1} [m]^{-2}$ should be the preferred unit for the ordinate axis of emission spectra.

Chemists have, on the whole, been the most reluctant to adopt the SI. It is not in the least remarkable to see in the preamble to school examination syllabuses phrases like "Metric (SI) units are to be used subject to the following conditions: the unit of mass will be the gram.† and the units of volume may be the cm^3 or the $dm^3 = 1$ litre". The reluctance is strange, in one way, in that concentrations in kilomoles per cubic metre would be numerically the same as the familiar moles per litre. This myopic determination to hold on to the past is a great pity since it tends to isolate chemistry from the rest of science and technology but despite pleas to the contrary, the chemical literature for the foreseeable future will continue to be in essentially cgs units.

In this literature, and especially when solutions or gases are being considered, it is the extinction coefficient ε_m which is discussed rather than the absorption coefficient α. The two are related by Beer's law

$$\alpha = \varepsilon_m c_m, \qquad (1.2.28)$$

where c_m is the molar concentration. The usual units for ε_m are therefore $(Np) (cm^{-1}) (l) (mol^{-1})$, though care is needed, for in much of the earlier literature decimal logarithms are used. The fundamental molecular properties are related more to the integral of ε_m over a band rather than to ε_m itself. The units for integrated extinction are usually $(neper) (cm^{-2}) (litre) mol^{-1}$ since the integration is nearly always with respect to wavenumber in units of cm^{-1}. A named unit for integrated extinction—the "dark"—is sometimes encountered. The dark is usually defined with dimensions $(cm^{-1}) (cm^2) (mmol^{-1})$ but this is identical to the units given above. Integrated absorption coefficients in units of $(neper) (cm^{-2})$ can be converted to darks by multiplying by $M\rho^{-1} 10^{-3}$ where M is the molecular weight and ρ the density. For a gas at atmospheric pressure at $0°C$ the multiplying factor is 22.4.

To complete this section, two further non-SI units which will be

---

† The mole which is an official SI quantity is actually defined in terms of the *gram* molecular weight—a farcical situation.

encountered occasionally should be mentioned. These are the Ångstrom unit (Å) which equals 10^{-10} m, i.e. 10^{-1} nm and the electron-volt (eV) which is equivalent to 241·8 THz (i.e. 8065·5 cm^{-1}). The Ångstrom unit is rapidly fading away but the electron-volt is so firmly established and so convenient that it has just to be accepted as the common currency of both theoretical and experimental physics. In this situation conversion factors are necessary. The relations for transmitting frequency (or energy) in one set of units into another are:

$$1\,eV = 1\cdot6021892[46] \times 10^{-19}\,J = 2\cdot417970 \times 10^{14}\,Hz$$
$$= 8\cdot065478 \times 10^{5}\,m^{-1} = 8\cdot065478 \times 10^{3}\,cm^{-1} = 1\cdot160450$$
$$\times 10^{4}\,K$$
$$1\,cm^{-1} = 100\,m^{-1} = 0\cdot1239852 \times 10^{-3}\,eV = 29\cdot979246\,GHz$$
$$= 1\cdot438786\,K$$
$$1\,THz = 3\cdot3356409 \times 10^{3}\,m^{-1} = 33\cdot356409\,cm^{-1} = 4\cdot13570 \times 10^{-3}\,eV$$
$$= 47\cdot99273\,K$$
$$1\,K = 0\cdot695030\,cm^{-1} = 69\cdot5030\,m^{-1} = 8\cdot61735 \times 10^{-5}\,eV$$
$$= 0.02083648\,THz$$

The centigrade and absolute temperature degrees are the same but the absolute zero is at $-273\cdot16°C$.

1.2.4 *Physical constants and conversion factors*

The principal universal constants used in spectroscopy [3] are listed in Table 1.2.

1.3 Division of the electromagnetic spectrum

It is conventional to divide the electromagnetic spectrum into named regions for purposes both of exposition and description. Thus, for example, one speaks of the optical or visible region and thereby conveys the essential quality of that region, viz. 420 to 750 THz, namely that radiation lying within it can be detected by eye. Divisions of this nature are necessary if the vast array of phenomena which occur are to be organised into some sort of order but they stem more from man's desire for tidiness rather than from any fundamental schisms dictated by nature and it soon becomes obvious that any system of division has its drawbacks. Thus optical spectroscopy, for example, is carried out throughout the near infrared, visible and near-ultraviolet regions. A reasonable compromise and one which has the advantage that it links radiofrequency and optical practice is to divide into decades—that is to use a linear logarithmic scale. This is shown schematically in Table 1.3. The lower frequency bands which appear in this Table, stretching from 300 Hz to 0·3 THz, have been named by international convention—they are the bands used for telecommunications. The higher frequency bands have not yet a

TABLE 1.2

The fundamental constants and their numerical values

Quantity	Symbol	Value
Velocity of light	c	$2.997\,924\,580\,(12) \times 10^8\ \text{m s}^{-1}$
Permittivity of free space	ε_0	$8.854\,187\,818\,(71) \times 10^{-12}\ \text{F m}^{-1}$
Planck's constant	h	$6.626\,176\,(36) \times 10^{-34}\ \text{J Hz}^{-1}$
	$\hbar$	$1.054\,5887\,(57) \times 10^{-34}\ \text{J s}$
Electronic charge	e	$1.602\,189\,2\,(46) \times 10^{-19}\ \text{C}$
Electronic mass	m_e	$9.109\,534\,(47) \times 10^{-31}\ \text{kg}$
Proton mass	m_p	$1.672\,648\,5\,(86) \times 10^{-27}\ \text{kg}$
Neutron mass	m_N	$1.674\,954\,3\,(86) \times 10^{-27}\ \text{kg}$
Boltzmann's constant	k	$1.380\,662\,(44) \times 10^{-23}\ \text{J K}^{-1}$
Avogadro's number	N	$6.022\,045\,(31) \times 10^{23}\ \text{mol}^{-1}$
Mass for unit atomic weight	$10^{-3}N^{-1}$	$1.660\,565\,(98) \times 10^{-27}\ \text{kg}$
Stefan–Boltzmann constant	σ	$5.670\,32\,(71) \times 10^{-8}\ \text{Wm}^{-2}\,\text{K}^{-4}$
Gas constant	R	$8.314\,41\,(26)\,\text{J mol}^{-1}\,\text{K}^{-1}$
Bohr magneton	μ_B	$9.274\,078\,(36) \times 10^{-24}\ \text{J T}^{-1}$
First radiation constant	$c_1 = 2\pi h c^2$	$3.741\,832\,(20) \times 10^{-16}\ \text{Wm}^2$
Second radiation constant	$c_2 = hc/k$	$1.438\,786\,(45) \times 10^{-2}\ \text{mK}$
Rydberg constant	$R_\text{H} = m_\text{e}e^4/8\varepsilon_0^2 h^3 c$	$1.097\,373\,177\,(83) \times 10^7\ \text{m}^{-1}$

The figures in parentheses give the uncertainties in the last two digits.

TABLE 1.3

The electromagnetic spectrum

Band name	Frequency	Wavelength
γ-Ray	3 EHz–30 EHz	100 pm–10 pm
X-Ray	300 PHz–3 EHz	1 nm–100 pm
Soft X-ray	30 PHz–300 PHz	10 nm–1 nm
Vacuum-ultraviolet	3 PHz–30 PHz	100 nm–10 nm
Optical	300 THz–3 PHz	1 μm–100 μm
Infrared	30 THz–300 THz	10 μm–1 μm
Far infrared	3 THz–30 THz	100 μm–10 μm
Submillimetre	300 GHz–3 THz	1 mm–100 μm
Extra-high frequency (EHF)	30 GHz–300 GHz	1 cm–1 mm
Super-high frequency (SHF)	3 GHz–30 GHz	10 cm–1 cm
Ultra-high frequency (UHF)	300 MHz–3 GHz	1 m–10 cm
Very high frequency (VHF)	30 MHz–300 MHz	10 m–1 m
High-frequency (HF)	3 MHz–30 MHz	10^2 m–10 m
Medium-frequency (MF)	300 KHz–3 MHz	10^3 m–10^2 m
Low-frequency (LF)	30 KHz–300 KHz	10^4 m–10^3 m
Very-low frequency (VLF)	3 KHz–30 KHz	10^5 m–10^4 m
Extra-low frequency (ELF)	300 Hz–3 KHz	10^6 m–10^5 m

settled nomenclature and the names suggested in the Table must be regarded as provisional. Thus the top two bands are very conveniently labelled as γ-rays and X-rays but in common usage these names are used to reveal the origin of the radiation rather than its frequency. Thus radiation emitted by excited nuclei or in the course of elementary particle reactions is referred to as γ-radiation whilst that coming from the sudden deceleration of electrons is called X-radiation. Within this definition it is perfectly possible, of course, to have a γ-ray photon whose energy is less than that of an X-ray photon!

Another example of popular usage being at variance with the desire for rational order comes from the EHF and SHF bands which are commonly lumped together under the portmanteau title "microwaves". In fact the situation is rather worse than this since the term is used to describe also radiation having both lower and higher frequency. Originally the touchstone for the use of the term was whether the wave-length was short enough for the radiation to be useful for radar but nowadays any coherent radiation which is produced by electron tube oscillators and whose wave-length is less than about 10 cm is liable to be labelled "microwave". One again has a paradox in that one can have radiation from a laser, which by definition is not microwave, whose wavelength is longer than that from a carcinotron which definitely is! The microwave region, which, from the definition given above, stretches from about a GHz or so to well beyond some hundreds of GHz, was intensively developed during the war years for military radar systems and since the protagonists were at some pains to keep their operating frequencies secret, the overall band came to be subdivided into smaller regions which were identified merely by a random alphabetical code. These codes have tended to stick but unfortunately there are several of them still in common use and this can be very confusing! A system in fairly general use in Great Britain and endorsed by the IEE, is shown in Table 1.4.

TABLE 1.4
The microwave bands

Band	Frequency (GHz)	Wavelength (mm)	Waveguide code UK	US(EIA)
L	1–2	300–150	WG6	WR650
S	2–4	150–75	WG10	WR284
C	4–8	75–38	WG12	WR187
X	7–12	43–25	WG16	WR90
J	12–18	25–17	WG18	WR62
K	18–26	17–12	WG20	WR42
Q	26–40	12–8	WG22	WR28
V	40–60	8–5	WG24	WR17
O	60–90	5–3	WG26	WR12

The place of the infrared in the general context of the electromagnetic spectrum is well brought out by the division system of Table 1.3 but when one comes to discuss infrared matters in isolation, a subdivision of the more than ten octaves becomes desirable. What one does, therefore, is to divide the infrared into the three regions: the far, the mid, and the near infrared. The dividing lines have been drawn at various places at various times, mostly on the grounds of the state of the art at the time or else because of the particular interests of the practitioners. Thus Smith Jones and Chasmar [4] in 1957, suggested the following:

near infrared	$0.75–1.5$ μm
mid infrared	$1.5–10$ μm
far infrared	$10–1000$ μm

This classification was based on the different types of detector which were then available: photoelectric detectors for the near infrared, photoconductive detectors for the mid infrared and thermal detectors for the far infrared. In the last fifteen years there has been an explosive development in detector technology and the divisions given above no longer seem so natural.

From the chemical spectroscopic point of view, the region between 3 and 40 μm is particularly important as it is the "finger print" region where the characteristic absorption bands of organic molecules occur [5]. It is reasonable therefore to define this region as the mid infrared. The region from 100 to 1000 μm is then consequentially the far infrared and the region from 0.7 to 3 μm the near infrared. These latter divisions do have some element of naturalness in their own right, for the absorption bands found in the near infrared arise from overtones and electronic transitions, not vibrational fundamentals and the interest of the physicist in the infrared, as opposed to that of the chemist, tends to be strongly localised in the 100 to 1000 μm region. We will, therefore, for the purposes of the present work, divide the infrared as follows:

near infrared	$0.75–3.0$ μm
mid infrared	$3.0–40$ μm
far infrared	$40–1000$ μm

but it must be stressed that this division, like all others, is subject to the provisos made earlier.

To round off this discussion of the divisions of the spectrum, it must be mentioned that spectroscopists use many terms which are so well entrenched that, despite the difficulties of fitting them into a completely logical system, they will probably continue as part of the common language of the art well into the future. Audiofrequency (AF) is strictly in the range 30 Hz–16 kHz which is detected by the human ear, but it is loosely taken to mean from 0 Hz (i.e. DC) to beyond 40 kHz and there is therefore considerable overlap with the lower radio frequency bands. The distinction which electronic engineers seem to draw is between oscillations propagated in conductors—which would be AF—or oscillations propagated through space in the form of electromagnetic waves which would be classified as radio frequency (RF). By tradition the usual symbol for audiofrequency is f but there does not seem to be any good reason for continuing this. Radio frequencies are currently held to extend up to about 600 MHz, the present high-frequency limit of commercial broadcasting (colour television on UHF). Beyond this limit, conventional, valve, transistor or integrated circuitry becomes unable to deal with the ultra-high frequency oscillations and one enters the world of microwaves. The low microwave region is usually considered to run from 1 to 30 GHz and the high microwave or millimetre-wave band from 30 to 300 GHz. These frequencies are generated by magnetrons, klystrons, backward-wave oscillators (BWOs or carcinotrons) and increasingly in recent years by solid-state devices such as Gunn and IMPATT diodes. The submillimetre band lies strictly from 300 GHz to 3 THz but the phrase is loosely used to denote the much wider range from 60 GHz to over 6 THz. In modern terminology little distinction is drawn between the phrases "submillimetre" and "far infrared". The former tends to be used by physicists and engineers, the latter by chemists and biologists. A similar phrase, "near-millimetre" is used by military technologists to denote the wavelength range from 3 to 0·3 mm. At much shorter wavelengths the terms vacuum ultraviolet and soft X-rays are used because the atmosphere is, at these wavelengths, heavily absorbing and special spectroscopic techniques are necessary.

1.3.1 *The relation of photometric to radiometric units*

The measurement of radiant intensity in power per square metre terms is virtually universal throughout the electromagnetic spectrum with the notable exception of the visible region. The reason for this is, of course, that light, being visible, used to see with and especially used for illumination, is a very special case. The important matter is not so much how intense is the source as determined by an absolute radiometer, but rather how bright is the source as perceived by the human eye. The two can differ hugely because of the strong spectral dependence of the sensitivity of the eye.

The standards of luminous intensity, luminous flux and illumination all have their origins in ancient material comparisons as is clearly brought out by such terms as "candlepower". These standards are, however, not easy to relate directly to the radiometric standards and the visible region is becoming something of a jungle from a standards point of view, with various groups using rather unrelated units. Thus lighting engineers are increasingly coming to use the SI approved units, the candela, the lumen, the lux etc. Laser engineers on the other hand use the joule and the watt whilst the engineers concerned with the development of light-emitting diodes use, of all things, the foot-lambert to define the brightness of their devices.

This is a confusing situation and since physical phenomena do not observe the divisions of the spectrum arbitrarily adopted by man, it is unfortunately necessary for infrared technologists [6] to have some acquaintance with these non-radiometric units and some ability to translate them, at least roughly, into radiometric terms. Modern photometry is related to an essentially non-material fundamental standard, the *candela* which is a modern version of the old *candle*. It is defined by the statement that one square metre of the surface of a black-body at the temperature of freezing platinum (2046 K) gives out 6 × 10⁵ candelas of light. The unit of luminous flux is the *lumen* which is defined as the radiant flux emitted into unit solid angle by a point source giving out one candela. In numerical terms therefore, 4π lumens equals one candela. The SI unit of illumination is the *lux* which represents a level of illumination equal to one lumen falling on one square metre. Some other units encountered are the *lambert*, equal to $1/\pi$ cd cm^{-2} (it is a unit of brightness), the *foot-lambert*, equal to $1/\pi$ cd ft^{-2} and the *foot-candle* (a unit of illumination) equal to one lumen per square foot: it is therefore equal to 10·76 lux.

From Table 2.1, it will be seen that a black body under the specified conditions emits roughly 2 W cm^{-2} of visible radiation, that is radiation having a wave number between $1·43 \times 10^6$ and $2·5 \times 10^6$ m^{-1}. However, not all of this is perceived with the same efficiency and one should calculate the integral of $S_{\bar{\nu}}$ multiplied by $V_{\bar{\nu}}$, where $V_{\bar{\nu}}$ is a measure of the spectral variation of the eye's sensitivity. Since this quantity varies from one person to another, a standard "eye" is introduced and $V_{\bar{\nu}}$ for this defined by international agreement. It is shown in Fig. 1.1

When the spectral curve of Fig. 2.12 is multiplied by that of Fig. 1.1 the integrated result between the limits of the visible region is found to be

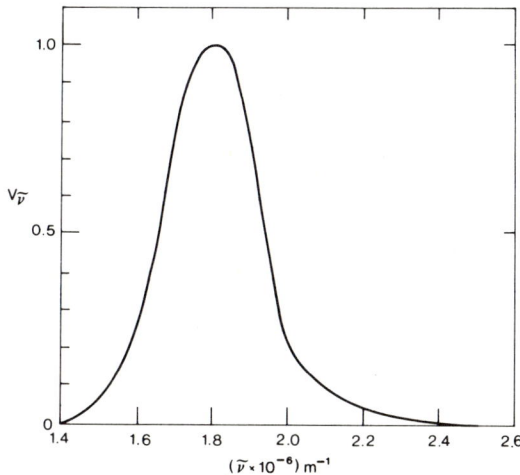

FIG. 1.1. Spectral variation of the sensitivity of the standard eye.

$0.0867 \text{ W cm}^{-2} \text{ sr}^{-1}$. This must by definition equal 60 lumens and therefore we have the result

$$692 \text{ lumens equals } 1 \text{ watt}$$

With monochromatic radiation, such as that from a laser, the number of lumens is equal to the power output of the laser multiplied by 692 and $V_{\hat{y}}$. As an example, 100 mW of the 632·8 nm radiation from a helium/neon laser would be equivalent to only 14 lumens. There would be five times as many lumens from a laser of the same power working at 555·0 nm. It is this strong variation of eye sensitivity which makes green-emitting electroluminescent diodes (GaP:N, $\lambda_{max} = 550$ nm) more attractive than the red ones (GaAs$_{0.6}$P$_{0.4}$, $\lambda_{max} = 655$ nm) despite the greater electrical efficiency of the latter.

1.4 The use of complex quantities in electromagnetic theory

1.4.1 *Complex amplitudes in propagation theory*

The use of complex quantities in electromagnetism has its origin in the treatment of AC currents flowing in LCR networks. A complicated inductive (L), capacitive (C) and resistive (R) network can be analysed by the use of differential equations and Kirchoff's Laws, but for all save the simplest networks, this soon becomes impossibly tedious. Fortunately it is readily shown that the form of solution of the differential equations may be generalised and for each circuit element a complex quantity, the *impedance*, may be introduced. The overall impedance (also in general complex) of the network may be found by simply treating the individual impedances as though they were resistances and combining them according to the normal series and parallel laws. Thus two impedances in series have a total impedance

$$\hat{Z} = \hat{Z}_1 + \hat{Z}_2. \tag{1.4.1}$$

and two in parallel have a total impedance

$$(\hat{Z})^{-1} = (\hat{Z}_1)^{-1} + (\hat{Z}_2)^{-1} \tag{1.4.2}$$

where the "hat" (^) is used to denote an explicitly complex quantity. One can therefore calculate the currents flowing in the various parts of the network immediately. If the incident voltage were of the form

$$\hat{V} = V_0 \exp i\omega t, \tag{1.4.3}$$

then the current will take the form

$$\hat{I} = I_0 \exp i(\omega t + \alpha). \tag{1.4.4}$$

The interpretation of equation (1.4.4) is that $\hat{I}$ represents a current of *magnitude* I_0 and *phase* (relative to $\hat{V}$) α. The mathematical reason why this complex approach works is that we can consider a real voltage $V = V_0 \cos \omega t$ as the real part of (1.4.3)—see Fig. 1.2. Now provided one is considering only *linear* operations, the operator "take the real part of" or "(Re)" can be taken outside all the brackets to be multiplied together and one can postpone doing it right till the last step. One may therefore use this highly convenient formalism quite freely and in practice the restriction to linear operations is not unduly oppressive since the only non-linear operation which arises in elementary circuit theory can also be treated by making a slight modification to the complex algebra. This operation is the calculation of the average power being dissipated in any circuit element and it is non-linear since it involves the evaluation of the average value of the product of voltage and current. In the non-complex approach this would be given by

$$P = \frac{\omega}{2\pi} \int_0^{2\pi/\omega} V_0 \cos \omega t \, I_0 \cos(\omega t + \alpha) \, dt \qquad (1.4.5a)$$

$$= \tfrac{1}{2} V_0 I_0 \cos \alpha. \qquad (1.4.5b)$$

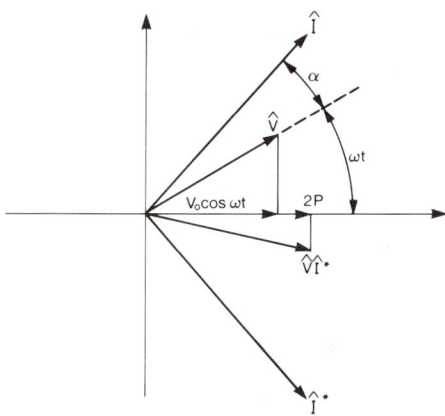

FIG. 1.2. Argand diagram representation of voltage, current and average power dissipation in an AC circuit.

In the complex approach one may write (see Fig. 1.2)

$$P = \tfrac{1}{2} (\text{Re}) [\hat{V} \hat{I}^*] = \tfrac{1}{4} [\hat{V} \hat{I}^* + \hat{V}^* \hat{I}], \qquad (1.4.6)$$

where the asterisk is used to denote the complex conjugate. The reason why this works is that the oscillatory terms in (1.4.5a) average to zero. Some practical illustrations of the use of this complex formalism (to analyse the performance of RC filters) are given in section 4.2.4.

The interaction of the electromagnetic field with matter is not fundamentally different at low and at high frequencies and although it may be convenient to think in terms of propagation in conductors at the low frequency end and wave propagation in free space or dielectrics at the other, it must always be remembered that this is solely a matter of practical convenience. It follows then that the complex algebra may be used to describe wave propagation and one will expect to derive the phase shifts and absorptive loss which the wave suffers directly from the complex field quantities. One may write in elementary terms a time-dependent

$$E(t) = E_0 \cos(\omega t + \alpha) \tag{1.4.7}$$

or a space-dependent

$$E(x) = E_0 \cos(2\pi \tilde{\nu} n x + \alpha) \tag{1.4.8}$$

electric field where n is the refractive index, $\tilde{\nu}$ the free space wavenumber $(\omega/2\pi c)$ and α an arbitrary phase angle to define the zero mark. In complex terms one may therefore write

$$\hat{E}(t) = E_0 \exp i(\omega t + \alpha) \tag{1.4.9}$$

and

$$\hat{E}(x) = E_0 \exp i(2\pi \tilde{\nu} n x + \alpha). \tag{1.4.10}$$

Many examples of the use of this complex notation to simplify the treatment of propagation problems will emerge later in the book.

For waves in free space, the intensity at a point is calculated from the simple complex relation

$$I = \tfrac{1}{2} c \varepsilon_0 \hat{E} \hat{E}^*. \tag{1.4.11}$$

This is quite analogous to (1.4.6). Equation (1.4.11) is much used in interferometry where one *adds* the amplitudes of the various waves and then calculates the resulting intensity. As a simple example, consider the two-beam interferometer where the two beams have the same amplitude and frequency but one has suffered a phase shift β. The total amplitude will then be

$$\hat{E}_{\text{total}} = \tfrac{1}{2} E_0 \exp[i\omega t] + \tfrac{1}{2} E_0 \exp[i(\omega t + \beta)] \tag{1.4.12a}$$

i.e.

$$\hat{E}_{\text{total}} = \tfrac{1}{2} E_0 \exp[i\omega t][1 + \exp(i\beta)]. \tag{1.4.12b}$$

We have hence

$$\hat{E}^*_{\text{total}} = \tfrac{1}{2} E_0 \exp[-i\omega t][1 + \exp(-i\beta)], \tag{1.4.13}$$

and therefore from (1.4.11)

$$I = \tfrac{1}{4} c \varepsilon_0 E_0^2 [1 + \cos \beta], \tag{1.4.14a}$$

i.e.

$$I = \tfrac{1}{2} I_0 [1 + \cos \beta]. \tag{1.4.14b}$$

The intensity thus varies from 100% of that of the full beam to zero as β goes from 0 to π. If β has its origin in a path-difference x introduced into one arm of the interferometer, one then has

$$I(x) = \tfrac{1}{2} I_0 [1 + \cos 2\pi \tilde{v} x], \tag{1.4.15}$$

which is the basic equation of Fourier spectrometry [7, 8].

The same ground can be covered in an alternative presentation, much favoured by microwave engineers, in which each complex quantity is represented by a vector in the Argand diagram. These vectors, whose magnitudes are given by their lengths and whose phases are denoted by the angles which they make with the abscissa axis are often called "phasors". For the Michelson interferometer, which is the optical equivalent of the microwave "magic tee", one can take the beam arriving from the fixed mirror as the origin of phase. The corresponding complex field will then be represented by a vector 00′ along the abscissa axis (Fig. 1.3) whose length will be $E_0/\sqrt{(2)}$. The field arriving from the moving mirror will then be represented by the phasor O′P with the same length but having a phase-angle $\beta = 2\pi \tilde{v} x$. From the geometry of Fig. 1.3, it immediately follows that the phasor of the combined field, i.e. OP, will have an amplitude $\sqrt{(2)} E_0 \cos \beta/2$ and that the intensity will therefore be $(1/2) I_0 [1 + \cos \beta]$ as before.

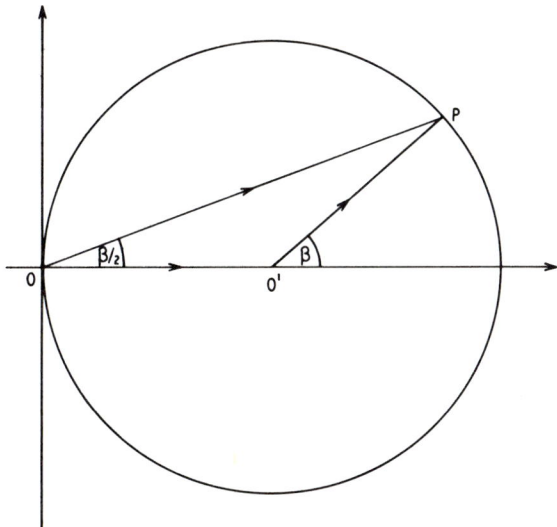

FIG. 1.3. Complex plane interpretation of the Michelson interferometer.

The use of complex quantities was introduced, as remarked above, to simplify the mathematics of AC circuit analysis and to deal with electromagnetic propagation in a more elegant fashion. However, much more can be gained by the use of this powerful formalism. As an example one might cite linear response theory which is nowadays widely used to analyse networks and to provide a conceptual framework for understanding the properties of dielectrics. A core concept here is the Dirac delta function defined by the twin requirements

$$\delta(t) = 0, \; t \neq 0,$$

and

$$\int_{-\infty}^{+\infty} \delta(t)\,dt = 1. \qquad (1.4.16)$$

Clearly it is a rather strange function [9] and for this reason it is usually regarded as the simplest member of the class of generalised functions [10] but it is of great value in modern analysis. Some mathematicians even hold that its use has provided them with the first really rigorous proof of Fourier's theorem! Physically, however, one can think of it as the limit of several much more normal functions as some defining parameter approaches zero. Thus the rectangle function (Fig. 1.4), defined by

$$(2D)^{-1}\,\Pi(x/2D) \begin{cases} = 0, & x < -D \\ = (2D)^{-1}, & -D < x < +D \\ = 0, & x > +D \end{cases} \qquad (1.4.17)$$

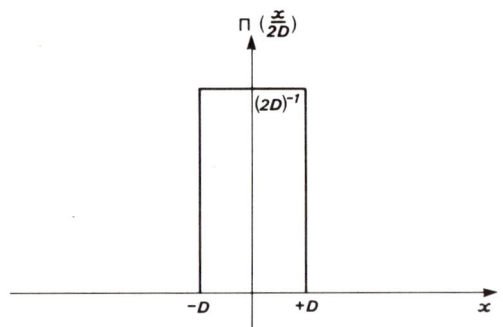

Fig. 1.4. The normalised rectangle function $(2D)^{-1}\,\Pi\,(x/2D)$.

can be considered to turn into the delta function as $D \to 0$. The delta function can thus be thought of in physical terms as a very brief, very intense pulse. Because of this aspect, it can be used to give a representation of instantaneous events. Thus if a function $F(t)$ is multiplied by $\delta(t - t_0)$ and the product integrated over the whole range of t, the result is the single ordinate $F(t_0)$. Another way of looking at this is to say that the delta function is a sampling operator. A related property is that when used in convolution (see section 1.6.3), it acts as a shift operator. Thus the convolution of $F(x)$ with $\delta(x - m)$ is $F(x - m)$. In other words convolution with $\delta(x - m)$ serves to shift the origin of coordinates along the abscissa axis by an amount m.

If a delta function spike is applied to a physical system at time $t = 0$, the system will develop a response (for example a voltage response if one is considering a network) and one may therefore define a response function which gives the time behaviour of the response. It is at this point that the complex analysis enters since it is far more convenient to do Fourier analyses in terms of complex exponentials than it is in terms of cosines and sines separately [11]. The Fourier transform of $R(t)$ yields the transfer function $\hat{T}(v)$, thus

$$\hat{T}(v) = \int_0^\infty R(t) \exp(2\pi i v t)\, dt \tag{1.4.18}$$

and this will tell us everything there is to know about the frequency response of the system. Its amplitude, for example, tells of the variation of attenuation with frequency whilst its phase gives the corresponding variation of phase-shift. However there is much more to it than just this. $R(t)$ is, by definition, a one-sided function since the Causality Principle, which is believed to be a universal law of nature, prohibits any physical system from showing a response *before* the stimulus is applied. The complex Fourier transform of a one-sided function (Appendix 1) is Hermitian and its real and imaginary parts are thus of necessity Hilbert transforms of one another (section 1.6.4). The complex representation therefore brings out immediately the remarkable result that, for all physical systems, the attenuation and phase shift are not independent of one another. The representation of physical properties such as refractive index, permittivity, conductivity etc. in complex form therefore takes on a natural significance. Within this natural formalism, simple relations often exist. Thus the complex relative permittivity

$$\hat{\varepsilon}_r = \varepsilon'_r + i\varepsilon''_r \tag{1.4.19}$$

is related to the complex refractive index

$$\hat{n} = n + i(\alpha/4\pi\tilde{v}) = n + ik \tag{1.4.20}$$

by the very simple formula

$$\hat{\varepsilon}_r = (\hat{n})^2. \tag{1.4.21}$$

This very convenient use of complex quantities does require, however, the adoption of some sign convention for the imaginary components. This is, in the final analysis, purely a matter of convenience since the positive sign corresponds to one sense of rotation of the vector in the complex plane, the negative sign to the other and there is no fundamental justification for a particular preference. The difficulty is that although one is indeed at liberty to choose the sign for any particular complex quantity, once one has done so, the signs for all the others become determined. Thus in electrical engineering, it is usual to write

$$V = V_0 \exp(j\omega t) \tag{1.4.22}$$

where j is used for the square root of minus one so that the symbol i can be retained for the current. This is a natural convention since it implies a vector rotating anticlockwise into the first quadrant. However, once it is adopted, the complex permittivity $\hat{\varepsilon}$ must have a *negative* sign for its imaginary component since the charge on the plates of a lossy capacitor lags *behind* the applied voltage by a phase-angle delta whose tangent is given by $\varepsilon''/\varepsilon'$. To keep Maxwell's relation equation (1.4.21) simple, one then has to adopt a negative sign for the imaginary component of $\hat{n}$ and this in turn leads inexorably to the use of a negative sign for the argument of the exponential in the electromagnetic propagation equation (1.4.10). Physicists generally find all this much too high a price to pay for the marginal advantage of the positive sign in equation (1.4.22) and they almost invariably nowadays adopt the positive convention. It has been pointed out, not entirely facetiously, that the two bodies of opinion and usage can be readily reconciled by the simple convention that

$$j = -i. \tag{1.4.23}$$

1.4.2 *Some standard results in complex algebra*

Some theorems in complex algebra find repeated usefulness throughout a work of this kind and it is helpful to have them gathered together in compact form.

1. The first general result is that the complex conjugate of a given complex number is found by replacing i everywhere by $-i$. Thus if

$$\hat{Z} = x + iy = \sqrt{(x^2 + y^2)} \exp\left[i \tan^{-1} (y/x)\right], \tag{1.4.24a}$$

then

$$\hat{Z}^* = x - iy = \sqrt{(x^2 + y^2)} \exp\left[-i \tan^{-1} (y/x)\right]. \tag{1.4.24b}$$

2. The square root of a complex number is given by

$$\hat{Z}^{\frac{1}{2}} = (x + iy)^{\frac{1}{2}} = \left[\frac{x + \sqrt{(x^2 + y^2)}}{2}\right]^{\frac{1}{2}} + i\left[\frac{-x + \sqrt{(x^2 + y^2)}}{2}\right]^{\frac{1}{2}} \tag{1.4.25a}$$

which can also be written

$$\hat{Z}^{\frac{1}{2}} = (x + iy)^{\frac{1}{2}} = (x^2 + y^2)^{\frac{1}{4}} \exp\left\{i \quad \tan^{-1} \quad \left[\frac{-x + \sqrt{(x^2 + y^2)}}{x + \sqrt{(x^2 + y^2)}}\right]^{\frac{1}{2}}\right\}. \tag{1.4.25b}$$

3. The reciprocal of a complex number is given by

$$\hat{Z}^{-1} = (x + iy)^{-1} = \frac{x}{x^2 + y^2} - i\frac{y}{x^2 + y^2}, \tag{1.4.26a}$$

which can also be written

$$\hat{Z}^{-1} = (x + iy)^{-1} = \sqrt{\left(\frac{1}{x^2 + y^2}\right)} \exp\left[-i \tan^{-1}(y/x)\right]. \quad (1.4.26b)$$

4. The commonly occurring form $(a - \hat{Z})/(a + \hat{Z})$ can be expressed in polar form thus

$$\frac{a - \hat{Z}}{a + \hat{Z}} = A e^{i\delta} \quad (1.4.27a)$$

where

$$A = \frac{(a - \hat{Z})(a - \hat{Z}^*)}{(a + \hat{Z})(a + \hat{Z}^*)} = \frac{a^2 - 2a|\hat{Z}|\cos\phi + |\hat{Z}|^2}{a^2 + 2a|\hat{Z}|\cos\phi + |\hat{Z}|^2}, \quad (1.4.27b)$$

and

$$\tan\delta = \frac{-2a|\hat{Z}|\sin\phi}{a^2 - |\hat{Z}|^2}, \quad (1.4.27c)$$

and where

$$\hat{Z} = |\hat{Z}|\exp[i\phi]. \quad (1.4.27d)$$

5. Trignometric functions of complex variables can be elucidated using De Moivre's theorem directly. One then has

$$\left.\begin{array}{l} \sin(x + iy) = \sin x \cosh y + i \cos x \sinh y \\ \cos(x + iy) = \cos x \cosh y - i \sin x \sinh y \\ \tan(x + iy) = \dfrac{\tan x + i \tanh y}{1 - i \tan x \tanh y} \end{array}\right\} \quad (1.4.28)$$

1.5 Some special functions used in infrared science and technology

1.5.1 *Functions which arise in Fourier transformation*

Apart from the standard elementary functions, square-root, exp., log., sin, cos, etc. which occur generally, there are some transcendental, higher transcendental and even odder functions which occur quite frequently in infrared theory and which it is convenient to gather together and describe in one place. The Dirac delta function, which can be regarded as a limiting case of the rectangle function, has already been introduced. A very useful extension of the delta function is the sampling, replicating or Dirac comb function $\sqcup\!\sqcup(x/D)$, sometimes called [11] "Shah", which consists of an infinite set of delta functions spaced uniformly with a separation D. The value of this function lies in its ability to provide an analytical expression for the operation of sampling. Sampling is an unfortunate necessity in the real world where one will have usually recorded continuous data but wish to process it in a digital computer. All one can do is to take a set of equispaced samples of the data and process this set, since clearly it would take an infinite time to process the infinite set represented by the continuous function. If one multiplies a continuous function by Shah and then integrates over the entire range of the abscissa variable, one will produce just the sampled set. One can then investigate in terms of various models, what will be the likely result of the computer processing a set of sampled data rather than the original continuous data.

A close relative of the rectangle function and one which can also be regarded as a generator of the delta function is the triangle function (x/D) which is defined by

$$\Lambda (x/D) = 1 - |x|/D, \qquad |x| \leqslant D$$
$$\Lambda (x/D) = 0 \qquad\qquad\quad |x| > D \qquad\qquad (1.5.1)$$

The rectangle function finds its principal application as an analytical expression for the operation, unfortunately unavoidable in the real world, of restricting the range of abscissa variable over which observations are taken. The triangle function is used instead of the rectangle function when a brutal chopping off of the data beyond $|x| > D$ would be undesirable and a progressive attenuation would be much preferred. This progressive attenuation can be ensured by several other functions, e.g.

$$f(x/D) = 1 - (x/D)^2, \qquad |x| \leqslant D, \qquad f(x/D) = 0, \quad |x| > D \quad (1.5.2)$$
$$f(x/D) = [1 - (x/D)^2]^2, \qquad |x| \leqslant D, \qquad f(x/D) = 0, \quad |x| > D \quad (1.5.3)$$
$$f(x/D) = \cos(\pi x/2D), \qquad |x| \leqslant D, \qquad f(x/D) = 0, \quad |x| > D \quad (1.5.4)$$
$$f(x/D) = \cos^2(\pi x/2D), \qquad |x| \leqslant D, \qquad f(x/D) = 0, \quad |x| > D \quad (1.5.5)$$

Weighting or tapering functions [7, 12, 13] of this kind, play an important role in modern data processing. They find widespread use, for example, in Fourier transform spectrometry [7] where the primary data (the interferogram), recorded in the time or distance domain, has to be Fourier transformed to give the desired spectrum in the frequency or wave-number domain. When used in Fourier transform spectrometry, expressions of this type are often called "apodising" functions because the subsidiary damped oscillations, that is the side-lobes or "feet" which accompany sharp features in the Fourier transform of a truncated interferogram, are much reduced in amplitude if the primary data set is first tapered. The spectrum is then said to be apodised (from the Greek αποδοσ = footless). Spectral lines in an apodised spectrum look much more like those produced by a conventional dispersive spectrometer and are therefore considered "more acceptable" but this "improvement" has to be bought at the price of a reduction in the resolving power, a point which will be taken up again in section 1.6.3. The improvement in spectral line-shape has its origin in the removal of the sharp discontinuity at $|x| = D$ but since, by their nature, the weighting functions are themselves discontinuous, it is worthwhile investigating the order of the discontinuity to see what effect this has on the ultimate line-shape. The functions (1.5.1), (1.5.2) and (1.5.4) are merely themselves zero at $|x| = D$, whereas (1.5.3) and (1.5.5) have also that their first derivatives are zero at the cut-off point. The latter pair certainly give better line-shapes and, for this reason, are very popular but no one so far seems to have felt it necessary to use still higher order functions. The point is nicely discussed by Norton and Beer [14].

Despite the discontinuities of the apodising functions, they nevertheless have continuous well-behaved Fourier transforms. The Fourier transform is defined and described in the next section, so here we will merely say that it gives the spectrum of the function, that is if one imagines the function synthesised from an infinite set of sine and cosine waves, it shows how the amplitude of a narrow band of these waves varies with wave "frequency". The Fourier transform of the rectangle function $\Pi(x/2D)$, is the *sinc* function defined by

$$2D \operatorname{sinc} (2\tilde{v}D) = 2D \sin (2\pi\tilde{v}D)/2\pi\tilde{v}D. \tag{1.5.6}$$

The sinc function is therefore of the form "sin x over x". It is shown in Fig. 1.5. The side-lobes of the sinc function are the fundamental cause of the similar lobes produced in the Fourier transformation of any truncated function $f(x)\Pi(x/2D)$. In the particular case of Fourier transform spectrometry, an interferometer irradiated with monochromatic light will yield an interferogram which is an endless cosine wave and if a truncated portion of this wave is Fourier transformed, the result will be a spectral "line" whose shape is given by (1.5.6). In this sense, the sinc function is the natural spectral window or instrumental line-shape for Fourier transform spectrometry. The Fourier transform of the triangle function (x/D) is the sinc^2 function

$$D \operatorname{sinc}^2 (\tilde{v}D) = D[\sin (\pi\tilde{v}D)/\pi\tilde{v}D]^2. \tag{1.5.7}$$

This is also shown in Fig. 1.5: the reduction in the amplitude of the side-lobes is evident. The sinc^2 function is the natural spectral window for a diffraction limited dispersive instrument and since dispersive instruments were the common currency of spectroscopy for most of the history of the art, this particular line-shape is the form instinctively preferred by spectroscopists. It should be stressed, however, that a preference is still merely a preference and suppressing information, as one does when one uses tapering, can hardly be justified. This point will be discussed in detail in section 1.6.3.

The Fourier transforms of the other tapering functions are more complex [7], but show the same general form with still smaller side-lobes.

1.5.2 *The Bessel functions* $J_n(x)$, $N_n(x)$, $H_n(x)$, $I_n(x)$, *and* $K_n(x)$

The Bessel functions are a class of higher transcendental functions which occur frequently in mathematical physics [15]. They arise, for example, in the solution of the differential equation

$$x^2 f''(x) + xf'(x) + (x^2 - n^2) f(x) = 0. \tag{1.5.8}$$

They also occur as the Fourier transforms of some algebraic functions, for example

$$J_0(x) = \mathscr{F}(1 - 4\pi^2 \tilde{v}^2)^{-\frac{1}{2}}, \tag{1.5.9}$$

where $\mathscr{F}$ means "the Fourier transform of" and they also occur in the expansions of some periodic functions of a function as Fourier series, for example

$$\cos[a\cos\omega t] = J_0(a) - 2J_2(a)\cos 2\omega t + 2J_4(a)\cos 4\omega t \text{ etc.} \tag{1.5.10}$$

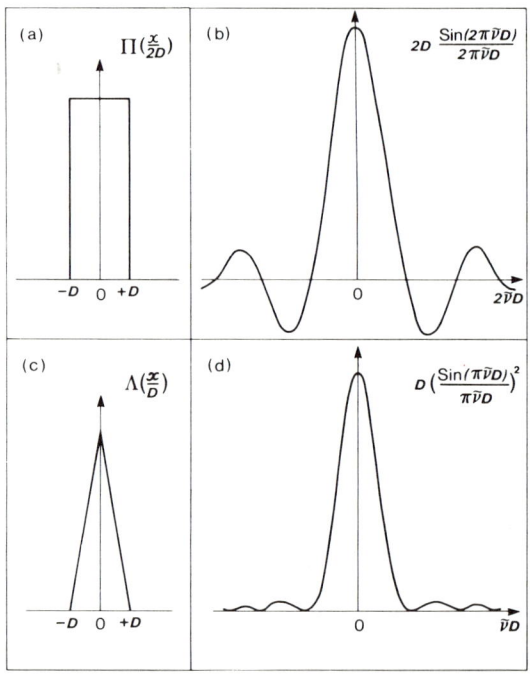

FIG. 1.5. The rectangle function (a), $\Pi(x/2D)$, and the sinc function (b), $2D \sin(2\pi\tilde{v}D)/2\pi\tilde{v}D$, are Fourier transforms of one another as are the triangle function (c), $\Lambda(x/D)$, and the sinc² function (d), $D[\sin(\pi\tilde{v}D)/\pi vD]^2$.

The parameter n which appears in the notation for the Bessel function $J_n(x)$ is called the order of the function and may have any value positive or negative, but for the great majority of physical applications it is either zero or else a small positive integer. The Bessel functions can be thought of as generalisations of the sine and cosine functions and can likewise be expanded in a power series.

$$J_n(x) = \sum_{s=0}^{s=\infty} \frac{(-1)^s}{s!(n+s)!} \left(\frac{x}{2}\right)^{n+2s}. \tag{1.5.11}$$

Like the sine and cosine functions, the Bessel functions oscillate but unlike them they are not periodic (Fig. 1.6) and their amplitudes show progressive attenuation. $J_0(x)$ has something in common with the sinc function and for large x approaches the asymptotic form

$$J_0(x) = (2/\pi x)^{\frac{1}{2}} \cos(x - \pi/4). \qquad (1.5.12)$$

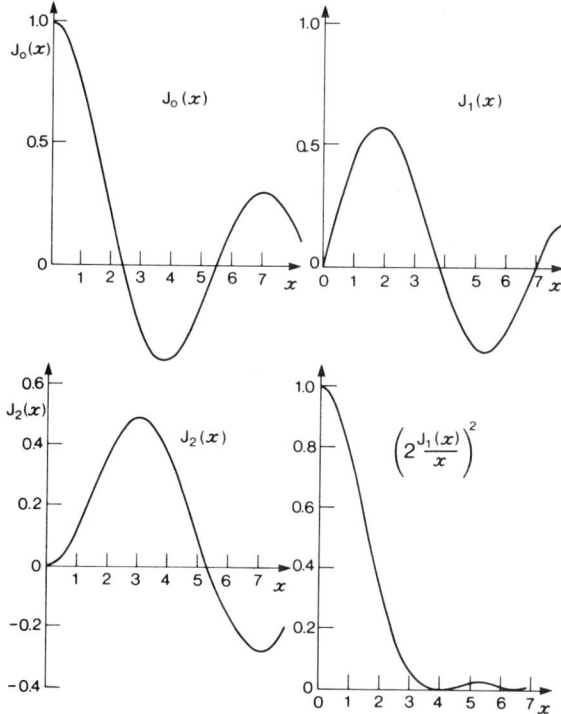

Fig. 1.6. Some low-order Bessel functions used in infrared science.

$2J_1(x)/x$ also resembles the sinc function. Its square (Fig. 1.6) is the celebrated Airy function which describes diffraction at a circular aperture and thus parallels in this respect the sinc function which describes that at a rectangular aperture. A treatment of diffraction in terms of these functions is given in section 3.2.

There are several relations connecting the Bessel functions just as there are for the trignometrical functions but possibly the most useful of these is the recurrence relation

$$J_{n-1}(x) + J_{n+1}(x) = (2n/x) J_n(x) \qquad (1.5.13)$$

which is used, amongst other things, in the practical computation of these functions. Equation (1.5.11) was originally derived on the understanding that the denominators were the ordinary factorial expressions or, in other words, that n was an integer. However, by invoking the extended factorial function $v!$ which is defined for any value of v by the relation

$$v! = \Gamma(1+v) \tag{1.5.14}$$

where Γ is the gamma function, the equation can be reinterpreted to *define* Bessel functions of any order. Of particular interest in the present context is the class of Bessel functions of half integral order, $J_{(2n+1)/2}(x)$. They are readily derived in series form by making use of the relation

$$\left.\begin{array}{l} (m+\tfrac{1}{2})! = (m+\tfrac{1}{2})\,(m-1+\tfrac{1}{2})! \\[2mm] \tfrac{1}{2}! = \tfrac{1}{2}\pi^{-\tfrac{1}{2}} \end{array}\right\} \tag{1.5.15}$$

It will be found however, (after some considerable manipulation!!) that the series can be rearranged into simple combinations of the familiar trignometric series, that is these particular Bessel functions degenerate into simple analytical forms. In particular

$$J_{1/2}(x) = \left(\frac{2}{\pi}\right)^{1/2} x^{-1/2}\,(\sin\,x) \tag{1.5.16a}$$

$$J_{3/2}(x) = \left(\frac{2}{\pi}\right)^{1/2} x^{-3/2}\,(\sin x - x\cos x) \tag{1.5.16b}$$

$$J_{5/2}(x) = \left(\frac{2}{\pi}\right)^{1/2} x^{-5/2}\left[(3-x^2)\sin x - 3x\cos x\right] \tag{1.5.16c}$$

By the application of (1.5.13) it will be seen that all the higher half integral order Bessel functions will also be simple analytical forms. The equations (1.5.16), after a little rearranging, readily yield the Fourier transforms of the commonly used weighting functions: in fact

$$(2\tilde{v})^{-1/2}\,J_{1/2}(2\pi\tilde{v}) = (\sin(2\pi\tilde{v})/2\pi\tilde{v}) = 2\,\mathrm{sinc}\,(2\tilde{v}) \tag{1.5.17a}$$

$$(2\tilde{v})^{-1}[J_{1/2}(\pi\tilde{v})]^2 = (\sin(\pi\tilde{v})/\pi\tilde{v})^2 = \mathrm{sinc}^2\,(\tilde{v}) \tag{1.5.17b}$$

$$4\,(\pi/2)^{1/2}\,(2\pi\tilde{v})^{-3/2}\,J_{3/2}(2\pi\tilde{v}) = (\pi\tilde{v})^{-2}\,(\mathrm{sinc}\,(2\tilde{v}) - \cos(2\pi\tilde{v})) \tag{1.5.17c}$$

$$15\,(\pi/2)^{1/2}\,(2\pi\tilde{v})^{-5/2}\,J_{5/2}(2\pi\tilde{v}) = 15\,(2\pi\tilde{v})^{-5}\left[\begin{array}{l}(3-(2\pi\tilde{v})^2)\sin(2\pi\tilde{v}) \\ -6\pi\tilde{v}\cos(2\pi\tilde{v})\end{array}\right] \tag{1.5.17d}$$

The right-hand sides of these equations are respectively the Fourier transforms of the rectangle function, the triangle function, the function given by (1.5.2) and that given by (1.5.3), when D is set equal to unity. The half-integral order Bessel function can therefore be used to describe apodised spectral windows, an application which is of some use in the more esoteric branches of Fourier transform spectroscopic theory.

In concluding this section it should be mentioned that there exists a set of functions called Neumann functions, or Bessel functions of the second kind, which are a second independent solution of Bessel's equation (1.5.8). These functions, usually denoted by $N_n(x)$, are (negatively) infinite at the origin and may therefore be used in the solution of those physical problems which involve a discontinuity at the origin. A common situation where Bessel and Neumann functions enter is the discussion of the propagation of electromagnetic radiation down a pipe (see Appendix 3). If in this case there is a central conductor, that is we have a coaxial line, then the Neumann functions will serve to describe the central discontinuity. It was mentioned earlier that the Bessel functions can be regarded as generalisations of the ordinary trigonometric functions and just as one normally combines these latter into the complex form

$$\exp(ix) = \cos(x) + i\sin(x), \tag{1.5.18}$$

so one may combine the two kinds of Bessel function into the complex forms

$$H_n^{(1)}(x) = J_n(x) + iN_n(x), \tag{1.5.19}$$

$$H_n^{(2)}(x) = J_n(x) - iN_n(x), \tag{1.5.20}$$

which are the Hankel functions. These are very useful in electromagnetic theory for ensuring the correct asymptotic behaviour of the field at infinity. Because $J_n(x)$ and $N_n(x)$ are strongly related to the trigonometric functions and may even be regarded as their generalisations, one can usefully consider the Bessel functions with imaginary arguments. These will then be the analogues of the sinh, cosh etc. functions. One generates in this way, the so-called *Hyperbolic Bessel* functions defined by

$$I_n(x) = i^{-n}J_n(ix), \tag{1.5.21}$$

$$K_n(x) = (\pi/2)i^{n+1}H_n(ix). \tag{1.5.22}$$

These functions find occasional use in infrared physics. Their series expansions are similar to those for the ordinary Bessel functions except that the factors of $(-1)^s$ are omitted.

1.5.3 *The harmonic oscillator and the Hermite polynomials*

The classical one-dimensional oscillator experiences a restoring force proportional to its displacement from its equilibrium position, i.e. $F = -kz$. It is readily shown that if the mass of the moving entity is m, then the frequency of oscillation is given by $\omega^2 = k/m$. It is also readily shown that the average potential energy is given by $V = \frac{1}{2}kz^2$. In terms of these quantities, the quantum mechanical wave equation [15] becomes

$$\frac{\hbar^2}{2m}\frac{d^2\Phi(z)}{dz^2} + \left[\mathscr{E} - \tfrac{1}{2}kz^2\right]\Phi(z) = 0, \tag{1.5.23}$$

which, with the substitutions

$$x = (m\omega/\hbar)^{1/2}z, \qquad \gamma = 2\mathscr{E}/\hbar\omega \tag{1.5.24}$$

where $\mathscr{E}$ is the total energy, becomes

$$\frac{d^2\Phi(x)}{dx^2} + \left[\gamma - x^2\right]\Phi(x) = 0. \tag{1.5.25}$$

This equation has solutions well behaved at infinity, only if γ is odd integral, i.e.

$$\gamma = (2n+1). \tag{1.5.26}$$

It follows that the energy is quantised and given by

$$\mathscr{E} = (n+\tfrac{1}{2})\hbar\omega. \tag{1.5.27}$$

With n integral, the solutions of equation (1.5.25) are the orthonormal functions

$$\Phi_n(x) = 2^{-(n/2)}\pi^{-(1/4)}(n!)^{-1/2}\exp(-x^2/2)H_n(x), \tag{1.5.28}$$

where $H_n(x)$ is the Hermite polynomial of order n. Care must obviously be exercised not to confuse these polynomials with the Hankel functions.

The Hermite polynomials are generated by the recurrence relations

$$H_{n+1}(x) = 2x\,H_n(x) - 2n\,H_{n-1}(x), \tag{1.5.29a}$$

$$\frac{d}{dx}H_n(x) = 2n\,H_{n-1}(x). \tag{1.5.29b}$$

They can also be generated by the successive derivatives of the $\exp(-x^2)$ function. Thus, if $y = \exp(-x^2)$, then

$$\frac{d^n y}{dx^n} = (-1)^n H_n(x)\,y. \tag{1.5.30}$$

The first few members of the series are given in Table 1.5. Some of these are illustrated in Fig. 1.7.

TABLE 1.5
Some low-order Hermite Polynomials

$$H_0(x) = 1$$
$$H_1(x) = 2x$$
$$H_2(x) = 4x^2 - 2$$
$$H_3(x) = 8x^3 - 12x$$
$$H_4(x) = 16x^4 - 48x^2 + 12$$
$$H_5(x) = 32x^5 - 160x^3 + 120x$$
$$H_6(x) = 64x^6 - 480x^4 + 720x^2 - 120$$

A non-linear molecule made up of N atoms has $3N$-6 normal vibrations and in practice it is found that each of these is well described by a harmonic oscillator model. Wave functions of the form (1.5.28) may therefore be used to describe molecular vibrations and hence to derive the selection rules for

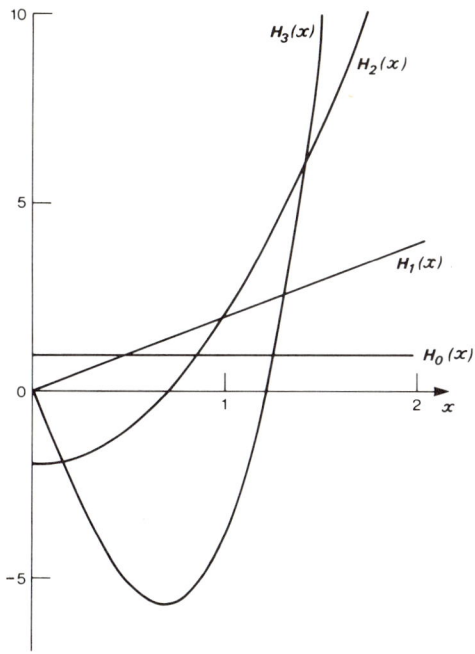

FIG. 1.7. The first four Hermite polynomials.

infrared absorption and the intensities to be expected in the infrared spectrum. Because of the form of the Hermite polynomials, and especially because of the alternating parity, the transition moment integral

$$[\mu]_{nm} = \frac{\partial \mu}{\partial x} \int_{-\infty}^{+\infty} \phi_n(x)\, x\, \phi_m(x)\, dx \qquad (1.5.31)$$

is only non-zero if $m = n \pm 1$. Therefore to first order, only fundamentals are expected in the infrared absorption spectrum of an ensemble of weakly interacting molecules.

The mathematical form of (1.5.28), in which the $\exp(-x^2)$ term ensures that the function is confined essentially to the region near the origin whilst a simple algebraic term determines its behaviour in that region, makes this relation useful in other connections apart from the quantum mechanical treatment of molecular vibrations. Thus the propagation of beam modes in open resonators (see sections 2.1.1 and 3.3) involves equations similar to (1.5.23). Because of this commonality, it is interesting to see the general shape of these functions. The first few functions, of the generic type $\phi_n(x) = \exp(-x^2) H_n(x)$ are therefore shown in Fig. 1.8.

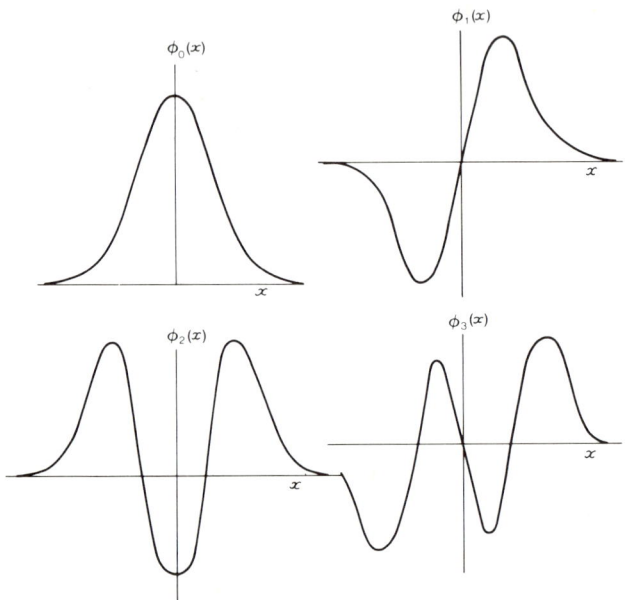

FIG. 1.8. The first few functions of the type $\exp(-x^2) H_n(x)$, drawn to arbitrary scale. In quantum mechanics the function itself has no direct physical meaning but its square gives the probability of the oscillator being in a small region surrounding the point. In electromagnetic propagation theory, the function gives the electric field amplitude and its square gives the beam intensity.

1.6 The use of integral transforms in infrared physics

1.6.1 *The Fourier transform*

The Fourier transform arose originally as the limiting case of the Fourier series (Appendix 5) and conceptually also in the evaluation of the coefficients of Fourier series. The idea was to resolve a given function into a set of cosine and sine waves which would add up everywhere to exactly reproduce the original function. This concept is very useful but it is even more useful when generalised. In this one thinks of two domains spanned by two different independent variables, say time and frequency. These two variables always have the characteristic that they have inverse dimensions (length and wave number are another pair) and if one has a function, say $F(t)$ in the first domain, the Fourier transform $f(v)$ in the other domain can be thought of as the "spectrum" of $F(t)$. Fourier transforms can be treated in terms of purely real functions and in terms of cosine and sine resolutions separately but it makes much more sense to start with the general complex form

$$\hat{f}(v) = \int_{-\infty}^{+\infty} \hat{F}(t) \exp(2\pi i v t) \, dt, \tag{1.6.1}$$

where $\hat{F}(t)$ and $\hat{f}(v)$ may be complex. The real power of the Fourier transform stems from Fourier's integral theorem which states that if (1.6.1) is true then

$$\hat{F}(t) = \int_{-\infty}^{+\infty} \hat{f}(v) \exp(-2\pi i v t) \, dv. \tag{1.6.2}$$

These equations are sometimes given in terms of ω (or equivalent variable) instead of v in which case factors of 2π enter. Rather than lose the symmetry of equations (1.6.1) and (1.6.2) it is conventional to write

$$\hat{f}(\omega) = (2\pi)^{-1/2} \int_{-\infty}^{+\infty} \hat{F}(t) \exp(i\omega t) \, dt \tag{1.6.3}$$

and

$$\hat{F}(t) = (2\pi)^{-1/2} \int_{-\infty}^{+\infty} \hat{f}(\omega) \exp(-i\omega t) \, d\omega. \tag{1.6.4}$$

With either form one can therefore readily pass back and forth between the two domains. This is very important physically because one often finds it more convenient to work in one domain rather than the other. Thus radio frequency engineers frequently prefer to study the time response of a system exposed to very brief pulses (that is to determine $F(t)$) rather than to laboriously record the response to sinusoidal inputs at varying frequencies (i.e. $f(v)$). Likewise infrared spectroscopists are increasingly finding it far preferable to record interferograms $I(x)$ in which mirror displacement, x, is the independent variable, rather than to have to record an infrared spectrum, i.e. $f(\tilde{v})$ directly. The methods of computing Fourier transforms in a high-speed digital computer are given in Appendix 5.

Some special cases of the Fourier transform are worth mentioning. Thus if $F(t)$ is real and symmetrical, only the real part of (1.6.1) is different from zero and the basic equation becomes

$$f_e(v) = \int_{-\infty}^{+\infty} F_e(t) \cos 2\pi v t \, dt, \tag{1.6.5}$$

where the subscript e means even. Likewise if $F(t)$ is real and odd, then only the imaginary part survives and

$$\hat{f}_0(v) = i \int_{-\infty}^{+\infty} F_0(t) \sin 2\pi vt \, dt, \qquad (1.6.6)$$

which can be divided through by i to give an equation in purely real terms. Sometimes, however, one of the pair of functions is purely real whilst the other is complex. In this case, we can divide the real function into two functions one even and the other odd:

$$F(t) = F_e(t) + F_0(t). \qquad (1.6.7)$$

Applying equation (1.6.1), we then have

$$\hat{f}(v) = \int_{-\infty}^{+\infty} F_e(t) \cos 2\pi vt \, dt + i \int_{-\infty}^{+\infty} F_0(t) \sin 2\pi vt \, dt \qquad (1.6.8)$$

and from this it follows at once that

$$\hat{f}(v) = \hat{f}^*(-v), \qquad (1.6.9)$$

where the asterisk indicates the complex conjugate. A function which satisfies equation (1.6.9) is said to be *Hermitian*. From the application of (1.6.2) it will be realised that the Fourier transform of an Hermitian function is pure real. The interesting physical consequences of the Fourier transformation of one-sided functions are discussed in Appendix 1.

Fourier transformation is basically a means for taking a problem defined in one domain into the conjugate domain where its solution may be easier or more illuminating. It follows that there will be conservation laws between the two domains and these will take the form of integral invariants. One of the most useful of these is given by Rayleigh's theorem [11], which states that if (1.6.1) and (1.6.2) are true then

$$\int_{-\infty}^{+\infty} |f(v)|^2 \, dv = \int_{-\infty}^{+\infty} |F(t)|^2 \, dt. \qquad (1.6.10)$$

In the particular case where the symbols refer to voltages, (1.6.10) merely states energy conservation but the theorem is generally true and it has several uses in quite diverse areas of optical physics and technology.

1.6.2 *The Hankel transform*

The Fourier transform is a member of the general class of *integral transforms* given by

$$\Psi(x) = \int_{-\infty}^{+\infty} \phi(y) \, K(x, y) \, dy \qquad (1.6.11)$$

where $K(x, y)$ is called the *kernel* of the transform. In the particular case of the Fourier transform the kernel is just $\exp(2\pi i xy)$. Another important integral transform, the Hankel transform, arises in connection with two-dimensional Fourier transformation. It involves Bessel functions which, as shown in section 1.5.2, can be regarded as generalisations of the trigonometric functions. One can define a two-dimensional Fourier transform in a manner entirely analogous to (1.6.1) via the relation,

$$\hat{f}(u, v) = \int_{-\infty}^{+\infty} \int_{-\infty}^{+\infty} \hat{F}(x, y) \exp 2\pi i [ux + vy] \, dx dy. \qquad (1.6.12)$$

Now in infrared systems, one always tries to work with cylindrically symmetric optics, not least because such an arrangement maximises energy throughput. It consequently makes more sense to work with cylindrical polar coordinates (r, θ, z) and the two-dimensional transform can then be reduced to a one-dimensional formulation since there will be no angular variation. Making the appropriate substitutions in (1.6.12) and invoking the standard identity [11]

$$J_0(x) = \frac{1}{2\pi} \int_0^{2\pi} \exp(-ix \cos \theta) d\theta \qquad (1.6.13)$$

leads to the result [11]

$$\hat{f}(q) = 2\pi \int_0^{\infty} \hat{F}(r) r \, J_0(2\pi qr) \, dr, \qquad (1.6.14)$$

where $q^2 = u^2 + v^2$ and $\hat{f}(q)$ is the Hankel transform of $\hat{F}(r)$. The importance of the Hankel transform arises from this experimental consideration and also from its reciprocity, for, entirely analogously to the case of the Fourier transform, one has that if (1.6.14) is true then

$$\hat{F}(r) = 2\pi \int_0^{\infty} \hat{f}(q) q \, J_0(2\pi qr) \, dq. \qquad (1.6.15)$$

In the case of the Hankel transform, the kernel is the zero'th order Bessel function. An important practical case of Hankel transformation arises with uniformly illuminated circular apertures for then $F(r)$ takes on the very simple form of the polar rectangle function $\Pi(r/2a)$ where a is the radius of the aperture. One then has that

$$\hat{f}(q) = 2\pi \int_0^a r J_0 (2\pi qr)\, \mathrm{d}r. \qquad (1.6.16)$$

The integral on the right-hand side of (1.6.16) is readily evaluated via one of the integral recurrence relationships for the Bessel functions, namely

$$x^{-n} J_n(x) = \int_{-\infty}^{+\infty} x^n J_{n-1}(x)\, \mathrm{d}x, \qquad (1.6.17)$$

so (1.6.16) becomes

$$\hat{f}(q) = (a/q) J_1 (2\pi aq). \qquad (1.6.18)$$

The use of this relation to describe diffraction at a circular aperture is outlined in section 3.2.

1.6.3 *Convolution, deconvolution and the convolution theorem*

Another very important transform—the convolution—arises in connection with the Fourier transform of a product. This operation is a frequent occurrence in infrared physics. It is involved, for example, in the consideration of the performance of a measuring instrument—that is the question of how faithfully it gives an account of the phenomena it is being used to investigate. A well-known case is the limitation of the resolving power of a dispersive spectrometer brought about by the need to use a finite slit-width. The equivalent for an interferometric system is the limitation of resolution due to the practical requirement of studying only a finite stretch of the interferogram. This is equivalent to multiplying the endless interferogram by a truncating or apodising function and then taking the Fourier transform of this product. The ability to deal with products comes from the remarkable *convolution theorem*, which states that if $f(v)$ and $g(v)$ have Fourier transforms $F(t)$ and $G(t)$ then [11]

$$f(v) * g(v) = \int_{-\infty}^{+\infty} F(t) G(t) \exp (2\pi ivt)\, \mathrm{d}t, \qquad (1.6.19)$$

where the symbol * means "convolved with". The definition of "convolved with" is the integral transform

$$f(v) * g(v) = \int_{-\infty}^{+\infty} f(v') g(v - v')\, \mathrm{d}v'. \qquad (1.6.20)$$

In this equation v' has the same dimensions and scans the same domain as does v—a different symbol is used because v is essentially parametric, i.e. it is a *fixed* value of abscissa about which the integral is to be evaluated. It is important to note that the convolution symbol * implies the dimension of the argument of f and g. Convolution is equivalent to first displacing one function, g, over the other h until the new origin of g, namely $v' = v$, is reached; g is then reflected about the ordinate at this point. The transposed and reflected function g is then multiplied by f and the resulting function integrated over the whole range of v'. It will be realised of course that one could take either g or f first and still get the same answer, i.e.

$$f * g = g * f. \tag{1.6.21}$$

In most infrared applications at least one of the functions is usually even in which case the convolution operation is somewhat simplified since the reflection operation is then redundant. The broadening of a spectral line by convolution with a rectangular slit function is shown in Fig. 1.9.

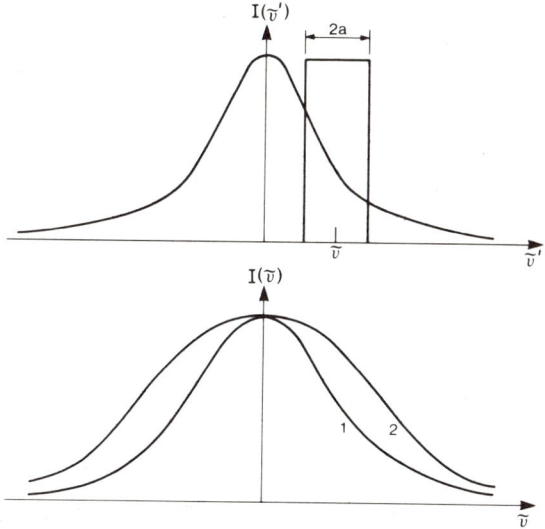

FIG. 1.9. Convolution of a Lorentz line shape function of half-width 2 with a rectangular slit function of total width $2a$. In the lower inset are shown the results for $a = 1$ and $a = 2$.

The Lorentz line function shown in the upper inset of Fig. 1.9 has the analytical form

$$I(\tilde{v}') = \frac{1}{1 + (\tilde{v}')^2}, \tag{1.6.22}$$

where the line centre frequency has been chosen as origin and the half-width is 2 units. It will be seen that the effects of the convolution only become noticeable when the width of the scanning function (i.e. $2a$) becomes comparable with or greater than the line width. In analytical terms, one can write the result of the convolution (from (1.6.20)) in the normalised form:

$$I(\tilde{v}) = \frac{\tan^{-1}(\tilde{v} + a) - \tan^{-1}(\tilde{v} - a)}{2 \tan^{-1} a}. \tag{1.6.23}$$

This function degenerates to (1.6.22) as a tends to zero and tends towards $\Pi(v/2a)$ as a tends to infinity. This illustrates the equivalence of the convolution operation.

Line broadening by real spectrometers (convolution with the *apparatus function* as it is sometimes called) is very frustrating and, since its exact origin is known, it is very tempting to enquire whether one could mathematically reverse the process, i.e. invoke *deconvolution*, and thereby get high resolution spectra from an instrument of only moderate resolving power. The question applies equally to all types of spectrometer but, since grating instruments and interferometric instruments derive their prime data in different domains, it is convenient to discuss deconvolution from the two different viewpoints first before showing how they are related.

The limiting apparatus function for a grating (or prism) spectrometer is, as will be discussed in Chapter 3, set by diffraction at the various apertures which the beam encounters within the instrument. It is usually of the sinc² form. However in practical infrared spectrometry it is most unusual to experience this fundamental limit. More commonly, energy shortage compels the spectroscopist to open up his slits and then the apparatus function becomes a simple triangle function whose span is determined by the slit widths and the geometry of the instrument. In either case, the observed spectrum will be given by

$$S_{\text{obs}}(\tilde{v}_i) = \int_0^\infty S(\tilde{v}) A(\tilde{v}_i - \tilde{v}) \, d\tilde{v}, \tag{1.6.24}$$

where $A(\tilde{v})$ is the apparatus function or spectral window. In principle, therefore, since one knows $S_{\text{obs}}(\tilde{v})$ and $A(\tilde{v})$ one ought to be able to calculate $S(\tilde{v})$. Unfortunately the general integral transform (1.6.24) does not have any simple inverse so the problem of deriving $S(\tilde{v})$ is far from trivial. One possible approach is to express $S(\tilde{v})$ as a Taylor's series and to integrate term by term. One then has

$$S_{\text{obs}}(\tilde{v}_i) = S(0) + \sum \left(\frac{d^n s}{d\tilde{v}^n} \right) \frac{1}{n!} \int_0^\infty \tilde{v}^n A(\tilde{v}_i - \tilde{v}) \, d\tilde{v}. \tag{1.6.25}$$

Hopefully this series will rapidly converge and then it becomes permissible to employ only a reasonable number of terms in the Taylor series. Next an equal number of regularly spaced ordinates of $S_{\text{obs}}(\tilde{v}_i)$ are selected from the experimental record, whereupon (1.6.25) becomes a simple matrix-vector equation with the transformation matrix square. The inversion to give the coefficients of the Taylor series is then rapidly carried out using a standard library programme in the computer. The difficulties are (a) that the matrix, made up of the successively higher integrals of the frequency weighted apparatus function, may be ill conditioned and (b) that the presence of noise, that is random uncertainty, on the selected ordinates may make the transform even more unreliable than the original. All this of course ignores the difficulty of evaluating all the integrals in (1.6.25), many of which cannot be expressed analytically. It will be realised therefore that computer deconvolution by this route is a laborious and uncertain operation and it can only be recommended when dealing with a simple spectrum made up of at most a few sharp lines and, furthermore, when the observed signal-to-noise ratio is high. Nevertheless such situations do sometimes occur and they especially tend to do so under circumstances where it is only the absolute peak height of a line which is of interest. When this is the case, this Taylor series approach can be helpful.

The interferometric spectroscopist will look at deconvolution in a quite different way. This is because his primary data are obtained in the time or path-difference domain (section 4.3) and he sees at once that the observation of lines broader than they ought to be and with the wrong profile is a consequence of terminating the observation of the interferogram before it has finished oscillating. Of course, in the strictest sense, the interferogram will *never* finish oscillating but most interferograms soon reach a point where either the oscillations have become less than the discrimination of the recording system and/or less than the noise level. In this case the termination will have little effect. On the other hand, when the spectrum contains sharp lines one may well find that one has insufficient mirror travel available and observation may have to cease whilst there is still meaningful information waiting to be gleaned. The experimentalist will usually take full advantage of the symmetry (or sometimes antisymmetry) of the interferogram to make the best of a bad job and will set the span of his mirror travel to cover just one side of the interferogram but even so he may run out of mirror travel prematurely. The situation is shown schematically in Fig. 1.10. One can represent the termination at $x = x_{max}$ analytically by writing

$$I_{obs}(x) = I(x) \prod \left(\frac{x}{2x_{max}} \right), \tag{1.6.26}$$

and one then obtains an inexact spectrum according to

$$S_{obs}(\tilde{v}) = \mathscr{F} \left[I(x) \prod \left(\frac{x}{2x_{max}} \right) \right], \tag{1.6.27}$$

a result which is identical with merely carrying out the integration over the limits 0 to x_{max}. Equation (1.6.27) is, however, a useful way of looking at the situation since by the use of the convolution theorem one has at once

$$S_{obs}(\tilde{v}) = S(\tilde{v}) * A(\tilde{v}), \tag{1.6.28}$$

where here $A(v)$ is the spectral window or apparatus function corresponding to rectangular truncation and it is hence the Fourier transform of the rectangle function, i.e.

$$A(\tilde{v}) = 2x_{max} \, \text{sinc}(2\tilde{v} \, x_{max}). \qquad (1.6.29)$$

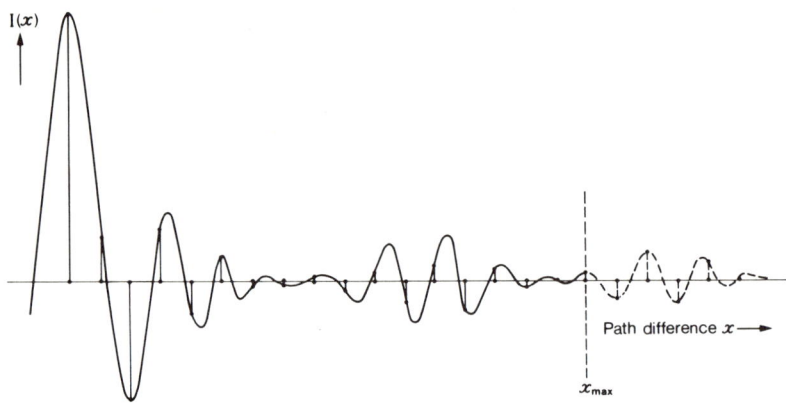

F$_{IG}$. 1.10. The interferogram which is produced by a two-beam interferometer is the autocorrelation function of the incoming radiant field and therefore contains all the information about the spectrum of the field. Unfortunately it can only be determined out to a maximum path-difference x_{max} which is set by experimental considerations. A reliable method for extrapolating beyond x_{max}, if available, would yield estimates of the undetermined ordinates and, when the set of augmented samples is fed into a discrete Fourier transformer, the resulting spectrum will show enhanced resolution and better line-shapes than would be available from straight Fourier transformation of the truncated interferogram.

This particular spectral window (Fig. 1.4) is most undesirable since the subsidiary oscillations (or side-lobes), of a strong line, can either be misidentified as real features or, perhaps even worse, mask weaker genuine features. When the spectrum contains a number of sharp lines, then the rich side-lobe structure (or "side-lobe rubbish" as it is sometimes called by frustrated spectroscopists!) is most annoying. Faced with this situation, the almost universal practice [16] is to resort to apodisation by multiplying the interferogram by a taper function $T(x)$ which is unity at $x = 0$ and which is zero at $x = x_{max}$. Some representative examples from the infinite set of taper functions have been discussed previously in section 1.5. When the interferogram is tapered, equation (1.6.28) still applies but $A(\tilde{v})$ is now the Fourier transform of $T(x)$. An important point to realise is that these spectral windows *are* apparatus functions, that is they are determined only by the apparatus and not at all by the spectrum under study. The window is therefore inflexible and unable to adapt to particular spectral data—thus if two lines are close together and of greatly differing intensities then the "rigidity" of the window, that is the unvarying placing of its side-lobes, can be a considerable nuisance since it may prevent the lines being resolved. Tapering can help slightly but not under the circumstances where help is most needed, that is when the data is sparse. More flexible approaches will be discussed later. One can analyse tapering from an information theory[17] point of view with some illuminating results. Abrupt truncation is equivalent to sending two false statements into the communication line (in this case the Fourier transforming computer). The first is that the ordinates beyond x_{max} are zero and the second is that there is a discontinuity at x_{max}. These two falsehoods are independent of one another and it is possible to achieve a trade-off. Thus by eliminating the discontinuity one can drastically reduce the subsidiary oscillations (a form of "ringing" closely related to the well-known Gibbs phenomena) at the price of only a moderate (twice) increase in the observed line-width. To illustrate this, the application of a triangular taper (equation 1.5.1) gives the sinc^2 spectral window which has a much weaker side-lobe structure (Fig. 1.5), but because of the reduction of information flow has an increased width for its central lobe. If this increase in line-width can be tolerated, then the consequent suppression of the interline side-lobe rubbish can be very valuable indeed and most computer programmes feature a tapering sub-routine with a choice between several taper functions.

The above outline contains all the basic ideas but in practice one needs also to bear in mind that it is impossible to perform continuous numerical integrations in a computer. What one has to do is to take a set of $(2M + 1)$ samples (usually equispaced) of $I(x)$ and to replace the integration by the summation

$$S(\tilde{\nu}) = \mathscr{F}_D[I(n\Delta x)] = \sum_{-M}^{+M} I(n\Delta x)\exp(2\pi in\tilde{\nu}\Delta x), \qquad (1.6.30)$$

where $\mathscr{F}_D$ means "discrete Fourier transform of". Equation (1.6.30) gives a continuous function $S(\tilde{\nu})$ from a sampled one—a fact which can be underlined by writing (1.6.30) in the equivalent form:

$$S(\tilde{\nu}) = \int_{-\infty}^{+\infty} \sqcup\!\sqcup\left(\frac{x}{\Delta x}\right)I(x)\exp(2\pi i\tilde{\nu}x)\,dx. \qquad (1.6.31)$$

The operation of sampling was analysed from an information theoretical viewpoint by Nyquist who showed that provided $S(\tilde{\nu})$ contained only zero ordinates for wavenumbers higher than $\tilde{\nu}_N = (2\Delta x)^{-1}$, then the reconstruction of $S(\tilde{\nu})$ from the sampled interferogram would be perfect over the range $-\tilde{\nu}_N < \tilde{\nu} < \tilde{\nu}_N$. This is the famous sampling theorem and the limiting or "folding" wavenumber (or frequency) $\tilde{\nu}_N$ is often called the Nyquist wavenumber (or frequency) in tribute to Nyquist's work. It also follows that the spectrum in the fundamental range will be replicated eternally throughout the domain of $\tilde{\nu}$ (or ν), a fact which can readily be established by noting that (1.6.30) is a Fourier series and is therefore periodic. If Nyquist's condition is *not* satisfied, that is if the interferogram is sampled too coarsely, then real spectral energy lying above $\tilde{v}_N$ will be "folded-back" or "aliassed" into the fundamental range leading to a loss of information in the spectrum. The use of (1.6.30) as it stands is nowadays called the "slow Fourier transform"—slow because of the immense amount of computer time required to calculate all the cosines and sines making up the complex exponential. In most modern work the "Fast Fourier Transform" or FFT based on the Cooley—Tukey algorithm (Appendix 5) is used and then $S(\tilde{v})$ is also only available at a set of discrete values for $\tilde{v}$. This is equivalent to replicating the interferogram throughout its domain and one has a symmetric situation which is sometimes called the periodogram approach. It is interesting in this connection to note that the use of a linear taper is unique in that it can be described succinctly by the matrix vector techniques of linear algebra. Thus if a column vector is defined as

$\mathbf{E} = \mathrm{col}\,[1,\ \varepsilon,\ \varepsilon^2\ \dots\ \varepsilon^{M-1}]$ where $\varepsilon = \exp\,(-2\pi i\Delta x\tilde{v})$ and a "correlation" matrix $\mathbf{R}$ whose elements $R_{ij} = I_{j-i}\ (= I \pm (j-i)\Delta x)$ then, with linear tapering,

$$MS(\tilde{v}) = \tilde{\mathbf{E}}\mathbf{R}\mathbf{E}^* \qquad (1.6.32)$$

where $\sim$ indicates transposition. Because of this unique feature, linear tapering has received considerable attention and the corresponding sinc^2 apparatus function has even come to be occasionally called by a special name "the Bartlett window" after Bartlett who carried out much of the initial work [18]. Matrices such as $\mathbf{R}$ whose elements are determined by the difference of their indices are often called Toeplitz matrices.

Having carried out the examination from the two different points of view, one is now in a position to compare the two different forms of spectroscopy. The dispersive spectroscopist does not ever observe true data since his raw output is already the true spectrum convolved with his apparatus function. The interferometric spectroscopist does observe true data but to get reasonable looking spectra he has to falsify the data before feeding it into the computer. At the end of the day, when both have a spectrum on paper, they are in equivalent positions. However this comparison does show how the deconvolution problem can be analysed. The interferometric spectroscopist can, if he wishes, remove the tapering and get an unapodised spectrum, but that is apparently as far as he can go to improve the resolution. The dispersive spectroscopist can see at once that some measure of amelioration may be possible since, by the convolution theorem, it follows that

$$I(t)R(t) = \int_{-\infty}^{+\infty} [S(v)^* A(v)]\exp(-2\pi ivt)\,dv, \qquad (1.6.33)$$

where $I(t)$ is the Fourier transform of $S(v)$ and $R(t)$ is that of $A(v)$. One then has

$$S(v) = \int_{-\infty}^{+\infty} \left[\frac{1}{R(t)} \int_{-\infty}^{+\infty} S_{\mathrm{obs}}(v)\exp(-2\pi ivt)\,dv\right]\exp(2\pi ivt)\,dt. \quad (1.6.34)$$

This equation will give a spectrum with enhanced resolution provided $R(t)$ has no zeroes and $S_{obs}(v) = S(v) * A(v)$ is known over the whole frequency range. Neither assumption, unfortunately, is commonly true but when used with care, abetted by good physical intuition, some measure of improvement can usually be achieved. Clearly, however, and especially when the effects of noise are considered, one would have to be very sanguine to expect dramatic improvement from the deconvolution of the spectra obtained from a grating instrument. Kauppinen and his colleagues have discussed this matter at some length [19] and have come to the conclusion that, with realistic noise levels, improvements in resolution by a factor between two and three may be possible. This approach, in common with all others of this class, does involve the assumption of a line-shape function. As a result, one can only expect to get improved estimates of line *position* by this technique since, of course, no information is then forthcoming about the true shape of the line.

The real problem that one is grappling with in deconvolution is that the interferogram (or its equivalent in dispersive spectroscopy) is not known over its entire range. To get a better result than that given by straightforward transformation requires interpolation, extrapolation, or both. The case of the truncated interferogram in FTS is the simplest to analyse so this case will be used for illustration but entirely equivalent arguments apply to all the others. At once it must be said that the whole of the traditional apodisation procedure looks most strange when examined from the vantage point of information theory. It is, after all, odd to assert that all ordinates after $x = x_{max}$ are zero when one knows full well that they are not! It is even odder to throw away perfectly good information as one does when one applies a taper. The information theoretician would analyse the situation by saying that the perfect reciprocity between the two domains, which applies in the ideal world, fails in the real world and whilst it remains true that knowing $S(\tilde{v})$ one can calculate $I(x)$ perfectly over the entire observed range of x, it does not work the other way round. If one feeds this information-depleted interferogram into a communication line (the Fourier transformer), one will get out a result which has impaired resolution and a distorted profile. By reducing the information flow still further with the introduction of a taper, one gets, necessarily, still worse resolution and the concomitant loss of information about the true line-shape. Obviously, what one wants to do is to find some way of maximising the information flow and minimising the insertion of the falsehoods which stem from the operator's prejudice. In other words we want a recipe for giving the best unbiased estimates for the experimentally undetermined ordinates which lie beyond x_{max}. Naturally we resort to probability theory and, since the conceptual basis of thermodynamics and of the probability theory of information flow are virtually the same, it is usual to use the language of thermodynamics to describe the situation. In particular the word "entropy" is used to mean a measure of information loss or ignorance gain. In our case, the information beyond x_{max} is totally lost so we seek a solution which maximises the entropy. The approach is, for this reason, usually called the *Maximum Entropy Method* (or MEM)[20]. It was invented by J. P. Burg in 1967 and a closely related approach called the "Maximum Likelihood Method" was introduced by Capon in 1969.

One begins the analysis of the MEM by considering a random event in which one has i possible outcomes and the probability of each event is p_i. The least prejudiced estimate of the state of knowledge of the actual outcome is obtained by maximising the quantity, $p_i \log p_i$ subject to all the known constraints. These include the obvious ones $0 < p_i < 1$ and $\sum p_i = 1$, but there may be others if some of the outcomes are weighted (loaded dice for example!). The use of the entropy function, $p_i \log p_i$, for the solution of probability distributions under a set of constraints was first clearly spelled out by Jaynes[21] and it is therefore sometimes called Jaynes's Principle but, conceptually at least, it is much older. It has been discussed very lucidly by Ables in some unpublished notes and its application to spectroscopy has been discussed by Lacoss[22] and by White[23]. The approach owes much to Bartlett[18] who showed how one could estimate the entropy of a spectrum by considering it to be produced from a "white" noise input to a linear filter whose power gain function was $S(v)$. Bartlett's result is that the entropy gain of the filter is

$$\Delta H = \int_{-\infty}^{+\infty} \log S(v)\,dv. \tag{1.6.35}$$

In practical interferometric spectroscopy where one works with a set of samples of the interferogram (that is the autocorrelation function of $S(v)$), $S(v)$ is constrained to be zero above the Nyquist frequency v_N, so equation (1.6.35) is frequently written

$$\Delta H = \int_{-v_N}^{+v_N} \log S(v)\,dv. \tag{1.6.36}$$

The integration over negative values of the frequency is retained for mathematical convenience but since $S(v)$ has to be an even function of frequency, (1.6.36) can be replaced by twice the integral over the physical domain $0 - v_N$. The situation now is clear, one wishes to maximise ΔH subject to the conditions that the spectral function which emerges gives all the sampled ordinates $I(n\Delta x)$ exactly, i.e. that

$$\int_{-v_N}^{+v_N} S(\tilde{v}) \exp(2\pi i \tilde{v} n \Delta x)\,d\tilde{v} = I(n\Delta x) \tag{1.6.37}$$

for *all* the observed values of n. This is a calculus of variations problem and it can be solved by introducing a Lagrangian multiplier for each of the constraint equations (1.6.37). Ables, in his notes (but see also [21]) shows that the final solution can be expressed in the form

$$S_{MEM}(\tilde{v}) = g/|\mathscr{F}_D(g_i)^2|^2, \; i = 0, 1, \ldots n, \tag{1.6.38}$$

where $g = \Delta x/\varepsilon$ and the sequence of the g_i is obtained as the solution of

$$\mathbf{R} \cdot \mathbf{E} = \mathrm{col}(1, 0, 0, \ldots 0) \tag{1.6.39}$$

where $\mathbf{R}$ is the Toeplitz correlation matrix introduced earlier. Toeplitz matrices have the property that they can be inverted very rapidly even if they are very large (e.g. $10^3 \times 10^3$). This is an important point since one would normally only consider the use of the MEM when one had run out of data points against one's will and in normal laboratory practice, that would mean when one had reached a rather large value of M. The discrete Fourier transform would be carried out using the very efficient Cooley-Tukey algorithm (see Appendix 5) so the whole operation in the computer can be quite quick even for large numbers of interferogram samples. The results which have been obtained so far are impressive and underline the conceptual soundness of the MEM approach.

Looked at in the interferogram domain, the MEM is a recipe for calculating further ordinates, beyond x_{max}, in terms of those already known. The matrix inversion, just mentioned, is then equivalent to having a recurrence or autoregressive relation which gives the further ordinates uniquely. It is an interesting point in the more esoteric reaches of Fourier transform theory that merely requiring that (a) the interferogram or correlation function be stable, that is that on the large-scale it falls in amplitude rather than rises, and (b) that all the ordinates of its transform be positive, suffices to give a unique solution. This sounds at first close to miraculous but on reflection it seems much more reasonable. Instinctively one feels that if one has observed an oscillating interferogram for a sufficient amount of path-difference, up to the experimental cut-off at x_{max}, then one could fairly reliably sketch in a continuation beyond x_{max} which would decay away at a reasonable rate to negligibly small values. The MEM provides a mathematically tight rule for carrying out this extrapolation. The Fourier transform of the extrapolated interferogram will be superior because more "information" is included and because there is no longer any sudden discontinuity. The equivalent treatment for the dispersive case is entirely similar except that interpolation, rather than extrapolation, is required to fill in the parts of the correlation function which are rendered zero, or very small, by the minima in the spectral window function. The MEM can be shown, from these arguments, to give a spectral window which is not fixed but which rather adapts itself to give the best fit to the data. For this reason, it is often called a "data-adaptive" method of spectral analysis.

The use of the MEM is still very new in infrared spectroscopy so such things as noise analysis for it are, as yet, in their infancy. It seems that one can either violate the principles of the method by considering noise signals propagated through a *fixed* data analysis system—in other words to interpret the constraints of equation (1.6.37) to force $S(\hat{v})$ to give the noisy $I(x)$ exactly—or more rationally merely to require that the interferogram function produced from $S(\hat{v})$ should pass close to the observed (but experimentally uncertain) ordinates. In either case it appears from what has been done so far that the MEM is no worse, from a noise point of view, than the traditional approach. One potential disadvantage of the MEM is that it is fundamentally non-linear. In the traditional method, the spectrum produced from the sum of two correlation functions is simply the sum of their individual spectra and signal and noise propagate through the mathematical transformation without interaction. In the MEM there is the possibility of "cross-talk" but again, from what has been done so far, it appears that the problem is minimal and the MEM has even been referred to as a "quasi-linear" technique.

A related topic to convolution and deconvolution is "mathematical filtering". This is the term used to describe the process of altering the spectral distribution of power mathematically by invoking logical operations in a digital processor rather than physically with the help of an absorptive or an interference filter. It is nowadays an enormously important topic in communication engineering because of the rapid advance in the use of digital techniques for virtually all forms of information transfer. The theory of digital filters [24] has even become a subject in its own right. In infrared spectroscopy, digital filtering was originally introduced to deal with the dynamic range problems which arose with the early computers and their associated digital peripherals. The idea was that if one had a broad-band input spectrum whose correlation function (that is the interferogram) might swamp the dynamic range of the system, one might wish to multiply the spectral function by a suitably located rectangle function which, by restricting the spectral range, would lead to the observation of an interferogram whose dynamic range was reduced but which would still contain all the spectral information of interest. This multiplication in the frequency (or wavenumber) domain could be ensured by convolution with a sinc function in the time (or path-difference) domain. Modern digital equipment is much better, from a discrimination point of view, and mathematical filtering is no longer so necessary, but it is still occasionally of value when one is trying to squeeze the last drop of information out of a narrow range of an interferometrically observed spectrum.

1.6.4 *The Laplace and Hilbert transforms*

The Laplace transform defined by

$$\mathcal{L}_p[F(t)] = \int_0^\infty F(t)\exp(-pt)\,\mathrm{d}t \qquad (1.6.40)$$

where p is a complex variable is a close relative of the Fourier transform and, in fact, when p is pure imaginary, it degenerates into a simple half range Fourier transformation. Often, though, in practical problems p is pure real. The Laplace transform is much used in electrical engineering because of its power to deal with differential equations which may be stubborn in one domain but much more tractable in the other. It does however, unlike the Fourier transform, suffer from the drawback that there are often ranges of the variable p over which the integral does not exist. But, given that the integral does exist for a given range of p then the Laplace transform may be obtained from a known Fourier transform by simply setting $p = i\omega$, that is $2\pi i v$. For example, since the Fourier transform of the rectangle function $\Pi(t)$ is sinc v, i.e. $(\sin \pi v)/\pi v$, it follows that the Laplace transform will be given by

$$\mathscr{L}_p[\Pi(t)] = \sin(ip/2)/(ip/2) = \sinh(p/2)/(p/2), \qquad (1.6.41)$$

the last step following from (1.4.28). As another example, the Laplace transform of $\exp(-t/\tau)$ will be $(p + \tau^{-1})^{-1}$ and hence the half-range Fourier transform will be $(i\omega + \tau^{-1})^{-1}$. This function and some of its relatives are widely used in the analysis of dielectric phenomena (see section 5.5.2) and of line-shapes (see Appendix 4). The close relationship between the Laplace and the Fourier transforms is well brought out by the existence of an exact analogue of the convolution theorem:

$$\mathscr{L}_p(f_1 {}^*\!f_2) = \mathscr{L}_p(f_1)\mathscr{L}_p(f_2). \qquad (1.6.42)$$

Another integral transform which occurs reasonably frequently in infrared work is the Hilbert transform $\mathscr{H}(f(t))$. This is a special case of convolution, viz. convolution with the $(\pi t)^{-1}$ function which itself is the sine Fourier transform of the signum function

$$\begin{aligned} \operatorname{sgn}(v) &= +1, & v &> 0 \\ \operatorname{sgn}(v) &= 0, & v &= 0 \\ \operatorname{sgn}(v) &= -1, & v &< 0 \end{aligned} \qquad (1.6.43)$$

It follows that the Hilbert transform is then defined by

$$\mathscr{H}(f(t)) = \pi^{-1} \int_{-\infty}^{+\infty} [f(t')/(t'-t)]\,\mathrm{d}t'. \qquad (1.6.44)$$

Its application in connection with the Causality Principle and the consequent derivation of the Kramers–Kronig relations is described in Appendix 1.

1.7 Coordinate transformations in connection with the solution of partial differential equations

The partial differential equations which arise in electromagnetic theory are often referred to one coordinate system (usually the Cartesian), whereas the geometry of the experimental arrangement is more naturally described by another. One needs therefore to transform the equation to this more appropriate system. The operation can be illustrated by a rather common example, the Cartesian to cylindrical coordinate transformation

$$(x, y, z) \rightarrow (r, \phi, z), \tag{1.7.1}$$

where

$$x = r \cos \phi \quad \text{and} \quad y = r \sin \phi. \tag{1.7.2}$$

In this particular case, since $z \rightarrow z$, we need only consider the transformation

$$(x, y) \rightarrow (r, \phi). \tag{1.7.3}$$

The method of linking the two systems is to use the formula for a total differential. Thus if we have a function $U(x, y) = U(r, \phi)$, then we may write

$$dU = \left(\frac{\partial U}{\partial x}\right)dx + \left(\frac{\partial U}{\partial y}\right)dy \quad \text{and} \quad dU = \left(\frac{\partial U}{\partial r}\right)dr + \left(\frac{\partial U}{\partial \phi}\right)d\phi. \tag{1.7.4}$$

By substituting the differentials from (1.7.2) and identifying terms, it follows that

$$\frac{\partial U}{\partial x} = \cos \phi \left(\frac{\partial U}{\partial r}\right) - \frac{\sin \phi}{r}\left(\frac{\partial U}{\partial \phi}\right), \tag{1.7.5a}$$

$$\frac{\partial U}{\partial y} = \sin \phi \left(\frac{\partial U}{\partial r}\right) + \frac{\cos \phi}{r}\left(\frac{\partial U}{\partial \phi}\right). \tag{1.7.5b}$$

By applying the same operation to $(\partial U/\partial x)$ and $(\partial U/\partial y)$ it follows (after some intricate symbol manipulation) that

$$\frac{\partial^2 U}{\partial x^2} = \cos^2 \phi \left(\frac{\partial^2 U}{\partial r^2}\right) + \frac{\sin^2 \phi}{r^2}\left(\frac{\partial^2 U}{\partial \phi^2}\right) - \frac{2\sin \phi \cos \phi}{r}\left(\frac{\partial^2 U}{\partial r \, \partial \phi}\right)$$
$$+ \frac{\sin^2 \phi}{r}\left(\frac{\partial U}{\partial r}\right) + \frac{\sin 2\phi}{r}\left(\frac{\partial U}{\partial \phi}\right), \tag{1.7.6a}$$

$$\frac{\partial^2 U}{\partial y^2} = \sin^2 \phi \left(\frac{\partial^2 U}{\partial r^2}\right) + \frac{\cos^2 \phi}{r^2}\left(\frac{\partial^2 U}{\partial \phi^2}\right) + \frac{2\sin \phi \cos \phi}{r}\left(\frac{\partial^2 U}{\partial r \, \partial \phi}\right)$$
$$+ \frac{\cos^2 \phi}{r}\left(\frac{\partial U}{\partial r}\right) - \frac{\sin 2\phi}{r}\left(\frac{\partial U}{\partial \phi}\right). \tag{1.7.6b}$$

The differential equations that arise in practice are always symmetrical in x, y and z and therefore involve either the operator

$$\nabla = \frac{\partial}{\partial x} + \frac{\partial}{\partial y} + \frac{\partial}{\partial z} \tag{1.7.7a}$$

or more commonly the operator

$$\nabla^2 = \frac{\partial^2}{\partial x^2} + \frac{\partial^2}{\partial y^2} + \frac{\partial^2}{\partial z^2} \tag{1.7.7b}$$

and then one has, for the present case,

$$\nabla^2 U = \frac{\partial^2 U}{\partial r^2} + \frac{1}{r^2} \frac{\partial^2 U}{\partial \phi^2} + \frac{1}{r} \frac{\partial U}{\partial r} + \frac{\partial^2 U}{\partial z^2}. \tag{1.7.8}$$

The use of this identity in the solution of Helmholtz's equation is illustrated in section 3.6.2. Spherical polar coordinates are the other common system and for these the same methods eventually yield

$$\nabla^2 U = \frac{1}{r^2 \sin \theta} \left[\sin \theta \frac{\partial}{\partial r} \left(r^2 \frac{\partial U}{\partial r} \right) + \frac{\partial}{\partial \theta} \left(\sin \theta \frac{\partial U}{\partial \theta} \right) \right.$$
$$\left. + \frac{1}{\sin \theta} \frac{\partial^2 U}{\partial \phi^2} \right]. \tag{1.7.9}$$

Chapter 2
The Emission and Absorption of Infrared Radiation

2.1 Introduction and the properties of cavities

We begin by imagining a quantised system, e.g. a molecule, atom, bound electron, etc. which possesses two states of energies, $\mathscr{E}_1$ and $\mathscr{E}_2$ ($\mathscr{E}_1 < \mathscr{E}_2$). These two states have a finite transition moment (see for example, equation 1.5.31) connecting them, and therefore when the system goes from state 2 to state 1, a quantum of radiant energy, that is a photon,

$$h\nu = \mathscr{E}_2 - \mathscr{E}_1, \tag{2.1.1}$$

can be emitted of frequency ν. Likewise, in the presence of an electromagnetic field of frequency ν, a photon can be absorbed by the system which is then taken from state 1 to state 2. These processes will be discussed in considerably more detail later, but here it will be stressed that the emission process seldom occurs for a completely isolated system although in principle it can do so. Only in the ultra-low pressure conditions of deep space can one envisage completely spontaneous emission to occur, and the recent discovery [25] of numerous maser emissions from cosmic clouds has made us realise that even there it is rather improbable. All realisable systems are therefore bathed in radiation and interact with their surroundings. Photons which are emitted may even return and be reabsorbed after a round trip which may involve other entities, the walls of the container or both. The emitter is thus always enclosed by a cavity, real or virtual, and to understand the processes of emission of infrared radiaton, both coherent and incoherent, it is essential to investigate the fundamental properties of cavities.

The basic concept of a cavity containing and interacting with radiation takes on different physical realisations for different ranges of frequency, as shown in Fig. 2.1. At audio- or radio frequencies one has an LCR circuit, at microwave

60

frequencies one has to resort to a hollow metallic resonant cavity, whilst for millimetre wave frequencies and above, one uses the open-resonator or Fabry–Perot interferometer. These are, however, all formally equivalent and, since the analysis of one will apply just as well to any of the others, we choose to analyse the resonant circuit because this is easier mathematically. The complex impedance (section 1.4) of the circuit will be

$$\hat{Z} = (i\omega C)^{-1} + R + (i\omega L) \tag{2.1.2}$$

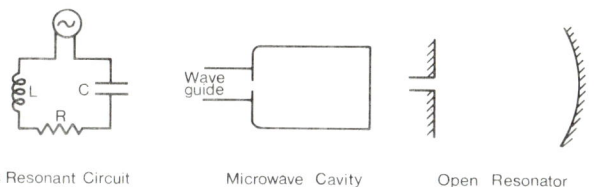

FIG. 2.1. Various forms of resonant "cavity".

which, on rationalisation followed by reciprocation, gives the conductance:

$$\hat{Z}^{-1} = R\{R^2 + [(1 - \omega^2 CL)/\omega c]^2\}^{-1} + i[(1 - \omega^2 CL)/\omega c]\{R^2 + [(1 - \omega^2 CL)/\omega c]^2\}^{-1} \tag{2.1.3}$$

The real and imaginary parts of $\hat{Z}^{-1}$ are shown in Fig. 2.2. The peak or resonance in the real part at $\omega_0^2 CL = 1$ becomes sharper and sharper as $R \to 0$ and at the same time the imaginary part approaches closer and closer to a discontinuity. Behaviour of the type shown in Fig. 2.2 is very common throughout electromagnetic physics, thus the variations of the absorption coefficient and the refractive index of a material in the neighbourhood of an isolated resonance look very similar to the curves shown in Fig. 2.2. The reason for this common occurrence lies in the necessary connection between the real and imaginary parts of functions such as (2.1.3), brought about by the causality condition (Appendix 1).

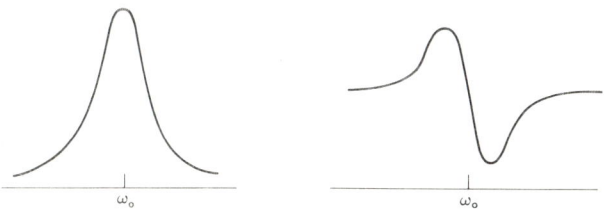

FIG. 2.2. The real (left insert) and imaginary (right insert) components of the complex conductance of an AC resonant circuit.

If one carries out a full electromagnetic treatment of these cavities, in other words sets up the relevant Maxwell equations and solves them for the given boundary conditions, one gets almost the same result for the AC circuit, just slightly modified by the small contribution to the loss of power due to radiation of electromagnetic waves of frequency $\omega_0 = (CL)^{-\frac{1}{2}}$. In the case of either of the other types, one finds that there is now an infinite set of resonant frequencies and for each there is a unique distribution of electric and magnetic fields. Each of these is referred to as a *cavity mode* and the set of modes represents the only electromagnetic patterns and frequencies which can be sustained in the cavity. This point will be taken up again in the next chapter when propagation is being discussed, but it is worthwhile to slightly anticipate that treatment and to remark that, as is usual in the solving of differential equations, any linear combinations of the solutions is also a solution. It follows therefore that the cavity will be able to sustain many modes of oscillation simultaneously. If the modes are closely spaced on the frequency axis, then any time and space variation of the field can be accommodated by a suitable combination of the cavity modes. This means that any arbitrary wave incident on, or arising in, the structure will be sustained and propagated and one has arrived at the situation described by geometrical optics. This is a most important concept for the infrared worker to grasp: "optical" propagation is to be regarded as the limit of a properly constructed electromagnetic approach as the number of possible modes tends to infinity.

Returning now to the consideration of the real part of (2.1.3), it will be seen that as the cavity losses are reduced, the resonance for a given mode gets sharper. If the losses are reduced sufficiently, the "lines" making up the mode spectrum become resolved from one another and the cavity will only sustain, or in modern terminology "oscillate on", a discrete set of frequencies. This is the situation which prevails in lasers. On the other hand, when the resonances are broad and densely spaced along the frequency axis, any frequency can be present in the cavity and the nature of the cavity will no longer affect the spectrum of the radiation contained in it. This is the situation which holds in a black-body cavity. The treatment of black-body or "full-radiator" radiation will be given in detail later, but in the present context it should be remarked that this treatment, as is almost universal, assumes both of the conditions given above, i.e. that the body is black and that it is much larger than the wavelength being radiated in order that its modes will be densely spaced. The situation of black bodies whose dimensions are of the same order as the wavelength has been discussed by Balthes [26].

2.1.1 *Infrared Fabry–Perot cavities*

The Fabry–Perot cavity consists of two mirrors, either plane or curved, facing one another and separated by a distance d. There are two basic versions, the interferometer where d is variable and the etalon where d is fixed. The instrument was developed by its authors and by Benoit in 1913 [27] as a high resolution spectrometer for metrological purposes. It was used, for example, in a precise determination of the metre in terms of the red line of cadmium. For this purpose and for the resolution of very close doublets, this multiple-beam instrument has the advantage over the two-beam Michelson interferometer in that the fringe pattern is sharper, that is it has a higher "finesse". Fabry–Perot interferometers and etalons are still widely used for high-resolution spectroscopy in the visible region [28] and the scanning form [29] is finding very useful applications in the near mm-wave emission spectroscopy of dense plasmas but the main use of the device, in the infrared, is to provide the cavities or resonators of lasers and to provide the frequency selective elements to be used, within the main cavity, to force the laser to operate on a chosen line. One can think of the operation of a Fabry–Perot in several ways but one useful one is in terms of a ray suffering endless (and usually attenuated) journeys back and forth between the reflectors. In the laser case, the medium between the plates shows gain rather than loss so the ray gains in intensity with each round trip and eventually can become very intense indeed. This is why one refers to the cavity as a resonator. An exact theory of the Fabry–Perot interferometer would require an analytical solution of Maxwell's field equations under the boundary conditions set by the geometry of the instrument. Such a recipe would appear to be impossibly difficult to follow through and, although numerical approximations could be derived using high-speed computers, most workers feel this would be a sterile approach. It is usual, therefore, to use analytical approximations which are chosen to be suitable for the particular wavelength region under consideration. At very short wavelengths the ideas can be used of geometrical optics, suitably modified where appropriate by the incorporation of linearised diffraction theory. At very long wavelengths an approximate electromagnetic theory based on pseudo-boundary conditions can be used—this essentially incorporates the first order diffraction effects into the theory. Finally, in the intermediate region where both approaches are inappropriate the so-called quasi-optical theory can be used. This has at its base the idea of gaussian beams propagating through a system but it can also be developed as shown by Fox and Li [30] in terms of diffraction at a series of regularly spaced apertures. The Fabry–Perot with the beam passing back and forth between the two mirrors is formally equivalent to a linear array of such apertures with the beam passing in just one direction. Fox and Li [30], using numerical methods, showed that stable solutions existed—in other words there were quasi-modes, and their theory is now very widely used not only in the infrared for the analysis of laser cavities, but also at microwave frequencies for the analysis of the properties of open resonators. The equivalent theory in terms of gaussian beams has been given by Kogelnik and Li [31]. The actual distribution of radiant flux which prevails at any moment will in general be a linear combination of the allowed modes. As the frequency increases, the modes become more and more closely spaced along the frequency axis and the cavity or resonator will be able to sustain more and more modes simultaneously. Eventually at very high frequencies any arbitrary field configuration can be sustained and propagated and the situation described by geometrical optics exists. The infrared-submillimetre- microwave spectral region therefore spans the range from that where the classical theory of the Fabry–Perot may be applied down to that where wave-optics or even electromagnetic theory is the only appropriate technique.

The Fabry–Perot cavities used for lasers can be made in several different configurations. The only essential requirement is that each mirror be sufficiently highly reflecting for the losses per pass not to exceed the (usually small) gain per pass. The four commonest forms are the plane-mirror cavity, the concave-mirror cavity, the plano-concave cavity and the convex mirror cavity. These are illustrated in Fig. 2.3. The concave-mirror cavity is of particular interest because of its stability. Thus if the mirrors are not exactly aligned there may nevertheless exist a pseudo-axis and the rays will be constrained to pass back and forth between the mirrors several times before finally wandering off.

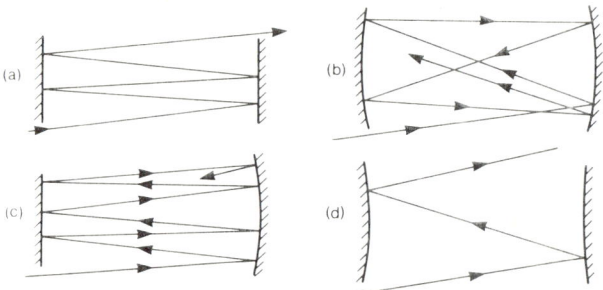

FIG. 2.3. Fabry–Perot resonators used in laser construction. The plano-concave arrangement shown in (c) can be regarded as a special case of the concave–concave arrangement—the plane mirror serving to provide an image of the curved one.

The stability or otherwise of a resonator can be defined analytically by utilising the "walk-off" ideas illustrated in Fig. 2.3. The result is that a resonator will be stable provided [31]

$$0 < (1 - d/r_1)(1 - d/r_2) < 1, \qquad (2.1.4)$$

where d is the intermirror separation and r_1 and r_2 are the mirror radii. A plot of $y = (1 - d/r_1)$ against $x = (1 - d/r_2)$ will therefore display areas of stability which are bounded by the two axes and the rectangular hyperbola

$$xy = 1. \qquad (2.1.5)$$

Special points on this diagram correspond to some simple arrangements of the resonator. Thus the *confocal* resonator which has $r_1 = r_2 = d$ and which has the feature that the two mirrors share a common focus, corresponds to the origin of the plot. It will be seen immediately that such a resonator is highly metastable since it corresponds to a singular point on the exact intersection of the stable and unstable areas. This exactly confocal arrangement is thus always avoided in laser work. It is also to be avoided in most other applications of the Fabry–Perot resonator since it has the additional undesirable feature of a highly degenerate mode spectrum. The avoidance is easily achieved merely by making d slightly larger or slightly less than $r_0 = r_1 = r_2$. The plane-parallel resonator $(r_1 = r_2 = \infty)$ and the spherical resonator $(r_1 = r_2 = d/2)$ correspond to the two points where the line $y = x$ cuts the hyperbola. Both therefore are on the verge of incipient instability and are thus usually avoided in high gain or frontier laser work. From all this, it will be readily seen that for the great majority of laser experiments and laser arrangements the nearly confocal resonator is very attractive. However a direct consequence of the stability is that when the mirrors *are* aligned perfectly there is a wide range of angles for rays inside the cavity which lead to the rays being essentially trapped. Nearly confocal cavities therefore give, as mentioned above, high gain, but the other side of the coin is that not only are there many axial modes present in the output, there are also many non-axial modes. The wavefront emerging from the laser will therefore have a far from uniform profile—it will certainly not be gaussian (see later). When a smooth spatial profile is required the convex mirror cavity may be used. For this both r_1 and r_2 (usually equal) are negative and the cavity is therefore absolutely unstable as clearly shown in Fig. 2.3(d). This sort of cavity is sometimes called just "the unstable resonator" though from what has gone before it will be realised that there are other types of unstable resonator. The unstable resonator is difficult to align and gives a lower gain but apart from its desirable near- and far-field patterns, it is also attractive because of the small mode-volume. In simple terms one can understand this as a consequence of gain being confined to a small region near the axis where paraxial rays are trapped. The small mode volume can be an important consideration in the case of optically pumped lasers where the gas to be pumped may be costly—an isotopic form for example—and also in those cases where it is necessary to make the most efficient use of the available pumping power. Unstable resonators are also used in frequency metrology work where the presence of solely longitudinal modes is a great convenience.

Fabry–Perot resonators have, as mentioned above, much in common with microwave resonant cavities and, although there are clearly defined boundary conditions for the latter and only asymptotic ones for the former, it still makes sense to talk of the mode patterns of the Fabry–Perot resonator. As usual one can approach a description of the mode phenomenon via the initial use of the ray approximation. For the plane mirror resonator which consists of two plane-parallel mirrors of diameter $2a$ facing each other and separated by a distance d, a ray launched parallel to the axis will bounce back and forth indefinitely merely being attenuated at each reflection by the imperfect reflectivity of the mirrors. However a ray launched at an angle (Fig. 2.3(a)) will eventually wander off and the apparent attenuation will be much higher. The zig-zagging of an off-axis ray is exactly the same concept that is used to describe propagation in a metallic waveguide and one can think of the plane-parallel resonator as a kind of waveguide sustaining very high order modes. The longitudinal mode number will thus be very large ($l = 2d/\lambda$) and the "waveguide" will be nearly cut-off for modes of about this number. As we are near cut-off, the transverse wavelength will be very long, of the order $2a$. An arbitrary wavefront originating inside the cavity will have, in general, mode components lying above the cut-off and these will not be able to propagate transversely. There is thus a weak reflection of some of the radiant energy back into the cavity and hence a certain measure of weak containment.

One can approach the same problem from the geometric optics point of view using some of the diffraction approximations discussed in section 3.2. A parallel beam of radiation of wavelength λ will be diffracted at a circular aperture of diameter $2a$ into a half-angle of tangent (λ/a). This diffracted beam will be intercepted by an identical aperture, set at a distance d, if $(\lambda/a) = (a/d)$. We are led therefore naturally to the classical concept of the Fresnel number

$$N_F = a^2/\lambda d, \tag{2.1.6}$$

which, in the case just mentioned, would be unity. With this particular configuration, the second aperture would just fill the first Fresnel zone when viewed from the first aperture. If N_F is less than unity, the diffraction losses will be large and a geometrical optics approach inappropriate. If, on the other hand, N_F is large the diffraction losses soon become negligible and geometrical optics proves perfectly adequate. For all practical lasers, N_F will be of the order of unity or larger but this will depend somewhat on the operating wavelength, getting larger as the wavelength falls. Thus for a far-infrared laser operating at 337 μm, the tube length would be usually about 3 m, the mirror diameter would be about 5 cm and N_F would have a value of about 0·6. For a near-infrared laser, on the other hand, operating say at 3·39 μm, the mirrors might have a diameter of 2 cm and the total length might be much less than 3 m and N_F would therefore be greater than 10. The quantitative calculation of diffraction loss in terms of the Fresnel number is, in general, a difficult problem but if N_F is considerably larger than unity some simplifications are possible. For the plane-mirror case, the fractional loss per transit may be written [32].

$$\Delta I/I = 0 \cdot 129\, N_F^{-3/2}. \tag{2.1.7}$$

Successful operation of infrared lasers usually requires that $\Delta I/I$ be less than 1 % so N_F values in excess of 5 are required. Plane-mirror cavities are acceptable therefore for near-infrared lasers but are not attractive propositions for far-infrared work. They also have the disadvantage, in comparison with near-confocal cavities, that they are difficult to align. Open resonators of all these types tend to sustain nearly plane waves so it is usual to describe their modes in transverse electromagnetic (i.e. TEM) terms. The TEM modes require three mode labels to designate the longitudinal (l) and the two orthogonal transverse mode numbers (mn). These mode numbers have the same significance as they do in microwave work, namely specifying the number of nodes of the field along the given direction. The resonant frequency of the cavity will depend on N_F and on the labels of the TEM$_{mnl}$ mode. The result, due to Weinstein [33], is

$$v_R = v_0 \left[(l+1) + \frac{2}{\pi} U^2_{n(m+1)} \cdot \frac{M(M+0\cdot824)}{[(M+0\cdot824)^2 + 0\cdot824^2]^2} \right], \quad (2.1.8)$$

where $v_0 = c/2d$, $M = (8\pi N_F)^{1/2}$ and $U_{n(m+1)}$ is the $(m+1)^{\text{th}}$ zero of the Bessel function J_n. From what was said above, plane parallel resonators would always have large values of N_F so the second term in (2.1.8) will always be negligible in practice. One can therefore say, to acceptable accuracy, that the resonant frequencies of the plane parallel resonator are just integral multiples of v_0.

The resonator with spherical mirrors is much more attractive from the infrared laser point of view since the calculations reveal diffraction losses orders of magnitude less. For $N_F > 1$, the result [32] is

$$\Delta I/I = 2^4 \pi^2 N_F \exp(-4\pi N_F). \quad (2.1.9)$$

It follows that for all intents and purposes diffraction losses can be ignored when using a properly designed spherical resonator. The basic reason for these much lower losses is that the mirror surfaces get closer and closer as one moves further off axis. The number of modes which are cut-off increases and the fraction of the power which is reflected back into the cavity increases. The same point in ray language is illustrated in Fig. 2.3(b), where it will be seen that a ray initially off-axis tends nevertheless to get trapped in the resonator. This type of resonator, as mentioned earlier, is very easy to align and will work quite well even if not exactly aligned.

The detailed operation of the spherical resonator cannot be discussed in terms of an exact electromagnetic theory because of the absence of properly defined boundary conditions but if, as is usually the case, the diameter of the mirrors is very much greater than the wavelength, and N_F consequently very much greater than unity, then the diffraction effects at the edges can be dealt with very effectively by the use of the "beam-wave" theory of Goubau and Schwering [34] as summarised by Kogelnik and Li [31] and nicely illustrated for the infrared case by Martin and Le Surf [35]. The result is the normalised gaussian beam

$$\hat{E}_{mn} = E_0 F_{mn}(x, y) \left(\frac{w_0}{w}\right) \exp\left(-\frac{r_z^2}{w^2}\right) \exp\left[-i\left((2\pi\tilde{v}z - \phi) - \frac{\pi\tilde{v}r_z^2}{r}\right)\right],$$
$$(2.1.10)$$

which is shown schematically in Fig. 2.4.

The quantities which appear in equation (2.1.10) are defined as follows: r_z is the radial distance from the z axis, w_0 is the so-called "beam-waist", i.e. the value of r_z where the gaussian function has fallen to $1/e$ of its initial value for $z = 0$, w is the value of the beam-waist for other values of z, i.e.

$$w^2 = w_0^2 \left(1 + \frac{\lambda^2 z^2}{\pi^2 w_0^4} \right). \qquad (2.1.11)$$

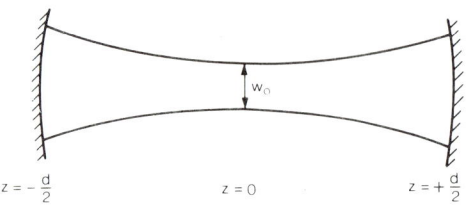

$$z = -\frac{d}{2} \qquad\qquad z = 0 \qquad\qquad z = +\frac{d}{2}$$

FIG. 2.4. Gaussian beam in a spherical mirror resonator.

r is the radius of curvature of the wavefront given by

$$r = z \left(1 + \frac{\pi^2 w_0^4}{\lambda^2 z^2} \right), \qquad (2.1.12)$$

and ϕ is a phase angle defined by

$$\phi = (m + n + 1)\arctan(z\lambda/\pi w_0^2). \qquad (2.1.13)$$

The initial function $F_{mn}(x, y)$ defines the form of the transverse modes. It can be specified in two ways, firstly in rectangular coordinates as

$$F_{mn}(x, y) = H_m \left(\frac{\sqrt{2}x}{w} \right) H_n \left(\frac{\sqrt{2}y}{w} \right) \qquad (2.1.14)$$

where $H_n(x)$ is a Hermite polynomial (section 1.5.3) or else in terms of the more appropriate functions for the cylindrically symmetrical case

$$F_{mn}(x, y) = L_m \left(\frac{2r_z^2}{w^2} \right) = \sum_{s=0}^{s=m} \frac{m!\,(-2r_z^2/w^2)^{m-s}}{(m-s)!\,(m-s)!\,s!}, \qquad (2.1.15)$$

where $L_m(x)$ is a Laguerre polynomial and m is the radial mode number. There are also solutions which are not cylindrically symmetric and for these it is necessary to use the associated Laguerre polynomial $L_m^n(x)$ where there is a second index, the azimuthal mode number. The first few Laguerre polynomials $L_m(x)$ have the forms $L_0(x) = 1$, $L_1(x) = 1 - x$, $L_2(x) = 1/2[x^2 - 4x + 2]$, $L_3(x) = 1/6[-x^3 + 9x^2 - 18x + 6]$ etc. The correspondingvariations of mode intensity with radial distance are sketched in Fig. 2.5.

The non-cylindrically symmetric modes all have a line passing through the centre along which the field is zero. They are not however as important as the cylindrically symmetric variety in the operation of real lasers since, unless there is some element present in the cavity to fix them, they are basically unstable.

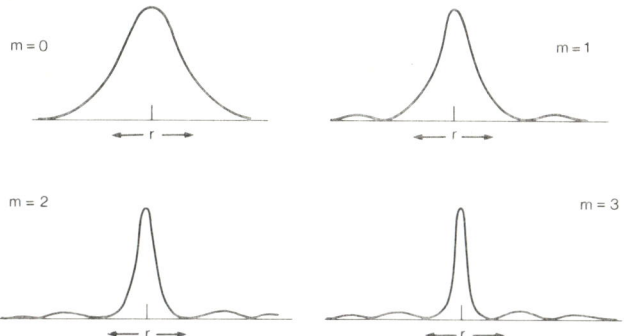

FIG. 2.5. Cylindrically symmetrical modes of a spherical mirror resonator.

The operation of the cavity is specified in terms of a single parameter w_0. This can be immediately related to the actual parameters of the cavity (its length d and the radius of curvature of its mirrors r_0) by making the assumption that the radius of curvature of the field, r, will match r_0 at the mirrors so that there will be a good fit. If this is done, then for the fully confocal case (i.e. $d = r_0$):

$$w_0 = \sqrt{\left(\frac{\lambda d}{2\pi}\right)}. \tag{2.1.16}$$

It is interesting to put some numbers into this equation. With $\lambda = 1$ mm and $d = 20$ cm, $w_0 = 5.64$ mm, showing how far away we are from the point focus of geometrical optics! Using equation (2.1.11) it follows that w at the mirror surface is 8 mm. The beam waist at the mirror is what laser engineers call the spot-size. A practical millimetre wave resonator would have a mirror diameter of 20 cm or so and hence it will be realised that, even at these long wavelengths, the fields at the mirror edges are small and diffraction losses very small also. In these circumstances the fractional loss per pass can be defined as

$$\mathscr{L} = \frac{\int_a^\infty \hat{E}_{mn} \hat{E}_{mn}^* \, dr}{\int_0^\infty \hat{E}_{mn} \hat{E}_{mn}^* \, dr}. \tag{2.1.17}$$

The higher-order modes have bigger ordinates further out and hence their losses are higher. Roughly speaking the losses go as $(m + 1/2)$, so if the loss in the fundamental mode were $1/2\%$ per pass, it would be 1% for $m = 1$, 2.5 for $m = 2$ etc. The laser gain per pass tends to be only a few per cent so the non-axial modes usually appear only weakly in the spectrum of the laser (Fig. 2.6). They can be suppressed altogether if desired either by the use of diaphragms or else by the use of coupling schemes which favour the axial modes. Thus hole coupling in which some radiation escapes through a small hole drilled in the centre of one of the mirrors will suppress the non-cylindrically symmetrical modes altogether since they have a node at the centre and will discourage the $m = 0$ transverse symmetrical modes which, relative to the $m = 0$ modes, have less of their intensity there.

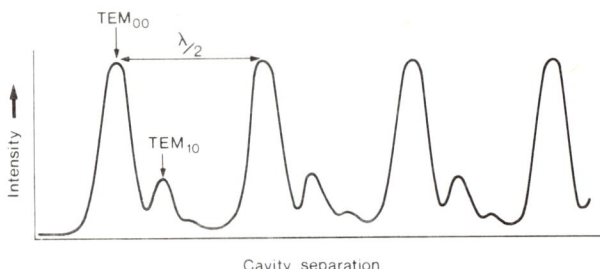

FIG. 2.6. Laser mode spectrum produced by varying the length of the resonant cavity.

The resonance condition is found by insisting, as a proper boundary condition, that either one or other component of $\hat{E}_{mn}$ vanish at the mirrors on axis (i.e. at $r = 0$). This will be so when the argument of the second exponential function in (2.1.10) is any integral multiple of $\pi/2$, i.e. when

$$\pi \tilde{v} d - \phi = (l+1)\lambda/2 \quad l = 0, 1, 2, \text{ etc.,} \tag{2.1.18}$$

that is when $2\tilde{v}d = (l+1) + 2[(m+n+1)/\pi] \arctan (z\lambda/\pi w_0^2)$

which, from (2.1.12), may be written

$$2\tilde{v}d = (l+1) + 2\frac{(m+n+1)}{\pi} \arctan \left[\sqrt{\left(\frac{d}{2r_0 - d}\right)} \right]. \tag{2.1.19}$$

Using the identity

$$2 \arctan \sqrt{\left(\frac{a}{2b - a}\right)} = \arccos \left(1 - \frac{a}{b}\right) \tag{2.1.20}$$

we may substitute to derive the final result that the cavity is resonant at any frequency v, which satisfies the equation

$$2\tilde{v}d = v/v_0 = (l+1) + \left(\frac{m+n+1}{\pi}\right) \arccos \left(1 - \frac{d}{r_0}\right). \tag{2.1.21}$$

The modes are therefore spaced by the constant amount $c/2d$ but they have a constant offset and this offset is different for the various transverse modes, that is the modes suffer dispersion as do the modes in a waveguide. This is why the transverse modes are separated from the much stronger axial modes in Fig. 2.6. The spacings are not absolutely constant however since the basic equation used for all the work (2.1.10) is in fact an approximation (section 3.3) and there are slight shifts, but under all normal circumstances these are too small to be worth considering. One interesting consequence of equation (2.1.21) is that when $d = r_0$, i.e. when we have the confocal condition, the resonance equation becomes

$$2\tilde{v}d = v/v_0 = (l+1) + \left(\frac{m+n+1}{2}\right).$$ (2.1.22)

Thus if m or n increases by 2 or their sum does likewise nothing will change, that is the mode spectrum becomes highly degenerate. As mentioned earlier it is usual to avoid this condition in most applications of Fabry–Perot resonators.

The performance of a Fabry–Perot resonator can be further analysed along the lines given above, but an alternative approach based on an equivalent set of approximations can be used to derive some important further results in a rather simple way. This approach is that used in section 3.5.2 to analyse multiple beam interference of plane-waves in a plane parallel slab. Using this method it may readily be shown that the transmissivity and reflectivity of a Fabry–Perot interferometer defined by mirrors of reflectivity R are given by

$$\tau = (1 + F \sin^2 \delta/2)^{-1},$$ (2.1.23a)

$$\rho = F \sin^2 \delta/2 (1 + F \sin^2 \delta/2)^{-1},$$ (2.1.23b)

where $\delta = 4\pi\tilde{v}d$, and the parameter F, sometimes referred to as the "cavity finesse", is equal to $4R/(1-R)^2$. This latter quantity should be carefully distinguished from the "fringe finesse" $\mathcal{F}$, given by

$$\mathcal{F} = \Delta v/\delta v,$$ (2.1.24)

where δv is the half-width in frequency terms of a fringe and Δv is the frequency spacing between fringes. For a high-performance (i.e. $R \approx 1$) interferometer, these two types of finesse are related by

$$\mathcal{F} = \pi\sqrt{F}/2.$$ (2.1.25)

Using the same approach, it follows that at equilibrium, i.e. after start up transients have died down, the intensity in the cavity I_c is related to that incident, I_i, by

$$I_c = \frac{I_i(1+R)}{(1-R)(1+F \sin^2 \delta/2)} = \left(\frac{1+R}{1-R}\right) I_i \tau.$$ (2.1.26)

The power in the cavity can therefore vary by orders of magnitude as the cavity goes into and out of resonance: it is this aspect which justifies the name of Fabry–Perot resonator. This feature, i.e. high energy storage and high internal field strength at resonance, is vital to the operation of the non-linear Fabry–Perot etalons discussed in section 3.4.7 and by Abraham and Smith [36].

Looked at from inside the cavity, the overall transmission is equivalent to a normalised mode gain (or response) G, but since there is no way of distinguishing absorptive loss in the mirrors from transmissive loss when observed from inside, one naturally defines an effective reflectivity R_e and an effective cavity finesse F_e which take both into account. The effective reflectivity is then defined by

$$R_e = 1 - \Delta P/P, \tag{2.1.27}$$

where P is the power present in the cavity and ΔP is the power loss per round trip. From what has been said above it will be realised that for a cavity of reasonable diameter and with the reflecting surfaces gold-plated it is easy to achieve ρ values in excess of 0.98. The cavity is resonant when G is a maximum that is when $\cos \delta = 1$ and hence

$$2\tilde{v}d = l + 1. \tag{2.1.28}$$

This is similar to (2.1.21) but with the second (correction) term missing. However from (2.1.23) it is more obvious that the cavity still has a finite response away from resonance; in fact the cavity response oscillates between $G_{max} = 1$ at resonance to $G_{min} = [(1 - R_e)/(1 + R_e)]^2$ at the halfway points. The frequency values at which the response is exactly half way between the extreme values are found from the solution of

$$G = 1/2 \left[1 + \left(\frac{1 - R_e}{1 + R_e} \right)^2 \right]. \tag{2.1.29}$$

These are sometimes known as the 3 dB points since the cavity response is commonly reckoned from $G_{min} = 0$ rather than the true zero and 3 dB in power represents a factor of two in attenuation. The values are given by

$$\delta_{1/2} = \arccos \left(\frac{2R_e}{1 + R_e^2} \right), \tag{2.1.30}$$

from which the halfwidth δv is given by

$$\delta v = \frac{1}{\pi} \left(\frac{c}{2d} \right) \arccos \left(\frac{2R_e}{1 + R_e^2} \right). \tag{2.1.31}$$

From these equations it follows that

$$\frac{\delta v}{\Delta v} = \frac{1}{\pi} \arccos \left(\frac{2R_e}{1 + R_e^2} \right), \tag{2.1.32}$$

where Δv is the mode spacing namely $(c/2d)$. The LHS of (2.1.32) is the reciprocal of the externally observed finesse and gives, therefore, as mentioned earlier, a measure of the absolute sharpness of the fringes. It varies from 0.5 when R_e is zero to zero when R_e is one. A value of zero corresponds obviously to infinitely sharp fringes. Rearranging equation (2.1.32) gives

$$\cos\left[\pi\frac{\delta v}{\Delta v}\right] = \left[1 + \frac{(1 - R_e)^2}{2R_e}\right]^{-1}. \tag{2.1.33}$$

Now Fabry–Perot resonators are usually only used under high performance conditions so we can regard both δv and $(1 - R_e)$ as small quantities and expand both sides of (2.1.33) to first order. This gives

$$\frac{\delta v}{\Delta v} = \frac{1 - R_e}{\pi}, \tag{2.1.34}$$

an important equation since it relates the width of the resonance δv to the losses in the cavity $(1 - R_e)$. One can therefore measure δv and thus infer $(1 - R_e)$. This is the basis of a valuable way of measuring dielectric loss at microwave frequencies since if a sheet of lossy dielectric is placed in the cavity the resonances will broaden and the extra width gives immediately the loss in the dielectric. A lucid account of this method has been given by Cullen and Nagenthiram [37]. Some modern applications of it are discussed by Clarke and Rosenberg [38].

Microwave engineers usually use a different nomenclature to describe the performance of a cavity, this is its *Quality Factor*, Q for short. This is defined as

$$Q = \frac{\text{Energy stored in the cavity}}{\text{Energy dissipated per radian}} \tag{2.1.35}$$

which can therefore be written

$$Q = \omega_0 \frac{\text{Energy stored in the cavity}}{\text{Energy dissipated per second}}. \tag{2.1.36}$$

The exact evaluation of this quantity needs some knowledge of the geometry of the cavity and of how the radiant energy is distributed within it, but one can get a surprisingly good answer from considering very simple models. Thus if one has a cylindrical cavity of length d, bounded by end plates (mirrors) of area A and supporting the propagation of plane waves of intensity I, then the energy stored in the cavity is IAd/c and the energy dissipated per second is $IA(1 - R_e)$; therefore

$$Q = \frac{\omega_0 d}{c(1 - R_e)} = \frac{\pi v_0}{(1 - R_e)}\left[\frac{2d}{c}\right] = \frac{v_0}{\delta v}. \tag{2.1.37}$$

Q is thus alternately defined as the centre frequency of the resonance divided by its half-width. This is a good definition of Q for the metallic closed cavities used in the lower microwave region but Fabry–Perot resonators have an infinite set of resonant frequencies (spaced by $c/2d$). Applying the straight definition of centre frequency divided by half-width can therefore give some very large numbers. Thus if we have a cavity 1 m long with $(1 - R_e)$ equal to 0·01 (i.e. $R_e = 99\cdot0\%$) then the Q of the first resonance at 150 MHz will be 314 but that for the 200th resonance at 30 GHz will be over 62 000. Very high

Q's of this order are commonly quoted for millimetre and submillimetre Fabry–Perot resonators.

It was remarked earlier that plane-parallel resonators are unattractive unless the Fresnel number is high, that is, in practice, that the wavelength is short. Nevertheless the first submillimetre lasers to be discovered [39] featured plane-parallel cavities! The explanation of this apparent paradox lies in the need to confine the active medium of the laser (usually a low-pressure gas) in a suitable vessel, usually a glass tube. Dielectric reflection of an off-axis ray at the tube walls will return the ray into the active medium and will therefore reduce the apparent losses. In other words the tube not only contains the gas but acts as a hollow dielectric waveguide. Much of our knowledge of this topic stems from the careful work of Kneubuhl and his colleagues [40] and of Veron and his colleagues [41] who investigated the mode patterns of glow discharge lasers and also from that of Hodges and Hartwick [42] who introduced the optically pumped waveguide laser.

2.2 Thermal radiation and the properties of black bodies

2.2.1 *General properties of thermal radiation*

Infrared radiation is not in any fundamental way different from other kinds of electromagnetic radiation, but it is *par excellence* the form for which the interaction with matter is dominated by thermal processes. Thus the preferred sources of incoherent infrared are thermal in character, usually hot bodies, whereas thermal sources are only used in other regions of the spectrum when there is nothing better available. A thermal source may have a spectrum which depends on the nature of the material from which it is constructed, but this is usually undesirable and one aims to make a "black" source, that is one which gives out a broad-band continuous spectrum whose profile is determined both qualitatively and quantitatively by a single parameter—the thermodynamic temperature of the source. The characteristics of the black-body spectral curve are that it rises quite steeply to a maximum and then falls off relatively slowly—exhibiting a rather long tail (Fig. 2.7). It is this tail which is responsible for the well-known phenomena associated with the visual observation of hot bodies. Thus one speaks of dull red-heat, red-heat, white-heat etc. (Fig. 2.8). A particularly important case is provided by the sun. Its photosphere temperature of 6000 K corresponds to black-body radiation peaking in the near infrared (0·85 μm), but having a long tail which, extending well into the visible and ultraviolet regions, gives us daylight and the essential actinic radiation which plants need for photosynthesis. It is not possible to heat any solid object to this sort of temperature without its vapourising, so in the manufacture of incandescent lamps for illumination one does the best one can and chooses the most refractory metals that are available (tungsten for example) and heats them to the highest temperature possible. At these temperatures, 3000 K, the metal has a significant vapour pressure and metal distils onto the cold glass envelope. Eventually the light output is seriously attenuated by absorption in the resulting metallic film. This troublesome effect can be considerably alleviated by introducing iodine into the lamp bulb. This combines with the metal film to make a volatile iodide which is promptly decomposed back to metal again on the extremely hot filament. One can therefore afford to run the lamp at a higher temperature but to do this it is also necessary to have a fused quartz envelope since there is a great risk of an ordinary glass envelope

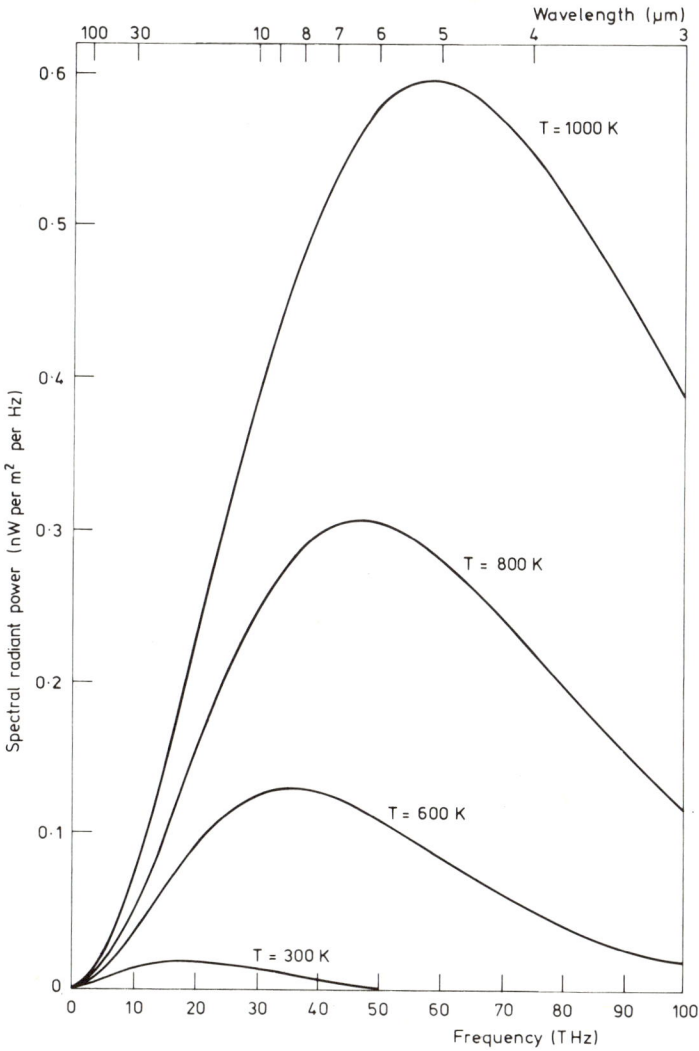

FIG. 2.7. Emission spectra of black-body sources at various temperatures.

softening under the considerable temperatures which will then prevail. These *Quartz–Halogen* lamps are important objects of commerce since they are used for the headlights of cars. Nevertheless they are still very inefficient sources of light, only a small fraction ($< 10\%$) of the electrical power going in coming out in the form of visible radiation. Conversely, however, they are excellent sources of near-infrared radiation since their output peaks in the $1.7\ \mu m$ region.

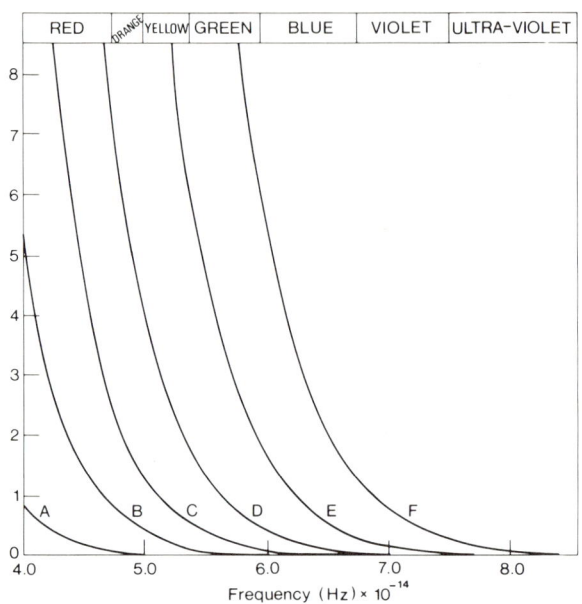

F$_{IG}$. 2.8. Total hemisphere visible emission from hot black bodies (in units of $10^{-12}\ Wm^{-2}\ Hz^{-1}$) at various temperatures. $A = 1273\ K$, "yellow-hot"; $B = 1453\ K$, "white-hot"; $C = 1573\ K$, "brilliant white-hot"; $D = 1700\ K$, not defined; $E = 1850\ K$, "blue white-hot"; $F = 2000\ K$, not defined. The associated colours come from the IES Lighting Handbook, 1st Edition, 1947, Appendix A-35, Table A-16. They should be regarded, however, as purely subjective. Thus when an astronomer refers to the sun as being "yellow", he is using the word in a quite different sense.

The position of the peak in the black-body curve is given by Wien's displacement law:

$$\tilde{\nu}_{max} = 196.1\ T \qquad (2.2.1a)$$

or in units of cm^{-1}

$$\tilde{\nu}_{max} = 1.961\ T \qquad (2.2.1b)$$

where $\tilde{\nu}_{max}$ is the wavenumber at which maximum power per unit wavenumber interval, per unit area is radiated and T is the absolute temperature. This law is often remembered in the approximate form "the peak emission occurs at a wavenumber in cm^{-1} equal to twice the absolute temperature of the source". In frequency (THz) terms, Wien's law becomes

$$v_{max} = 0.05879\, T, \tag{2.2.2}$$

where v_{max} is the frequency at which maximum power per unit frequency interval per unit area is radiated. Our subjective feeling that thermal radiation is confined to the infrared has its origin in this law because our normal experience is restricted to a narrow range of medium temperatures. Thus the surface of the Earth with a temperature near 300 K is cooled at night by the loss of thermal radiation (maximum at $588\,cm^{-1}$) into space and is warmed by receiving thermal radiation from the sun (T = 6000 K, $\tilde{v}_{max} = 12\,000\,cm^{-1}$) during the day. All the hot bodies which we come across in our normal lives have temperatures between these two bounds and must therefore have the bulk of their thermal radiation confined to the infrared.

In cosmological situations a very much wider range of temperature is encountered and examples of thermal radiation are known covering most of the electromagnetic spectrum. The current "Big-Bang" theory of the origin and evolution of the Universe indicates that at a time 15 000 000 000 years ago the Universe was in a highly condensed state in which temperatures of the order 10^{10} K prevailed [43]. For such temperatures, the equivalent frequencies are hundreds of exaHertz and the thermal photons, which would be gamma rays, could materialise as electron–positron pairs. As time advanced [44], the primeval fireball rapidly expanded and the temperature quickly dropped. The radiation correspondingly shifted to longer wavelengths but it remained strongly coupled to the material component because this would be completely ionised and therefore would be a *plasma*. Plasmas absorb very strongly electromagnetic waves whose frequencies are less than the characteristic electron plasma frequency

$$v_p = (Ne^2/4\pi^2 m\,\varepsilon_0)^{1/2} \approx 9N^{1/2} \tag{2.2.3}$$

where N is the number of electrons per cubic metre. Regarded from the outside therefore a plasma looks something like a metal, but does not have a 100% reflectivity because of the form of its optical constants (see later). The absorption inside the plasma is so great that the plasma is perfectly black and the material component and the radiative component will be in equilibrium. However, as the temperature of the plasma, almost entirely composed of an equal mixture of protons and electrons, fell through the value $T = 1.579 \times 10^5$ K, which corresponds to the ionisation energy of hydrogen (13.6 eV), a dramatic change occurred. The electrons and protons combined to form neutral hydrogen atoms and the expanding gas became transparent at all frequencies save those of the line spectrum of hydrogen. The matter and the radiation became essentially uncoupled and each set off on its separate evolutionary path. The hydrogen gas condensed, as is well known, to form galaxies and eventually stars. The temperature of the thermal radiation continued to fall as the Universe continued its expansion. This can be looked at in two different but equivalent ways. In classical terms, the temperature drops because the energy density is falling: in relativistic terms, the temperature drops because the radiation is cooled by the cosmological red-shift. Throughout the expansion, however, the radiation will preserve its black-body form. At the present epoch, the cooling has reached the stage at which the effective temperature of the cosmic black-body radiation is only 2·9 K. Such radiation (Fig. 2.9) is essentially confined to the microwave and submillimetre regions of the spectrum. The presence of an all-pervading microwave background was not suspected until quite recently, despite several pointers such as the finite population of the upper component of the split ground state of the CN radical in deep space. It fell to Penzias and Wilson [45] working with a horn antenna at the Bell Telephone Laboratories to make the crucial observation that despite every precaution their instrument showed an irreducible noise signal which, if it were thermal in origin, corresponded to a temperature of about 3 K. For this seminal discovery they were awarded the 1978 Nobel Prize for physics. The thermal nature of the radiation was subsequently confirmed by the observation of the expected peak at 6 cm^{-1}, in the millimetre waveband [46].

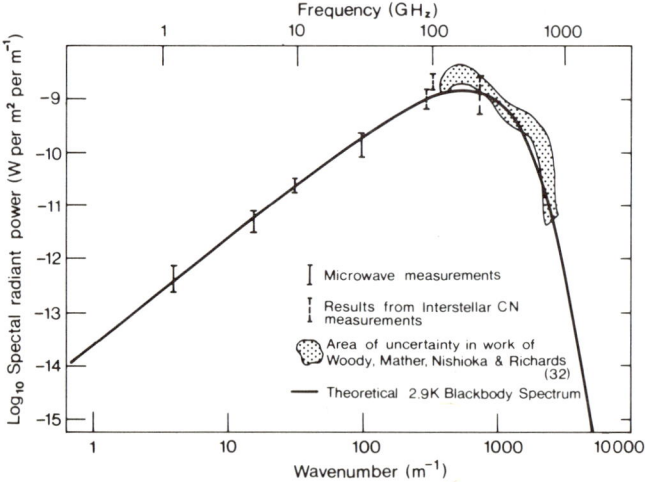

Fɪɢ. 2.9. The theoretical spectrum and some experimental observations of the cosmic background radiation. The data are taken from reference [46].

The whole Universe is therefore filled with 2·9 K thermal radiation and only in exceptional circumstances can anything be at a lower temperature than this. The principal exception occurs in cryogenics when specimens in metal dewar vessels and thereby shielded from the background can be cooled to small fractions of a degree absolute. Natural examples do, however, occur in molecular gas clouds in interstellar space where stimulated emission can depopulate upper states and lead to an effective temperature less than the background. This is sometimes known, rather fancifully, as the "cosmic refrigerator effect" or less fancifully as "the antimaser effect" [47]. The other cosmic situation where temperatures less than 2·9 K are apparently encountered is that involving "black holes". A black hole is an object so massive that it has collapsed completely under its own gravitational field. It is currently thought that stars more than a few times the solar mass cannot avoid this fate when their stocks of nuclear fuels have become exhausted. There is intense theoretical activity at the moment addressed to the question of what ultimately happens to the matter in a black hole since the tensor field equations of Einstein appear to indicate that the star continues to contract down to a point—a singularity!! What is reasonably certain is that when the star passes through the so-called Schwarzschild radius

$$R = 2GM/c^2 \qquad\qquad (2.2.4)$$

(where G is the universal constant of gravitation $(6·672·10^{-11}\ Nm^2kg^{-2})$ and M is the stars mass) which corresponds exactly to the classical radius at which the escape velocity would be that of light—no electromagnetic radiation can leave the object. It has become a one-way hole in space and to an outside observer is perfectly black. Its radiative temperature would therefore appear to be zero. The work of Penrose and Hawking[48] has shown that this is not quite true and that it is meaningful to ascribe a thermodynamic temperature to a black hole. Basically this is because when one considers the quantized electromagnetic field one must consider the creation of virtual electron/positron pairs near the Schwarzschild surface or "event horizon" as it is sometimes called. If one member of the pair is captured by the black hole, it will go "over the horizon" and disappear from the observable universe. The other member of the pair will therefore appear apparently from nowhere, but will obviously be interpreted as having been emitted by the black hole. Because of these quantum field effects, a black hole will have an effective thermodynamic temperature and the theory of Hawking [49] shows that this temperature is inversely proportional to the mass. The heaviest black holes are therefore the coolest. For most black holes, this temperature is less than that of the cosmic background and the 2·9 K radiation is pouring down "the drain". It is interesting to speculate [50], however, on the far future of the Universe when the adiabatic cooling will have lowered the temperature of the background to a value *below* that of the black holes. When this situation prevails, the holes will *lose* radiation to space, their masses will fall, their temperatures will rise and inevitably they will evaporate.

Apart from black holes and their thermal properties there will be very many more fascinating phenomena for infrared astronomers to study once stable observing platforms are available on large satellites. It is essential to work in space because the Earth's atmosphere is virtually opaque throughout most of the infrared, but once on the satellite and armed with a good infrared telescope, it will be possible to study protostars and especially those (such as MWC 349) which appear to be in the early stages of the formation of a planetary system. Large gas clouds, such as M 42 in Orion, which have already yielded so much information [25, 51] at millimetre and longer wavelengths, will surely yield still more when the submillimetre and infrared regions are available for systematic investigation.

2.2.2 Thermal radiation in cavities

We have made several references so far to "black bodies", it is now time to discuss their properties in more detail and to say how they may be realised in practice. A black body is, by definition, an object which absorbs all the radiation incident upon it. A body may be, of course, "black" at one wavelength and merely "grey" at another; it is however more convenient to develop the theory of completely black bodies first and to treat the complications due to imperfect blackness subsequently. We begin by noting that the concept of a black body, though an ideal, is not in the same category as frictionless surfaces, perfect simple-harmonic motion etc. Practical black bodies can be constructed which approach the theoretical ideal to any desired degree. These usually take the form of a spherical cavity with a very small hole drilled through so that some radiation can enter or leave the cavity. The hole has to be extremely small, compared to the surface area of the cavity, so that the amount of radiation entering or leaving the cavity will be negligible in comparison with that already present inside the cavity. In other words, a photon can readily enter the cavity but cannot so readily find its way out again. Problems do arise at very long wavelengths, where the wave nature of electromagnetic radiation prevents it being propagatable along a cylinder with diameter much shorter than the wavelength (Appendix 3). A very small hole is therefore impassable—i.e. it acts as a cut-off waveguide—to microwave radiation, and special techniques are required for the construction of a microwave black body. However in the infrared region, where wavelengths are always less than 1 mm, there is no difficulty in constructing practical black bodies.

If any body is inside such a cavity which itself is at a constant temperature it follows that, at equilibrium, the rate of energy gain by the body must equal the rate of energy loss. Thus if the surface radiance of the body is $I (W\,m^{-2})$ the cavity radiation density incident upon the surface I_c (W m^{-2}) and the absorptivity a, we have

$$I = aI_c, \qquad (2.2.5)$$

The maximum value of I will obviously occur when a has its maximum value, namely unity: in other words the maximum radiant power from a body occurs when that body is a black body. If we now define an emissivity, ε, always less than unity, by the equation

$$I_{\text{obs}} = \varepsilon I_{\text{black body}}, \qquad (2.2.6)$$

it follows that $\varepsilon = a$; that is the better an absorber an object is, the better a radiator it will be. This is Kirchoff's law which is fundamental to any discussion of infrared emission. If ε is unity at all wavelengths the body is truly black but if it is constant at all wavelengths but less than unity the body is said to be "grey".

Equations (2.2.5) and (2.2.6) apply strictly to the total power at all wavelengths since they are derived using only the First Law of Thermodynamics. However by invoking the Second Law they can be shown to apply at each wavelength separately. The derivation of this result, especially when developed with full rigour, is somewhat lengthy but a simplified account can be given which recovers the essential results. Firstly one establishes that the total energy density and the spectral energy density, in a cavity, depend only on the temperature. To show this, one assumes that they do not and imagines two different kinds of cavity connected by a window which is transparent at some wavelengths and not at others. In the transparent region, by assumption, the energy densities are different, therefore energy will flow from one cavity to the other and the temperature of one will rise whilst that of the other falls. A difference of temperature has been brought about, therefore, without any expenditure of work and this violates the Second Law of Thermodynamics. Since no assumption has been made about the spectral location of the window transparency, nor about its width, the contradiction applies to all wavelengths and therefore the basic assumption is untrue for all wavelengths. It follows that the spectral energy density (i.e. the energy density having frequencies between v and $v + dv$) cannot depend in any way on the character of the cavity walls but depends only on the temperature. From this it further follows that it will not change if an object at the same temperature is inserted into the cavity and hence that

$$I(v) = a(v) I_c(v) \qquad (2.2.7)$$

for all v. We thus have the general result that

$$\varepsilon(v, T) = a(v, T). \qquad (2.2.8)$$

The emissivity then is identically equal to the absorptivity, so to consider the quality of any object as a radiator it is necessary to discuss the factors contributing to the absorptivity. The energy transmitted by a specimen will equal the energy incident minus the sum of the energies absorbed and reflected at the surfaces. Thus

$$a = 1 - \tau - \rho \qquad (2.2.9)$$

where τ is the transmissivity and ρ the reflectivity. It is noticed at once that for metals, where $\tau = 0$, $a = 1 - \rho$ which can be very small indeed. Polished metallic surfaces are therefore rather poor radiators. For many systems, gases being prime examples, τ is a rapidly varying function of v so the system will be "black" in narrow frequency bands. However τ will nowhere be identically zero, so, provided the system is thick enough, it can become "black" anywhere. To discuss this further and especially for those cases where the concentration of absorbers is constant, it is usual to introduce the absorption coefficient $\alpha(v)$ defined by Lambert's law†

$$\tau = \exp[-\alpha(v)l], \qquad (2.2.10)$$

where l is the specimen thickness. For those cases where the concentration can vary the discussion is in terms of Beer's law:

$$\tau = \exp[-kC_m], \qquad (2.2.11)$$

where k is a constant and C_m the concentration. These two laws are frequently combined to give

$$\tau = \exp[-\varepsilon_m(v)C_ml], \qquad (2.2.12)$$

where $\varepsilon_m(v)$ is the extinction coefficient which is an intrinsic property of the absorbing entities. It will be seen from (2.2.12) that, no matter how small $\varepsilon_m(v)$ is and no matter how small C_m is, τ can approach zero provided l is large enough. In extreme cases, which occur for example in studies of the line spectra of the interstellar medium, C_m will be minute and $\varepsilon_m(v)$ for a particular line may be vanishingly small because the absorption is forbidden by some selection rule. In these cases, l may well have to be of galactic dimensions before the line can be "black".

The "blackness" or otherwise of a source is often discussed in terms of its optical density or optical thickness. The optical density is defined to be

$$(\text{o.d.}) = -\ln I/I_0, \qquad (2.2.13)$$

where I is the transmitted intensity and I_0 the incident intensity. It is thus roughly equal to $\alpha(v)l$ if reflection effects can be ignored. The optical thickness is in general usage a somewhat looser term, but one may make it more precise by defining a standard length l_0 which is the thickness for which $\tau = \exp(-1)$. One would then have $l_0 = [\alpha(v)]^{-1}$ and it could be said that a specimen was optically thin if l was much less than l_0 and optically thick if l was much greater than l_0. These concepts are very important throughout infrared physics and not just only in the theory of sources. Thus in quantitative infrared spectroscopy, one is faced with the problem of designing an experiment so that the errors in determining either $\alpha(v)$ or C_m (if $\alpha(v)$ is known) are minimal. If the experiment were being done by absorption spectroscopy, it might be thought that it would be desirable to have an optically thin specimen, for then $\alpha(v)l$ would be small and

$$\alpha(v)l = \ln\tau \simeq \Delta I/I_0, \qquad (2.2.14)$$

---

†This form of Lambert's law can be regarded as a definition of τ which otherwise is defined as the ratio I/I_0 corrected for all reflection effects at the interfaces.

where $\Delta I = I_0 - I$. In this regimen $\alpha(v)$ would be linearly proportional to a scale deflection and the experiment would be very simple and especially easy to automate. However by differentiating (2.2.14),

$$\frac{d[\alpha(v)l]}{\alpha(v)l} = \frac{1}{\tau \ln \tau} d\tau, \qquad (2.2.15)$$

and by differentiating this once more and setting the result equal to zero it will be seen that the percentage error in $\alpha(v) l$ is a minimum when $\ln \tau$ is unity. For the best accuracy one should therefore choose l so that it equals l_0, i.e. $[\alpha(v)]^{-1}$, whereupon one has, of course, a specimen which is optically neither thick nor thin. In the laboratory, this result is often approximated by the rule of thumb that the specimen thickness should always be chosen so that the average transmission over the region of interest is 50%. More quantitative guidance can be drawn from Fig. 2.10 which shows [52] how the percentage error in α varies with τ for given percentage errors in τ.

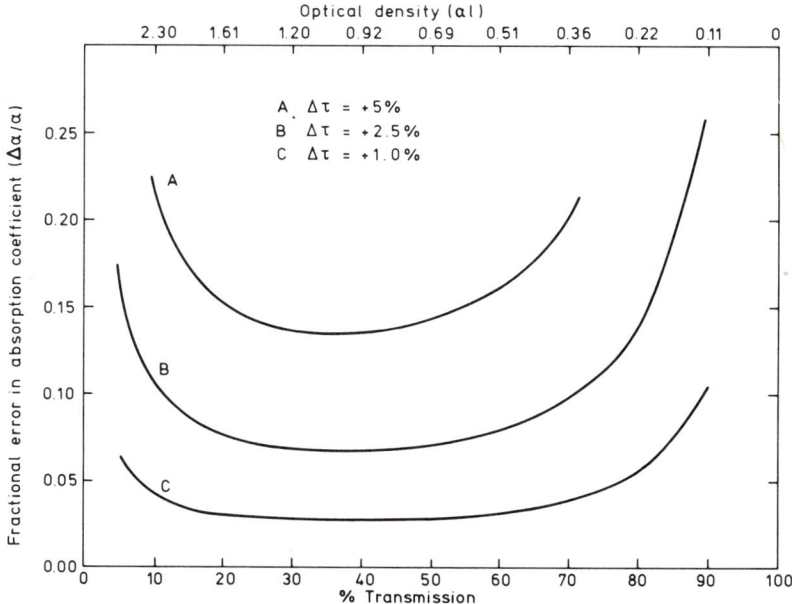

FIG. 2.10. Fractional error in absorption coefficient for given fractional errors $\Delta \tau$ in transmission.

Similar considerations apply to an emission experiment where one might, for example, be attempting to determine the concentration of an emitting species from measurements of the radiant intensity at its characteristic frequencies. The radiant power at a given frequency v is a linear function of $C_m l$

provided this product is small enough. The radiant power levels will, however, necessarily be small and the final signal-to-noise ratio poor. If $C_m l$ becomes large compared to $[\varepsilon_m(v)]^{-1}$, there will be adequate signal, but the radiant intensity will be approaching the black-body limit and will be a highly non-linear function of $C_m l$. The precision of determining C_m will therefore be poor. Once again there exists an optimum regimen where the best compromise is found of signal-to-noise ratio *vis-à-vis* linearity. This compomise occurs for the same range of optical thickness as was found optimal for the absorption experiment, i.e. neither thick nor thin. An example of how the emission spectrum of a heated solid can vary with thickness is shown in Fig. 2.11 which gives results [53] for polyethylene terephthalate films at thicknesses of 0·013 (A), 0·026 (B), and 0·052 (C) mm at a temperature of 100°C. These spectra are to be compared with that of a blackened metal plate (D) which serves as a reference "black-body" spectrum at the same temperature. The Figure also includes reflection and transmission spectra recorded at room temperature and it will be seen that, as expected, the strong emission features correspond exactly with the strong absorptive and reflective features. The comparisons cannot be taken very much further than this because of the differences in temperature and because of the differences in resolution in the various spectra, very wide slits having to be used to get an acceptable signal-to-noise ratio in the

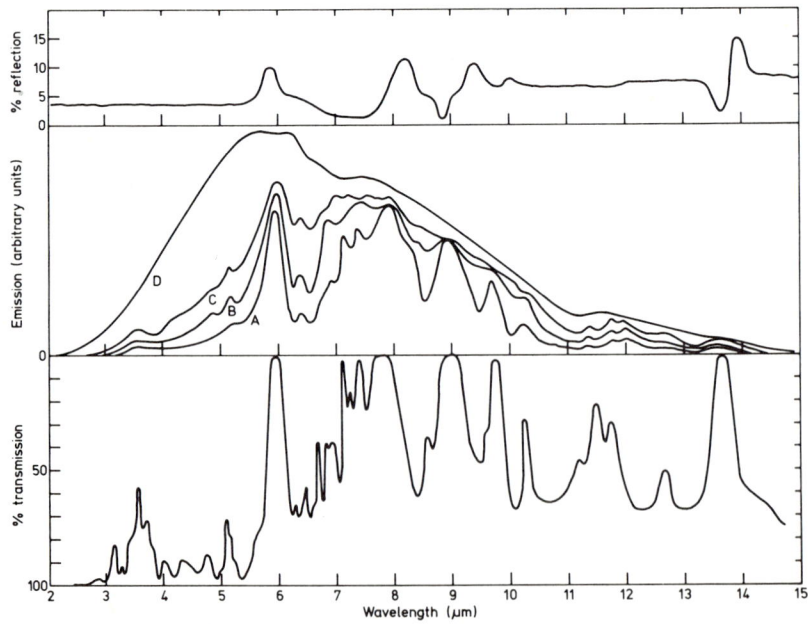

FIG. 2.11. Reflection, emission and transmission spectra of thin films of polyethylene terephthalate films (unpublished work of H. A. Willis and M. E. A. Cudby).

emission spectrum. Nevertheless, the approach to "grey body" radiation as the thickness increases is clearly demonstrated. Curve D should peak at $\lambda = 7\cdot77\,\mu\text{m}$ (equation 2.2.44b) since the spectrometer used was linear in wavelength rather than in wavenumber. The failure to do so probably arises largely from the falling emissivity at longer wavelengths of the "black" coating on the reference surface which is a consequence of its increasing transparency. The Figure clearly also shows that the most intense bands reach the black-body limit first, but that eventually, at a sufficient thickness, the discrete features are lost and the specimen radiates everywhere as a more or less black body.

2.2.3 *Mathematical theory of black-body radiation*

So far we have discussed the physics of black bodies and approximations to black bodies in general terms, but now it is necessary to develop the mathematical theory in some detail. The derivation was originally given by Planck in the early years of this century but important contributions were later made by Einstein, especially the concept of stimulated emission. A black body is assumed to be made up from a very large assembly of quantum oscillators, each having its own characteristic frequency, but since we have shown that the "blackness" holds at each frequency separately we may discuss a set of oscillators which have just the two levels $\mathscr{E}_1$ and $\mathscr{E}_2$ with $\mathscr{E}_2 - \mathscr{E}_1 = h\nu$. The final black-body equation can be derived then if necessary by considering all possible such sets.

We consider a closed cavity at a temperature T within which there is an equilibrium spectral energy density U_ν. The problem is to calculate U_ν as a function of ν and T. The method we use is to imagine that we have some probe two-level oscillators of energy separation $h\nu$ and that we introduce sufficient of these into the cavity that statistical methods may be assumed valid. We can now calculate how the population of the oscillators will divide itself between the two possible states, the lower N_1 and the upper N_2, by two independent methods, statistical thermodynamics and radiative equilibria respectively, and by obliging the two results to be the same we shall be able to derive the form of U_ν. At thermal equilibrium the populations of the two levels will be given by the Maxwell–Boltzmann equation:

$$N_2 = N_1 \exp(-h\nu/kT) \qquad (2.2.16)$$

This equilibrium must be maintained by a balance of stimulated absorption which takes oscillators from the lower state $\mathscr{E}_1$ to the upper state $\mathscr{E}_2$ and stimulated emission plus spontaneous emission which do the reverse. The stimulated emission is in-phase and coherent with the radiation field but the spontaneous emission, which can be thought of as random stimulated emission due to zero-point fluctuations of the vacuum, is not.

The rates of transfer of oscillators from one state to the other are determined by the Einstein coefficients. The chance per second that an oscillator will spontaneously pass from the upper to the lower state is given by the Einstein coefficient

$$A_{21} = \frac{16\pi^3 v^3}{3h\varepsilon_0 c^3} \left(\int \psi_1 \mu_{21} \psi_2 d\tau \right)^2.$$ (2.2.17)

Where ψ_1 and ψ_2 are the wave functions of the two states and μ_{21} is the dipole moment operator. The Einstein coefficients for stimulated absorption B_{12} and for stimulated emission B_{21} are equal to one another and are given by

$$B_{12} = B_{21} = \frac{2\pi^2}{3h^2\varepsilon_0} \left(\int \psi_1 \mu_{21} \psi_2 \, d\tau \right)^2.$$ (2.2.18)

We thus have that the chance per second that an oscillator will transfer from the upper state to the lower state is

$$R_{21} = A_{21} + U_v B_{21},$$ (2.2.19)

where U_v is the spectral energy density function. The chance of transfer per second from the lower to the upper state is given by

$$R_{12} = U_v B_{12}.$$ (2.2.20)

At equilibrium the number of oscillators leaving and entering each state per second must be equal, i.e. it is necessary that

$$N_2 R_{21} = N_1 R_{12}.$$ (2.2.21)

Substituting from equations (2.2.16), (2.2.19) and (2.2.20) one has

$$U_v = \frac{A_{21}}{B_{12}} \frac{1}{\exp(hv/kT) - 1},$$ (2.2.22)

and finally from (2.2.17) and (2.2.18)

$$U_v = \frac{8\pi h v^3}{c^3} \frac{1}{[\exp(hv/kT) - 1]}.$$ (2.2.23)

This equation in the form

$$U_v = \frac{8\pi v^2}{c^2} \frac{hv}{[\exp(hv/kT) - 1]}$$ (2.2.24)

has a simple physical interpretation. The first term is the number of modes per unit volume per unit frequency interval and the second is the average energy per mode, which is obtained by multiplying the mode-energy, i.e. hv, by the chance that the mode is excited, i.e. $[\exp(hv/kT) - 1]^{-1}$. This latter expression is the celebrated Bose–Einstein distribution function which describes the statistics of integral spin particles, such as photons. These, in contradistinction to the more familiar "material" particles of half-integral spin, such as electrons, are not prohibited from having more than one particle in any given quantum state. The radiation in the cavity can therefore be regarded as a quantum "gas" made up of particles, the photons, several of which may have the same frequency and the same momentum at the same time. The degeneracy δ, of a given mode of the cavity, is then simply $[\exp(h\nu/kT)-1]^{-1}$ where ν is the mode frequency. At optical frequencies δ is very small and most modes are unexcited but at far-infrared frequencies δ may be appreciable. Care needs to be exercised therefore when transferring any concepts based on photon statistics from the optical region to the far-infrared region.

A particular case arises in the detection of photons by means of high sensitivity detectors such as photomultipliers and photoconductors. In these devices, electrons or other carriers are produced ideally at the rate of one per photon absorbed. The correlation between the photons in the radiant field can then be deduced from that of the photoelectrons. This can be measured using an electronic correlator fed from the detector output current. With thermal radiation at optical frequencies, the photons behave like independent particles which obey classical Poisson statistics. At far infrared wavelengths, however, where δ may be appreciable, bunching of the photons is observed and there is an increased chance of detecting two photons which are much closer than a coherence interval (section 2.3) apart.

The division of particles into two fundamentally different classes—the bosons of integral spin and the fermions of half odd-integral spin is a unique feature of quantum mechanics. It has its origin in the fundamental postulate that the behaviour of any system is determined by the properties of a physically unobservable underlying wave function ψ. The square of ψ is, however, an observable property — it gives the probability that the system will be in a particular condition at any given time or alternatively if one is considering an individual particle what is the chance of finding that particle at a particular point in space. If the system contains two identical particles, then clearly an interchange of these cannot affect any physically observable property so there are only two possible responses of the wave function to the interchange of identical particles. One can have $(\psi \rightarrow -\psi)$ which is the symmetry of fermions, or one can have $(\psi \rightarrow +\psi)$ which is the symmetry of bosons. The Pauli exclusion principle can be defined in a much more concise fashion within this formalism. It is equivalent to demanding that all physical total wave functions, that is the product of all the component wave functions, be always antisymmetric. Systems corresponding to symmetric total wave functions do not exist. It follows that no two fermions in a system can have the same set of quantum numbers. There is no such restriction on bosons. Boson "gases" are always liable therefore to undergo a so-called "Bose-condensation" in which all the particles collapse into a common state where they all have the same momentum and all move completely coherently. The remarkable properties of the low temperature (He II) phase of liquid helium below 2·19 K form a well-known example. Here, because of the complete coherence, the liquid develops a high heat conductivity and loses all viscosity: it is a superfluid. This phase transition is a direct consequence of the helium atom having zero electronic and zero nuclear spin. The collapse of all the photons in a laser cavity into a common quantum state when the laser is operating above threshold is another example of a Bose condensation. Laser light (in which one might have $\delta = 10^{15}$!) is therefore fundamentally different from ordinary light just as superfluid helium is different from ordinary liquid helium. It ought to be mentioned in passing that *pairs* of fermions will have integral spin and can therefore behave like bosons. Thus below 2 mK, liquid ^{3}He develops superfluid properties because the ^{3}He atoms, of nuclear spin $I = \frac{1}{2}$, can condense in pairs and thus produce a compound boson fluid. Likewise pairs of electrons can combine into so-called Cooper pairs under low-temperature conditions when phonon coupling (that is the linking of the electrons via the mutual distortion of their host lattice) can overcome the electrostatic repulsion. The resulting pairs can then move through the lattice with zero ohmic loss. This gives the phenomenon of superconductivity which forms the basis of some very sensitive infrared detectors (section 4.4.4).

The quantum ideas provide, as is well known, a resolution of the dilemma that the classical version of (2.2.24), derived as usual by letting h tend to zero, namely

$$U_\nu = (8\pi\nu^2 kT)/c^3, \qquad (2.2.24a)$$

predicts an "ultraviolet catastrophe" since the power radiated will increase without limit as ν increases. This result was first derived by Rayleigh and Jeans using the classical equipartition principle and to them there was no escape from the paradox. The quantum form relieves the dilemma since, at a sufficiently high value of the frequency, the denominator will start to increase at a very much faster rate than does the numerator and U_ν will tend to zero. The Rayleigh–Jeans result (2.2.24a), however, often holds very well in the far infrared and is very convenient for numerical calculations.

All that remains to do to derive the final radiation formula is to remove the unnecessary restriction that the spectral line corresponding to the transition from 1 to 2 is infinitely narrow, i.e. that it is a delta spike. This is simply achieved by multiplying (2.2.23) or (2.2.24) throughout by the differential increment $d\nu$. $U_\nu d\nu$ then has the physical meaning that it is the energy per m^3 lying in the band of frequency from ν to $\nu + d\nu$. The important practical matter, however, is more often the power coupled out of the cavity rather than the power density within it. This is readily derived by imagining that one has a small hole, of area A, in the cavity walls. The hole is so small that the power escaping through it is negligible and the power loss does not therefore affect the equilibrium within the cavity. In one second, the amount of energy which has flowed out through the hole will be $U_\nu d\nu Ac$ and this will be distributed over the total 4π of solid angle (section 2.2.4). The rate of energy loss per unit area per unit solid angle will be given by

$$F(v, T)dv = \frac{U_v c dv}{4\pi} = \frac{2hv^3}{c^2} \frac{dv}{\exp(hv/kT) - 1}. \qquad (2.2.25)$$

This is Planck's function. In wave number terms it becomes

$$F(\tilde{v}, T)d\tilde{v} = 2hc^2\tilde{v}^3[\exp(hc\tilde{v}/kT) - 1]^{-1}d\tilde{v}. \qquad (2.2.26)$$

These various radiation formulae are sometimes required in terms of the entire emission into the observable 2π hemisphere. To achieve this one needs to know how radiance varies with angle of observation. This is discussed in the next section where it is shown that the total radiant flux is only half the brightness. The total radiated power is obtained by multiplying the RHS of equation (2.2.25) or (2.2.26) by π and not 2π as might be thought at first. That is

$$I(\tilde{v}, T)d\tilde{v} = 2\pi hc^2\tilde{v}^3[\exp(hc\tilde{v}/kT) - 1]^{-1}d\tilde{v}. \qquad (2.2.27)$$

It is very convenient to cast the quantitative expression for Planck's Law into a general form independent of particular frequencies and temperatures. To do this one introduces the dimensionless parameters

$$x = (hc\tilde{v}/kT), \qquad (2.2.28a)$$

$$y = \frac{h^2 c I(\tilde{v}, T)}{2\pi k^3 T^3}. \qquad (2.2.28b)$$

In terms of these Planck's law (2.2.27) becomes

$$y = x^3(\exp(x) - 1)^{-1}, \qquad (2.2.29a)$$

or in differential form

$$ydx = x^3[\exp(x) - 1]^{-1}dx. \qquad (2.2.29b)$$

The convenience is now evident in that a single plot of the function (2.2.29b), as presented in Fig. 2.12, will serve for *any* source temperature. All that is necessary is to relabel the axes with the aid of the relations (2.2.28a) and (2.2.28b). In fact it is possible to buy special slide rules based on (2.2.29b) which can be used to calculate the radiant power at any frequency for any temperature. However, modern electronic calculators, especially programmable calculators, are even more convenient for this purpose, and are rapidly replacing the slide-rules in common usage. The calculation is carried out by first computing x from the numerical form

$$x = 1{\cdot}438\,786 \times 10^{-2}\,\tilde{v}/T, \qquad (2.2.30)$$

where $\tilde{v}$ is the wave number in m^{-1}. One next calculates y using (2.2.29) and then evaluates $F(\tilde{v}, T)$ from the relation

$$F(\tilde{v}, T) = 1{\cdot}256\,31 \times 10^{-10}\,yT^3. \qquad (2.2.31)$$

As an example of this, if $\tilde{v}$ were $10\,000\,m^{-1}$ (i.e. $100\,cm^{-1}$), and T were $1000\,K$ then x would be $0{\cdot}1439$ and y would be $0{\cdot}0192$. $F(\tilde{v}, T)$ would therefore be

$2\cdot418 \times 10^{-3}$ W m^{-2} m^{-1}. One can of course work just as well in terms of other units: thus to get the spectral power per Hz, one merely divides (2.2.31) by c to produce

$$F(v, T) = 4\cdot1906 \times 10^{-19} \, yT^3. \tag{2.2.32}$$

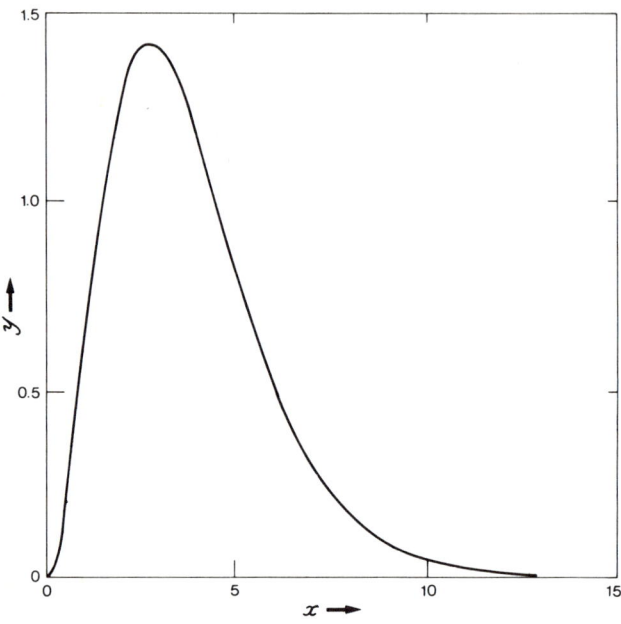

FIG. 2.12. Radiant power function for a black-body source. The abscissa and ordinate axes are given in terms of reduced variables in order that the curve be general rather than apply to any particular value of the temperature.

It is worth pausing for a moment to consider the implications of the above example. The effective area of source which a spectrometer can utilise is usually no more than 1 cm^2. If as much as 1 steradian of solid angle could be used and if the spectrometer were working to a resolution of 10 m^{-1}, then the total available power would be $7\cdot7 \times 10^{-7}$ W. This is ample enough signal for a sensitive detector, but if the power were attenuated by instrumental losses and especially by being lost at the jaws of the slits of a monochromator, then the signal could rapidly become inadequate. It was such considerations of energy limitation that led experimentalists to introduce Michelson interferometers (see section 6.6.3) for the purpose of carrying out far-infrared spectroscopy since these instruments do not have any slits to attenuate most of the precious energy. They also have a muliplex advantage because they observe all the spectral frequencies all the time unlike the dispersive instruments which, being sequential, observe only one spectral component at a time. Because of these advantages interferometric spectrometers give excellent performance in the

100 cm^{-1} region and with cryogenically cooled detectors they have even been used down to 2 cm^{-1}, that is 60 GHz!

The spectral power available from a black-body source compares very poorly with that available from laser sources. Thus an optically pumped laser (section 6.2.5) working at 100 cm^{-1} can readily deliver a few milliwatts in a spectral bandwidth of less than 10^5 Hz. In the same bandwidth, a very hot black body could do no better than about 1 µW. This fact is sometimes translated into temperature terms and figures of 10^7 K quoted for the effective laser temperature! It is all rather misleading because the laser, being essentially a fixed frequency device, ought not to be compared in this way with the hot-body source which can be used over a huge range of frequency. However when widely tunable lasers become readily available the comparison will become valid and the attractions of a source which has achieved a super-high effective temperature by concentrating all its power into the desired frequency whilst radiating none at undesired frequencies will be very obvious. However, at the moment, tunable lasers are bulky, costly, expensive and limited in range so there is still much for the humble black-body source to do.

Differentiation and integration of the basic function (2.2.29) yield two very well-known results. By differentiation, the wavenumber (or frequency) where the power per unit wavenumber (or frequency) is a maximum may be found. $F(x, T)$ is a maximum for that non-zero value of x which satisfies the transcendental equation

$$1 - x/3 = \exp(-x), \tag{2.2.33}$$

namely $x = 2{\cdot}821\,439\,37$. Bearing in mind the definition of x, namely $x = h\nu/kT$, one arrives immediately at equation (2.2.2) in the form

$$\nu_{max} = 2{\cdot}821\,44\,(k/h)T = 5{\cdot}8788 \times 10^{10}\,T. \tag{2.2.34}$$

Integration is rather more difficult since the integral of the RHS of (2.2.29) cannot be expressed in terms of elementary functions. However, as is often the case, the definite integral over the full range from zero to infinity can be evaluated in simple terms. In fact

$$\int_0^\infty \frac{x^3 dx}{\exp(x) - 1} = \frac{\pi^4}{15}. \tag{2.2.35}$$

We have, therefore, by substitution,

$$I(T) = \int_0^\infty F(\nu, T)\,d\nu = \frac{2\pi h}{c^2}\left(\frac{kT}{h}\right)^4 \int_0^\infty y\,dx = \left(\frac{2\pi^5 k^4}{15h^3 c^2}\right)T^4 = \sigma T^4. \tag{2.2.36}$$

This is the *Stefan-Boltzmann law* and σ is *Stefan's constant* equal to $5{\cdot}670\,32 \times 10^{-8} \text{ W m}^{-2}\text{ K}^{-4}$. This law had been earlier derived by purely macroscopic thermodynamic arguments and it was its successful derivation in terms of a microscopic theory which persuaded many physicists at the time that Planck, despite his revolutionary ideas, was in fact correct. To illustrate the quantitative aspect of this law, consider the high-pressure mercury arc lamp (section 6.1) which is commonly used as a source of far-infrared radiation. This is essentially a quartz cylinder of total surface area 10 cm^2 (i.e. 10^{-3} m^2) which is run off an electrical power supply from which it draws 125 W. If all the dissipated power were to be given off in the form of thermal radiation, that is if there were no conductive or convective losses, then the equilibrium temperature of the quartz surface would, from (2.2.36), be 1219 K. In practical use, of course, there are non-radiative losses and the observed temperature is closer to 1000 K.†

In practical work it is often necessary to calculate the radiant power not over the entire range of frequency, but only over a restricted range. This cannot be done analytically, as mentioned above, so recourse is often taken to Tables of the integral. The special slide rules, also mentioned earlier, usually feature as an extra bonus a scale which is essentially this integral. Nowadays programmable calculators often feature an integrating routine and this can be used to calculate the integral directly. However, it is still of some value to have access to the tabulated function and this is given in Table 2.1. The quantity given in this Table is

$$I_i = \int_0^{x_i} \frac{x^3 \mathrm{d}x}{\exp(x) - 1} . \tag{2.2.37}$$

To obtain the radiant power between wave numbers $\tilde{v}_i$ and $\tilde{v}_j$ one computes x_j and x_i from (2.2.28a), looks up I_j and I_i in the Table and calculates their difference, divides this by 6·494 (i.e. $\pi^4/15$) and then multiplies the answer by σT^4 to give the final power per unit area. Sometimes, however, all that is required is the fraction of the total power which is radiated between two specified wave numbers: this can be obtained from the first three steps. As an example, consider the sun with photosphere temperature of 6000 K. If we wished to know the fraction of its output which is radiated as light, we would have

$$\tilde{v}_{\min} = 1\cdot43 \times 10^6 \text{ m}^{-1}, \quad \text{hence} \quad x = 3\cdot43,$$

$$\tilde{v}_{\max} = 2\cdot50 \times 10^6 \text{ m}^{-1}, \quad \text{hence} \quad x = 5\cdot99.$$

From the Table, the difference between the two corresponding values of I_i is 2·34 and therefore the visible fraction is 36 %. It is interesting to note that the sun's power is divided almost exactly between the infrared and the higher frequencies.

---

† It is assumed here that the emissivity of fused quartz is unity. This is an excellent approximation over most of the infrared region.

Many works on radiation physics are written with wavelength as the operating variable—this stems from historical reasons, since the early workers actually measured wavelength and therefore naturally tended to use it for their abscissae. To convert from frequency to wavelength, it is noted that

$$\tilde{\nu} = 1/\lambda,$$
(2.2.38)

TABLE 2.1

The total radiated power integral $I_x = \int_0^x x^3 [\exp(x) - 1]^{-1} dx$

x	I_x	x	I_x	x	I_x	x	I_x
0·0	0·0	3·6	3·3747	7·2	6·0624	10·8	6·4598
0·1	0·0003	3·7	3·5045	7·3	6·0895	10·9	6·4623
0·2	0·0025	3·8	3·6315	7·4	6·1150	11·0	6·4646
0·3	0·0080	3·9	3·7561	7·5	6·1391	11·1	6·4667
0·4	0·0183	4·0	3·8771	7·6	6·1618	11·2	6·4687
0·5	0·0344	4·1	3·9949	7·7	6·1831	11·3	6·4706
0·6	0·0571	4·2	4·1094	7·8	6·2032	11·4	6·4723
0·7	0·0871	4·3	4·2205	7·9	6·2221	11·5	6·4739
0·8	0·1249	4·4	4·3281	8·0	6·2398	11·6	6·4754
0·9	0·1707	4·5	4·4322	8·1	6·2564	11·7	6·4768
1·0	0·2248	4·6	4·5328	8·2	6·2721	11·8	6·4781
1·1	0·2871	4·7	4·6299	8·3	6·2868	11·9	6·4793
1·2	0·3576	4·8	4·7234	8·4	6·3005	12·0	6·4792
1·3	0·4360	4·9	4·8134	8·5	6·3134	12·1	6·4802
1·4	0·5221	5·0	4·9000	8·6	6·3255	12·2	6·4811
1·5	0·6155	5·1	4·9831	8·7	6·3369	12·3	6·4820
1·6	0·7158	5·2	5·0628	8·8	6·3475	12·4	6·4828
1·7	0·8226	5·3	5·1391	8·9	6·3574	12·5	6·4836
1·8	0·9353	5·4	5·2122	9·0	6·3667	12·6	6·4843
1·9	1·0534	5·5	5·2821	9·1	6·3754	12·7	6·4850
2·0	1·1763	5·6	5·3488	9·2	6·3835	12·8	6·4856
2·1	1·3054	5·7	5·4125	9·3	6·3911	12·9	6·4861
2·2	1·4362	5·8	5·4732	9·4	6·3983	13·0	6·4866
2·3	1·5692	5·9	5·5310	9·5	6·4049	13·1	6·4872
2·4	1·7081	6·0	5·5860	9·6	6·4111	13·2	6·4876
2·5	1·8444	6·1	5·6383	9·7	6·4169	13·3	6·4880
2·6	1·9848	6·2	5·6880	9·8	6·4223	13·4	6·4884
2·7	2·1263	6·3	5·7352	9·9	6·4273	13·5	6·4888
2·8	2·2683	6·4	5·7800	10·0	6·4321	13·6	6·4891
2·9	2·4104	6·5	5·8225	10·1	6·4365	13·7	6·4894
3·0	2·5521	6·6	5·8627	10·2	6·4406	13·8	6·4897
3·1	2·6931	6·7	5·9008	10·3	6·4443	13·9	6·4900
3·2	2·8330	6·8	5·9369	10·4	6·4479	14·0	6·4902
3·3	2·9715	6·9	5·9710	10·5	6·4512	15·0	6·4918
3·4	3·1082	7·0	6·0033	10·6	6·4543	16·0	6·4925
3·5	3·2424	7·1	6·0337	10·7	6·4571	∞	6·4939

and hence

$$d\tilde{\nu} = -(1/\lambda)^2 d\lambda. \qquad (2.2.39)$$

Planck's Law therefore becomes

$$F(\lambda, T)d\lambda = 2hc^2\lambda^{-5}[\exp(hc/\lambda kT) - 1]^{-1}d\lambda, \qquad (2.2.40)$$

where $F(\lambda, T)$ is the radiant intensity per *unit wavelength interval*. It should be noted carefully that the wavelength enters this expression as the *fifth* power whereas, of course, the wavenumber appears in (2.2.26) merely as the *third* power, the reason being the form of (2.2.39). An immediate consequence of this difference is that $F(\tilde{\nu}, T)$ and $F(\lambda, T)$ have their maxima in different parts of the spectrum. The maximum in $F(\lambda, T)$ is found as before by differentiating the core function, which now (with $z = \lambda kT/hc$) is

$$y = \frac{z^{-5}}{\exp(z^{-1}) - 1}. \qquad (2.2.41)$$

The maximum is at that value of z which satisfies the equation

$$\exp(1/z) = 5z(5z - 1)^{-1}, \qquad (2.2.42)$$

namely $z_{max} = 0.201\,405\,236$. From the definition of z, we have at once

$$\lambda_{max} = (hc/kT)z_{max} = (2.8978 \times 10^{-3})T^{-1}. \qquad (2.2.43)$$

which on rearranging becomes

$$\lambda_{max}T = 2.8978 \times 10^{-3} \quad \text{in metres}, \qquad (2.2.44a)$$

$$\lambda_{max}T = 2.8978 \times 10^{3} \quad \text{in } \mu m. \qquad (2.2.44b)$$

These are the forms in which Wien's displacement law was first given and in which it is still commonly quoted. The *wavenumber* corresponding to λ_{max} is

$$\tilde{\nu}_{max}(cm^{-1}) = 3.451\,T. \qquad (2.2.45)$$

Clearly (2.2.45) and (2.2.1b) are incompatible but the incompatibility arises entirely from the definition of $F(\tilde{\nu}, T)$ and $F(\lambda, T)$ as intensity per unit *frequency* interval and per unit *wavelength* interval respectively. There can be no physical discrepancy and if one were to calculate the total power being radiated between two points in the spectrum, one would get the same answer whether one used (2.2.26) or (2.2.40). Nevertheless, there does remain the point that the two functions have their maxima at different spectral locations and this can cause difficulty unless the terms are carefully defined. As an example it is commonly asserted that the sun radiates maximally in the blue-green region of the visible. This is certainly true if $F(\lambda, T)$ is quoted since, from (2.2.44b), λ_{max} would be 0.483 μm (Fig. 2.13). This fact has even led some biologists to suggest that the human eye has a peak of response in the green (Fig. 1.1) in order to match the peak in the solar spectrum!! It need hardly be pointed out that in frequency terms the solar maximum occurs at 353 THz (0.84 μm) and that it is the balance of photon energy and photochemical activation energies which dictates a maximum of eye sensitivity near 0·5 μm. However it should be stressed, in biological contexts, that it is the distribution of solar power at the Earth's *surface* which is important and it is then true that the combination of infrared absorption and ultraviolet scatter plus absorption leads to the situation where the bulk of the solar power, which does reach the ground, is concentrated in the visible region.

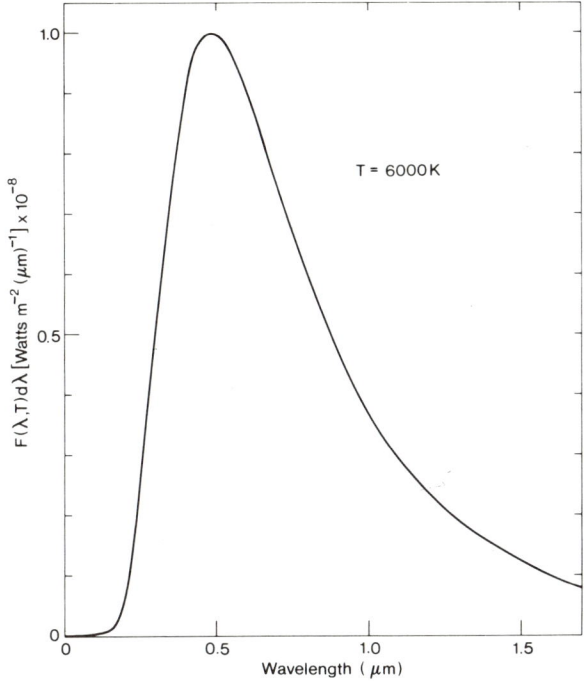

FIG. 2.13. Radiant power from the sun in wavelength terms.

The question of which presentation is the better is certainly a very contentious one. In practice it is influenced by the temperature of the source, by the spectral response of the detector and by the dispersion characteristics of the spectrometer. Early workers did their measurements in the visible region with the eye as the principal detector and they naturally tended to use $F(\lambda, T)$ since this emphasises the visible. However plots of $F(\lambda, T)$ versus λ, for example Fig. 2.13, cannot be meaningfully drawn for small values of the abscissa, i.e. the ultraviolet, X-ray regions etc., and cannot be drawn practically for large values of abscissa where the important infrared radiation lies. It is for this reason that one often seen plots of $F(\lambda, T)$ featuring only the region of the maximum. $F(\lambda, T)$ is, therefore, not a useful way of describing the infrared radiation from hot-bodies. Also one must bear in mind that wavelength, unlike frequency, is a medium-dependent quantity and that it is frequency not wavelength which is the fundamental quantity (cf. section 2.1.1) in radiation physics and therefore the presentation in wavenumber terms is always preferable.

It is particularly apposite under the condition that $h\nu \ll kT$ for then, with the usual approximation that

$$\exp(x) = 1 + x, \tag{2.2.46}$$

one has analogously to (2.2.24a),

$$F(\bar{\nu}, T) = 2\pi ckT\bar{\nu}^2. \tag{2.2.47}$$

This is the Rayleigh–Jeans radiation formula which, as mentioned earlier, was derived by classical arguments in pre-quantum days. It would clearly be true at all wavenumbers and there would be an "an ultraviolet catastrophe" if h were to be zero, that is in the classical limit. Obviously one cannot plot the spectrum in the Rayleigh–Jeans limit in wavelength terms—at least not on a finite piece of paper! To give some idea of the physical significance of these ideas, we recall that for the commonly used source temperature of 1000 K, the Rayleigh–Jeans condition demands that $\bar{\nu}$ be less than $600\,\mathrm{cm}^{-1}$, i.e. lie in the far infrared. In the Rayleigh–Jeans limiting region, just as for every other region, there is always an increase in power output as the source temperature rises, but here the output only varies linearly with temperature. This can be illustrated dramatically by the observation that in a far-infrared spectrometer, the source is only three times brighter than the spectrometer walls! This fact dictates the positioning of the chopper in a far-infrared instrument since the detector can be constrained to "see" the source alone only if the chopper is immediately in front of the source. In optical spectrometry, on the other hand, where the background is dark, the chopper can be placed anywhere that is convenient. Another problem with far-infrared spectrometry is that one will have to operate the source at as high a temperature as possible to get a good signal, but then the amount of mid- and near-infrared radiation will increase very rapidly (roughly as the fourth power of the temperature) and there will be a severe "stray light" problem.

In the opposite limit, i.e. when $h\nu \gg kT$, the radiation formula reduces to

$$F(\bar{\nu}, T) = 2\pi hc^2\bar{\nu}^3 \exp(-hc\bar{\nu}/kT), \tag{2.2.48}$$

which is Wien's law. This too had been derived by classical arguments before the advent of quantum theory and in fact Planck was led to his epoch-making discovery by the search, using interpolation theory, for a function which would behave like (2.2.47) for low wave numbers and like (2.2.48) for high ones. The quality of an infrared source, at a fixed frequency, is determined by whether it is operating in the Rayleigh–Jeans, the Wien or the intermediate regime. Thus one can imagine a detector observing a very narrow band of frequencies centred on ν_0. As the source temperature starts to rise, one will be in the Wien regime with the peak frequency less than v_0 and the power detected rising very rapidly with source temperature. As the temperature reaches the value where the peak coincides with v_0, one will be in the intermediate region and for higher temperatures with the peak lying at higher frequencies, one will be in the Rayleigh–Jeans regime with the power rising slowly with source temperature. This is illustrated in Fig. 2.14. The choice of source is set by striking the balance between the practical difficulties of operating the source at a very high temperature and the need to get enough signal to get an acceptable signal-to-noise (S/N) ratio. In the far infrared, Rayleigh–Jeans operation is inevitable so there is no need to indulge in expensive and capricious sources just because they are marginally hotter than the cheap and reliable ones. Also when carrying out careful measurements in the far infrared it may be desirable to reduce the risk of stray light by working at a lower temperature. In the mid infrared, it is always possible to achieve the intermediate regime but in the near infrared one will always be in the Wien regime. One will therefore always seek to operate a near infrared source as hot as possible. These concepts can be quantified in terms of an equivalent temperature $T_e = hv/k$ and a scaled temperature $T_s = T/T_e$ and then in terms of a dimensionless power output U, one can write

$$U = [\exp(T_s^{-1}) - 1]^{-1}. \qquad (2.2.49)$$

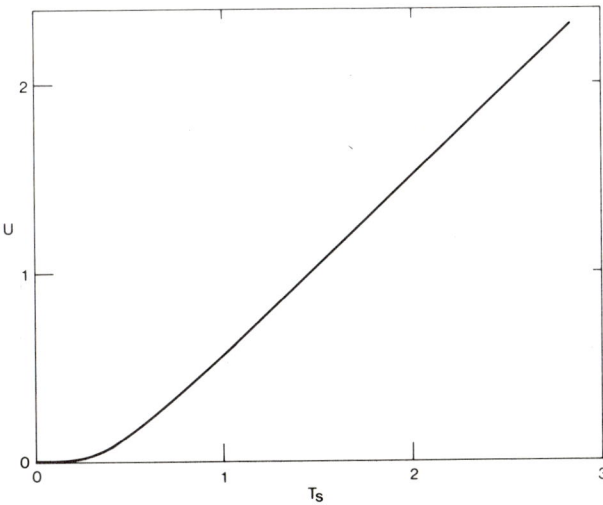

FIG. 2.14. Radiant output from a black-body source at a fixed frequency as a function of temperature: this latter is plotted in terms of the scaled quantity defined in the text.

A useful rule of thumb for remembering equivalent temperatures is that room temperature corresponds to 6.25 THz that is 208·5 cm^{-1}. The function (2.2.49) is plotted in Fig. 2.14. It will be seen from this Figure that for T_s much less than unity the plot is strongly curved—this is the Wien region. For T_s unity or larger, the plot is sensibly linear, this is the Rayleigh–Jeans region. As mentioned earlier, the source temperature should always be as high as possible consistent with practical design requirements and acceptable stray-light limits, but it will be realised from Fig. 2.14 that temperatures less than $0.4\,T_e$ should be avoided. This is not too onerous a requirement for far- and mid-infrared work, but for near-infrared work it may be difficult to satisfy. Thus at $2\,\mu m$ (i.e. $5000\,cm^{-1}$) the temperature would have to be 3000 K. This is the reason for the remark made earlier that in near-infrared spectroscopy one is nearly always working in the Wien regime.

The Rayleigh–Jeans and Wien approximations may be used to side-step a difficulty which arises in the use of Table 2.1. This is that it is hard to make accurate interpolations for small and large values of x. However, making the usual approximations, the integral (2.2.37) can be expressed in closed form, thus:

$$I = \int_0^x = x^3/3, \quad x \ll 1, \qquad (2.2.50)$$

$$I = \int_{x_1}^{x_2} = [6 - \exp(-x)[x^3 + 3x^2 + 6x + 6]]_{x_1}^{x_2}, \quad x_1, x_2 \gg 1, \qquad (2.2.51)$$

and by evaluating these expressions between the desired limits arrive at good estimates.

2.2.4 Radiative transport in extended media

The mathematics and concepts which have been used to describe the equilibria inside cavities and the radiation by black-body sources can be used just as well to describe the propagation of infrared radiation through an absorbing/emitting medium. Suppose we have such an extended medium, as in Fig. 2.15.

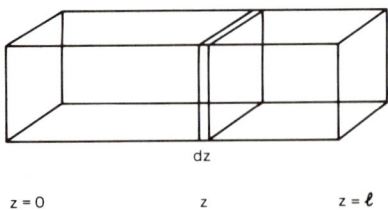

z = 0   z   z = ℓ

FIG. 2.15. Configuration adopted for the radiative transport calculation.

We consider propagation in the z direction and choose an elemental slice of thickness dz situated at a distance z into the medium. For simplicity we again assume a two-level system and further assume that the transition is reasonably sharp (i.e. $v_0 \gg \Delta v$) and described by a profile or line-shape function (section 2.5) $\phi(v)$. The profile function is normalised so that the area under it is unity and it has therefore dimensions of time T. Applying equations (2.2.19) and (2.2.20) we can write

$$dS_v = \left[N_2 A - (N_1 - N_2) \, U_v B 4\pi \right] \left(\frac{hv}{4\pi} \right) \phi(v) \, dz, \qquad (2.2.52)$$

where S_v is the spectral intensity. We have therefore

$$\frac{dS_v}{dz} = \left[N_2 A - (N_1 - N_2) \, S_v 4\pi B/c \right] \left(\frac{hv}{4\pi} \right) \phi(v). \qquad (2.2.53)$$

It should be carefully noted that the spontaneous contribution, being isotropic, is divided by 4π to give the observed intensity, but the stimulated emission and absorption which only take place in the beam direction are not. The physical interpretation of equation (2.2.53) is that there is gain from spontaneous emission and loss from the excess of stimulated absorption over stimulated emission. Equation (2.2.53) can also be written

$$\frac{dS_v}{dz} = \frac{N_2 A h v \phi(v)}{4\pi} - \alpha_v S_v, \qquad (2.2.54)$$

where α_v is the absorption coefficient given by

$$\alpha_v = h\tilde{v}(N_1 - N_2) \, B\phi(v). \qquad (2.2.55)$$

Equation (2.2.54) is the general radiative transport equation. To apply it to particular systems, it is solved with the appropriate boundary conditions. Direct integration of (2.2.54) yields the equation

$$\frac{\alpha_v S_v - N_2 A h v \phi(v)/4\pi}{\alpha_v S_v^0 - N_2 A h v \phi(v)/4\pi} = \exp(-\alpha_v l), \qquad (2.2.56)$$

which describes the general case where the extended medium is emitting and where there is also an external source of spectral intensity S_v^0 located at $z = 0$. Two special cases may be picked out. Firstly, when the frequency is high enough or the temperature low enough for A to be assumed negligible in comparison with $\alpha_v S_v$. One now clearly *must* have an external source and in fact the set-up is the familiar one used for absorption spectroscopy. Equation (2.2.56) in this limit becomes

$$S_v = S_v^0 \exp(-\alpha_v l). \qquad (2.2.57)$$

This is usually known as Lambert's law, though there is some controversy as to who actually first proposed it. The second special case of interest is the opposite limit, i.e. when the medium *is* emitting and there is no external source. Equation (2.2.56) then becomes

$$\alpha_v S_v = \left[N_2 A h v \phi(v)/4\pi \right] \left[1 - \exp(-\alpha_v l) \right], \qquad (2.2.58a)$$

which, from (2.2.55), (2.2.22) and (2.2.25), can be written

$$S_v = F(v, T) \left[1 - \exp(-\alpha_v l)\right]. \tag{2.2.58b}$$

This equation shows clearly that as the specimen thickness increases, the emission comes more and more to be just that expected from a black body at the same temperature. This quantitative demonstration of the approach to the black-body limit is the justification for the qualitative argument given earlier and illustrated in Fig. 2.11. In the opposite limit, i.e. in the optically thin condition, one has, when one applies the usual approximation to (2.2.58b),

$$S_v = F(v, T) \alpha_v l, \tag{2.2.58c}$$

and the intensity becomes proportional to the thickness. This is the situation which prevails for most of the lines produced by electrical discharges in Geissler tubes. The gas is optically thin because of the negligible population of the highly excited states which are involved in the transitions.

The intermediate case of equation (2.2.56), i.e. when we have both an external source *and* an emitting medium, leads to a much more complicated analysis since the received signal will now be made up of two components. In the laboratory this situation is usually avoided by modulating the radiation from the hot source and then detecting only at the modulation frequency. In this way the DC signal from the warm specimen is ignored and, provided it is not high enough to overload the detector, it will have no effect and equation (2.2.57) can be used as it stands. In astrophysical situations, however, this obviously cannot be done and both sources of radiation must be taken into account. Astrophysicists usually prefer to use the spectral brightness B_v rather than the spectral intensity S_v, but for most intents and purposes these two quantities can be taken to be identical. Astrophysicists also prefer to use the *brightness temperature* of a source. This is a purely phenomenological quantity which defines the observed brightness at a given frequency in terms of that of an equivalent black body. It is *not* a thermodynamic temperature and therefore, in general, will be expected to vary with frequency. Radioastronomers who usually regard broadband radiation as coming from "noise" sources prefer to use the concept of *antenna temperature* to describe the received power from an astrophysical object. This is because the noise power generated per unit bandwidth in an object at a temperature T is simply related to the temperature (section 4.2) and radioastronomers actually measure noise power. The antenna temperature is related to the source temperature by the equation

$$T_A = T_s \left(\frac{hv}{kT}\right) \left[\exp(hv/kT) - 1\right]^{-1}, \tag{2.2.59}$$

so at the relatively low frequencies used by radioastronomers (i.e. in the Rayleigh–Jeans region) it is equal to the brightness temperature. In terms of spectral brightness and brightness temperature, equation (2.2.56) may be written

$$B_v = B_v^0 (T_m) + [B_v^0 (T_s) - B_v^0 (T_m)] \exp(-\alpha_v l), \qquad (2.2.60)$$

where $B_v^0 (T)$ is the spectral brightness of an ideal black body having a temperature T and at a frequency v. T_m and T_s are the brightness temperatures of the medium and of the background source respectively. What might possibly be observed for the case of a sharp line and for the two situations $T_s > T_m$ and $T_s < T_m$ is shown in Fig. 2.16. For the former case the feature is seen in absorption, whereas for the latter it is seen in emission. Behaviour of this nature is commonly observed in the study of astrophysical objects [25], but complicating effects are often present as well. Thus apart from the expected variation of brightness temperature with frequency, there are unexpected variations between the source brightness temperature and the medium brightness temperature which lead to features being seen in absorption in one spectral region and in emission in another. It may even happen, by an unfortunate coincidence, that $T_s = T_m$ when, from (2.2.60), it will be noted that the spectral features will not be seen at all! Various examples which have been observed in the microwave and millimeter-wave regions have been discussed by Winnewisser and his colleagues [25].

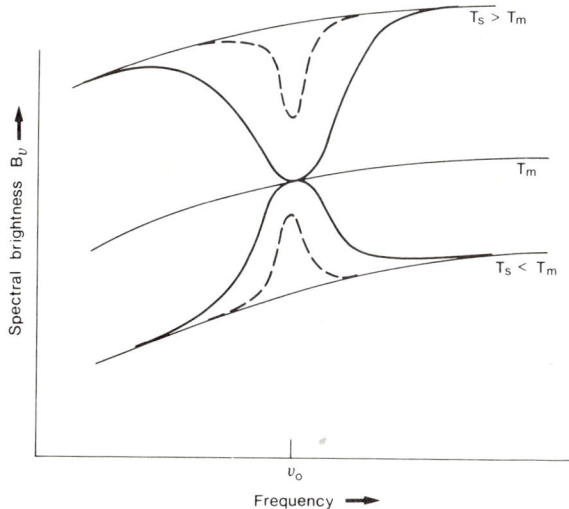

FIG. 2.16. The observation of a discrete feature in the spectrum of an extended medium which has an additional broad-band source behind it. The optically thick and thin cases are illustrated for two values of the source temperature T_s with respect to the medium temperature T_m.

So far we have tacitly assumed that the detector in the system has an identically zero temperature. Now, whilst experimentalists are often at pains to operate their detectors at as low a temperature as possible, the final operating temperatures which they achieve are nevertheless finite, and a proper discussion of radiative transfer must acknowledge that the detector is a part of the overall system. If one is observing an extended warm medium with a detector which is cooler then there will be a net flow of thermal radiation to the detector and the spectral features of the medium will be seen in emission. Conversely, if the detector is warmer the features will be seen in absorption. This discussion applies, of course, *only* to thermal signals; it would not be a useful approach, for instance, to a treatment of sharp-line spectroscopy of electrically excited gases unless one were prepared to accept highly artificial values for the effective temperatures in the gas. Some nice examples of this type of spectroscopy have come from observations on the stratosphere made from high-flying aircraft or from balloons. The traditional method is to use the sun, or sometimes the moon, as a "hot" source and this would certainly be advisable for mid- and near-infrared work. At far infrared wave-lengths, however, one can use either a "warm" detector at $\sim 300\,\text{K}$ to observe "absorption" spectroscopy of the cooler ($\sim 220\,\text{K}$) stratospheric gases, or else use a liquid helium-cooled detector to observe signals in emission. The latter approach is preferable because there is a greater difference of temperature involved, but both approaches have the advantage over the use of the sun as a source, in that observations can be taken at any time—at night, for example—and in any direction. Harries [54] has given some elegant examples of stratospheric spectra observed with a helium-cooled detector.

A particularly interesting case of the general radiative transfer problem is provided by those media in which the population is inverted so that $N_2 > N_1$. Now both terms in (2.2.53) are positive and there is gain all the time. However, the two terms are very different in their physical effects since the first, representing spontaneous emission, is incoherent, whilst the second, representing stimulated emission, is coherent. One can again distinguish two cases, that where the spontaneous emission is dominant and we have amplification of spontaneous emission (sometimes abbreviated to ASE), and secondly that where the stimulated emission is dominant and we have the basic condition for maser or laser action. These words, as is well known, are acronyms for *M*icrowave (or *L*ight) *A*mplification by the *S*timulated *E*mission of *R*adiation. The spontaneous emission is proportional to the cube of the frequency (2.2.17) and therefore seldom sets a limit to laser or maser action in the infrared or microwave regions. The situations where stimulated emission is dominant can again be divided into two subcases: firstly where the medium is optically thin, that is where the modulus $|\alpha_\nu|$ of the now negative absorption coefficient is less than l^{-1}, and secondly the converse where the medium is optically thick. The first case is only of astrophysical interest where it has the effect of making subtle changes in the observed line intensities and line-shapes, but the second, being the practical condition for observable maser or laser action, is of widespread interest. In astrophysics, for example, it accounts for the well-known OH, H_2O and SiO masers at 1·665, 22·2 and 43 GHz respectively, and in the laboratory, where it is readily satisfied for a wide range of media (section 2.4), it accounts for the large number of infrared lasers which are available. With α_v negative, equation (2.2.54) leads, in the optically thick limit, to

$$S_v = F(v, T) \exp(|\alpha| l). \qquad (2.2.60)$$

It follows that, ignoring the effects of the competing spontaneous process, one can get an arbitrarily high spectral intensity (or brightness) provided one has a long enough path length. This explains how it is possible to get maser emission from interstellar clouds, for although $|\alpha|$ will be very small because of the low number density of the molecules in the cloud ($10^8 \, \mathrm{m}^{-3}$), the path lengths can be very large indeed to compensate, in fact of the order $10^{18} \mathrm{m}$. In the laboratory, where the "straight through" path lengths available to the experimenter can usually be measured in metres, behaviour of this type is rare—it is usually called "superradiant emission" and the phenomenon itself, "superradiance". In laboratory practice, therefore, it is usual to place the gain medium inside a Fabry–Perot resonator of one of the kinds mentioned earlier, and the beam will then traverse the medium very many times. The effective length will now be the intermirror separation multiplied by the Q of the cavity. In this way the optically thick condition can be realised, provided that the gain per pass exceeds the losses per pass. More details of laser theory will be given in section 2.4, and an account of the operation of infrared lasers will be given in section 6.2.

Before concluding this account of the radiative transport phenomenon, it should be pointed out that in many experimental situations the simple theory given so far has to be modified to take account of the presence of interfaces and of the finite resolution of the spectroscopic system. The interfaces cause an apparent loss by reflection, either single or multiple, and one needs to introduce a multiplicative term which is a function of the frequency, the specimen length and the reflection coefficient at the interface. More details are given in section 3.5. In the particular case of a hot extended dense medium, the reflection losses at the interface will be manifest as an apparently imperfect emissivity, and the body will be "grey" rather than truly black even when optically thick. The finite resolution has two effects. The first and more minor of these is that S_v is no longer a truly observable quantity, since it has to be regarded as the limit of observations of intensity at progressively reducing bandwith. This is why Lambert's law is often quoted in the form previously given as equation (1.2.22). The second and more serious consequence is that the spectrometer is not able to follow the spectral profile $\phi(v)$ if this is rapidly varying, as it will be for gaseous samples. This means that one cannot determine α_v, only its average value over a finite bandwidth. This is a matter of some consequence since one would like to be able to use equation (2.2.55) to determine B and hence the dipole moment elements. The solution to this dilemma is to add to the gas under investigation a moderate pressure of a non-absorbing foreign gas. This will often prove sufficient to pressure-broaden the line enough so that its profile can be followed by the spectrometer. The area under the α versus $\tilde{v}$ curve can then be integrated and, provided the line is reasonably sharp:

$$\int \alpha(v)\,dv = h\tilde{v}_0\,(N_1 - N_2)\,B \int \phi(v)\,dv = h\tilde{v}_0\,(N_1 - N_2)\,B. \qquad (2.2.61)$$

This is the basis of the method introduced by Wilson and Wells [55] for determining matrix elements of vibrational transitions using a grating infrared spectrometer. Nowadays there are several other methods available based on interferometric spectrometers, but the pressure-broadening method is still occasionally used. It depends, of course, on the principle of spectral stability, that is that the integral on the LHS of equation (2.2.61), usually called the line strength, is independent of the line width. This is certainly true for narrow lines.

2.2.5 Solid-angle and Lambert's cosine law

So far, we have considered only the simple case of the test object immersed in completely uniform surroundings. In practical cases of thermal equilibrium, one has to consider non-uniform surroundings and then one must introduce a quantity to define the amount of radiation received from an area A situated at a distance d and then go on to specify the effect of a finite angle of inclination between the normal to the source-plane and the direction of observation.

The first point is covered by the introduction of the quantity "solid-angle". One defines this by imagining a sphere of radius r constructed about the test point as centre. An area A of the sphere's surface then subtends a solid angle of A/r^2 steradians at the test point. By definition, therefore, the totality of space surrounding a point subtends a solid angle of 4π steradians at the point. If there is effectively radiation only from and into a hemisphere, then the solid angle will be reduced to 2π steradians. This definition is readily extended to the two practical cases of interest, the flat disc and the sphere by means of the geometrical constructions shown in Fig. 2.17.

In both cases the solid angle is given by the same expression

$$\Omega = (1/r^2)2\pi r^2\,(1 - \cos\theta) = 2\pi(1 - \cos\theta) \qquad (2.2.62)$$

which in terms of the parameters for the two cases becomes

$$\Omega = 2\pi\{1 - [1 + (a/d)^2]^{-\frac{1}{2}}\}, \qquad (2.2.63a)$$

and

$$\Omega = 2\pi\{1 - [1 - (R/r)^2]^{+\frac{1}{2}}\}. \qquad (2.2.63b)$$

If $a \ll d$ or $R \ll r$, then the appropriate expression may be simplified by means of the usual approximations and one has, for example,

$$\Omega = \pi a^2/d^2 = A/d^2. \qquad (2.2.64)$$

This equation is very widely used and, by a simple conceptual extension, it is also used when A is not a circular disc or the apparently circular projection of a sphere. In fact it is the normal practice to define the solid angle subtended by a plane (or apparently plane) irregular area A situated at a distance d just as A/d^2. This relation will nearly always prove reliable, but occasionally the approximations made in deriving it have to be borne in mind. In particular it will fail for close approach of the test point to the radiating/receiving area A. The approximate form is frequently used by illumination engineers who even go so far as to state "the inverse square law of illumination" but, from what has been said above, it will be realised that such a "law" would only apply strictly to point sources.

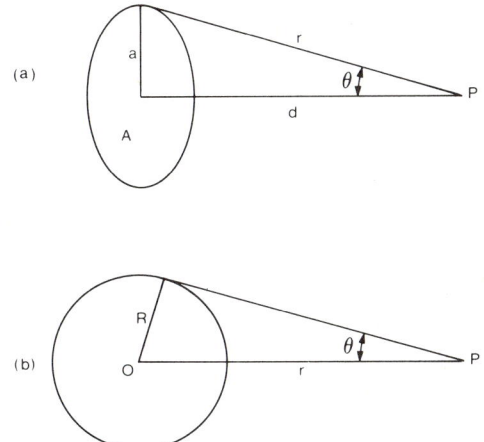

Fig. 2.17. Two definitions of solid angle: (a) that subtended by a circle of radius a situated at a distance d; and (b) that subtended by a sphere of radius R whose centre is at a distance r from the test point.

The problem of the effect of angle of inclination on the radiation exchanges was investigated in the last century by Lambert. He was concerned with the analysis of practical illumination arrangements in which considerable angles of inclination occur between the sources (lamps) and the receivers (walls of a room etc.). Such large angles do not usually occur in infrared instrumentation but, nevertheless, in the more refined analyses it is necessary to take the non-normal irradiation into account. This might be wise, for example, if one were designing a high aperture spectrophotometer or interferometer. Lambert solved his problem by imagining (Fig. 2.18) that he was observing a uniformly bright circular disc of radius a. If the disc were inclined to the direction of observation, then it would appear to be an ellipse with one semi-axis still the same (i.e. a) but with the other fore-shortened to $a\cos\theta$. The area of an ellipse is given by

$$A = \pi ab, \qquad (2.2.65)$$

where a and b are the semi-axes. The effective area of the inclined disc, as seen by the observer, is therefore $\pi a^2 \cos \theta$. Lambert's cosine law follows. This states that if the brightness of a surface (defined as the energy radiated normally from a surface per second, per unit area, per unit solid angle) is B_0, then the apparent brightness seen from a point which makes an angle θ with the normal will be

$$B(\theta) = B(0) \cos \theta. \qquad (2.2.66)$$

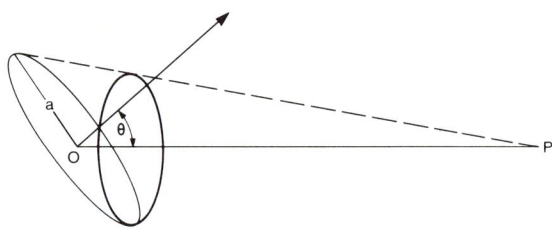

FIG. 2.18. A circle of area πa^2 observed obliquely at an angle θ appears in projection to be an ellipse of area $\pi a^2 \cos \theta$.

Surfaces which obey this equation, either intrinsically as sources or extrinsically as reflectors are said to be "Lambertian surfaces". Lambertian, or perfectly diffuse, surfaces are relatively common and if the angle θ is restricted to values less than 50° they are very common. The reason for the restriction is concerned with electromagnetic phenomena observed in dielectric and metallic sources at high angles of observation (section 3.8) but if the source is truly black, Lambert's law becomes a purely geometric law—that the energy flowing from a source through a given area is proportional to the solid angle subtended by the area at the point—and therfore applies reigidly. By integration of (2.2.66) with respect to θ, it may readily be shown that the total power radiated, over the whole 2π steradians of the half sphere, by a black source will be

$$W = \pi B_0, \qquad (2.2.67)$$

and therefore that the average flux density per unit solid angle is half the brightness.

An important consequence of Lambert's law is that perfectly diffuse sources or reflectors appear uniformly bright regardless of the angle of observation. Thus if one considers the sun, the area of photosphere observed in a solid angle $d\theta$ at an angle θ to the line of normal observation is proportional to $(\cos \theta)^{-1}$ and this exactly counters the diminution of brightness brought about by Lambert's law and consequently the sun appears to be a uniformly illuminated disc. Very accurate measurements on the sun do, however, indicate a perceptible limb darkening from which it may be concluded that the sun does not radiate as a perfect black body: explanations of this have been put forward in terms of the radial variations of density which must occur in the vicinity of the photosphere. From the spectral viewpoint as well, the sun is not a perfect black body. Its brightness temperature (i.e. the temperature of an ideal black body which would give the same output power in the same small frequency interval centred on the same frequency) varies from a minimum value of about 4500 K in the far infrared to its commonly quoted value of 6000 K in the visible region. The average value, in terms of total power output, is approximately 5500 K.

A more extreme case of non-diffuse reflection has been suggested as an explanation of the paradox that Pluto, from its observed brightness, appears to be a small planet, yet it was discovered following an analysis of apparent residual perturbations in the orbits of Uranus and Neptune which, if they were gravitational in origin, could only have been caused by a large massive ninth planet. However, if by some means the surface of Pluto were covered with a material which reflects specularly rather than diffusely, it would be observed as a point of light (in fact the image of the now exceedingly remote sun) with no perceptible disc, even if it were a large body comparable in size to the other outer planets! Unfortunately for this ingenious theory, Pluto has recently been shown to be essentially a double planet with a moon, Charon, of the same order of size and from an analysis of their relative motion, it is clear that both objects are small (diameter of Pluto $\sim$ 4000 km, diameter of Charon $\sim$ 2000 km). The reported perturbations of the orbits of Uranus and Neptune are probably either fictitious or else due to an as yet undiscovered tenth planet of considerable mass.

2.2.6 *Some practical examples of thermal equilibrium*

Although the sun gives out quite intense quasi-monochromatic radiation (for example, Lyman α at 121·5 nm), the bulk of its output is continuous thermal radiation and the temperatures achieved by the various planets, satellites and other objects in the solar system are determined almost exclusively by the balance between thermal radiation arriving from the sun and thermal radiation departing into the depths of space. The only significant exception appears to be in the case of the giant planets where for Jupiter, Saturn and Neptune there is good evidence that more heat is given out than arrives from the sun. This presumably is due to the energy released by continuing gravitational contraction. A contraction of only 1 mm per year would explain the disparity for Jupiter. For the remaining planets, however, their surface temperatures are determined by thermal equilibrium alone. From the point of view of life on Earth, this example of thermal equilibrium is clearly the most important of all!

We begin a detailed analysis by considering the simplest possible case, a black disc of negligible thermal conductivity situated at a distance r from the sun and inclined normally to the axis connecting the centre of the sun and the centre of the disc. The geometry of the situation is shown, in essence, in Fig. 2.17(b). The solid angle subtended by the photosphere at a point on the sunward side of the disc is $2\pi[1 - \cos\theta]$, that is

$$\Omega = 2\pi\{1 - \sqrt{[1 - (R_s/r)^2]}\}, \qquad (2.2.68)$$

where R_s is the radius of the sun (0.696×10^6 km). A point on either side of the disc will only radiate into and receive radiation from 2π of solid angle, so the radiative equilibrium equation for the sunward side will be

$$\{1 - \sqrt{[1 - (R_s/r)^2]}\} T_s^4 + \sqrt{[1 - (R_s/r)^2]} T_B^4 = T^4, \qquad (2.2.69)$$

where T_s is the effective solar brightness temperature (5500 K) and T_B is the temperature of the cosmic background (3 K). The corresponding sub-solar temperatures T at the various planetary distances are shown in Table 2.2. The dark side temperatures are obtained by putting $T_s = T_B$, that is by replacing the sun by an equivalent piece of sky. The answer of course is that $T = T_B$ for all distances. If an attempt is now made to extend this treatment to the planets themselves, several complicating features have to be taken into account. One must consider

a. non-zero thermal conductivity;
b. varying spectral emissivity;
c. absorption and scattering effects in the planetary atmospheres.

TABLE 2.2

Planet	Distance (10^6 km)	T_{ave} (T_{max})	$T_{sub\text{-}solar}$	Albedo	T_{ave}^{calc}
Mercury	57·9	? (677)	507	0·1 (?)	415
Venus	107·8	700	372	0·5 (?)	263
Earth	149·7	280 (337)	315	0·2	251
Mars	227·7	245 (280)	256	0·1	209
Jupiter	777·3	125	138	0·33	105
Saturn	1426	90–95	102	0·33	77
Uranus	2869	57	72	0·33	55
Neptune	4495	55	58	0·33	44
Pluto	5900	37 (?)	50	0·33 (?)	38 (?)

The non-zero thermal conductivity effect is immediately manifest on, say, the dark side of the moon where the temperature, though very low ($-130°C$), is still high compared to that of the cosmic background. The remaining effects show up in Table 2.2, in the discrepancies between T_{max} and $T_{sub\text{-}solar}$. For the terrestial planets where one is measuring the temperature of a solid surface, the discrepancies are positive, whereas for the giant planets, where one is measuring the temperature of the top of a dense and very cloudy atmosphere, the discrepancies are negative. We turn first to consider the effects of clouds and make the simplest assumption, namely that we can deal with these by introducing an effective or averaged albedo a. Heat will be transferred within the planet's atmosphere both by winds and by the planet's rotation and the heat loss over the full 4π radians of solid angle can therefore be averaged to give the simple radiation balance equation

$$\{1 - \sqrt{[1 - (R_s/r)^2]}\} \, (1 - a) \, T_s^4 + \sqrt{[1 - (R_s/r)^2]} \, T_B^4 = 2T_{ave}^4 - T_B^4.$$
$$(2.2.70)$$

No albedo term is included for the cosmic background radiation because it is assumed that all the incident radiation is absorbed. This assumption does not, however, lie within the range of experimental test since the cosmic background term does not make any significant contribution to the heat balance in any accessible region of the solar system. The cosmic background term is retained in the equations given above so that the equations will still make physical sense in the limit of very large distances from the sun. The T_{ave} values given in the last column of Table 2.2 are in good agreement with those derived [56] using a much more sophisticated model. The remaining discrepancies for Jupiter, Saturn and Neptune are believed to be real (due to continuing contraction perhaps) but why Uranus should be exceptional is unknown.

When we come to consider the terrestrial planets, it is obvious that the values of T_{ave}^{calc} given by equation (2.2.70) are in poor agreement with those that are in fact observed. To explain these discrepancies, we need to consider in detail two very important effects, which for reasons which will emerge later are conveniently known as the "greenhouse effect" and the "solar panel effect". Both effects lead to a trapping of solar radiation and to equilibrium temperatures higher than would otherwise be expected [6]. Both have inverses and, when these apply, equilibrium temperatures are observed *lower* than would be expected. The greenhouse effect applies, as is clear from the epithet, to the glasshouses used to raise early crops and for the cultivation of tender plants in high latitudes. The near-infrared radiation from the sun is transmitted by the glass so the heat goes in but the mid- and far-infrared radiation from the contents cannot get out because the glass is opaque at these wavelengths. For good measure glass is also highly reflecting in the long wavelength infrared, due to "reststrahlen" effects and this causes the thermal radiation to be reflected back rather than be absorbed in the thin glass and then be lost by conduction to the cold outside. A further property of glass, not relevant in the context of infrared physics but highly relevant to the economics of greenhouses, is the transparency which it displays in the near ultraviolet. This permits the essential actinic radiation, which the plants need for photosynthesis, to be transmitted to them. Glass is almost an ideal material for the construction of greenhouses, its only drawback being its fragility. Polyethylene is becoming more and more widely used since it is cheap, flexible and easy to handle but it is not so good from a radiation point of view since it is relatively transparent in the mid infrared and is not strongly reflecting. The transmission spectra of glass and polyethylene are compared in Fig. 2.19. Polyethylene "tunnels" should therefore not be regarded as being true greenhouses but should rather be regarded as a method for protecting plants from cold winds and for maintaining the moist atmospheres in which plants can thrive. Research continues in the plastics industry to find a material which would be cheap and easy to use, like polyethylene, but which would also be more opaque in the mid infrared. No ideal candidate has emerged as yet but several types of polyester seem suitable for small-scale operations where the basic cost of the material is not the limiting factor. This research is very important since it would be difficult to exaggerate the economic importance of greenhouses. A large fraction of the food supply raised in northern countries comes from them and many autumn, winter and spring crops could not be grown in these countries without their use. Roughly speaking, the availability of a greenhouse is equivalent to a $10°$ southerly shift in latitude!

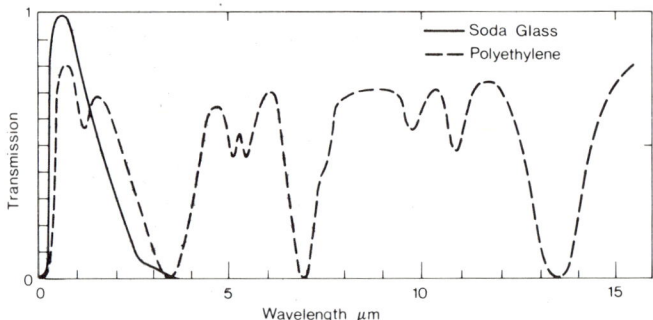

FIG. 2.19. Transmission spectra of soda-glass and polyethylene for sample thicknesses of 1 mm. The relatively poor performance of polyethylene in the visible region is due to scattering in this rather opalescent material.

The Earth's atmosphere is relatively transparent in the near infrared but much more heavily absorbing in the mid and far infrared. There will therefore be a marked greenhouse effect at the surface of the Earth due to the extensive atmosphere above. A transmission spectrum through the atmosphere at sea level is shown in Fig. 2.20. This corresponds to a path length of about 300 m in the horizontal direction but since the scale height of the atmosphere (i.e. the height at which the pressure has fallen to $1/e$ of its value at sea level) is much larger, namely 6 km, the effective vertical path length will be almost the same. It follows that apart from the 9–13 μm window, the atmosphere is for all intents and purposes quite opaque throughout the infrared beyond 5 μm. The 9–13 μm window is not, of course, completely transparent and additional filling-in occurs because of absorption near 9.6 μm due to the layer of ozone in the stratosphere, but nevertheless there is considerable heat loss through this window. One can quantify the effect by introducing the apparent sky

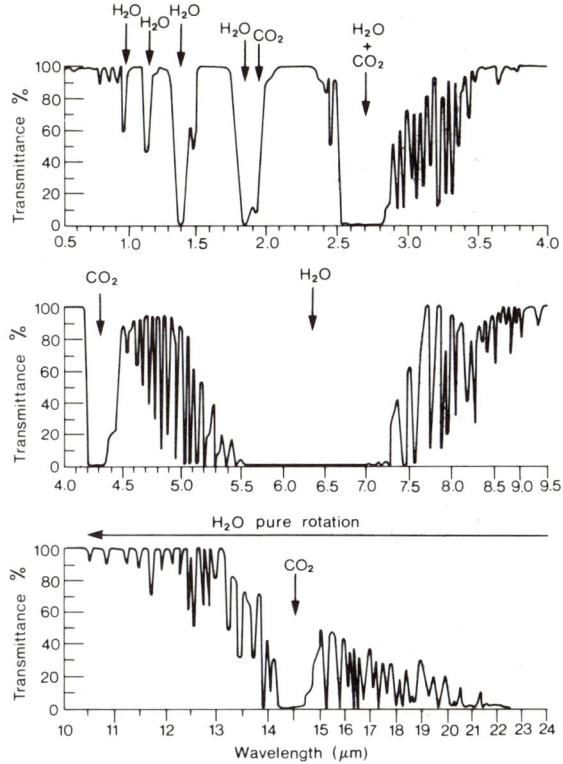

Fig. 2.20. Transmission spectra along a 300 m horizontal path at sea level through an atmosphere containing about 6 mm of precipitable water at a temperature of 26°C.

temperature and this is shown in Fig. 2.21. The observed sky temperature is close to that expected for a 280 K blackbody for much of the range: this is due, as mentioned above, to the "blacking out" of the CO_2 and H_2O bands. In the window regions, however, the temperature is much cooler and since these regions correspond to the peak of the spectral emission from the ground underneath, they are very significant in determining the average sky temperature which is only 230 K. The prevailing surface temperatures under cloudless conditions can be calculated by replacing T^4 in Equations such as (2.2.69) and (2.2.70) by $(T^4 - 230^4)$. In this way maximum temperatures of about 62°C and minimum temperatures of about -43°C are found. Temperatures of this order have indeed been reported in sub-solar equatorial deserts and for Antarctic winter nights. Modified in this way the thermal balance equations yield calculated average temperatures of about 285 K which is in fair agreement with observation. The residual discrepancies arise almost exclusively from the effects of clouds but these are very difficult to calculate in quantitative terms. Qualitatively one can say that a cloud, being made up of water droplets or ice crystals will be very opaque indeed in the 10 μm region and will therefore close

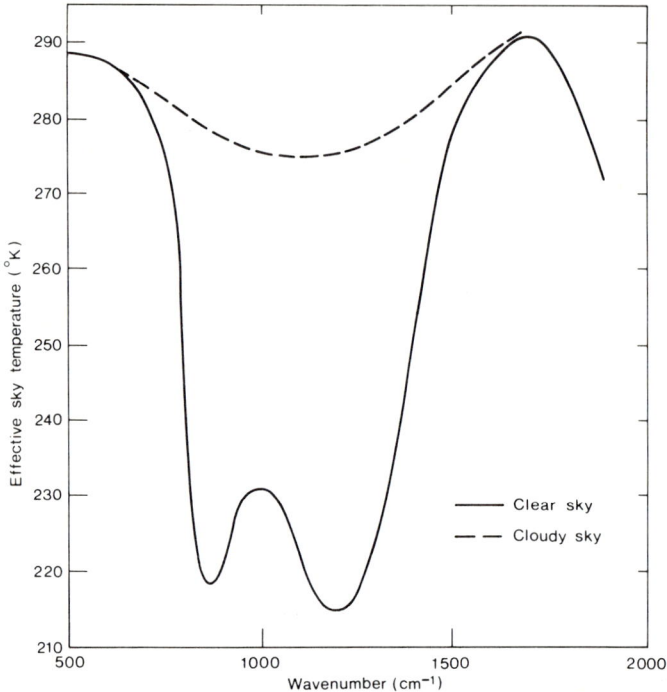

FIG. 2.21. Variation of sky temperature with wavenumber. The peak near 1000 cm^{-1} is due to stratospheric ozone. Absorption by the ozone tends to fill in the 10 μm window where the major atmospheric absorbers (H_2O and CO_2) are relatively transparent. The ozone is effectively "black" at band centre with an ambient temperature of about 230 K.

up the window and prevent the ground temperature on a winter's night reaching low values. Conversely, by scattering incident sunlight back into space they lower summer daytime temperatures. The effects of clouds have to be interpreted in a statistical sense since the very strong convective currents (winds) in the Earth's atmosphere can lead to wide differences in temperature for otherwise identical places under otherwise identical conditions. One gets thus an averaged effect which is the major component of the albedo correction but the fluctuations about this average can be considerable. The greenhouse effect leads to average temperatures on the Earth which are about twenty degrees warmer than one might expect otherwise. This obviously makes any possible changes in the composition of the atmosphere a matter of concern. A case in point is the rise in the amount of CO_2 in the atmosphere due to the massive burning of fossil fuels that has occurred this century. Increased CO_2 means higher average temperatures and these could lead to melting of the polar ice-caps and to a dramatic increase in the sea level. One theory of the origin of ice ages holds that they are due to a varying greenhouse effect due to a varying amount of CO_2 in the atmosphere. We are at the moment in an interglacial warm period within an overall ice age that began about two million years ago, and clearly the concentration of CO_2 and any other gas which absorbs in the 10 μm region in our atmosphere is of more than academic interest [57].

On Venus the greenhouse effect has run away to give cataclysmic results. The Venusian atmosphere is mostly CO_2 and it exerts at the planet's surface a pressure more than twenty times larger than that which prevails on Earth. The trapping of the incoming solar radiation leads to surface temperatures in excess of 400°C! Now since Venus and the Earth are otherwise so similar, it is not too fanciful to imagine that they originally had very similar compositions. On Earth, most of the CO_2 has become tied up in limestone rocks or else dissolved in the oceans, whereas on Venus it is free in the atmosphere. It is quite possible that Earth-like planets face a catastrophe situation in that if they are situated too close to their primary the equilibrium

$$CaCO_3 \rightleftharpoons CaO + CO_2 \qquad (2.2.71)$$

will be displaced to the right, the greenhouse effect will be enhanced, the temperature will rise, the equilibrium will be displaced still further and so on to disaster. In the high-temperature regime there will be no liquid water to remove CO_2 by dissolution and the large quantities of water vapour in the atmosphere will intensify the greenhouse effect. Eventually ultraviolet radiation from the primary will photolyse the H_2O into hydrogen and oxygen and the hydrogen will escape into space but the CO_2 will still be sufficient to keep the planet's temperature very high.

The second effect to be considered which may perturb straightforward radiative equilibrium is called, for convenience, the solar panel effect because it is vital to the operation of the panels used in solar heating systems. What one needs here is a surface which is black at near-infrared wavelengths but which is "white" or at least "pale grey" at mid-infrared wavelengths [6]. One will then have once more the situation where heat flows in but cannot so readily flow out. It was shown above that the maximum temperature which a true black body could achieve through solar irradiation on Earth would be 67°C (temperatures of about 60°C have in fact been reported). This is hot, but not hot enough for the solar heating engineers who would ideally prefer to raise steam so that conventional generating equipment can be used. In addition these high temperatures would only apply in deserts near the equator. For the densely populated northern countries, much lower temperatures (about 30°C) would prevail. However with the use of panels whose surfaces have been specially painted to absorb the near infrared solar radiation transmitted by the atmosphere but not to radiate very effectively at longer wavelengths, much higher temperatures can be achieved. Water circulated through channels in the panels can be made hot enough for the domestic water supply. Still higher temperatures can be achieved by mounting the panel beneath a thick sheet of glass so that the solar panel effect is augmented by the greenhouse effect and by the suppression of convective heat loss. Near boiling temperatures have been achieved in Britain on sunny days in summer using this combination and the systems can be made efficient enough to heat water significantly even when there is 100% cloud cover. In warmer climes and where the skies are clearer much higher temperatures have been reached. Thus values as high as 130°C have been reported for installations in the western deserts of the USA and when these installations are combined with concentrators, that is high aperture sun following focussing optics, it becomes possible to build solar powered generating stations. The combination of the solar panel effect and the greenhouse effect is certainly potent as may have been brought home painfully to anyone who may have left their car out in the summer sun with its windows wound up and who then have re-entered the car in swimming gear and sat down on a black leather seat!

The inverse solar panel and greenhouse effects also have their uses. Thus there is an increasing use of near-infrared reflecting coatings for the windows used in office blocks so that the offices will not over-heat in summer. This is especially important in high rise buildings where, for safety reasons, opening of the windows may not be permitted. In the middle east people wear white clothes out of doors,† and in space, astronauts wear special highly reflecting clothes when they leave the safety of their spacecraft. In both cases the equilibrium thermal temperature may well be above blood heat and only the inverse solar panel effect makes survival possible. In the context of planetary temperatures, the solar panel effect must have something to do with the reported high (677 K) subsolar temperatures on Mercury since this planet has only a vestigial atmosphere.

2.3 Coherence aspects of infrared sources

The coherence of a radiant field is a measure of the correlation in space and time of the field fluctuations. It can be analysed within a rigorous quantum mechanical framework [59] but nearly all the phenomena which are encountered in the infrared can be interpreted just as well within the simpler classical formalism [60] which considers merely the fluctuations of the electric field strength $E(t)$. These fluctuations are usually stationary and ergodic

---

†This argument is basically true but it has to be admitted that Bedouins seem able to manage in *black* robes and that the native goats have *black* hair! Undoubtedly convective and conductive effects have also to be taken into consideration [58].

(section 4.2) so one can normally replace the ensemble averages by simple time (or space) averages and thus write the correlation in the form

$$\gamma(\tau) = \frac{\langle\, E(t)\,E(t+\tau)\,\rangle}{\langle\, E(t)\,E(t)\,\rangle}. \tag{2.3.1}$$

Naturally there are strong connections with noise theory, as spelled out in section 4.2, and with spectroscopy theory, since the spectrum of the radiation is simply related to the Fourier transform of $E(t)$.

The radiant fields which are encountered in practice range from the nearly perfectly monochromatic and hence almost fully coherent to the nearly completely "white" and hence almost fully incoherent. Experimental investigations of partially coherent fields are, however, usually restricted to the narrow-band end since coherence phenomena are then more readily demonstrated. One is dealing thus with fields which range from the quasi-monochromatic to the fairly narrow-band polychromatic. In the ideal monochromatic case, it is convenient to introduce a complex electric field $\hat{E}$ and to regard the observed field as its real component. In the physically more realistic case where the radiation has a finite band-width, this approach is still very convenient so one seeks an equivalent formalism. The first step is to introduce the spectrum of the field fluctuations as a set of amplitudes $a(v)$ each being associated with a random phase $\phi(v)$. This spectrum is, of course, restricted to positive only values of v, but for mathematical convenience one resolves it into the sum of two functions, one even, $a_e(v)$, and the other odd, $a_o(v)$, each of which spans the entire range from $-\infty < v < +\infty$. One then has

$$a(v) = 2a_e(v), \quad v > 0,$$
$$a(v) = 2a_o(v), \quad v > 0,$$
$$a(v) = a_e(v) + a_o(v). \tag{2.3.2}$$

The real polychromatic field can then be written

$$E'(t) = \int_{-\infty}^{+\infty} a_e(v)\cos\left[2\pi vt + \phi(v)\right]dv = \int_{0}^{\infty} a(v)\cos\left[2\pi vt + \phi(v)\right]dv, \tag{2.3.3}$$

which suggests that we introduce a complementary field

$$E''(t) = \int_{-\infty}^{+\infty} a_o(v)\sin\left[2\pi vt + \phi(v)\right]dv = \int_{0}^{\infty} a(v)\sin\left[2\pi vt + \phi(v)\right]dv, \tag{2.3.4}$$

and combine the two fields to produce

$$\hat{A}(t) = E'(t) + iE''(t) = \int_{0}^{\infty} \hat{a}(v)\exp\left[2\pi ivt\right]dv, \tag{2.3.5a}$$

where

$$\hat{a}(v) = a(v)\exp[i\phi(v)].$$ (2.3.5b)

$\hat{A}(t)$ is then said to be the analytic signal associated with the real field $E'(t)$. It is analytic because the definitions (2.3.3) and (2.3.4) compel $E'(t)$ and $E''(t)$ to be Hilbert transforms of one another (section 3.4.5 and Appendix 1). The principal merit of the analytic signal is that it gives an immediate access to the observable quantity at optical frequencies namely the "envelope" of the electric field fluctuations. The envelope will be changing at a rate given by the bandwidth Δv, whereas the field fluctuations themselves will be occurring at a rate determined by the mean frequency $\bar{v}$, and since for quasi-monochromatic radiation the ratio of these two quantities will be very small, the envelope variations may well lie in an experimentally accessible region. Even so, it is still only the intensity (rather than the amplitude) variations which can be detected, so one applies Rayleigh's theorem to equations (2.3.2), (2.3.3) and (2.3.4), to give the results

$$\int_{-\infty}^{+\infty} |\hat{A}(t)|^2 \, dt = \int_0^\infty |\hat{a}(v)|^2 \, dv$$

$$= 2\int_0^\infty [E'(t)]^2 \, dt = 2\int_0^\infty [E''(t)]^2 \, dt$$

$$= 2\int_0^\infty [a_e(v)]^2 \, dv.$$ (2.3.6)

The intensity of the field can be calculated uniquely, therefore, in terms of the square-modulus of the envelope function, that is the analytic signal. The analytic signal itself can be derived from a given $E'(t)$ by inverting (2.3.3) to give $a_e(v)\exp[i\phi(v)]$, multiplying this by 2 and suppressing negative frequency amplitudes to give $\hat{a}(v)$ and finally Fourier transforming this quantity over the half-range. The inverse operation is quite straightforward and, including for completeness the full optical time dependence, one may write,

$$E(t) = (1/2)[\hat{A}(t)\exp(2\pi i\bar{v}\tau) + \hat{A}^*(t)\exp(-2\pi i\bar{v}\tau)],$$ (2.3.7)

where τ is the total delay, i.e. $(t - nz/c)$. This equation degenerates, as it should, into a simple cosine wave if $\hat{A}(t)$ becomes constant, that is in the truly monochromatic case.

The experimental measurement of coherence requires, as implied above, a method for slowing down the ultrafast oscillations of the radiant field so that the fluctuations are shifted into an accessible frequency region. The problem is the same as that faced by the optical spectroscopist (section 4.3) and the solutions available are the same, namely the use of either heterodyne beating or interferometric techniques. Heterodyne techniques have already been used to measure the high coherence of laser fields [61] and they can be very convenient, especially when ultra-stable lasers are available to act as the local oscillators, but since the time coherence of most optical fields can be measured in nanoseconds or less, the interference technique is more generally useful. It is particularly so for weak radiant fields. The simplest form of interferometry involves the division of the beam into two split beams which traverse different path-lengths and which are then recombined to give an interference pattern which displays fringes. The pattern, so produced, is stationary but, as will emerge later, one can obtain quantitative measures of the coherence from the contrast of the fringes. The classical way of doing this experiment, especially for the case of visible radiation, is to use the double slit arrangement of Young, which is shown schematically in Fig. 2.22.

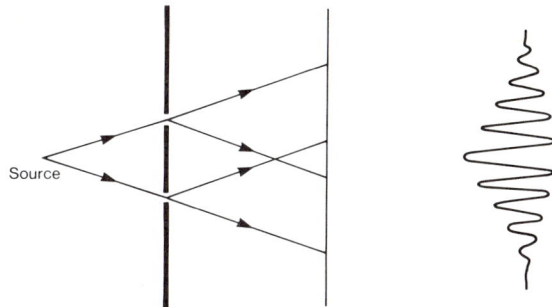

Fig. 2.22. Young's double slit experiment used to define the mutual coherence of two points on a wavefront. The left inset shows the optical set-up and the right shows the sort of fringes which might be observed.

In this experiment, we imagine a wave-front (ideally plane) arising from a source S and incident on an opaque screen which has had two slits cut through it. A subsequent screen will then display the interference pattern (if any). One can assume for simplicity that the intensities at the two slits are equal and that the space between the two screens is non-absorbing and non-dispersive. The correlation, at a point Q on the receiving screen, of the two analytic signals arriving from the slits 1 and 2 will be defined in terms of the combined analytic signal

$$\hat{A}(Q, t) = \hat{A}(1, t - t_1) + \hat{A}(2, t - t_2), \qquad (2.3.8)$$

where $t - t_1$ and $t - t_2$ are the times of flight of the two beams from the two slits. Using equation (3.1.3) and invoking the ergodicity and stationarity, it follows that the intensity at Q will be given by

$$I(Q) = 2I + \hat{\Gamma}_{12}(\tau) + \hat{\Gamma}_{12}^{*}(\tau), \qquad (2.3.9)$$

where τ is now used for $t_1 - t_2$ and

$$\hat{\Gamma}_{12}(\tau) = (1/2)\varepsilon_0 c \langle \hat{A}(1, t - t_1)\hat{A}(2, t - t_2) \rangle \qquad (2.3.10)$$

is the mutual coherence of the field fluctuations reaching Q from S via the two slits. The mutual coherence is physically an intensity and it may therefore be normalised, by dividing by I, to give a mutual coherence coefficient $\hat{\gamma}_{12}(\tau)$. From the definition this is also the average cross-correlation coefficient of $\hat{A}(1, t)$ and $\hat{A}(2, t)$. One can therefore finally write

$$I(Q) = 2I[1 + (1/2)\{\hat{\gamma}_{12}(\tau) + \hat{\gamma}_{12}^*(\tau)\}], \qquad (2.3.11)$$

that is

$$I(Q) = 2I[1 + |\hat{\gamma}_{12}(\tau)| \cos \psi]. \qquad (2.3.12)$$

Here ψ, which is the argument of $\hat{\gamma}_{12}(\tau)$, is in general a complicated function of the nature of the source, the geometry of the system, the spectral distribution of the radiation and of the distribution of phases amongst the spectral components. It is seldom calculated and then only for the quasi-monochromatic case where acceptable approximations may be made. However this is no great hardship in the present context since ψ determines only *where* the fringes will be formed: the quality of the interference is set by $|\hat{\gamma}_{12}(\tau)|$ and this quantity is therefore immediately accessible to experiment. If it is unity, we have a perfect *interference field*, the fringes will show 100% modulation (or contrast) and the field is fully coherent. If, on the other hand, it is zero, there will be no fringes, the intensity will be found everywhere merely by adding the intensities incident from each slit and we have an *intensity field*. In practice, one usually has an intermediate situation, that is $0 < |\hat{\gamma}_{12}(\tau)| < 1$ and one has a partially coherent field. If near a point Q, in the interference pattern produced from such a field, the maximum fringe intensity is $I_{max}(\tau)$ and the minimum is $I_{min}(\tau)$, then, from (2.3.12), it follows that

$$|\hat{\gamma}_{12}(\tau)| = \frac{I_{max}(\tau) - I_{min}(\tau)}{I_{max}(\tau) + I_{min}(\tau)}. \qquad (2.3.13)$$

Alternatively, one can say that the receiving screen is uniformly irradiated by an incoherent intensity I_{incoh} superposed on which there is a coherent irradiation I_{coh} which shows perfect fringes. Equation (2.3.13) can then be rewritten

$$|\hat{\gamma}_{12}(\tau)| = \frac{I_{coh}}{I_{coh} + I_{incoh}}. \qquad (2.3.14)$$

The modulus of $\hat{\gamma}_{12}(\tau)$ obviously depends on the two variables d_{12}, the distance apart of the two slits and τ the delay time. One could therefore plot it in three dimensions with $|\hat{\gamma}_{12}(\tau)|$ as the ordinate and with d_{12} and τ as abscissae. The two sections through this plot along the abscissa axes give respectively $|\hat{\gamma}_{12}(0)|$, the mutual spatial (or transverse) coherence, often written $|\hat{\mu}_{12}|$, and $|\hat{\gamma}_{11}(\tau)|$, the time or longitudinal coherence. It is found that for many practical situations both these functions have Gaussian dependences on their arguments so it is possible to define $1/e$ widths for the curves and thus derive numerical values for the coherence times or distances.

The transverse coherence is mostly of importance when one is dealing with quasi-monochromatic radiation or else with radiation which has a small angle of divergence. When either of these conditions is satisfied the powerful Van-Cittert and Zernike theorem [60] can be brought to bear on the problem of calculating the amplitude and phase of the complex spatial coherence $\hat{\mu}_{12}$. This theorem relates the area of coherence around a fixed point P_1 on a screen illuminated by an extended quasi-monochromatic source of mean wavelength $\bar{\lambda}$ and area A set at a distance D from the screen, to the diffraction pattern which would have been produced had a spherical wave converging to a focus at P_1 been interrupted and diffracted by an opaque aperture equal in size and position to the source. The theorem states that the correlation of the radiant field fluctuations at a movable point P_2, relative to those at P_1, is given simply by the ratio of the amplitudes in the diffraction pattern at P_2 relative to P_1. Diffraction theory (section 3.2) shows that the pattern on the screen will be determined by the Fourier transform of the source intensity function. A common situation is where the source is circular of radius r and uniformly illuminated. The diffraction pattern is then the two-dimensional Fourier transform of the disc function, in other words the Airy function $[J_1(x)/x]$. The spatial coherence function is thus

$$\hat{\mu}_{12} = [2J_1(v)/v] \exp(i\psi), \tag{2.3.15}$$

where

$$v = 2\pi r d_{12}/\bar{\lambda}D \quad \text{and} \quad \psi = -\pi d_{12}{}^2/\bar{\lambda}D. \tag{2.3.15a}$$

Born and Wolf [60] consider two special cases of (2.3.15). Firstly a disc, centred on P_1, and small enough to be essentially coherently illuminated. For this they suggest that a fall in the amplitude of the Airy function by 12% would be acceptable. This occurs when $v = 1$ and it then follows that

$$d_{12} \leqslant D\bar{\lambda}/2\pi r, \text{ i.e. } \leqslant 0 \cdot 16 D\bar{\lambda}/r. \tag{2.3.16a}$$

Secondly they consider an area on the screen which is of maximal size to maintain positive correlation everywhere within the disc. Such a disc encloses exactly the first Airy lobe and the value of d_{12} will be set by the first zero of the Airy function. This occurs at $v = 3 \cdot 83 \ldots$ and hence

$$d_{12} \leqslant 3 \cdot 83 D\bar{\lambda}/2\pi r, \text{ i.e. } \leqslant 0 \cdot 61 D\bar{\lambda}/r. \tag{2.3.16b}$$

These two relations can be used to specify various experimental conditions. Thus in the Young's slit experiment, the intensities in the fringe pattern will be given by

$$I(Q) = 2I\{1 + |\hat{\gamma}_{12}(0)| \cos[(2\pi c\tau/\bar{\lambda}) - \beta]\}, \tag{2.3.17}$$

where β is a small phase-offset. If the slits are spaced by a distance less than the value of d_{12} given by (2.3.16a), then $|\hat{\gamma}_{12}(0)|$ will be close to unity and the fringe contrast will be good. If, on the other hand, the spacing is greater than that given by (2.3.16b) no fringes will be obtained. Equations (2.3.16) form therefore a good recipe for setting up the experiment.

In the infrared and millimetre-wave regions, because $\bar{\lambda}$ is large, coherence effects are often obvious, especially if lasers are being used. One can discuss two contrasting examples. Firstly millimetre-wave astronomy where D/r will be large and one can write (2.3.16a) in the approximate form

$$d_{12} \leqslant 0.16\bar{\lambda}/\alpha, \qquad (2.3.16c)$$

where α is the angle subtended by the object under study. Clearly for an observational angle of even as much as $1'$ of arc ($\alpha = 0.0003$ radians) d_{12} will be of the order of metres and for smaller observing angles it will be still bigger. This means that large "dishes" (i.e. paraboloid reflectors) can be used to collect the energy effectively without one having to give up the use of the very sensitive coherent detectors which are now available to the astronomer (section 4.4.4). The second example is provided by the use of a laser coupled to a Mach–Zehnder interferometer to determine the optical constants of a specimen [62]. The Mach–Zehnder configuration is used here to prevent difficulties due to radiation being returned to the laser cavity. The almost parallel radiation from the laser can be imagined to originate in a sphere of diameter of the order of the wavelength so the effective value of D/r will be very large indeed. This is an alternative way of understanding the very low divergence angles observed with laser radiation. It follows that the transverse coherence will be very high and it is then not usually necessary to have the interferometer in perfect adjustment to get good fringes. Indeed it is even possible to get interference when one mirror image has been flipped over with respect to its companion! There is thus an obvious hazard in that beams which have managed to reach the image plane by some unexpected and inadvertent path can nevertheless interfere. It is just this possibility which bedevils attempts to determine the absolute power of a laser source and which detracts from the absolute status of spot-frequency dielectric measurements made in the infrared using laser sources. The difficulty in the first case is that if the measurement is to be absolute, it has to be made with respect to an exactly characterised source, that is a black body under identically the same conditions. The possibility of several kinds of long-path interference makes it impossible to be sure that the measuring equipment has the same antenna pattern (section 3.1) for the two types of source. In the second case, one would want the laser measurements to be absolute so that they could be used to calibrate the relative measurements made with dispersive Fourier transform spectrometers. The realisation that it is no easier to make absolute laser measurements than it is to make absolute interferometer measurements has led to this approach being abandoned and the present practice is to do both but to regard the results of each as having equivalent status. Each can then serve as a check on the other and any discrepancies can be used to help track down the almost inevitable systematic errors.

The time (or longitudinal) coherence of a radiation field can be defined by imagining the two points considered so far to approach one another closer and closer until they eventually coalesce. One thus derives the self-coherence at a point, 1, viz. $\hat{\gamma}_{11}(\tau)$. This quantity, from its definition, is the autocorrelation coefficient of the field fluctuations at a single point in space but for two *different* instants of time. The development of the concept owes much to the pre-war work of van Cittert but it has been extensively discussed and extended in the more recent work of Janossy [63]. To get to grips with the physical meaning of self-coherence, it is very helpful to have a model for the electric field fluctuations, which is capable of encompassing the whole range of phenomena from near perfect coherence to near total incoherence. Two such models are commonly used, both based on the idea of a wave-packet. Firstly, one can have a wave-packet in which the oscillations gradually build up to a maximum and then gradually die down again. In analytical terms one might express this in the form

$$E(t) = E(0)\exp\left(-(t/\tau_c)^2\cos(2\pi vt)\right),\qquad(2.3.18a)$$

which clearly shows that as the coherence time τ_c goes from zero to infinity, the wave train changes from a delta-spike to an infinitely long perfect cosine wave. The second model is that of a wave train of constant amplitude but of finite length. Analytically this has the form

$$E(t) = E(0)\,\Pi(t/\tau_c)\cos(2\pi vt)\qquad(2.3.18b)$$

Again τ_c is an exact measure of the coherence and obviously the coherence must then be zero for delay times in excess of τ_c.

The experimental determination of $\hat{\gamma}_{11}(\tau)$ requires, as mentioned above, an interferometric technique but the Young's double-slit arrangement, though convenient for purposes of exposition and for the experimental determination of the transverse coherence, does not give directly the longitudinal coherence. The ideal instrument for this latter measurement is the moving version of the two-beam interferometer introduced by Michelson and illustrated schematically in Fig. 2.23. In this instrument an incident beam is divided in amplitude into two partial beams, one of which suffers a time delay with respect to the other before they are finally recombined to give the interference pattern in the image plane. The delay time τ can be changed by altering the path difference in the interferometer and this can be ensured by moving one of the mirrors along an axis perpendicular to its reflecting surface. The observed interference function will then be

$$I(x) = (1/2)c\varepsilon_0\left\langle(E(t)+E(t+\tau))^2\right\rangle,\qquad(2.3.19a)$$

which clearly reduces to

$$I(x) = I(0)\left[1+|\hat{\gamma}_{11}(\tau)|\right].\qquad(2.3.19b)$$

The interference function is thus simply related to $\hat{\gamma}_{11}(\tau)$ and hence can be regarded as the autocorrelation function of the field fluctuations. Its general form can be derived by considering either of the models introduced above. Thus if one takes the truncated cosine wave, then, as the delay time increases, the overlap between the two partial beams gets less and less and eventually

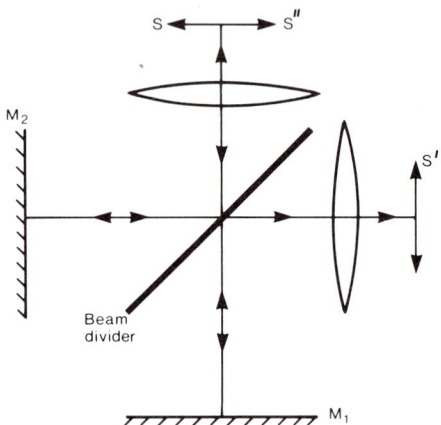

F$_{\text{IG}}$. 2.23. Schematic diagram of the Michelson interferometer. Two images S′ and S″ are produced from an extended source S. Interference occurs in the two image planes but this is complementary so that the sum of the intensities in S′ and S″ for all corresponding points equals that of the original points in S. Energy is conserved.

becomes zero. The self-coherence coefficient $|\hat{\gamma}_{11}(\tau)|$ therefore falls steadily from a value of unity down to zero. This behaviour is always found. The envelope of $I(x)$, on the ground scale, falls inexorably as x increases and the scale-length of this decline is a direct measure of τ_{c}. It follows immediately that when the interferometer is irradiated with quasi-monochromatic radiation of line-width Δv then the interferogram will be damped with a scale-time τ_{c} equal to $(\Delta v)^{-1}$. In more complex cases (Fig. 2.24), the interferogram will show fine structure but there will always be, on the large scale, an over-all damping. The fine structure can be thought of as arising from interaction between the various damped cosinusoids which each spectral component contributes to the over-all interferogram. In simple cases, the interferogram develops recognisable "beats" as shown in Fig. 2.24. In this connection it is interesting to note that a continuous-wave laser operating on a single longitudinal mode will have a self-coherence coefficient which will only significantly depart from unity at very large values of τ. If, however, there is present, in the output, more than one longitudinal mode (see next section), then, since these will differ in frequency (being, in fact, spaced along the frequency axis with a separation $c/2d$), beats will result in any interference pattern produced by the laser. In particular, at certain values of τ the fringe contrast will vanish. This is clearly not desirable in any application (holography for example) where a high degree of coherence over considerable spatial separations is required. It is also, of course, not acceptable in spectroscopic work where one is attempting to resolve spectral intervals smaller than $c/2d$. For these applications, single-mode lasers must be used despite their cost and relatively lower power outputs.

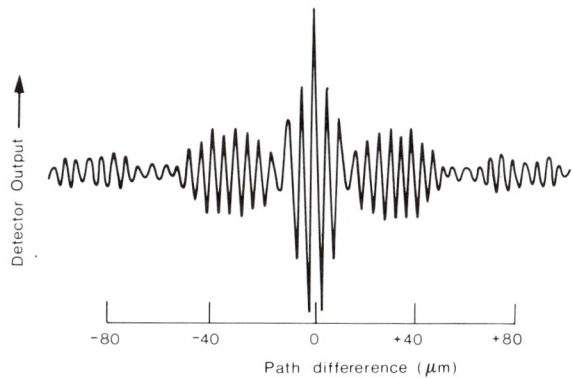

F$_{IG}$. 2.24. Zero-path-difference ("white-light" fringe) pattern produced by a Michelson interferometer irradiated by continuous broad-band radiation. The "beat" pattern arises from the presence of a strong absorption band in the polypropylene which was used for the beam divider. This band essentially divides the spectral range into two regions each of which contributes a damped cosine wave to the interferogram. The beat pattern is formed as these two waves come into and out of phase with one another (J. W. Fleming, unpublished work).

The interpretation of the fine structure in the interferogram to reveal the spectral detail is the province of Fourier transform spectrometry (section 4.3) but in the context of coherence measurements it is obviously not necessary to leave the time domain. All that is required is a determination of the rate of the overall damping. Experiments of this kind were first performed by Michelson [64] but he naturally did not interpret their results in coherence terms and, moreover, he was unable to do much about the fine structure either since he had no computers available to help him. Michelson worked in the visible region and the method that he derived to extract the best available information from the interferogram is consequently called that of the "visibility of fringes". It is really only suitable for simple spectra made up of at most a few very sharp lines and, when this is the case, one may usefully introduce a quantity $\mathscr{V}(x)$, the visibility at a path-difference x, defined by

$$\mathscr{V}(x) = \frac{I_{\max}(x) - I_{\min}(x)}{I_{\max}(x) + I_{\min}(x)}, \qquad (2.3.20)$$

where $I_{\max}(x)$ and $I_{\min}(x)$ are the intensities of a "bright" and a "dark" fringe respectively in the neighbourhood of x. From (2.3.13), it follows that

$\mathscr{V}(x) = |\hat{\gamma}_{11}(x)|$ since $x = c\tau$. In passing it should be remarked that the visibility of fringes can be applied to spectroscopy in some specially simple cases. It can be used, for example, to determine the shapes of isolated lines [7] and the fine structure of close doublets [64]. In both cases the fringes are visible for considerable excursions of path-difference. If, by contrast, one were studying broad-band or polychromatic radiation, the fringes would be closely confined to the zero-path-difference position. With the eye as detector, one would see the beautiful phenomenon of "white-light-fringes". An example of the infrared equivalent is shown in Fig. 2.24. To get such interferograms with maximum contrast, it is necessary to have the interferometer in perfect alignment, that is to have the two images, one from each mirror, in perfect register. This is because of the very small spatial coherence of broad-band radiation from an extended source. One has thus the opposite side of the coin to the situation discussed earlier when laser sources were being considered. The radiation, whose interferogram is shown in Fig. 2.24, is produced from a black-body source but is confined to the region 900–1100 cm^{-1} by the use of absorptive filters. Its bandwidth is therefore 200 cm^{-1} and we would consequently expect a coherence length of about 50 μm. Ignoring the more rapid fluctuations, due to the discrete structure of the corresponding spectrum, it will be seen that the $(1/e)$ point is indeed reached at a path-difference of about 50 μm. This is a nice example of the use of the Michelson interferometer as a coherence meter. With still broader-band radiation, the coherence length will be even shorter. Thus with a perfect black body and a flat response detector, the spectral width will be that of the Planck curve. From (2.2.29),

$$x_{1/2} = 4 \cdot 25411 \ldots \qquad (2.3.22)$$

and hence, from the definition of x,

$$\Delta v_{1/2} = 8 \cdot 8641 \ldots 10^{10} \, T. \qquad (2.3.23)$$

A conventional infrared source ($T \approx 1000 \, \text{K}$) will therefore have a coherence time of the order of 10^{-14} s and a coherence length of about 3 μm. If one could somehow or other hear the fluctuations due to the radiant field from a black body they would sound like random noise, but not truly "white" noise because of the finite width of the Planck curve. If on the other hand, one could hear the field oscillations from a stabilised single-mode laser, one would hear a clear musical note. This analogy is a very useful way of summarising the classical theory of longitudinal coherence.

The classical theory can be extended, however, in several ways which cannot be represented by simple analogies. It has been shown [65], for example, that broad-band radiation can be coherent and conversely that narrow-band (but *not* monochromatic) radiation can be incoherent. It has also been shown that there exist hierarchies of correlation functions, analogous to (2.3.1) but involving higher and higher powers of the field fluctuation function $E(t)$. So far only the next member of the series, the correlation of intensities, has been used experimentally; this in the celebrated work of Hanbury-Brown and Twiss [66], who used a pair of photomultipliers and an electronic correlator to make a very much superior version of Michelson's stellar interferometer.

The classical theory was naturally concerned with the quasi-monochromatic radiation produced from gas-discharge sources and with the rather broader band polychromatic radiation produced by the use of narrow-band transmission filters in conjunction with a thermal or quasi-thermal source. If such a source were circular with radius r, then from (2.3.16b) it follows that at a distance D, one would have an area of spatial coherence A_c, defined by

$$A_c = \pi d_{12}^2 = \pi (0\cdot61)^2 D^2 \bar{\lambda}^2/r^2. \tag{2.3.24}$$

The power from this quasi-monochromatic source of bandwidth $\Delta \nu$ passing through A_c, would, from (2.2.25) be

$$P = (2\delta h\nu^3/c^2)\Delta\nu\pi r^2 A_c/2\pi D^2, \tag{2.3.25}$$

which, from (2.3.24), may be written

$$P = h\nu\Delta\nu\pi\delta (0\cdot61)^2 = h\nu\pi\delta (0\cdot61)^2 (c/l_c), \tag{2.3.26}$$

where l_c is the coherence length. The energy in a cylindrical volume of base A_c and length l_c will be Pl_c/c, which must of course equal $Nh\nu$, where N is the number of photons in the cylinder. It follows then that

$$\delta = N/\pi (0\cdot61)^2 = 0\cdot86 N \approx N. \tag{2.3.27}$$

One is led naturally therefore to the concept of a coherence volume

$$V_c = A_c l_c, \tag{2.3.26}$$

which, from its definition, contains just δ photons. The significance of the coherence volume is that, if one has an extended source, then roughly speaking one can get interference effects if one recombines radiation from two points both of which lie within V_c.

The quantum theory of coherence, as mentioned earlier, has so far found few applications in infrared physics, but it is worth sketching out its bare bones at least. One starts with a cavity and then quantises all the allowed modes of that cavity. The resulting wave functions of the quantised field represent stationary, or boson, states of the system. The quanta, that is the photons, then fill these states but, being bosons, there is no restriction on the number of them which may be in any given state. In the simplest way of looking at the system, the coherency may be defined in terms of the number of photons per boson state. If there is a high multiple occupancy, one has a high degree of coherence and so on. The quantity δ—the degeneracy of the radiation—is thus a direct measure of the field coherence. For all normal thermal sources, δ is small but for lasers it can be very large indeed because of the Bose condensation. Laser radiation is thus fundamentally different from non-laser radiation; it is not just simply a matter of relative linewidth. To obtain some idea of the order of magnitude of δ one can do a rough "back of the envelope" calculation by assuming that one has a total energy $\mathscr{E}$ in the form of a "ball" bouncing back and forth in a cavity which has a length d and which is terminated by two mirrors, one perfect and the other of reflectivity R_e. The energy loss in one round trip will then be $(1 - R_e)\mathscr{E}$ and since this takes place in a time $2d/c$ one has

$$P = (1 - R_e)\mathscr{E}c/2d, \qquad (2.3.27)$$

for the power escaping through the imperfect mirror. Taking as an example a single mode visible region laser with $(1 - R_e) = 10^{-4}$, length 1.5 m and giving an output power of 1 W, it follows that $\mathscr{E}$ will be 10^{-4} J and if the frequency of the laser is 600 THz then the number of photons in the single mode within the laser cavity will be of the order 10^{15}. Even for lower power lasers at longer wavelengths and with less perfect cavities one is still going to encounter huge values for the degeneracy parameter δ.

The quantum theory of coherence also leads naturally to the idea of a coherence volume but here it is the combination of the ordinary three-dimensional space V_c with a three-dimensional momentum space V_m into a six-dimensional phase-space

$$V_p = V_c V_m, \qquad (2.3.28)$$

which is significant. The coherence volume then corresponds to one cell in the quantum mechanical phase-space and this contains one photon. In very rough language one can say that interference is only possible when a photon interferes with itself! The average phase-space per mode for thermal radiation turns out to be h^3. When there are very large numbers of photons in the field, for example intense radiation at long wavelengths, the quantum theory becomes equivalent to the classical one. This is a familiar observation because quantum theory is not used to describe the radiation from broadcasting stations. In this limit the unit cells corresponding to each photon do not overlap. In the opposite limit, which would be encountered with lasers in the optical region, the unit cells do overlap and phenomena are in principle possible which cannot be described in classical terms. However, to be fair, even here there is an "optical equivalence" theorem [63] which can furnish some insight into the connection between the classical and the quantum descriptions. As an example the classical theory would indicate that Twiss–Hanbury–Brown intensity correlations would not be expected either for a continuous-wave klystron oscillator or for a continuous-wave laser. The quantum version would be that for a perfect cosine wave, the amplitude is exactly defined and hence, from the Uncertainty Principle, the photon arrival statistics must be completely random. It follows by inverting the argument that if one has an optical field which gives perfect 100 % contrast Young's fringes but which gives zero intensity correlation, then that field corresponds to a monochromatic constant amplitude signal. These points will be taken up again in Chapter 4 in connection with noise fluctuations in the electromagnetic field.

2.4 Stimulated emission in the infrared

2.4.1 *General introduction*

Stimulated emission devices can be divided into two main classes: these are, (1) the primary kind in which broad-band incoherent energy is absorbed and some fraction of it converted into narrow-band coherent energy, and (2) the secondary kind in which an already existing coherent beam from a primary device is converted into another coherent beam at a different wavelength. The first primary devices to be operated worked in the microwave region and were called MASERS, an acronym derived from *M*icrowave *A*mplification by the *S*timulated *E*mission of *R*adiation. Originally, as their name implies, masers were used just as high-gain and in fact very low noise amplifiers, but since any amplifier can be turned into an oscillator by the incorporation of positive feedback, they can also be used as sources. However, this remains an infrequent use, since microwave practitioners have no shortage of high-power, monochromatic, tunable sources (section 2.5) and they could afford to ignore the possibility of tuning their maser sources by varying the applied magnetic field. Instead, they concentrated on optimising the performance of masers as single-frequency amplifiers. Infrared practitioners, up to 1960, were less fortunate since they had neither tunable monochromatic sources nor stimulated emission amplifiers. So, when the optical equivalents of masers (originally and quite logically called "optical masers") came along they were very welcome indeed, despite their lack of tunability, because they were the only coherent devices available. The term "optical maser", however, proved unpopular and it was soon replaced by the equivalent acronym LASER where the L stands for "light". Some attempt was made to introduce further acronyms such as IRASER which would apply to the infrared devices but these terms never caught on and the modern practice is to call anything which looks like a laser, simply a laser! Like their microwave counterparts, lasers can be both oscillators and amplifiers but the trend seems to be to reserve the term solely for the oscillators.

In a primary laser, a suitable medium, solid, liquid, or gas, is enclosed in a Fabry–Perot cavity of one of the types discussed earlier and is then excited by exposure to broad-band energy. This might be, for example, the radiation from a flash-tube or else it might be transferred energy due to collisions with the electrons in an electrical discharge. By some mechanism (see later for details) a population inversion results and a wavelet, launched into a cavity mode by spontaneous emission, grows in amplitude, by stimulated emission, as it advances through the medium. In simple terms, the population inversion produces the gain and the cavity provides the feedback. The result is oscillation and the production of an intense highly directional monochromatic and coherent beam.

A very important point with primary lasers is that the stimulated emission is always in competition with the spontaneous emission. The stimulated emission, however, depends on the radiation density (equation 2.2.19) whereas the spontaneous emission does not; so provided there is a population inversion, a threshold of excitation will exist beyond which the stimulated emission will dominate and the laser will "lase". The practical problem is to make this threshold accessible. One can, without much loss of generality, assume a Gaussian line-shape, whereupon (cf. equation 2.6.11) the normalised profile function becomes

$$\phi(v) = \frac{2}{\Delta v} \sqrt{\frac{\ln 2}{\pi}} \exp\left[-\frac{\ln 2 \, (v_0 - v)^2}{(\Delta v)^2} \right]. \tag{2.4.1}$$

Substituting in equation (2.2.55) and invoking the ratio of B/A from equations (2.2.17) and (2.2.18) gives

$$\alpha_{max} = \frac{(N_1 - N_2)c^2}{4\pi^2 v^2 \tau \Delta v} \sqrt{\frac{\ln 2}{\pi}}, \tag{2.4.2}$$

where τ the natural (or in lasing circles the "fluorescence") lifetime is equal to A^{-1}. In a cavity where one will have absorptive and scatter losses in the medium (quantified by a medium absorption coefficient α_m) and losses at the mirrors (quantified by an imperfect reflectivity R_e), one would have that, relative to its initial value, the radiant intensity after one round trip would be altered by the factor

$$R_e^2 \exp(-2\alpha_m d) \exp(-2\alpha_{max} d). \tag{2.4.3}$$

For the threshold to be exceeded, this factor must clearly be larger than unity. Taking logarithms, one then has

$$(N_2 - N_1) > \frac{4\pi^2 v^2 \tau \Delta v}{c^2 d} \left(\frac{\ln 2}{\pi}\right)(\alpha_m d - \ln R_e). \tag{2.4.4}$$

The experimentalist will obviously try to make α_m as small as possible and R_e as close to unity as possible and he will also make his laser as long as possible to get a large value for d, but the remaining three parameters, namely v, τ, and Δv are, for any given system, fixed by nature. In the Gaussian case, the line-width parameter will be proportional to the frequency (equation 2.6.2) so over all, the RHS of (2.4.4) will vary as the cube of the frequency. This v^3 dependence of the minimum population inversion clearly favours long wavelengths for laser action. It is very easy indeed to make far infrared lasers and it is not surprising that this region of the spectrum is the most densely populated with practical devices. However, as one goes progressively through the infrared, the visible and into the ultraviolet, it gets harder and harder and we still do not have practicable X-ray lasers despite their technical attractions. The dependence on the observed line width, Δv, indicates that the sharper the line the easier it will be to get it to lase. This is an important point in the operation of gas lasers where the lines are naturally sharp but it is of major importance in the operation of rare-earth ion (for example Nd^{3+}) lasers where, due to shielding of the emitting $4f$ electrons, the observed fluorescence lines are surprisingly sharp. The dependence on τ, that is essentially the reciprocal of the line-strength, would appear to indicate, on first examination, that one should look for the most intense line one can find. However the competition with the spontaneous emission has also to be taken into consideration. At long wavelengths there is no problem and it is found, as expected, that the available lasers work on strongly allowed transitions. Into the near infrared, especially when high-power operation is considered, it is better to seek a less strongly allowed line and to compensate by turning up the level of excitation. Thus the ruby and neodymium lasers both depend on lines which are first-order forbidden and which are only made weakly active by crystal-field effects. Even in the mid infrared this tends to be true. Thus the powerful CO_2 laser operates on a difference band which again is first-order forbidden (section 5.2.3). Increasing the excitation does result in the problem that the energy transfer from the broad-band "pump" to the narrow-band lasing emission is very inefficient, seldom reaching much above 10%, so a very efficient population inversion mechanism is needed to compensate. In the ultraviolet, for example, the most efficient lasers, the so-called "excimer" type, work with molecules such as KrF, XeF, Xe_2, etc. which have stable excited states but completely dissociative ground states. The population inversion is therefore total! In the infrared such extremes are not necessary and many adequately efficient population inversion mechanisms are known (section 2.4.3) and more are being discovered, so there is no practical problem in making a wide range of very adaptable infrared lasers. Even so the gain-per-metre figures actually achieved are usually small, of the order of a few per cent. The round-trip gain ΔI_g, in these circumstance, can be approximated by

$$\Delta I_g = I_0 [\exp(2\alpha_0 d) - 1] \approx 2I_0 \alpha_0 d, \qquad (2.4.5)$$

where α_0, the small-signal gain coefficient, is usually numerically equal to α_{max}. The round-trip losses will be given by

$$\Delta I_l \sim I_0 [\alpha_m d + (1 - R_e)], \qquad (2.4.6)$$

so the engineer has to design his cavity such that

$$\alpha_m d + (1 - R_e) < \alpha_0 d. \qquad (2.4.7)$$

This equation is sometimes given in terms of the concept of the cavity "lifetime" τ_c, i.e. the time required for the power level in the unpumped cavity to fall to one half of its initial value. This is

$$\tau_c = d/c [(1 - R_e) + \alpha_m d], \qquad (2.4.8)$$

and one derives immediately the result

$$\alpha_0 c \tau_c > 1. \qquad (2.4.9)$$

The engineer's job is thus to make τ_c longer than a microsecond, or, in equivalent terms, to make the total cavity losses less than a per cent or so.

2.4.2 *Principles and operation of infrared lasers*

To keep the cavity losses down, one will invoke careful design, as discussed in section 2.1, and will use very highly reflecting materials, e.g. gold, for the mirror surfaces, but there still remains one very serious problem. This is that the active medium seldom completely fills the cavity so that there are interface losses which, even with the most transparent window materials, will be of the order of 4% per interface (section 3.5). A laser with two windows would have reflection losses of the order of 16% and laser action would be quite impracticable. Fortunately, dielectric reflection is sensitive to both the angle of incidence and to the polarisation of the incident wave (a fact taken advantage of by the manufacturers of polarising sunglasses) and it is possible to set the windows at a special angle—called the Brewster angle—for which the parallel polarised reflection coefficient is zero. Visible region lasers feature Brewster windows extensively (Fig. 2.25(a)), but in the infrared, where there are materials problems, it is usual to avoid them whenever possible. One way of doing this with a gas laser is to have the mirrors of the cavity inside the discharge vessel (Fig. 2.25(b)). This solves the interface problem but unfortunately introduces further difficulties, for example, attack on the mirror surfaces by the discharge or by the discharge products.

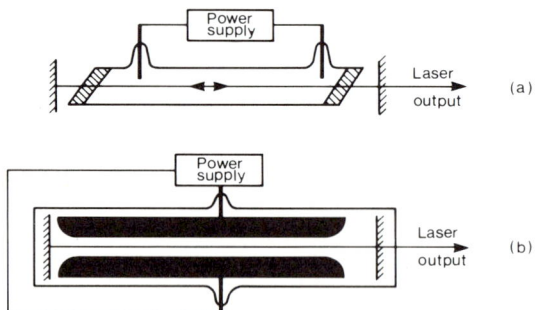

Fig. 2.25. Two types of gas laser. In (a) there are external mirrors and Brewster windows and the electrical excitation is by longitudinal discharge. In (b) the mirrors are internal and the excitation is by a transverse discharge.

For solid-state lasers, the interface problem can be solved by finishing off the faces of the gain medium at the Brewster angle. If the medium is isotropic, for example a cubic crystal or else a glass, then the even simpler expedient may be adopted of merely inclining the medium so that its faces intercept the beam at the Brewster angle. One or other of these solutions is usually chosen for those lasers, working in the near infrared, which feature transition or rare-earth ions dispersed, at a low concentration, in an inert matrix. For the other main category of solid-state laser, namely the semiconductor laser, one can actually turn the reflections at the interface to advantage. These materials have high refractive indices and the reflection is often strong enough for the assembly of an apparently mirrorless cavity. In this case, the opposing faces of the crystal, which are in fact the cavity "mirrors", have to be finished off very flat and very parallel to one another.

There are several ways available for coupling radiation out of the cavity but, since the coupling represents another form of cavity loss, it has to be kept small and in some cases has to be adjustable. At optical frequencies, it is usual to employ "through the mirror" coupling in which one of the multilayer dielectric-coated mirrors has a reflectivity slightly less than unity whilst the other is virtually perfectly reflecting at the laser frequency. The effect now is that every time the beam is reflected from the imperfect mirror, a small amount of radiation will be transmitted to the outside world. The method can still be used even when the mirrors are within the discharge vessel, provided the coatings will stand up to the conditions. It is sometimes chosen, for example, for the output coupling of CO_2 lasers where the mirrors would be multilayers on a germanium substrate. An alternative way of getting power out of the cavity is to have a small hole drilled through the centre of one of the otherwise perfect mirrors. This technique is known, aptly enough, as "hole-coupling" and is particularly appropriate to the case of far-infrared lasers with internal mirrors. The mirrors can thus be cheap, even fabricated from bulk metal, with perhaps an evaporated gold coating to finish them off, but at very long wavelengths there may be difficulty in getting the radiation through the hole which, at these wavelengths, looks like a "cut-off" waveguide. Another approach is to use a dielectric film beam-divider inside the cavity: this couples radiation out by Fresnel reflection. The reflectivity will vary with the film thickness and with its refractive index, because of the multiple-beam interference in the plane-parallel film (section 3.5.2), so there is the possibility of adjusting the degree of coupling. The fourth common form of output coupling is to use a pair of wire grids (section 3.10) as a Fabry–Perot interferometer. The reflectivity of the combination will depend on the spacing of the grids so the degree of coupling is, in principle, variable. These four types of output coupler are illustrated schematically in Fig. 2.26. For all these simple types it is not easy to vary the degree of coupling whilst the laser is running, but if a third mirror is added to a conventional beam-splitter coupled laser, this is readily achieved, as shown in Fig. 2.27(a). The combination of the beam-divider and the mirrors M_2 and M_3 makes up a Michelson interferometer and, by varying the position of M_3, the power output can be varied from zero up to twice what it would be if M_3 were not there at all. This assumes, of course, that the amount of power being extracted is negligible in comparison with the energy flow within the cavity. It is also interesting to consider the alternative arrangement shown in Fig. 2.27(b). This is essentially a double cavity sharing a common mirror. Gain is only high when a mode of the long cavity coincides with one of the short, and therefore by adjusting the position of the mirror M_3 it is possible to achieve single-mode operation of the laser. The arrangement was first described by Fox and Smith [67] who, however, used "through the mirror" coupling: it is often referred to, in the literature, as the Fox–Smith mode selector.

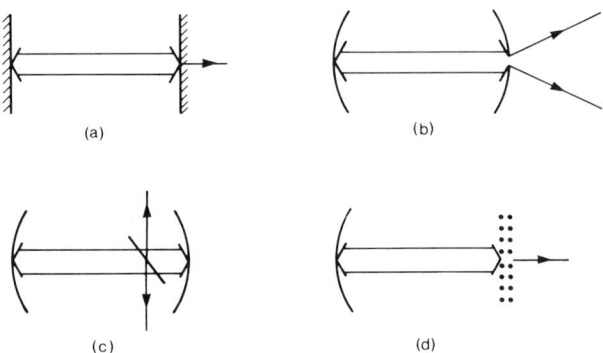

FIG. 2.26. Various forms of output coupling for infrared lasers: (a) "through the mirror" coupler; (b) hole-coupler; (c) beam-divider coupler; (d) wire-grid Fabry–Perot interferometer coupler.

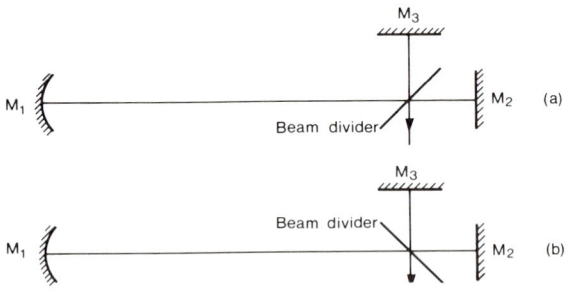

FIG. 2.27. Two forms of output coupling from a three-mirror beam-divider coupled laser.

There is a fifth way of extracting power from the cavity which is quite different from anything described so far. This is to take advantage of the inevitable diffraction losses round the cavity mirrors. It is attractive because the power extraction does not perturb the operation of the laser but, since it is only the very high order modes (section 2.1) which have appreciable intensity at the mirror edges, the far-field pattern of a diffraction-coupled laser will be unacceptably structured. This would not matter, of course, if one were merely using it in the near field to produce a power level reference for a main beam which was being coupled out conventionally and this application can therefore be quite useful when lasers are being used for quantitative measurements.

The output coupling schemes which give the best approximation to a clean Gaussian TEM_{00} beam are undoubtedly the "through the mirror" and the beam-divider arrangements, especially if internal diaphragms can be used to suppress off-axis modes. Partially transparent mirrors are widely used to couple power out of CO_2 lasers and the beam-divider format is very popular for far-infrared lasers. However, in the far infrared, low power levels may present a problem and it may be necessary to use hole-coupling even though the divergent beam contains many higher order transverse modes (Fig. 2.5). In this case the far field pattern will be far from smooth and the regions where the beam is focussed will show much structure.

Apart from power level and beam quality, the other parameter affecting the choice of output coupling is the desired state of polarisation. In the infrared, polarisation of the laser beam is the normal state of affairs since the $\Delta J = \pm 1$ selection rule leads naturally to the production of a linearly polarised beam. But real lasers may not appear to produce such a beam since, with cylindrical symmetry, the plane of polarisation can rotate randomly leading to an apparently unpolarised output. Even slight departures from full cylindrical symmetry may serve, however, to lock the plane of polarisation in a preferred direction. Thus Yamanaka et al. [68] have found that with an HCN laser, a set of fine parallel scratches on the surface of the plane mirror in a plano-concave cavity will ensure 100% polarisation parallel to the scratch direction. The beam-divider output coupler automatically produces 100% polarisation in the plane of Fig. 2.26(c). The reason for this is that the perpendicularly polarised component is so strongly reflected by the beam-divider that the cavity losses become too high for the laser to work, whereas for the parallel polarisation, where the encounter with the beam-divider is close to the Brewster angle, the losses will be low. The Fabry–Perot coupler will also give perfect polarisation if the reflectors are made from parallel wire grids or from mesh with rectangular elements. The hole-coupled laser and the "through the mirror" coupled laser do not have any inbuilt polarisation bias so if a polarised output is desired a polarisation sensitive element has to be incorporated in the cavity.

The non-axial modes in the laser output arise in several ways, but one prime cause is the electric field distribution in the laser cavity at axial resonance. In this, as shown in Fig. 2.28(a), a standing wave pattern is set up and there will be nodes of the field spaced at intervals of $\lambda/2$ along the tube axis. At these nodes, the field is, of course, zero and there can be no stimulated emission into longitudinal modes. The nodes are therefore growth points for the transverse modes which can feed on the unused gain because their field patterns are not constrained to be zero in these regions of the laser tube. One could suppress the transverse modes and incidentally get increased power in the desired longitudinal modes if one could eliminate the standing wave pattern. One attractive way of doing this is to use an interesting variant on the Fabry–Perot cavity, viz. the ring-resonator, as shown in Fig. 2.28(b). In the ring-laser cavity, the radiation flows continuously round a loop and there is thus no standing wave pattern. Unfortunately, the radiation can just as well flow either way round the loop and, if it is flowing in both directions at once, the standing wave pattern will promptly reappear. It is found experimentally, however, that if some measure of asymmetry is present in the cavity, then the radiation will prefer to flow in one direction only. The resulting absence of nodes and consequent attenuation of the non-axial modes makes it feasible to operate the laser on a low number of purely longitudinal modes. In favourable circumstances, it becomes relatively easy to get the desired single-mode operation. This design is often employed for this reason, in the near-infrared colour-centre lasers (section 6.2.7). The necessary asymmetry is readily introduced by pumping with a beam which flows in just one direction through the active medium.

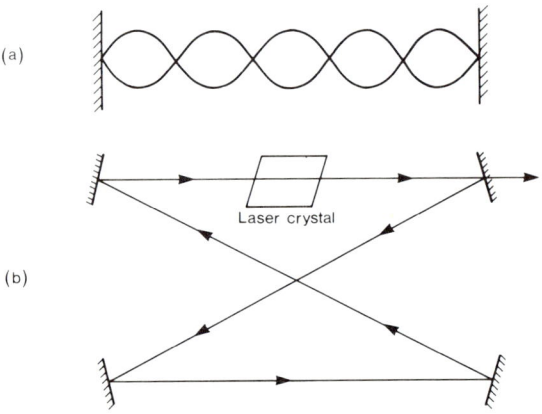

Fig. 2.28. Principle of the ring laser. In the upper inset is shown for comparison the field distribution in a conventional cavity.

In longitudinal mode operation of the laser, it is necessary, of course, to secure a match of the cavity resonance with that of its enclosed active medium. If the cavity resonances are spaced finely enough and the medium resonance is broad enough there is no problem. This is usually the case with visible region lasers where the spontaneous line width is commonly many times the cavity mode spacing. In the infrared, on the other hand, where the line width of the medium may be of the same order or even less than the cavity mode spacing, it may be necessary to tune the cavity, by altering its length, to ensure a match. The cavity resonances have of necessity to be sharp in order that the cavity losses be low; in practical terms they will have a half-width of about 0·5 MHz, so for all intents and purposes one can regard the set of cavity resonances as a

Dirac comb with a spacing given by the reciprocal of the round-trip time, i.e.

$$\Delta v_c = c/2d. \qquad (2.4.10)$$

For a one metre laser this would be 150 MHz. The need for cavity tuning will thus arise whenever the width of the medium resonance gets to be less than a few hundred MHz. The width of the gain region is determined by the degree of excitation and by the underlying spontaneous line-width. It is only with gas lasers that the spontaneous line-width is liable to be of the order 100 MHz or less so we need only consider this case. The basic theory of gas-phase line-shapes in the infrared is given in section 2.6 where it is shown that the observed widths arise from a combination of pressure-broadening which is homogeneous and Doppler-broadening which is heterogeneous. The pressure-broadening is linearly proportional to the pressure at low pressures and also depends on the dipole moment. For a typical polar molecule ($\mu = 1$ Debye), one would expect to find a value of the order of $10\,\mathrm{MHz\,torr^{-1}}$. Doppler broadening depends on the molecular mass, the frequency and the temperature. Gas lasers usually have a translational temperature close to ambient even when they are running strongly, so one can calculate (see section 2.6 for details) the expected Doppler broadening on this assumption. In fact, since the broadening only depends on the square root of the temperature (see equation 2.6.13) one will not be far out even if the effective translational temperature is higher. Some typical results are given in Table 2.3. Glow-discharge lasers tend to work at pressures between 1 and 10 torr and with effective kinetic temperatures near ambient. This latter is a rather rough approximation but, from the arguments given above, it will not affect seriously the conclusions which are that for millimetre and submillimetre wavelengths the pressure broadening will dominate whilst at near infrared wavelengths the Doppler broadening will dominate. At mid-infrared wavelengths, both types will contribute and the overall line-shape will be given by a convolution of the two primitive components. This is sometimes called the Voight profile. Optically pumped lasers usually operate at pressures which are an order of magnitude

TABLE 2.3
Doppler-widths for some technically important gas lasers

Laser	Wavelength μm	Doppler half-width MHz
He/Ne	0·6328	1,314
I	1.315	251
He/Ne	3·392	245
CO	5·300	133
CO_2	10·6	53
H_2O	27·97	31
HCN	336·7	2
CH_3F	496·0	1·3

less, so even in the far infrared they tend to be Doppler limited. The overall conclusion is that for mid- and far-infrared operation the line width resulting from the combined effects will always be less than 100 MHz, so some form of cavity tuning will usually be required.

The spontaneous emission profile gives, of course, only an upper limit to the width of the stimulated emission region. As the level of excitation is progressively increased one goes smoothly from a situation where there is loss everywhere on the profile to one where there is gain everywhere. The first part of the profile to show gain is the peak and the last parts are the far wings. This is shown schematically in Fig. 2.29. It is assumed here that the line is homogeneously broadened, but the same general conclusions would apply to the heterogeneous case apart from the possible presence of saturation (or "Lamb") dips (see later). For most infrared lasers, the degree of excitation is only moderate so out of a possible line width of say 50 MHz, one might find that laser action was confined to ± 15 MHz each side of the line centre frequency. Infrared gas lasers usually have to be long because, as mentioned above, the gain per metre figures are usually small, but even with a 2-m laser ($\Delta v_c = 75$ MHz) cavity length adjustment would still be needed to ensure a match with emission features only 30 MHz wide. As the cavity length is smoothly adjusted, the power output will show peaks when a cavity mode comes into resonance with the line-centre frequency and each peak will be followed by an extended minimum during which there is mismatch. An example of this sort of behaviour has already been given in Fig. 2.6. Cavity length adjustment is not at all convenient experimentally since one often has the mirrors inside the laser jacket and the micrometer adjustment will then have to be taken through a vacuum seal. However, the other side of the coin is that one can take advantage of the extra degree of freedom to both select individual longitudinal modes and suppress transverse modes—this latter because of the occurrence of mode dispersion (Fig. 2.6). High-pressure gas lasers and nearly all the near-infrared solid-state lasers have very large spontaneous line widths so cavity adjustment is not necessary, but on the other hand obtaining single mode output is difficult.

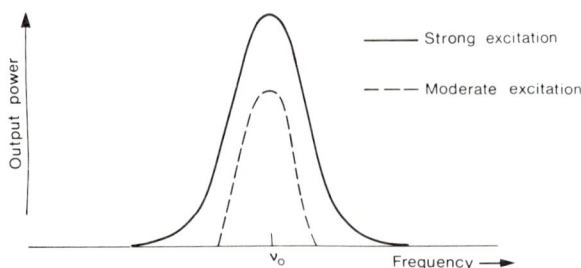

FIG. 2.29. Stimulated emission from a homogeneously broadened line at two values of the relative excitation.

The nature of the line broadening mechanism is of prime importance in any discussion of the intracavity interaction of the radiation with the active medium. The crucial point is that for homogeneous broadening *all* the molecules contribute to *all* the profile, whereas in heterogeneous broadening selected groups of molecules contribute to only narrow regions of the profile. Thus, in the case of Doppler broadening, molecules may be labelled according to their resolved component of velocity along the beam direction. The observed feature can then be regarded as made up of a large number of much narrower contributing features each having a centre frequency given by the Doppler-shifted fundamental molecular frequency. The width of these Doppler components will be the natural or homogeneous width so all molecules with Doppler shifts lying within the homogeneous width can be regarded as members of an equivalent set. Conversely those whose shifts do not, belong to non-equivalent and essentially non-interacting sets. A probe wave of frequency v_0 propagating through a medium whose absorption/emission is characterised by a Doppler-broadened profile will thus interact with all molecules whose Doppler-shifted frequencies lie in the band $v_0 \pm \Delta v_h$, but it will not interact with all the other sets. This leads at once to the very important concept of "hole-burning".

Hole-burning arises whenever one has intense monochromatic radiation propagating through an absorbing medium whose line shape is inhomogeneously broadened. Absorption of energy from the beam will transfer a small group of oscillators between the two states of the transition but leave the population distribution of all the other oscillators unaffected. If the beam intensity is high enough, the population distribution within the interacting group may even become equalised, that is the transition may become fully saturated. The commonest example of this phenomenon arises with Doppler broadening and, in this case, a plot of the population function $N(v)$, that is the number of molecules having velocity components lying between v and $v + \delta v$, against v will then show a diminution, at a velocity corresponding to the Doppler-shifted absorption frequency, for a ground-state molecules and a compensating augmentation for upper-state molecules. One then says that a "hole" has been burnt in the ground-state velocity distribution function which would otherwise be a featureless Gaussian function (section 2.6.2). Now inside a laser cavity the beam is usually passing in both directions simultaneously so it follows that *two* holes will be burnt in the distribution since the beam will interact with a different set of molecules going in one direction than it does when going in the other. This is illustrated schematically in Fig. 2.30. If the laser frequency is v_L and the cavity is tuned to a frequency v_c, then holes will be burnt at the velocities $\pm v$, where v is given by

$$v = c[(v_c - v_L)/v_L] \qquad (2.4.11)$$

From this equation, it will be seen that when the cavity is tuned into exact resonance with the laser frequency, the same set of molecules will interact with the beam travelling in both directions: in other words the two holes will coincide. The set for which this is true is composed of those molecules travelling normally to the beam direction. With the holes coincident, saturation effects can become immediately evident since the population equalisation produced by the intense wave propagating in one direction will be

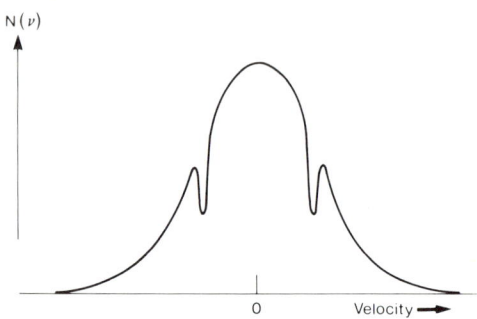

$N(\nu)$

0   Velocity

F$_{IG}$. 2.30. Hole-burning in the Doppler profile of molecules in a laser cavity. The velocity indicated on the abscissa axis is the resolved component in the beam direction.

made manifest by a lack of gain on the reverse trip. By contrast, when the cavity is off-tuned by more than the homogeneous line width, the laser beam will experience gain in both directions. The result is that the cavity mode pattern shows minima or "dips" at the exact resonance positions. This phenomenon was analysed by Lamb [69] and the minima are consequently usually referred to as "Lamb dips". A laser mode spectrum showing well-developed Lamb dips is shown in Fig. 2.31.

Very similar phenomena are encountered when laser beams execute a double-passage through an external cell. The double traverse is readily achieved by fitting the cell at the far end with a plane mirror and the result can be investigated by the simple expedient of using a dielectric beam-divider, inclined at 45°, to deflect the returning beam towards a suitable detector. If the external cell contains a gas which has a sharp inhomogeneously broadened resonance virtually coincident with the laser frequency, then, as one tunes the laser by altering the cavity length, one would normally expect to trace out the familiar absorption profile as the laser output frequency sweeps over the absorption band. If, however, the laser power is high enough to saturate the gas on one pass, then a sharp spike of enhanced intensity will be observed exactly at line centre. This spike is the "inverse" Lamb dip. It arises because the saturated gas cannot absorb the returning beam when this is exactly resonant, whereas it can when the returning beam is off-tuned by more than the homogeneous width. The width of the inverse spike is again given by the convolution of all homogeneous processes and can thus always be made less than the Doppler width. One has, therefore, an immediate method which in fact is still the most significant for overcoming the limitations imposed by Doppler broadening on high-resolution spectroscopy. Some examples are quoted in section 4.6.

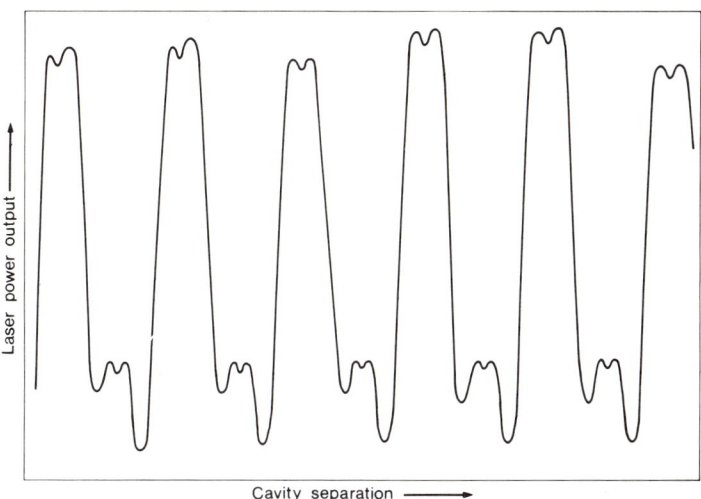

Fɪɢ. 2.31. Cavity mode spectrum of a water-vapour laser operating at 118.6 μm (unpublished work of C. C. Bradley). The stronger features arise from the TEM_{00} mode and the weaker from the TEM_{01} mode. Both sets show clear Lamb dips.

Lamb dips of both kinds can also be used to stabilise the laser output frequency. The frequency of an ordinary or "free-running" gas laser can wander considerably due to thermal variations of the cavity length and to fluctuations in the plasma conditions. Crude stabilisation can be ensured by using invar rods to connect the two ends of the laser and by using stabilised power supplies. Finer control can be obtained by having in addition an intracavity etalon to restrict the gain to a very narrow spectral region. However the highest degree of stabilisation demands that the laser output frequency be "locked" either to a fundamental frequency standard (e.g. the harmonically multiplied output from a caesium "clock") or else to a molecular or atomic absorption line. The inverse Lamb dip provides not only a very sharp feature to give good frequency control, it also has a very convenient form for servo-loops to lock onto. Lamb dip stabilised He/Ne and Argon ion lasers are nowadays the *de facto* frequency and wavelength standards for the visible and near-infrared regions of the spectrum.

The Lamb dip is a non-linear phenomenon and its observation requires reasonably high powers but the observation of other non-linearities and the successful outcome of various other laser-based experiments may require very much higher power, megawatts or more. Fortunately, power levels of this order are only required for the brief intervals determined by the reciprocals of the relaxation times, that is microseconds or less. One is not therefore called upon to have a private generating station; all that is required is a pulsed laser! A low-pressure pulsed gas laser will usually deliver some kilowatts of power in pulses some microseconds long, this being roughly equivalent to the tens of milliwatts obtainable from the laser in its continuous-wave regime. Higher pressure lasers, for example the TEA CO_2 lasers described in section 6.2.2, tend to be self-quenching so the pulses are shorter (~ 100 ns) and the peak powers are correspondingly higher (~ 10 kW). To obtain still higher peak powers the laser pulses have to be constrained to a still shorter time interval. This can be achieved, provided the relaxation times are long enough, by having the cavity gain high only for this very brief time. The stimulated emission will now only occur during the fleeting moments of high gain and the stored energy which would normally have been emitted gently over a long period of time becomes compressed into intense emission over a very much shorter period. This method of generating high peak-power, very short laser pulses, is called "Q-switched", "giant pulse" or "cavity spoiled" operation. There are two basic ways in which it can be brought about: active and passive. In the active methods one can arrange for one of the cavity mirrors to be rotated at high speed about an axis lying in its reflecting surface [70]. The cavity, as a high gain entity, will then be formed only for some nanoseconds. The same effect can be achieved by using an intracavity rotating prism. The passive methods rely on the use of saturable absorbers within the cavity. This kind of absorber has the property that its absorption coefficient decreases as the power level rises (section 2.6.3). In laboratory jargon, they are said to "bleach". What then happens is that after excitation of the laser the losses due to the intracavity absorber are too high to allow stimulated emission to occur. The absorber, however, picks up energy from the spontaneous emission and its absorption coefficient starts to fall. The stimulated emission then begins and the absorption coefficient falls still more rapidly. This is clearly a run-away situation and eventually all of the laser power will be radiated in a single very intense pulse. The laser then "goes out" and the saturable absorber returns to its normal condition ready for the cycle to begin again as the continuing excitation starts to rebuild the population inversion. Saturable absorbers are frequently used to Q-switch near and mid-infrared lasers which, for the reasons given earlier, have long-lived upper levels. Thus sulphur hexafluoride gas can be used to Q-switch the CO_2 laser [71]. Q-switching of longer wavelength lasers is not, however, effective [72]. The reason for this lies in the very rapid rates of rotational relaxation which are found for the gas phase. These are such that population inversions can be thermalised away in times much less than a microsecond and effective energy storage is no longer possible. Some of the transitions in long-wave glow discharge lasers appear on first sight to be vibrational in character but, as will be explained in the next section, they owe their lasing activity to perturbations which give them the character of rotational transitions and the rate of equilibration is very rapid. Attempts to Q-switch the HCN laser for example, have proved very disappointing [72].

Pulsing or Q-switching lasers which have wide gain profiles can lead to some very remarkable phenomena. Thus at a high level of excitation the laser may be able to give out several longitudinal modes simultaneously. If these modes are radiated with a fixed phase relationship to one another the output of the laser breaks up into a regularly spaced series of exceedingly brief pulses. The laser is then said to be "mode-locked". The physical reason for this is that the output of the laser in the frequency domain is an attenuated Dirac comb so its output in the time domain, which must be the Fourier transform, is an endless stream of pulses each having a finite width. The pulses are spaced by the round trip time $(2d/c)$ and the width of each is determined by the number of modes which are phase-locked to one another. The more there are, the briefer the pulses. For gas lasers in the near infrared one can have the pulses spaced by 10 ns with each having a length of 100 ps, but for solid-state, dye and colour-centre lasers, both these times can be very much reduced and true picosecond pulses are available. These have given us an extremely valuable way of studying very fast processes. Essentially the laser acts as an ultrafast flashgun! The picosecond pulses can be extremely intense and the selection of single mode-locked pulses is one of the best ways of starting the oscillator/amplifier chains used in ultra-high-power laser research. The operation of mode-locking can be analysed just as well in the time domain where one can think of the radiation in the cavity as taking the form of a tight "ball" of energy which bounces back and forth between the two cavity mirrors. The pulse of energy which escapes has then a width equal to the temporal dimension of the ball. From this point of view, one can bring about mode-locking by stimulating the laser in synchronism with the bouncing ball in rather the same way that one can build up the amplitude of a swing by timing the pushes well. This is active mode-locking. The same result can also be brought about by the use of non-linear elements, and in particular saturable absorbers, within the cavity. The absorption coefficient of the saturable absorber falls as the power rises so it tends to select for intense pulses in the laser output. For this and similar reasons, most of the giant pulses obtained from near-infrared and visible lasers actually consist of a large number of somewhat irregular and certainly very much briefer mode-locked pulses. When studied with very fast detectors they are seen to have very ragged and very irreproducible profiles.

The propagation of the beam through the active medium is determined not only by the absorption but also by the refractive index and these two must, of necessity (see section 3.4.5 for details), be related to one another. Thus as the absorption coefficient switches from positive to negative, the form of the dispersion curve flips over too. One has in general, therefore, a time-varying refractive index and this can lead to some interesting effects since the quantity d appearing in all the laser equations is the optical distance between the cavity mirrors not the physical distance. With pulsed lasers in particular one will have the situation that, as n varies, the cavity will come into and go out of resonance several times during the build-up of the pulse. The result is that the emitted pulse of radiation is found to break up into a series of much shorter, essentially self Q-switched pulses. Another effect is that the refractive index usually varies with radial distance inside the tube, especially when the excitation is longitudinal. This can lead to self-focussing or self-defocussing of the beam which can again give pulse break-up and the generation of non-axial modes. The frequency output of the laser can likewise be influenced by variations of refractive index. Thus if the laser is operating on a given cavity mode but the effective cavity length is varying, then the laser will try to stay in that mode as long as possible (mode-pulling) before jumping catastrophically to the next available mode (mode-hopping) with a sudden change of output frequency. This is only of academic interest in the case of pulsed gas lasers because no one would seriously think of using them as constant or steadily varying frequency sources. Similar effects are encountered, however, for the various broad-band tunable lasers which are starting to become available and these can restrict the resolution which these otherwise very welcome devices can give.

Before concluding this brief outline of some of the factors of importance in the dynamics of the active plasma, it is essential to point out that the ideal system of two, or three levels is seldom encountered in practice because all the levels involved in the inversion will usually be connected to several others by allowed transitions. If the overpopulated (or suprathermal) state is connected to two possible lower states, then operation on one transition will seriously degrade the performance on the other. That is we have competitive emission as shown in Fig. 2.32.

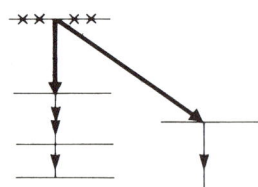

FIG. 2.32. Competitive (heavy arrows) and cascade (light arrows) emission in a laser system.

A well-known example is provided by the He/Ne laser where the 3·39 μm line comes from the same upper level as does the 0·6328 μm line and has the higher transition probability. If one wants the laser to work on the visible line, the infrared one has to be suppressed. This is in practice readily achieved by fitting the laser with glass Brewster windows since glass is completely opaque at 3·39 μm. The opposite type of phenomenon is often encountered when a laser is operating strongly on one of its transitions and the rapid flow of population brings about a secondary inversion further down the energy level ladder. One then gets cascade emission. There are two basic types: (a) population of an upper state by the laser action as shown in Fig. 2.32, and (b) depopulation of a lower state, an example of which is given in Fig. 2.38. Cascade emission is useful technically since one piece of laser equipment can be induced to give out several different lines and it is also useful from a more fundamental point of view since the various cascade lines can often give vital clues when one is attempting to assign the transitions in a newly observed laser system.

2.4.3 Population inversion mechanisms

So far we have seen how the interaction of the cavity with its active contents can lead to observed external laser action. The question remains of how the contents were made active in the first place, in other words how the population inversion was brought about. Population inversions are not remarkable in themselves, for if there is no radiative pathway between the two levels, then in a non-equilibrium situation any arbitrary population distribution between them is possible. Such situations are not, of course, of immediate interest because of the lack of an allowed transition but, as will emerge later, they can often be exploited to bring about a population inversion in a different system in which the two levels involved *are* connected by an allowed transition. It is therefore worth considering non-connected systems in some detail and in particular the classic example provided by helium gas through which an electrical discharge is flowing.

The electrons forming the current can collide with the helium atoms either elastically or inelastically. In the latter case, they either transfer the helium atom from its ground state to an electronically excited state, or more drastically they ionise it forming the He^+ ion, possibly also in an excited state. The ions eventually recombine with electrons and the now neutral atoms cascade down the ladder of allowed energy states and finally once more reach the ground-state. In doing so, they give out the characteristic emission spectrum of helium. Excited neutral atoms can do very much the same thing. These cascades are strongly allowed so that in the quasi-equilibrium condition one would expect to have a population distribution which was quasi-thermal and characterised by an effective excitation temperature T_{ex}. Because the excitation energies, several eV, are very much larger than kT_{ex}, the excited states will have negligible populations and the gas will be optically thin for all save the ground-state resonance lines. This scenario is to a large extent true but a major complication is introduced by the presence of certain levels which are not strongly connected to any lower level and atoms arriving in them have to stay there until they can be de-excited in some non-radiative way, for example by collision. The reason for this complication lies in the selection rules for changes of electron spin. Helium has two distinct sets of states, the singlet in which the spins of the two electrons are antiparallel (i.e. $S = 0$) and the triplet in which the electron spins are parallel (i.e. $S = 1$).† Electromagnetic transitions between these two different types of state (intersystem transitions) are extremely rare in the case of helium because of the strong selection rules which apply. The lowest triplet state 2^3S_1 is therefore indefinitely stable in the gas phase since a transition to the ground state would not only violate the electromagnetic selection rules, it would also violate all the collisional selection rules. One thus has the remarkable situation that a state, whose energy (19·72 eV) is quite close to the ionisation limit (24·6 eV), can only be de-excited by collisions with the walls of the discharge vessel. Under the discharge conditions, large concentrations of 2^3S_1 atoms can build up and as a result the gas is found to absorb strongly its own emission line $2^3P_2 \rightarrow 2^3S_1$ at 1·0829 μm. In fact this line behaves as a resonance line, for example displaying self-reversal. For all intents and purposes then, helium gas through which an electrical discharge is flowing behaves as though it were made up of two quite separate gases. These are sometimes referred to as ortho and parhelium respectively. A further complication to the simple scenario is that within each system it is possible to have one or more metastable states. Thus the first excited state of the singlet system, 2^1S_0 at 20.65 eV, is not connected to the ground state by any allowed dipole transition since there are no intervening P states and the radiative lifetime of this state will be very long indeed, in fact being measured in seconds. Under the discharge conditions, considerable concentrations of this species will also build up.

When a discharge is passed through a mixture of helium and neon, each gas will be excited in the manner sketched out above but now a new form of interaction is possible since metastable atoms of one component can lose their excitation by collision with ground state atoms of the other component. This type of collision is, of course, the only one which can happen with reasonable probability since, as mentioned above, the gas is optically thin and only ground-state and to a lesser extent metastable atoms are present in any numbers. The collision of a 2^3S_1 metastable helium atom with a ground state neon atom has a large cross-section since neon has a 2s (Paschen notation) state which is virtually resonant in energy with the helium metastable. There is thus a very effective transfer of neon atoms to the excited 2s state but now, in contradistinction to the helium case, there are lower levels, the 2p, which are connected by allowed transitions and, since these levels will be virtually unpopulated, a strong population inversion occurs leading to laser action in the 1·15 μm region (see Fig. 2.33). The 2^1S_0 state of helium is also virtually resonant with a neon state, the 3s, and again there is effective population transfer leading to laser action at 3.39 and 0·6328 μm (Fig. 2.33).

A similar form of resonant collisional transfer is involved in the operation of

†The spectroscopic notation and conventions used in this section are defined in more detail in Chapter 5.

the CO_2 laser, in which an electrical discharge is passed through a mixture of CO_2 and N_2 containing also some helium. All the vibrationally excited states of the nitrogen molecule are metastable because the molecule has a centre of symmetry and the transition matrix elements have, of necessity, to be zero. Electronically excited states of N_2, produced by electron collision, can,

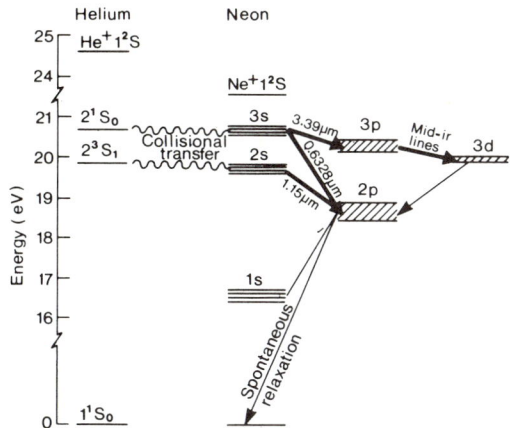

FIG. 2.33. Origin of the infrared and visible laser action in a He/Ne mixture. The various states of neon consist of close multiplets and, under conditions of strong excitation, several fine structure lines may be obtained in stimulated emission. However, under normal conditions only a single member of each set is available, for example $3s_2 \to 2p_4$ at 0·6328 μm.

however, decay to the ground-state by allowed transitions and the photons given off in these cascades account for the beautiful colours produced by electric discharges through nitrogen. A transition to the true ground-state is, however, unlikely because of adverse Franck–Condon factors and the molecules more usually finish up in vibrationally excited states where they must stay until they can be de-excited by a collision or, much less likely, undergo a quadrupole transition down to the true ground-state. In the discharge, therefore, high populations of nitrogen molecules in their first (and higher) vibrationally excited states build up. These can collide with CO_2 molecules and transfer them to the nearly resonant (00^01) excited state as shown in Fig. 2.34. Transitions from this state to the lower (and virtually unpopulated) (10^00) and (02^00) Fermi resonance pair (section 5.2.3.1) give rise to the well-known and technically very important laser emissions at 10·4 and 9·4 μm. Practical details of CO_2 lasers are given in section 6.2.2.

The explanation just given for laser action in terms of resonant collisional transfer is undoubtedly true, but it is not the whole story. Small additions such as transfer from not quite resonant states and in the He/Ne case, transfer of the excitation to other members of the multiplet, will account for the extra lines observed under conditions of strong excitation but there does remain the fact that both pure neon and pure carbon dioxide will lase if excited strongly enough. This can only involve selective excitation by electronic collisions aided perhaps by spin non-conserving collisional transitions. A good account has been given by Patel [73]. It follows that the conventional picture of the

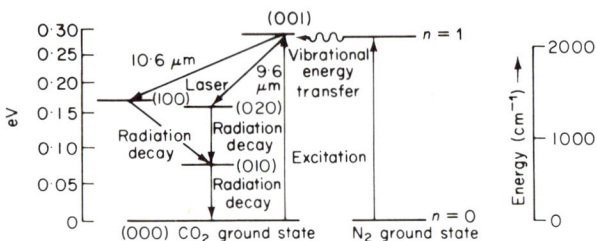

FIG. 2.34. Origin of the mid-infrared laser action in a CO_2/N_2 mixture. For both transitions, the entire ro-vibrational band can lase so a large number of individual lines is available (see Fig. 6.1).

electron stream as being equivalent to very high temperature ($\sim 10\,eV$ $\sim 10^5\,K$) black-body radiation is only approximately correct. The inelastic collisions with the atoms or molecules will introduce structure into the black-body curve and this can "feed-back" resulting in population distortions.

The phenomenon of pure gas lasing is fairly important in the present context since most of the lines so produced lie in the infrared. Thus pure helium gives lines at $1·9548$ ($4^3P_2 \to 3^3D_3$) and $2·0608$ ($7^3D_3 \to 4^3P_2$) μm in the near infrared and lines at $95·8$ ($3^1P \to 3^1D$) and $216·3$ ($4^1P \to 4^1D$) μm in the far infrared. The $2·0608$ μm linses from the sequence

$$He(1^1S) + e \ (+ \text{kinetic energy}) \to He(7^1P) + e$$
$$He(7^1P) + He(1^1S) \to He(7^3D) + He(1^1S) \qquad (2.4.12)$$

which populates the upper level, and the reaction

$$He(4^3P_2) + He(1^1S) \to He_2{}^+ + e \qquad (2.4.13)$$

which depopulates the lower level. Atoms in the 4^3P state arise principally from electron collision

$$He(2^3S) + e \to He(4^3P) + e \qquad 2.4.14$$

so any mechanism, the prsence of an impurity for example, which will reduce the concentration of the metastable 2^3S atoms will enhance laser action at $2·0608$ μm. These examples of pure gas lasing underline the general point that there exist very many potential lasing systems and that if they are hit hard enough all of them will eventually lase. For most of them, however, the threshold is too high for convenient operation and it is necessary to modify the system, as exemplified by the addition of helium or nitrogen, to bring the threshold down. It is low threshold lasers which are of interest in practical science and technology and this justifies the detailed study of their inversion mechanisms.

Another important class of laser in which a probably inherent tendency to lase is brought within the realm of practical thresholds is that of the transition and rare-earth ion lasers. Of this class, the chromium and neodymium ion lasers are the most important. These feature a fluorescent ion (Cr^{3+} and Nd^{3+} respectively) dispersed dilutely in an inert matrix. All transition and rare-earth ions absorb weakly in the visible and near-ultraviolet regions of the spectrum due to transitions within their incomplete d or f shells. Such transitions would be forbidden for the isolated ion because they violate the $\Delta l = \pm 1$ selection rule, but in the non-spherical (and in fact octahedral) crystal field of the surrounding matrix ions the five- or seven-fold degeneracy is slightly removed and $d \rightarrow d$ and $f \rightarrow f$ transitions become weakly allowed. Several well-known gemstones owe their value to colours induced by this mechanism. Examples include ruby (Cr^{3+} in alumina), saphire (thought to be a variant on ruby but with Fe^{2+} and Ti^{3+} ions present as well) and emerald (chromium ions in beryl $Be_3Al_2Si_6O_{18}$). The differing colours of apparently similar gemstones may arise from a different coordination or from a different oxidation state. The brilliance of gemstones, both natural and artificial, is somewhat enhanced by fluorescence effects in which higher frequency light is absorbed and then re-emitted at a longer wavelength. The long wavelength fluorescence lines of ruby near 694·3 nm are remarkable for their narrowness. In rough terms, this phenomenon, which is also observed for other transition and rare-earth ions, can be ascribed to the screening of the d or f electrons responsible for the emission, by the outer filled shells. The optical electrons are therefore relatively insensitive to outside influences, for example phonon-decay in the surrounding lattice, and the spectral lines are sharp. The absorption bands at higher frequency, on the other hand, are broad and diffuse. One can classify the electronic states of the absorbing ion in terms of the irreducible representations of the point group O_h which describes the symmetry of the cubic or octahedral field due to the matrix ions at the "guest" ion. It turns out for ruby that it is the ^2E states which are sharp and the $^4F_1$ and $^4F_2$ which are broad. The invention of the ruby laser followed naturally from Maiman's studies [74] of these states and of the flow of energy between them. The basic process is that broad-band blue/green or ultraviolet radiation from a flash-tube is absorbed very efficiently by the two broad F-states and then some of the absorbed energy is transferred to the ^2E state by radiationless transitions mediated by the lattice (Fig. 2.35(a)). If the flash is intense enough, it is possible to produce a population inversion between the ^2E and the ground states. Laser action then becomes possible and, since the ^2E state is slightly split by the trigonal component of the not quite perfectly cubic crystal field, two lines are observed, at 694·4 nm ($R_1, E \rightarrow {}^4A_2$) and at 693·0 nm ($R_2, 2A \rightarrow {}^4A_2$) for room-temperature operation. Both lines shift with temperature (R_1 for example going from 693·4 nm at $-180°C$ to 695·6 nm at $+200°C$) and with pressure and they furthermore split in axial magnetic fields so the experimentalist has some degrees of freedom to work with. The ruby laser is capable of producing high power and, moreover, is readily Q-switched, but it does have a relatively high threshold because the inversion is with respect to the ground state (i.e. it is a "three-level" laser). Nevertheless it is widely used in infrared technology, for example as the pump for near-infrared dye-lasers and for the driver of a difference frequency generator. From a purely infrared viewpoint, however, the most important solid-state laser is that involving the Nd^{+3} ion in various matrices. Here both states are excited (i.e. it is a "four-level" laser) and the difficulty of an inversion with respect to the ground state is avoided. This laser with the matrix either YAG (Yttrium Aluminium Garnet) or glass can not only be run continuous wave, it can also produce ordinary, Q-switched or mode-locked pulses of very high power. The combination of a neodymium/glass master oscillator with an amplifying chain composed of similar elements is a favoured candidate for the high-power laboratories working on laser fusion. So generally useful is this laser that it is not too fanciful to speculate that were its emission to lie in the visible region instead of the near infrared at 1·06 μm it would reign supreme as the number one solid-state laser. Its mechanism is indicated schematically in Fig. 2.35(b).

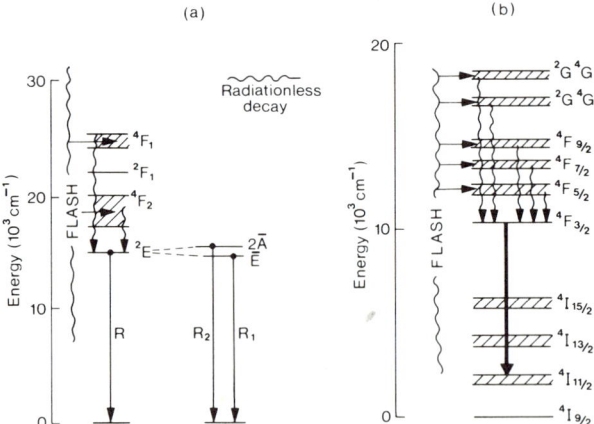

FIG. 2.35. Population inversion mechanisms for the Cr^{3+} (ruby) and the Nd^{3+} (YAG or glass) lasers.

Chemical lasers have always been attractive propositions because chemists have known for a long time that, in a gas-phase reaction, the products are often formed selectively in excited states so one has a built-in mechanism for providing a population inversion [75]. A well-known example occurs in the reaction of fluorine gas with hydrogen:

$$H_2 + F_2 = 2HF^*, \tag{2.4.15}$$

where the HF molecules are formed in vibrationally excited states. If the two reactants are separately introduced into a tube placed inside an optical resonant cavity and the products rapidly removed, then inside the tube a population inversion will exist and the vibration/rotation bands of HF can be obtained in stimulated emission. This gives a very useful set of laser lines near 2·7 μm [76]. Another very useful laser line, at 1·315 μm, is produced via the photo-dissociation reaction of perfluoropropyl iodide,

$$C_3F_7I + hv = C_3F_7(\to C_6F_{14}) + I^*(5p, {}^2P_{3/2}), \qquad (2.4.16)$$

followed by stimulated emission on the magnetic dipole transition

$$5p, {}^2P_{3/2} \to 5p, {}^2P_{1/2}. \qquad (2.4.17)$$

The iodine laser [77] is capable of delivering very-high-power pulses and for this reason is a strong candidate for the driving laser in a thermonuclear reactor (section 7.4.4).

Diode lasers are particularly important sources of coherent infrared radiation since, by a suitable choice of the basic material and of its stoichiometry, it is possible to obtain emission anywhere from the near ultraviolet out to as far as 30 μm. The output is not as spectrally pure as is that from a gas laser and it is, moreover, often highly structured spatially because of the presence of several modes, but it is nevertheless possible to obtain single-mode lines with a spectral width as narrow as 0·0001 cm^{-1}. The lasers can be tuned coarsely by the choice of operating temperature, by the application of a magnetic field or else by the application of hydrostatic pressure, and they can then be tuned finely by varying the driving current. Direct high-resolution infrared spectrometers using these diodes as their sources are therefore very attractive and they are rapidly establishing themselves, especially for pollution monitoring applications. Diode lasers being electrically driven are readily modulated and are thus very suitable as the sources for infrared optical-communication systems. The diode itself is formed at the junction of two blocks of a semiconductor, one doped to be p-type and the other doped to be n-type. When the diode is forward biassed, electrons are forced across into the p-region where they can combine with holes and have their energy converted into quanta of radiation with a frequency closely equal to the band-gap energy divided by h. The situation is illustrated in Fig. 2.36, where it will be seen that one naturally has an inverted population. At low current densities, however, the emission is not coherent and the device acts as the very familiar light-emitting diode or LED. LEDs are enormously important commercially as their ubiquitous use in pocket calculators and other forms of display testifies. The ability to change the operating wavelength by changing the band gap has proved invaluable. Thus, by alloying gallium arsenide with phosphorus or aluminium, it is possible to shift the wavelength of the well-engineered gallium arsenide LED from the near infrared at 0·85 μm into the red region of the visible, at 0·655 μm, and thus obtain a technically more useful product. LEDs like their laser counterparts are widely used in near-infrared fibre-optic communication systems and the freedom to choose the operating wavelength is a most important consideration since one can thus avoid those wavelengths where the glass absorbs.

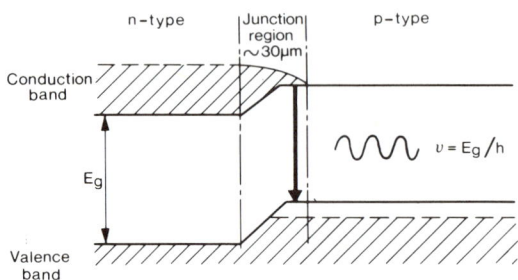

FIG. 2.36. Origin of the stimulated emission from a semiconductor diode.

Most work in the field of light emitting diodes has been done with GaAs and it was noticed quite early on that at low temperatures and high current densities, the line width of the incoherent emission from the junction region narrowed and the beam tended to be confined to the plane of the junction. These are the signs of incipient laser action and when the faces of the crystal, at right angles to the junction, were polished plane parallel to one another, so as to form a cavity, laser emission was, in fact, obtained. Nearly all the other light emitting diodes are now available in lasing forms and, as remarked above, with them it is possible to cover a wide range of the near and mid infrared. Further details of the construction and operation of these sources are given in section 6.2.9. The population inversion mechanism for diode lasers is not as easy to visualise as it is for the other lasers because one is not dealing with discrete states. The process is illustrated schematically in Fig. 2.36. A more quantitative account is available in a paper by Lax [78].

A rather different type of population inversion mechanism to those discussed so far is provided by the far-infrared glow-discharge molecular gas lasers. [79]. Several pure gases, e.g. HCN, H_2O, SO_2, H_2S and some of their isotopic variants emit in the far infrared when gently excited by an electric discharge. The small signal gain is rather low, always less than 2dB m^{-1}, but the lasers are easy to make and use and have found widespread application for these reasons. The inversion mechanism rests on the phenomenon of resonance perturbation as discussed in section 5.2.3.1. The requirement for it to be possible is that two ro-vibrational states having the same parity and J value should be in accidental coincidence. To the harmonic oscillator approximation there is no significance in the close approach and the two levels would not interact; but when cubic and quartic terms are incorporated into the molecular hamiltonian, interaction can take place and the levels can mix. Each observed state then becomes, to first order, a mixture of the two unperturbed forms. One will get, therefore, not only the normal intrastate transitions, i.e. the normal rotational spectrum, but also interstate transitions which, though formally vibrational, owe their intensity to rotational matrix elements. Under the discharge conditions it is easy to have different effective temperatures for the two states involved and, although within each there will be a thermal distribution of population, there will not necessarily be a thermal distribution between them. In fact there may well be a population inversion between the two states and the transition between them, which owes its intensity to the perturbation, will then show gain. The best known example occurs for HCN, for which the two states (11^10) and (04^00) are close in energy. The two states have slightly different rotational constants and because of this the slight mismatch in energy for the rotationless level is exactly cancelled at $J = 10$ and a strong Coriolis perturbation results [80]. This is illustrated schematically in Fig. 2.37. In the discharge, for reasons that are not at all well understood, the (11^10) state is highly populated whereas the (04^00) state is virtually un-populated. The $J^+(10) \rightarrow J^-(9)$ interstate transition can therefore be had in stimulated emission and the result is a powerful line at 337 μm (891 GHz). The weaker perturbation at $J = 11$ leads to a much weaker line at 311 μm. When these lines are radiating, the cascade lines also shown in the figure become possible and most of these have, in fact, been observed. This group of lines is very useful indeed because they lie in the $30\,cm^{-1}$ atmospheric window. Hocker and Javan [81] have measured the time frequencies of some of the transitions shown in Fig. 2.36, using the technique of heterodyne beating (section 7.6). They find, for example,

311 μm	$v = 964.3134\,GHz$	
335 μm	$v = 894.4142\,GHz$	estimated error $\pm\ 0.001\,GHz$
310 μm	$v = 967.9658\,GHz$	
337 μm	$v = 890.7607\,GHz$	

From these figures, the separation between the two states $J^-(11)$ of (11^10) and $J^-(9)$ of (04^00) is found by the two paths to be identical, in elegant agreement with the Ritz combination principle. There is therefore no doubt that the assignment of the lasing transitions and the mechanism for emission given in Fig. 2.37 are both correct but other factors must also be involved in the laser action. Thus HCN is unstable in an electrical discharge and its concentration will be very low. What little there is will, after excitation by electron impact, relax to the ground state presumably through a large number of available channels. Only a small number of these will include (11^10) so the number of molecules in the emitting state, at any given moment, must likewise be small. This does not agree with the observed power output. There must be some particular feature of the chemistry of the discharge which preferentially channels molecules through (11^10).

The water vapour laser is similar to the HCN laser in its operation and quite similar in its mechanism, except that both Fermi and Coriolis resonance are involved in generating the required finite transition moment between the inverted states. The assignment of the very large number of lines available from the H_2O laser has been worked out by Benedict and his colleagues [82]. The

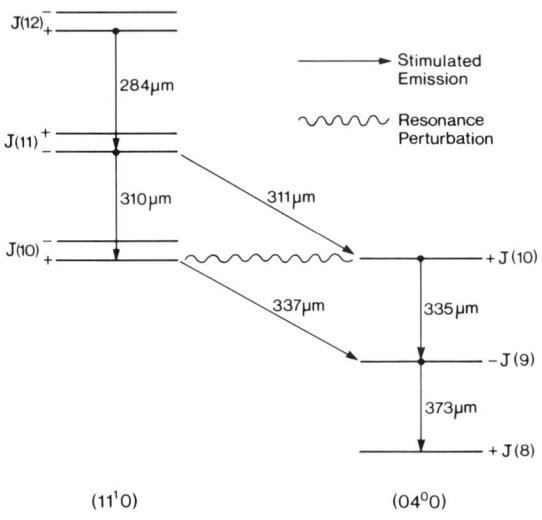

FIG. 2.37. Rotation–vibration energy levels of HCN near 2900 cm^{-1} above the ground state. The l-type doubling of the (11^10) state is shown greatly exaggerated (after Hocker and Javan [81]).

inversion mechanism has been discussed in some detail by Pollack [83]. Part of the very complex energy level diagram for H_2O and the origin of some of the laser lines are illustrated in Fig. 2.38.

Another increasingly important method for producing a population inversion is the use of an already existing laser to selectively pump a different system. At first sight this might not sound like a practical proposition because a virtually exact coincidence is required between the laser frequency and that of the absorption line of the system to be pumped. In practice, however, it is found to work well and very many molecules have been found to give what are essentially stimulated fluorescence lines when pumped with the powerful CO_2 or N_2O lasers. The reason for this is that a ro-vibrational band of a moderately complex polyatomic molecule will contain a very large number of lines and, since the laser will also have quite a few lines of reasonable power, the chance of a coincidence is greatly improved. In addition it should be noted that most organic molecules have several absorption bands in the 10 μm region and these are accompanied by combination and hot bands; the result is a virtual forest of closely spaced lines. One has also available isotopic variants of the pumping lasers and now also the so-called "sequence-band" lasers [84]. On closer examination, therefore, it is no longer surprising that a vast range of low-pressure vapours, when confined in a Fabry–Perot resonator and simultaneously pumped with intense CO_2 or N_2O radiation, give out laser emission at longer wavelengths. Lists of more than 1200 lines have been given

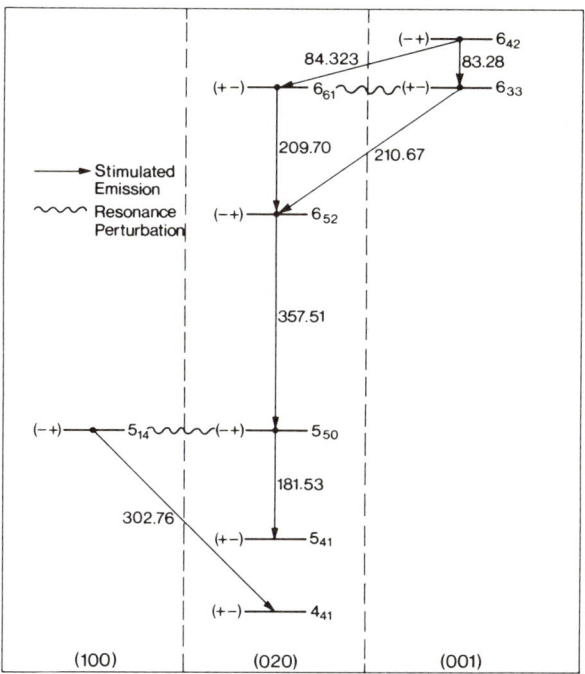

Fig. 2.38. Resonance perturbations and the origin of some of the laser lines from water vapour. The transition frequencies are in cm^{-1} and the term symbols are defined in section 5.2.2.

by various authors [85]. The origin of the useful lines at 496 (B, b), 541 (C, c) and 452 (A, a) μm from optically pumped CH_3F is shown in Fig. 2.39. The $CO_2 P(20)$ line pumps [86] two K components (K = 1 and K = 2) of the J = 12 to J = 12 Q branch line and this produces the primary transitions B and b at 20.1584 and 20.1570 cm^{-1} respectively. When these lines are running strongly, the cascade transitions C and c at 18.4804 and 18.4793 cm^{-1} can be obtained in stimulated emission. Finally, the strong pumping by P(20) can produce "holes" in the ground-state thermal distribution and these can lead to the transitions A and a at 22.1286 and 22.1776 cm^{-1} appearing. All these lines appear either because the pumping overpopulates the upper level or else because it depopulates the lower level. Similar diagrams and remarks apply to the other cases except that often the exact assignment is not known. It is interesting to observe, in this connection, that unlike the glow-discharge laser, no perturbations are involved in the operation of optically pumped lasers, so time-frequency measurements can give very accurate information about the rotational constants. This information complements rather than competes with microwave data because one is often considering vibrationally excited states or else high J values of the ground state and microwave spectroscopy is then severely restricted in its applicability.

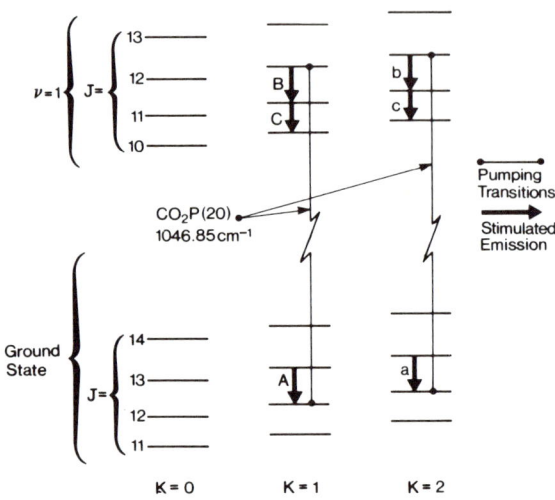

FIG. 2.39. Stimulated fluorescence and laser action in optically pumped methyl fluoride [86].

The theory given above is undoubtedly basically correct but to account for the finer details of laser operation some refinements are required. Firstly, one should take thermal population of the excited states into account when relating incident power to resulting population inversion. For the case of far-infrared lasers, as exemplified in Fig. 2.39, this will have negligible effects, but for the important mid-infrared optically pumped lasers where the transition is back to the ground vibrational state it will play an important role. Mid-infrared lasers therefore require much harder pumping and for this reason they are usually operated in the pulsed mode. Secondly one must take account of the effects of molecular collisions which can rapidly thermalise away the population inversion. At low pressures these will be infrequent but the laser gain will be low because of the few molecules in the beam. At high pressures the number of molecules will be high but the degree of inversion will be small and the gain will once again be low. There exists, therefore, an optimum pressure and this is found to fall as the laser frequency falls. Far-infrared continuous-wave optically pumped lasers thus operate at very low pressures ($<$ 1 torr). Pulsed lasers or mid-infrared lasers can operate at much higher pressures— several torr. Looking at the matter more closely, one now knows [87] that molecular collisions are subject to fairly strong selection rules so that in certain pressure regimes one might get the inversion of a given level transferred to neighbouring levels by means of collisions and new laser lines might appear starting on levels not apparently reached directly by the pumping. Thirdly it should be noted that exact coincidences of pump and pumped transitions are rare and that in the cases of several pulsed lasers where the frequency mismatch is very large ($\sim$ 1 GHz) the phenomenon of stimulated Raman emission [88] is probably encountered. This is a very interesting conclusion because the Raman process is theoretically very efficient since no population inversion is required. Both kinds of optically pumped laser are efficient, however, when compared with the glow discharge type, since gains in the small signal limit of the order 10 dB m^{-1} for the pulsed variety and 20 dB m^{-1} for the pulsed variety are readily realised.

2.5 Electronic sources of microwave and far infrared radiation

2.5.1 *Introduction*

Whenever a charge is accelerated or decelerated, it necessarily emits electromagnetic radiation and, depending on the circumstances, this radiation can be in any region of the spectrum. A well-known example is provided by the Bremsstrahlung ("braking" radiation) which is always observed when energetic particles lose energy in electric and magnetic fields. Quantum mechanically, the particle begins in an unbound state and finishes in one, so bremsstrahlung is often said to arise from "free-free" transitions. Galactic bremsstrahlung lies in the radio frequency and microwave regions but that observed from modern high-energy particle accelerators may lie in the ultraviolet or even the X-ray region. Bremsstrahlung radiation, although it is broad-band and incoherent, does not have the spectral characteristics of black-body radiation, tending rather to vary more slowly with frequency. Intense sources of this type of radiation are now available as by-products of the construction of the ultra-high-energy (many GeV) proton and electron synchrotrons in several research centres and the desirable spectral characteristics have led many spectroscopists to propose experiments to be carried out with the synchrotron radiation. In particular, it is likely that these machines will provide the brightest available sources at the two ends of the normally used incoherent spectrum, namely the far-infrared and the X-ray regions.

If the charges, by contrast, are being accelerated or decelerated by *periodic* fields, then the output will be quasi-monochromatic and coherent. The most well-known example of this is found for the currents flowing in a resonant circuit (see, for example, Fig. 2.1). If one sets up an amplifier with a tuned circuit used to connect its input and output terminals then feedback will occur only at the resonant frequency and one has made a single frequency oscillator. Oscillators based on electron-tube or transistor amplifiers coupled with conventional LC circuits can be used up to about 500 MHz but beyond this they run into difficulties caused by the inherent capacities and inductances of the components themselves and by the transit time effects. Oscillators for the microwave region and beyond have therefore to rely on very different methods of construction but the underlying physics is the same, the interaction of a beam of electrons with a periodic electromagnetic field. Sources of this type are the well-known klystron, magnetron and backward wave oscillator. Klystrons and magnetrons cannot be made for frequencies much above 150 GHz because the resonant "circuit" providing the periodic field is a metallic cavity and its dimensions soon become prohibitively small [89]. Backward-wave oscillators on the other hand, in which the periodic field is provided by an extended "slow-wave" structure, can be used as high as $25 \, \text{cm}^{-1}$ (750 GHz) and the recent development of the gyrotron and of several types of free-electron laser promises sources which will be able to deliver high power at submillimetre wavelengths. These developments are particularly welcome because the other options for providing coherent near-millimetre wave radiation based on the downwards extension of optical techniques rapidly become inefficient because they encounter fundamental limitations. Manley and Rowe [90], for example, show that any non-linear reactance providing long-wave radiation of wavelength λ_s from short-wave pump radiation of wavelength λ_p will have a maximum efficiency given by the ratio of λ_p to λ_s. For the optically pumped laser, this Manley–Rowe relation is readily interpreted in quantum language since one short-wave photon in produces at most one long-wave photon out. A millimetre-wave optically pumped laser will thus have a maximum efficiency (assuming CO_2 laser pumping) of about 1 %! Power levels of more than a milliwatt or so are therefore impracticable. The devices based on electron beams, on the other hand, in which the cavities are large compared with the wavelength, can readily provide kilowatts of continuous-wave power and megawatts of pulsed power. There is therefore considerable interest on the part of infrared spectroscopists and technologists as they see these coherent, powerful sources being developed for the bottom end of their frequency region.

Similar developments are also occurring in the field of solid-state sources. Gunn diodes have been established as microwave sources for some years now, but recently with the introduction of new materials such as InP and of new fabrication techniques, their upper frequency limit has been rising. At the moment InP Gunn diodes are working at frequencies above 100 GHz but this should soon be extended to at least 250 GHz. New types of solid-state source such as the IMPATT and TUNNETT diode which are based respectively on impact avalanche ionisation and on tunnelling across an insulating narrow gap and which have inherently high oscillation frequencies (~ 100 GHz) are very promising, especially since they have very non-linear characteristics and thus give useful power levels in their first few harmonics [91].

Microwave coherent sources are tunable over a restricted range (say 10 % of the centre frequency) and, furthermore, can readily be phase-locked through a suitable multiplication chain, down to a fundamental standard of frequency. One can thus carry out, quite routinely, ultra-high-resolution spectroscopy with very high absolute accuracy [92]. It was this aspect which for so long distinguished microwave spectroscopy as a separate discipline from infrared spectroscopy to which it is otherwise strongly related. The separation is now happily ending and the resemblances rather than the differences between the two disciplines are tending to be emphasised. The common points which link the two might be listed as follows.

a. The development of intense infrared lasers has provided coherent sources for the infrared region as well and the emergence of the new art of laser-microwave double-resonance [93] in which it is possible to transfer the high-resolution capabilities of the microwave source up into the infrared has welded strong connections.

b. Considerable attention is now being paid to the possible use of infrared lasers as fundamental metrological standards. This is discussed in detail in section 7.6, but for the moment it may be summarised by saying that the problem is to determine the absolute time frequencies of a chain of lasers, spanning the infrared, in terms of the fundamental microwave standard of time provided by the caesium clock [94]. Once these are known, it will be possible to dispense with a length standard altogether and use only the defined unit of time together with a defined velocity of light in vacuo based on the best experimentally determined value, to define the laser wavelength. The technology of the frequency determinations covers smoothly the microwave, millimetre-wave and infrared regions and makes a nonsense of any arbitrary divisions.

c. Molecular spectroscopy aims to provide a set of parameters to define the energy levels of the molecule in question. To derive these to the highest precision, it is necessary to use experimental data from all regions in which the molecule absorbs in its ground state. Microwave and infrared data must necessarily therefore be combined and a division based only on the coherent/incoherent barrier has always been rather ridiculous. This separation used to be justified by the almost coincident division into rotational spectra on the one hand and vibrational spectra on the other, but this too is artificial since we can often determine the frequencies of vibrationally excited states from the temperature dependence of the intensities of corresponding rotational lines. This can be an invaluable procedure if the state in question is not connected to the ground state by an allowed transition and is thus otherwise unobserved. A further point uniting the two forms of spectroscopy is that the wavelength or frequency calibration of infrared spectrometers is often best done by invoking microwave data. As an example Fourier transform spectrometers are usually calibrated in the far infrared by means of the pure rotational spectrum of carbon monoxide. The positions of the high J (that is far infrared) lines can be calculated to high precision from the molecular parameters deduced from the low J lines via microwave spectroscopy. All one needs to do therefore is plot a graph of measured frequency against true frequency for CO and then use this to calculate the true frequencies of the lines observed for other substances.

Infrared and microwave science and technology are therefore rapidly being unified and it is essential for a modern infrared practitioner to be familiar with at least the basics of microwave sources and detectors. In what follows a brief outline will be attempted of the physics of modern "electronic" sources of microwave and higher frequency radiation.

2.5.2 *Klystrons, backward-wave-oscillators and other vacuum tube sources of microwave radiation*

For a circuit to oscillate at microwave frequencies, the LC tuned part must transform into a resonant cavity and the transit time effects must be overcome in some way. A most ingenious way of achieving both objectives is represented by the Klystron [95]. A simple klystron has an electron gun and two loosely coupled identical cavities all inside a common evacuated glass envelope. The stream of electrons from the gun passes through the first cavity which acts as a "buncher", that is it breaks up the continuous stream of electrons into a series of bunches. It does this because the oscillating field in the cavity will pull back electrons which are ahead of the field and accelerate those which are lagging. The electrons therefore tend to be forced into bunches which are in phase with the field. In alternative language one can say that the electron stream is velocity modulated. When the bunches pass through the second cavity, "the collector", they transfer their energy to the resonant modes of this cavity and electromagnetic radiation at one or more cavity resonant frequencies can be coupled out via a suitable waveguide. Modern klystrons, however, feature only a single cavity which acts as both buncher and collector—the stream of electrons having passed through the cavity in one direction is reflected back along its path by a negatively charged reflector electrode, and therefore passes through the cavity again but in the opposite direction. This form of microwave tube is called a *reflex klystron*. Its construction is illustrated schematically in Fig. 2.40.

The output frequency of a klystron is determined primarily by the dimensions of the resonant cavity and to a lesser extent by the mode on which it is resonating. However, when set on a given mode the frequency can be swept by as much as 1 % of the mode centre frequency merely by varying the reflector voltage. Tuning rates are of the order of 10 MHz V^{-1}. This tunability is very convenient since it enables one to frequency modulate the klystron and also to

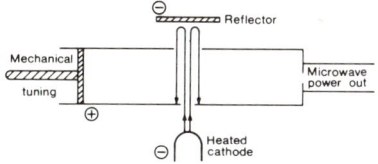

FIG. 2.40. Schematic of the construction of a reflex klystron.

lock its output frequency by means of a phase-lock loop. If larger excursions of frequency are required, as they will be say for microwave spectroscopy, then one must use a combination of mechanical and electrical tuning. With this combination it is possible to sweep a given mode across the entire tuning range which might typically be $\pm 20\%$ of the band centre frequency. X-band klystrons, for example, can be tuned all the way from 8 GHz to 12 GHz. Klystrons are available for all the radar bands and these tubes can deliver considerable power—some watts. It therefore becomes feasible to use harmonic multiplication in a crystal diode to generate the overtones of the fundamental frequency. Roughly speaking one gets about a factor of 100 attenuation at each multiplication so the power available falls off rapidly with harmonic number. Nevertheless Gordy at Duke University [96] and Kilp at Mainz [97] have managed to use harmonically multiplied klystrons to carry out submillimetre spectroscopy. One interesting point here is that whereas a 1% swing of the fundamental frequency, say 10 GHz, is only 100 MHz, when one considers a 1% swing on the multiplied frequency one gets a substantial range of the spectrum, for example nearly 0.1 cm^{-1} at 300 GHz. The shortest wavelength klystrons currently available work at 2 mm but these are very expensive, have short life times and, moreover, deliver rather low power. For these reasons, the much more powerful longer wavelength klystrons (4 mm and 8 mm) are usually used as the drivers for the multiplication chain. An example of a high resolution submillimetre spectrum obtained using these techniques by Gordy and his colleagues [98] is shown in Fig. 2.41.

A second type of widely used microwave tube is the backward-wave oscillator or BWO [99]. This is a development of the travelling wave tube which is in common use as a microwave amplifier. The principle of the backward-wave oscillator is illustrated in Fig. 2.42.

Again one has an electron beam produced by an electron gun but this time there is present an axial magnetic field to focus the beam. On their way to the positively charged collector, the elctrons pass through a metallic helix which is normally held at the same potential (300–2500 V) as the collector. This helix is a microwave transmission line of effective length equal to several wavelengths and fields applied to it will bunch the electrons just as in the klystron case. The magnetic field is there to prevent the electrons from being attracted to the positively charged helix. If a microwave field is applied to the helix through a

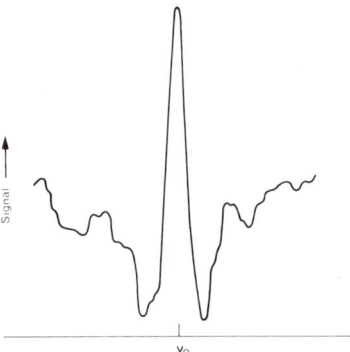

FIG. 2.41. The $2_{02}-2_{11}$ line of H_2O at $v_0 = 752.033\,44$ GHz $(25{\cdot}085\,135\text{ cm}^{-1})$. The line is presented in second derivative form and is about 5 MHz wide. (After P. Helminger, F. De Lucia and W. Gordy [98].)

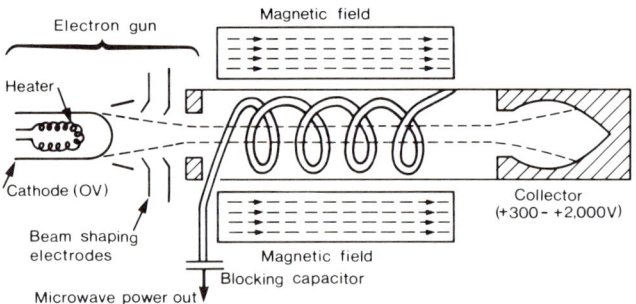

FIG. 2.42. Schematic of the backward-wave oscillator or BWO. The type illustrated is the so-called O-type which has the two fields parallel. There exists also an M-type in which they are perpendicular.

suitable DC blocking capacitor, then, at a certain critical collector voltage, the bunches of electrons will become synchronous with the field on the helix. One can loosely say that the field due to the bunches appears through the gaps in the helix and, interacting with the microwave field already present, transfers energy to it. There is thus amplification. In the absence of an external field, the components of the current noise which lie within the resonance bandwidth of the travelling wave system will be amplified and if the tube is set up appropriately feed back will occur and the tube go into oscillation. The situation is very similar to that of a laser which needs spontaneous initial emission (that is optical noise) to get it going.

When the tube is oscillating the wave generated is 180° out of phase with the electron bunches. This is why the tube is called a backward-wave oscillator. It is sometimes, more colourfully, called a carcinotron after the Greek word *carcinos* meaning a crab, because crabs are well-known for running backwards! BWOs are much more costly than klystrons because they are more intricate in their construction and also because they require very expensive permanent magnets. However, they can be electrically tuned over their entire range of oscillation—unlike the klystron—and it is much more practical to make extra-high frequency BWOs than it is to make the equivalent klystrons. Krupnov [100], for example, has used carcinotrons as high as 30 cm^{-1} (900 GHz). The other side of the tunability coin is that the output frequency is unfortunately much less stable and complex stabilising circuits are necessary. Details of the construction of these have been given by Krupnov [101]. In the latest form of his equipment TTL (i.e. *Transistor-Transistor Logic*) circuitry is used. This is very convenient since one can simpy dial up a starting frequency and the circuitry can then be used to print out an accurate abscissa scale for the spectrum as the frequency is scanned.

The third common type of microwave tube is the magnetron. This, however, because of the details of its construction, is basically a high-power source for the lower microwave bands. It finds extensive use in radar but does not nowadays compete with the cheaper and more readily available klystron for spectroscopic applications. The power available from klystrons is adequate for microwave spectroscopy and for microwave–infrared double-resonance spectroscopy. Even after several harmonic multiplications there is still sufficient power to give a good *Signal-to-Noise Ratio* (SNR) from the sensitive detectors now available in the submillimetre region. Klystrons also have the advantage that they start higher up the frequency scale so fewer multiplications are necessary. Magnetrons are thus seldom used in modern spectroscopy, but they have great historical interest since using an early form, Cleeton and Williams [102] in 1934 observed the first microwave spectrum, the inversion band of ammonia at 24 GHz.

2.5.3 Gyrotrons

One of the most remarkable developments in the last decade in the field of microwave tubes has been the introduction of the gyrotron. The first ideas in this area seem to have been evinced by Twiss [103] and Schneider [104] in the West but the full credit for the further development of the theory and especially for the creation of practical devices must go to the team in the Soviet Union led by Gaponov [105]. The gyroton or electron cyclotron maser depends for its operation on a phase-focussing effect which becomes manifest at relativistic energies. Electrons spiralling about a magnetic field B have a typical cyclotron resonance frequency

$$v_{cr} = eBc^2/2\pi\mathscr{E}, \qquad (2.5.1)$$

where $\mathscr{E}$ is the total energy of a particle of rest mass m_0 moving at a velocity v, namely

$$\mathscr{E} = m_0 c^2 (1 - v^2/c^2)^{-1/2}. \qquad (2.5.1a)$$

At low electron velocities, (2.5.1) therefore reduces to the simple classical result $v_{cr} = eB/2\pi m_0$ ($v_{cr} = 27\cdot992$ GHz T^{-1}), but at relativistic velocities where (v/c) is appreciable, the cyclotron frequency will no longer be independent of the velocity. The simple cyclotron is therefore limited in the energy to which it can accelerate particles and to overcome this limitation it is necessary to go over to the much more complicated synchro-cyclotron, or synchrotron for short, in which the frequency of the accelerating AC field is gradually reduced as the particle energy increases.

If the circling particles are subjected to an impressed electromagnetic field of a fixed frequency then the result is that the particles tend to bunch. This is because electrons whose initial phases are such as to lead to energy loss to the field will experience an increase in phase whilst those whose initial phase would favour a gain will experience a recession in phase. Both groups therefore reach the same equilibrium value of the phase which depends on (v/v_{cr}). When this ratio is greater than unity the work done by the field is negative for more than half the electrons, in other words there is gain. This is an exciting conclusion because it implies that it is, in principle, possible to make a maser which requires neither a high power pump nor cryogenic operation. The Soviet workers have shown [106] that this concept of radial (as opposed to the more familiar axial) bunching can be applied generally and there is a "gyro" equivalent of every normal microwave tube. One has, for example, gyro-klystrons gyro-carcinotrons, gyromonotrons and gyro-TWTs etc. In all these tubes the electrons follow a helical path as they go from cathode to anode— this contrasts, of course with the linear path followed by the electrons in a conventional microwave tube. The gyrotron has no periodic structure to sustain slow waves, its resonant cavity is just a hollow tube, it is the cyclotron resonance condition which maintains the synchronism. Several different designs for gyrotrons have emerged. A very successful Russian one is shown in Fig. 2.43.

The crucial part of the device is the cathode (1) which has a narrow emitting stripe (2); this, together with the first anode (3), gives the special spatial distribution of electrons and their velocities which are essential for successful operation. The electrons are then accelerated by a second anode (4) and then enter the resonant cavity (5) where the interaction occurs. The radiation is coupled out via an aperture (6), a beam collector (7) an output window (8) and a suitable waveguide (9). The device generates large quantities of heat so it has to be water cooled by suitable jackets (10) and (12). Magnetic fields are generated by the solenoids (11) and (13). Sections are insulated from one another by spacers (14). Devices of this type can produce pulses lasting several micro-seconds at power levels of 40 kW or alternately can be run continuous wave at power levels of the order 10 kW. These are most impressive figures.

The Soviet workers seem content to develop centimetre and millimetre wave gyrotrons for radar, communication and plasma heating purposes and seem only marginally interested in operating gyrotrons at sub-millimetre wave-

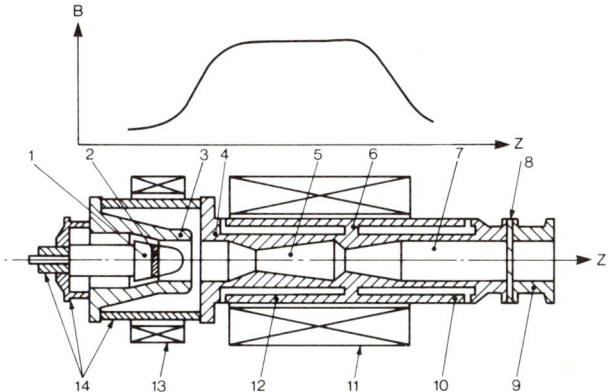

FIG. 2.43. Schematic of a Russian gyrotron operating at 8·9 mm (second harmonic) after Kisel *et al.* [107]. The device is about 20 cm long.

lengths. One can understand why they prefer to concentrate their efforts since the use of gyrotrons, for example, to augment the heating of tokamak plasmas may well be essential for reaching fusion temperatures [108], but it is nevertheless intriguing to speculate about the use of gyrotrons at higher frequencies [109]. The problems are two-fold. Firstly the cavity resonators and output waveguides become very small in diameter and this gives severe heating difficulties. Secondly intense magnetic fields are required to satisfy the cyclotron resonance condition (2.5.1). Various ways of using overmoded (and therefore larger) cavities have been considered and one Russian approach has been to invoke the "whispering gallery" modes which have their field intensities high only near the walls—this considerably reduces the heat dissipation problem. Nevertheless there are many difficulties in working with a system which has a crowded mode spectrum. The magnetic field requirement can be met either by using a superconducting solenoid for continuous-wave operation or else by using a pulsed conventional electromagnet for pulsed operation. Results so far have been reported only at 0·9 mm wavelength where 1·5 kW of continuous-wave power was obtained. Gyrotrons can apparently operate with very non-linear characteristics where the power drop-off with harmonic number is not catastrophic. One can therefore visualise considerable penetration into the infrared without collossal constructional expense and with only moderate magnetic field requirements. Another possibility for an infrared gyrotron is to use a quasi-optical, that is a Fabry–Perot cavity and to have the spiralling electrons enter it transversely. At the moment, however, it is still millimetre-wave gyrotrons which are in demand because of their unequalled power capabilities, but now that the USA and other countries have entered the gyrotron race [110, 111, 112], one can confidently expect that the upper frequency performance of gyrotrons will be raised considerably above its present limit of about 300 GHz.

2.5.4 *Synchrotron and other relativistic sources*

2.5.4.1. *Synchrotron radiation*

In Lawrence's cyclotron, electrons were accelerated every time they jumped from one "D" to the other. The accelerating energy came from an AC high-voltage supply connected across the "Ds" with a frequency exactly matching the rotation frequency of the electrons. At the sort of magnetic fields now available, this frequency would be of the order of some MHz but, of course, it was rather less in Lawrence's day. The circling electrons are subject to a radial acceleration and therefore radiate electromagnetic radiation at the fundamental rotation frequency. This radiation represents an energy loss mechanism for the cyclotron. Cyclotrons cannot be used into the relativistic regime, as they stand, as was discussed in the previous section, because the relativistic increase in mass slows down the electron bunches and they get out of synchronism with the driving field. This problem was overcome with the invention of the synchrotron where the frequency of the driving field is varied to match always that of the electron bunches. The only limits then to the energy to which the circling particles may be accelerated are set by cost, the size of the machine and energy loss by synchrotron emission. This synchrotron radiation, whilst it is a considerable nuisance to the high-energy particle physicists, has, nevertheless, some remarkable properties which make it very attractive to spectroscopists. It is for all intents and purposes a broad-band source which is much brighter than any available black-body source in the extremes of the spectral region of interest, namely the vacuum ultraviolet and the far infrared. Synchrotron radiation, since it comes from highly relativistic particles, is strongly depressed in directions out of the plane of the orbit and is almost perfectly plane polarised. It is therefore quite unlike cyclotron radiation where a normal "dipole" polar diagram is observed. The radiation is "sprayed out" tangentially from the electron bunches in a manner which has been compared with rain water being sprayed off the wheels of a racing bicycle [113]. Another consequence of the relativistic origin is that the harmonics of the fundamental frequency are radiated intensely, in fact to a first approximation the number of photons emitted in each harmonic is the same up to a cut-off frequency determined by the particle energy and the bending magnetic field strength. Thus if the electrons are moving on a circle of radius R with energy E, one has

$$v_c = 3c\gamma^3/4\pi R, \qquad (2.5.2a)$$

where

$$\gamma = (\mathscr{E}/m_0c^2). \qquad (2.5.2b)$$

The synchrotrons which are accessible to spectroscopists have $\mathscr{E}$ of the order of 1 GeV, so since the rest mass (i.e. m_0c^2) of an electron in energy units is 0·511 003 4 MeV, γ will be of the order 2000. The synchrotron will have an overall radius of some tens of metres but, since it is very difficult and very costly to maintain homogeneous magnetic fields over such distances, the electron beam bending is usually carried out in much smaller regions by the intense magnetic fields (~ 5T) provided by "wiggler" magnets. The wiggler regions are then separated by sections in which the electrons follow straight-line paths. The radius R which appears in (2.5.2a) is then the radius of curvature of the beam in the wiggler region. This might typically be 1 m and we would have $\nu_c = 5.7 \times 10^{17}$ Hz. This corresponds to a cut-off wavelength λ_c ($= 5.2 \times 10^{-10}$ m) lying in the X-ray region. At wavelengths shorter than λ_c the power rapidly falls with decreasing wavelength.

The radio-frequency fine structure to the emission is determined by the transit time of the electrons and by the structure of the beam. This is usually made up of one or more electron bunches of individual length 20 ns and each bunch will have a circulation time of about 300 ns. Radiation is only "seen" when "looking" into a curved part of the trajectory and when there is an electron bunch there to be seen. This is why the output radiation is necessarily pulsed. The corresponding pulses in frequency space are only separated by a few hundred MHz (~ 0.003 cm^{-1}) so for nearly all spectroscopic purposes the fine structure can be ignored and one has, as mentioned above, essentially ideal broad-band radiation. It must also be borne in mind, in this connection, that the finite length in time of the electron bunches leads to a broadening of the discrete features in the spectrum and this has the effect of still further smoothing out the spectral characteristic.

The angular spread of the beam in the plane of the orbit is wavelength dependent being almost negligible for X-rays, but 30° or so for the far infrared. One can use this phenomenon to achieve a rough spectral separation by means of stops placed in the beam. Alternatively one can use specially shaped mirrors to collect the beam very effectively for a given wavelength region. This is illustrated schematically in Fig. 2.44.

A very important adjunct to a synchrotron is an electron storage ring— sometimes referred to in the high-energy literature as an ESR. In this device, which is essentially a highly evacuated torus, electron bunches which have been accelerated in a synchrotron or else in a linear accelerator (LINAC) are stored by going round and round for long periods. The electron bunches lose energy by synchrotron emission and also by collisions with the residual gas atoms, but this loss can be compensated, to a large extent, by power supplied from an RF resonant cavity and the $1/e$ lifetime can be as much as 8 hours. Storage rings are typically of the order 16 m in diameter so the bunches take times of the order of

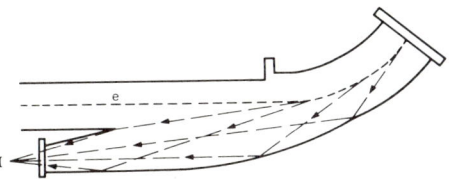

FIG. 2.44. Use of a curved mirror to collect very effectively the radiation from the synchrotron beam (after P. Lagarde [113]). The radiation is emitted when the electron path goes through one of the bending regions associated with a "wiggler" magnet.

3×10^{-7} seconds to complete a circuit. The low-frequency cut-off is therefore 3 MHz but since it is unusual to have only a single bunch in the storage ring, this cut-off and the corresponding spacing of the harmonics will increase. Storage rings are important in particle physics because by using two of them with beams circulating with opposite senses it is possible to bring about "head-on" collisions between the bunches, thus doubling the collision energy in the laboratory frame of reference. It was, in fact, by such "head-on" collisions between electrons and positrons that much of our knowledge of "charmed" quarks was discovered [114]. Electron storage rings have their attractions to spectroscopists also, since the long life-time means that they can do their experiments in relative peace without competing with the particle physicists for access to the main synchrotron and having to cope with the electrically very hostile environment. Lagarde has described [113] some infrared experiments with the electron storage ring at Orsay, called ACO (Anneau de Collision d'Orsay), which have given promising results.

The power output, per electron per unit horizontal angle, per unit vertical angle, at a wavelength λ and a vertical angle ψ, for a synchrotron or storage ring is [113]

$$P(\lambda, \psi) = \frac{27}{64\pi^4} \left(\frac{e^2 c}{\varepsilon_0 R^3} \right) \left(\frac{\lambda_c}{\lambda} \right)^4 \Delta\lambda \gamma^8 [1 + (\psi\gamma)^2]^2$$

$$\times \left[K_{2/3}^2(x) + \left(\frac{(\psi\gamma)^2}{1 + (\psi\gamma)^2} \right) K_{1/3}^2(x) \right] \qquad (2.5.3)$$

where the Ks are the modified Bessel functions discussed in section 1.5.2 and x is the dimensionless variable defined by

$$x = \left(\frac{\lambda_c}{2\lambda} \right) (1 + (\psi\gamma)^2)^{3/2}. \qquad (2.5.4)$$

Because of the strong depression into the horizontal plane, it is sensible to integrate over ψ, since all this angle can usually be collected, and then, in terms of the practical units used by the "machine" physicists, one has [115]

$$P(\text{W}) = 9{\cdot}81 \times 10^4 \frac{i(\text{mA})}{R(\text{m})^2} \mathscr{E}(\text{GeV})^7 \Delta\lambda(\mu\text{m})$$

$$\times \left[\left(\frac{\lambda_c}{\lambda} \right)^3 \int_0^\infty K_{5/3}(\eta) \, d\eta \right], \qquad (2.5.5)$$

where $\eta = (\lambda_c/\lambda)$. At long wavelengths where η will be small, the Bessel function may be replaced by the first term in its series expansion [15]:

$$K_{5/3}(\eta) = 2^{2/3} \Gamma(\tfrac{5}{3}) \eta^{-5/3} \qquad (2.5.6)$$

and after performing the integration one has the approximate (but quite good!) result

$$P(W) \approx 10^5 \frac{i(\text{mA})}{R(\text{m})^2} \mathscr{E}(\text{GeV})^7 \, \Delta\lambda(\mu\text{m}) \left(\frac{\lambda_c}{\lambda}\right)^{7/3}. \qquad (2.5.7)$$

The experimental conditions used by Lagarde [113], were $i = 70\,\text{mA}$, $R = 1\cdot1\,\text{m}$, $\mathscr{E} = 0\cdot54\,\text{GeV}$, $\lambda_c = 39 \times 10^{-10}\,\text{m}$, so at a wavelength of 100 μm with a bandwidth of 1 μm, P would be $4 \times 10^{-6}\,\text{W}$ into 1 rad of horizontal angle. The available collection angle (Fig. 2.44) is 30°, i.. $0\cdot53\,\text{rad}$, so the available power is about $2 \times 10^{-6}\,\text{W}$. This is about the same as is available from a good quality globar or mercury vapour lamp under reasonable conditions. However at longer wavelengths the synchrotron radiation rapidly becomes much more powerful than that from any available black body. This is shown very dramatically in Fig. 2.45.

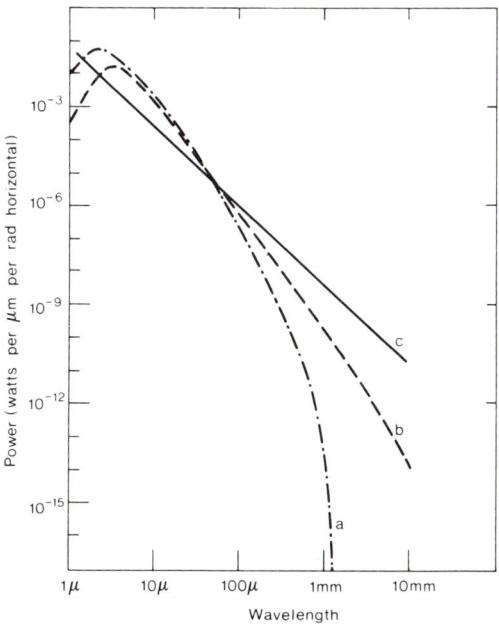

Fig. 2.45. Comparison between (a) a globar, (b) a HPK 125 mercury lamp and (c) the ACO storage ring, under the same spectroscopic receiving conditions (after Lagarde [113]).

The reason for the superiority is, of course, that the synchrotron radiation intensity is falling only as $\lambda^{-7/3}$ whereas that from a black body in the Rayleigh–Jeans region (equation 2.2.40) is falling as λ^{-4}. Synchrotrons and storage rings are therefore very attractive as sources for the energy starved far-infrared region. Higher current, higher energy machines will be better at even shorter wavelengths or alternatively provide a solution to those spectroscopic problems at long wavelengths which need high power. The angular dependence of the radiation on the wavelength was mentioned earlier and this is of some relevance in considering infrared experiments. In the relativistic regime x can be replaced by $2\pi R\psi^3/3\lambda$ for $\psi \neq 0$ and one can then use the asymptotic approximations for the Ks since this time the arguments are large. These approximations are given by [15]

$$K_n(x) = (\pi/2x)^{1/2} \exp(-x), \qquad x \gg 0, \qquad (2.5.8a)$$

so the vertical angle dependence will be given by a function of the form

$$P(\psi) \sim (\psi\lambda) \exp(-4\pi R\psi^3/3\lambda). \qquad (2.5.8b)$$

$P(\psi)$ will clearly fall rapidly as ψ increases unless λ is large enough to compensate. Some measure of beam spreading will therefore be expected in the far infrared but it will still be relatively small ($< 30°$) and should present no insuperable difficulties to the experimentalists whose equipment is normally constructed to accept such angles anyway. For all these reasons it will be seen that in the far infrared beyond 100 mm and especially for "difficult" experiments, the storage ring promises to be a very attractive source.

2.5.4.2 Free-electron lasers

Free-electron lasers, or FELs, belong loosely to the same general class of device which includes the gyrotron and the synchrotron. The common feature is emission from a relativistic electron beam and because of this commonality all these devices are in principle broad-band tunable. The FEL which couples the emission to a Fabry–Perot cavity gives out intense, moderately coherent, radiation and the absence of a condensed active medium is an enormous advantage since laser-damage problems are virtually absent. FELs have already been operated up to the megawatt level (pulsed) and because free electron generators of electromagnetic radiation are inherently highly efficient devices, there is good reason to expect even higher performance with up to 10% of the applied electric power emerging as radiation. The impact that this sort of power would have in fields such as laser fusion, plasma heating and military applications has stimulated much research in the USA and USSR. An excellent account of the work to date in the USA, together with a comprehensive theoretical analysis, has been given by Sprangle, Smith and Granatstein [116].

The basic principle of the FEL is stimulated back scattering of photons from the collective electronic ensemble, a phenomenon predicted theoretically by Kapitza and Dirac [117] in 1933. The observed photons, however, suffer an enormous Doppler shift since the electron beam is highly relativistic and low-frequency waves can be up-shifted into the microwave region and microwaves themselves emerge as infrared. There does not seem to be any limit to how far up in frequency one can go and Sprangle *et al.* [116] have even discussed visible region FELs. If the incident pump has a frequency v_0 then the scattered wave will have a frequency

$$v_s = v_0 \left[\frac{1 + (v/c)}{1 - (v/c)} \right] = v_0 \frac{[1 + (v/c)]^2}{1 - (v/c)^2} \approx 4v_0\gamma^2, \qquad (2.5.9)$$

where v is the axial velocity of the electron beam and, as before,

$$\gamma = [1 - (v/c)^2]^{1/2} = \mathscr{E}/m_0 c^2. \qquad (2.5.10)$$

The output frequency can therefore be changed merely by altering the electron energy. The pumping can be provided either by the real radiation from a laser, for example the CO_2 laser at $10.6 \mu m$ or else by the virtual radiation produced as the beam traverses the periodic magnetic field provided by an "undulator". Experiments to date, however, have used either the magnetic modulator alone or else an electromagnetic pump, together with magnetic modulator. Sprangle, Smith and Granatstein [116] give mathematical analyses of the various regimes under which the FEL can operate, but the basic physical mechanism is easy to visualise. The important point is that because the beam is highly relativistic it is essential to distinguish between the beam frame of reference and the laboratory frame of reference. In the beam frame, low-frequency longitudinal electrostatic waves, that is space-charge or density waves will exist, say of frequency v_1' where the prime signifies the beam frame. If now the beam encounters an intense pump wave of frequency v_0' and electric field strength E_0 then transverse oscillations will be induced which will couple with the density wave to produce currents at the sum frequency $v_+' = v_1' + v_0'$. The resulting back-scattered wave at frequency v_+' couples to the pump via the "velocity-cross-field" term in the Lorentz force equation to produce a sort of radiation pressure which bunches the electrons exactly in synchronism with the original density wave. There is thus a positive feedback mechanism present and oscillation results. In quantum terms one has stimulated emission at v_+'. An exactly similar argument applies to the forward scattering process but since v_-' is *downshifted* on transference back to the laboratory frame it is of no interest in the present context. The beam is propagating through a magnetic field, and there will thus be a defined electron cyclotron resonance frequency and this represents a possible mode for decay of the pump wave (the other is a space-charge beam mode). In the beam frame, because of the relativistic effects, the cyclotron resonance frequency may become equal to the pump frequency and then one can have stimulated resonance Raman scattering. This is undoubtedly an effective means for increasing the gain of the scattered wave. Another possible mode of operation depends on the relativistic analogue of stimulated Compton scattering and there are several others discussed in the article by Sprangle, Smith and Granatstein [116]. In general all these processes are occurring simultaneously but the one which is dominant depends on the electron beam parameters and on the pump frequency. Thus for the beam one defines a Debye shielding length, λ_D, sometimes loosely called the Debye length or wavelength by

$$\lambda_D = \left(\frac{kT\varepsilon_0}{Ne^2} \right), \qquad (2.5.11)$$

where N is the number of electrons per cubic metre. Collective Raman processes, i.e wave-wave scattering, will dominate when the pump wavelength, in the beam frame, is much greater than λ_D and single-particle stimulated Compton, i.e. wave-particle, processes will dominate when it is not. The Raman process is very suitable for high-power millimetre-wave production whilst the single-particle process is more suitable for infrared or visible generation.

Experimental demonstrations of free electron laser action have been few so far, mostly because elaborate and expensive equipment was necessary and because the experimentalists were mostly interested in checking out the fundamental theory. Elias et al. [118] in 1976 showed amplification of 10·6 μm radiation from a CO_2 laser in a single pass through a helical magnet containing a co-axial 24 MeV bunched electron beam. Using the same magnet Deacon et al. [119] increased the electron beam energy to 43 MeV and fitted two mirrors to make an optical cavity. They observed 7 kW (0·006 % efficiency) of 3·4 um radiation. Granatstein and his colleagues [120] in 1977 observed submillimetre radiation from the device shown schematically in Fig. 2.46.

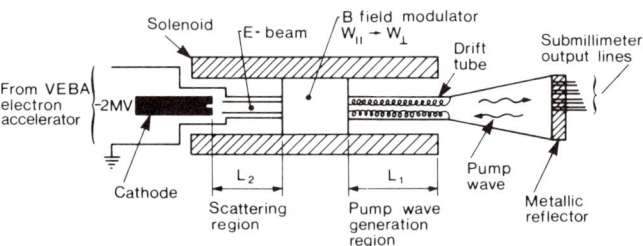

FIG. 2.46. Schematic of the apparatus used by Granatstein et al. [120] to produce intense submillimetre radiation from a free-electron laser.

A very high voltage (-2 MV), very short (50 ns) pulse applied to the cathode produced an annular beam of relativistic ($\gamma = 3.9$) electrons. When these electrons entered the non-adiabatic magnetic field region coupled to a drift tube, a pump wave was generated at 2 cm. This part of the apparatus is essentially a gyrotron. The 2-cm radiation was reflected from a metal place which carried output waveguides with internal dimensions less than 1 mm. These would be cut off of course for the 2-cm radiation. When this radiation after reflection re-encountered the beam in the scattering region, it was backscattered with a wavelength shift (2.5.9) to the 1/3 of a millimetre region and this radiation was coupled out via the waveguides. The experimental wavelength was found to be 400 μm in good agreement with the theory. The power output was of the order of 1 MW which must be a record for submillimetre power levels. The drawback to the simple apparatus shown in Fig. 2.46 is that pump wave generation and stimulated scattering take place in the same electron beam and it is therefore not possible to optimise both independently. New apparatus is therefore being developed in which the two functions are separated.

The simple undulator is composed of fixed permanent magnets but if one can use powerful electromagnets, it becomes possible to tune the output frequency magnetically instead of by varying the electron beam energy. This will be much more convenient experimentally and, of course, by keeping the beam energy constant it will be that much easier to keep it monoenergetic and thus attain better spectral purity. When the magnet field strength becomes appreciable (2.5.9) should be replaced by

$$v_s = 4 v_0 \gamma^2 [1 + e^2 B^2 / 8\pi^2 m_0^2 v_0^2]^{-1}, \qquad (2.5.12)$$

from which the dependence of v_s on B is evident. The output of an FEL will be so-called macropulses each about 1 μm long and with a repetition rate of about 25 Hz. Each macropulse will consist of a string of micropulses each about 60 ps long and spaced by about 2 ns. The spectral line width will not therefore be very sharp even with the Fabry–Perot cavity to help. The very best possible is about 0.01 % but practical machines will probably only realise 0.1%. At 100 μm, i.e. 100 cm^{-1}, the line width will therefore be 0.1 cm^{-1} which is not very good for a laser. Actually there is dispute whether an FEL is even a laser in any real sense since it does not involve stimulated transitions between defined quantum states. Nevertheless, and despite the broad emission line width, there is good evidence that the emitted radiation will be very coherent. This is entirely due to the higher-order quantum coherence effects briefly mentioned earlier. Present plans for FELs include one at the Science and Engineering Council's laboratories in the UK which will be fed from a 22–33 meV LINAC feature tunability through the range 40–170 μm where alternative tunable sources are rather scarce. Quite clearly these second generation forms of FEL will open up entirely new areas of sub-millimetre research since power levels of the order of tens of megawatts will be available.

Some close relatives of the FEL are provided by the various devices based on the Smith-Purcell effect [121]. This phenomenon, first observed in 1953, is the emission of radiation when a beam of electrons skims the surface of a metallic diffraction grating. The effect is normally weak and incoherent but, depending on the electron energy, the radiation can lie anywhere in the electromagnetic spectrum so, nevertheless, it is of considerable interest. The most promising approaches to a practical device are the Orotron, first invented in the USSR [122] but subsequently much developed in the USA [123], and the

Ledatron developed by Mizuno and his colleagues in Japan [124]. In both of these devices, the crucial element is the incorporation of the diffraction grating into one of the mirrors of a Fabry–Perot resonator. This has the effect of introducing feedback so that the intensity is enormously enhanced and also of introducing a longitudinally varying electric field which causes the electron beam to bunch. Because of the bunching, the radiation is produced more efficiently and it also becomes coherent. If in addition its frequency matches that of one of the resonator modes, one will have achieved an intense monochromatic coherent source. The modern orotrons, developed at the Harry Diamond Laboratories of the United States Army [125], have a Fabry–Perot comprised of a spherical upper mirror carrying the output waveguide and a lower cylindrical mirror in which is embedded the grating. An electron gun produces a sheet electron beam which is passed as close as possible to the grating surface and then collected by a positively charged collector on the other side. When the electron velocity becomes synchronous with the phase velocity of a slow wave propagating along the grating surface, build-up will occur and coherent radiation will be obtained. Orotrons are mostly being used, at the moment, as millimetre-wave sources because this is the spectral region where coherent tunable sources are most needed. The version developed at the Harry Diamond Laboratories works from 53 to 73 GHz giving a quasi-continuous wave output of about 100 mW. The electronic tunability is 0.25 MHz V^{-1} over a narrow range and the output line width is better than 0.4 MHz at 70 GHz. These are impressive figures and particularly so since there is no obvious obstacle to extending the range of operation up into the infrared [126].

2.5.5 *Solid-state electronic sources*

In microwave and millimetre-wave technology, solid-state devices are playing an increasing rôle. Solid-state detectors have been available for some years now, modulators and mixers are making an impact and now various types of solid-state source are appearing to challenge the klystrons and BWOs. These solid-state sources have the advantages of being small, lightweight, reliable and relatively inexpensive. The emission occurs from a narrow diode region embedded in a more extended semi-conducting structure having usually a rather complicated doping profile (section 6.5). These microwave emitting diodes, formally at least, have much in common with FELs and their relatives, the commonality stemming from the crucial feature of their operation, namely the velocity dependence of the mass of the carriers. In the free-space devices, this dependence arises from relativistic effects—in the solid-state devices it arises from the non-parabolicity of the bands. Gunn diodes have been used as routine microwave sources for some years [127, 128]. They give typically 1 Watt at 10 GHz down to 10 mW at 100 GHz. A relatively new device, based on *IM*Pact *A*valanche *T*ransit *T*ime effects and therefore known acronymously as the IMPATT diode offers the prospect of much higher powers at the higher millimetre-wave frequencies, thus IMPATTs can give 100 mW at 100 GHz and 10 mW at 300 GHz. Furthermore their characteristics are reasonably non-linear so measureable power is available at the harmonics which stretch well into the far infrared. Already IMPATTs have been used for dielectric measurements and spectroscopy in the region above 100 GHz and there is a good prospect that they may serve as the local oscillators for superheterodyne systems working above 200 GHz.

The physics of IMPATTs has been extensively discussed by Kuno [129]. The device is basically a simple p-n junction (section 6.5) reverse biassed to avalanche breakdown. The avalanche current is produced by knock-on ionisation and one can analyse it into its Fourier components which cover a very wide frequency band essentially continuously. The equivalent circuit of the whole device is made up of a resistance, a capacitance and an inductance in parallel. Kuno shows [129] that the resistive component becomes negative for frequencies in excess of the avalanche frequency given by

$$v_a = \frac{1}{2\pi} \left(\frac{2\alpha' J_0 v}{\varepsilon_0 \varepsilon_r} \right)^{\frac{1}{2}} \tag{2.5.13}$$

where α' is the number of ions produced per unit length per unit electric field strength, J_0 is the current density through the diode, v is the limiting carier velocity in the drift region which follows the avalanche region and ε_r is the relative permittivity of the material. Rough values for the parameters appearing in (2.5.13) are $\alpha' \approx 10^{-3}$, $J_0 \approx 10^8$ A m^{-2}, $v \approx 10^7$ m s^{-1} and $\varepsilon_r \approx 16$ and oscillation would therefore be expected for frequencies in excess of 20 GHz. There are two basic types of diode actually used for millimetre-wave work, p$^+$-n-n$^+$ called single-drift and p$^+$-p-n-n$^+$ called double drift. The double drift variety can be regarded as two single-drift diodes in series giving, therefore, twice the output power for the same current density. To get the best millimetre-wave performance, the doped layers should have very sharp transitions and the whole diode structure should cover a distance of less than a micrometre. To achieve this specification calls for the full resources of modern semiconductor device fabrication technology. The diode is, in fact, made by photolithographic and chemical etching techniques and is usually constructed with an integral heat sink commonly of gold but for the ultimate performance this may be made of better conductors such as silver, copper or even diamond. The diodes, together with their integral heat sinks, are very tiny: the diameter is typically 0·025 mm and when one bears in mind, as mentioned above, that the active layer may be only tenths of a micrometre, it will be realised that the power densities involved are very high ($\sim 10^{14}$ W m^{-3}). The diode is usually mounted in a cavity resonant at the frequency of interest so that only this frequency and its harmonics may be sustained. The driving waveform may be shaped to emphasise a particular harmonic and so one can partially compensate for the fall-off in power with harmonic number. Because the

IMPATT is in a cavity it can be phase-locked and so provide a useful way of introducing time-frequency calibrations into the infrared (section 7.6). So far only homojunction IMPATTs have been made and tested but there are good grounds for believing that heterojunction IMPATTs, for example Ge: Ga As, will give both higher power output and higher-frequency operation.

A close relative of the IMPATT is the TUNNETT [130] in which the mechanism for electrons to cross the biassed diode is quantum mechanical tunnelling. Physically the difference between the two is that whereas for an IMPATT the diode region might have a width of 200 nm, in a TUNNETT this width would be no more than 50 nm. The junction always represents capacitance and this in its turn implies high-frequency limitation. The TUNNETTs in which the junction is narrower can therefore work to higher frequency (say 300 GHz) but the efficiency is less. These two diodes and possibly some of the other variants such as the BARRITT [131] present interesting prospects for infrared work but for the immediate future their principal use will continue to be as invaluable coherent sources for the difficult near-millimetre wave spectral region stretching from 120 GHz to 300 GHz where there is a marked shortage of rival tunable coherent sources.

2.6 The absorption of infrared radiation

2.6.1 *Elementary theory in the weak signal limit*

To deal rigorously with the interaction between an electromagnetic wave and the absorbing medium, through which it is propagating, needs the full apparatus of quantum field theory, but since this is a formidably difficult branch of theoretical physics, it is seldom invoked in practice. Rather, one sets up a hierarchy of approximations each more general and correspondingly more difficult than its predecessor. Quantum field theory would then be the last resort to be used only when all else fails.

The simplest approach is the elementary classical theory which applies when the field intensity is so low that it has virtually no effect on the populations of the various states involved. This situation would arise, for example, with a very weak monochromatic beam or with virtually any broad-band source since with the latter the fraction of the total power lying within the absorption bandwidth would usually be negligible. One is therefore working in the weak signal limit. The next level of sophistication is the semi-classical theory which is capable of dealing with the situation where the field intensity within the bandwidth is high enough to produce significant population changes. The final stage of currently used theory, still strictly speaking semi-classical since the field is not quantised, considers the situation where the field intensity and coherence are both so high that the molecular energy levels themselves become perturbed.

In the elementary theory, a plane wave propagating through an absorbing medium (see Fig. 2.15) is considered and the very plausible assumption is made that what happens inside the differential increment dz cannot depend in any way on what has happened to the wave before reaching this point. In other words the attenuation in the differential increment depends only on the material properties of the increment and on the intensity incident upon it. The resulting theory is given in section 2.2.4 where it is shown that the variation of intensity I with depth of penetration into the medium z is given by

$$I = I_0 \exp(-\alpha z), \qquad (2.6.1)$$

where α is the absorption coefficient. This relation is usually known as Lambert's law. It is sometimes given in its differential form

$$\frac{dI}{dz} = -\alpha I \qquad (2.6.1a)$$

and both forms have been quoted and used several times in what has gone before. The practical application of Lambert's law is limited by the unavoidable presence of interfaces so that the observed signal I is not simply related to the incident signal I_0 because of the reflection losses and multiple beam interference effects. These are analysed in more detail in the next chapter. As mentioned in Chapter 1, the units usually quoted for α are "neper cm^{-1}".

If the medium is a pure liquid or a well-defined solid, α is a material constant and that is all there is to it. However, if one is considering, as one frequently is, a solution of an absorbing solute in a much more transparent solvent, α will vary with the concentration and we need some other quantity to define the system. The usual assumption is that α varies linearly with the concentration of the absorbing species—this is called *Beer's law*. The observed absorption coefficient divided by the concentration should therefore be a constant—the *extinction coefficient* usually denoted by ε. Experimentally it is found that a plot of α divided by concentration versus concentration is seldom a horizontal straight line but that it usually does approach this condition for low enough values of the concentration. The extinction coefficient has therefore to be regarded as a single molecule property which is obtained by extrapolating data, obtained at finite dilution, down to the limit of infinite dilution. The departures from Beer's law observed at high concentrations would be interpreted in this light, as evidence for interaction between pairs and ultimately higher groups of solute molecules. It is important to point out, however, that the limited success of Beer's law does not rule out infrared spectroscopy as a quantitative tool for determining concentrations in a mixture. All one needs to do is to plot absorbence versus concentration, for a given wavelength, and thus set up a calibration curve. These calibration curves unfortunately vary from solvent to solvent and indeed the infinite dilution extinction coefficients themselves are slightly solvent dependent. This does introduce difficulties into attempts to use infrared intensity measurements as a probe for investigating the electronic structure of molecules but, on the other hand, one might think of inverting the argument and try using these measurements as a direct probe of intermolecular interactions. Unfortunately, however, the results so far seem to show that this is a far more complex field than one might imagine at first sight and unambiguous interpretations seem hard to come by. From the purely practical vantage point it is important to note that trading concentration for path length with a view to keeping the transmission of the cell at its optimum value (section 2.2.2) can lead if one is not careful to some pitfalls in those cases which depart significantly from Beer's law behaviour.

2.6.2 *Gas-phase line-shape theory*

The lines which are observed in the absorption spectrum of a gas will be sharp but not infinitely sharp that is they will not be delta spikes. Their finite widths arise from the presence of three simultaneous effects. These are radiative damping, collisional broadening and Doppler broadening. The radiative damping arises because at least one of the two levels involved must necessarily have a finite lifetime and the energy of that level will, by the Uncertainty Principle, become indeterminate to the extent

$$\Delta \mathscr{E} \approx h/\tau, \tag{2.6.2}$$

where τ is the lifetime. The line will hence have a width

$$\Delta v \approx 1/\tau. \tag{2.6.3}$$

A very similar result comes from the classical argument that the wavetrain must be of finite length if it is to represent a finite energy flux, i.e. a quantum. The quantity τ is the natural lifetime of the upper state (usually) and by definition it is therefore the reciprocal of the Einstein A coefficient. The natural line width Δv is hence numerically equal to A. For this reason it is of some theoretical interest but in all experimental infrared and microwave spectroscopy it is not accessible to measurement because it is invariably much smaller than the other two contributions. Thus, for example, a pure rotational microwave transition at 10 GHz might have an A value of 10^{-8} Hz and this would be many orders of magnitude below the pressure broadening (~ 1 MHz) or the Doppler broadening (~ 30 kHz). Even at 1 μm where the natural line width will have gained enormously in comparison because of the cubic dependence on the frequency (equation 2.2.17), it will be only 300 kHz and this will be still much less than the Doppler width (~ 500 MHz). One can conclude, therefore, that natural or radiative broadening may safely be neglected in discussions of infrared spectra but of course this will not be true for spectroscopy at much higher frequencies. In high-energy particle physics, for example, the widths of the observed "resonances" are entirely determined by the lifetime of the "particle" involved.

Collisional broadening arises because the monochromatic wave-train from the emitter, or its equivalent for the absorber, is interrupted by a molecular collision. The Fourier transform of the wave train, i.e. the spectrum, will not now be a delta spike and the line will have a finite width and occupy a significant range of frequencies. This argument when translated into quantum mechanical terms gives explicit results for the line-shape functions. Some details of the alternative approach using dielectric response theory are given in Appendix 4. The important result which emerges from these treatments is that the line-shape is homogeneous—that is all the molecules absorb at all the frequencies where the profile has finite values. The line-shape itself gives, unfortunately, very little information about the nature of the molecular collision. The best one can do is to divide collisions into the two categories, weak and strong but what actually happens during either kind of collison is not forthcoming. The study of intermolecular interactions (including "collisions") has, however, taken a major step forward with the development of laser/microwave double resonance spectroscopy [132] which has shown how the exchange of quantised energy between molecules can take place through the weak but temporally extended torque applied when one of the pair passes by on a sort of "orbit" several Van der Waals' radii removed. This research has shown that head-on collisions are very rare and that our previous ideas of what constituted a "collision" were somewhat wrong.

Turning now to the explicit line-shape functions, the two most popular are the Lorentz [133],

$$\alpha(\tilde{v}) = \frac{2\pi\Delta N}{3hc\varepsilon_0} |\mu_{ij}|^2 \left[\frac{\tilde{v}\Delta\tilde{v}}{(\Delta\tilde{v})^2 + (\tilde{v}_0 - \tilde{v})^2} - \frac{\tilde{v}\Delta\tilde{v}}{(\Delta\tilde{v})^2 + (\tilde{v}_0 + \tilde{v})^2} \right], \quad (2.6.4)$$

and the Van Vleck–Weisskopf [134],

$$\alpha(\tilde{v}) = \frac{2\pi\Delta N\tilde{v}}{3hc\varepsilon_0 \tilde{v}_0} |\mu_{ij}|^2 \left[\frac{\tilde{v}\Delta\tilde{v}}{(\Delta\tilde{v})^2 + (\tilde{v}_0 - \tilde{v})^2} + \frac{\tilde{v}\Delta\tilde{v}}{(\Delta\tilde{v})^2 + (\tilde{v}_0 + \tilde{v})^2} \right], \quad (2.6.5)$$

In these expressions, ΔN is the population difference between the two levels concerned, $\tilde{v}_0$ is the basic wavenumber of the oscillator, $|\mu_{ij}|^2$ is the dipole moment matrix element and $\Delta\tilde{v}$ the line-width parameter. The two line-shape functions can also be written in the alternative forms:

Lorentz $\qquad \alpha(\tilde{v}) = \frac{2\pi\Delta N}{3hc\varepsilon_0} |\mu_{ij}|^2 \left[\frac{4\tilde{v}_0 \Delta\tilde{v} \tilde{v}^2}{(\tilde{v}_0^2 + (\Delta\tilde{v})^2 - \tilde{v}^2)^2 + 4\tilde{v}^2 (\Delta\tilde{v})^2} \right], \quad (2.6.4a)$

Van–Vleck–Weisskopf

$$\alpha(\tilde{v}) = \frac{2\pi\Delta N}{3hc\varepsilon_0} |\mu_{ij}|^2 \left[\frac{2\tilde{v}(\tilde{v}_0^2 + \tilde{v}^2 + (\Delta\tilde{v})^2)}{\tilde{v}_0 \left[(\tilde{v}_0^2 + (\Delta\tilde{v})^2 - \tilde{v}^2)^2 + 4\tilde{v}^2 (\Delta\tilde{v})^2 \right]} \right]. \quad (2.6.5a)$$

The Lorentz is an attractive function since it is zero at both $\tilde{v} = 0$ and $\tilde{v} = \infty$ as it should be and its peak, given by

$$\tilde{v}_{max} = \sqrt{(\tilde{v}_0^2 + (\Delta\tilde{v})^2)}, \quad (2.6.6)$$

will lie close to $\tilde{\nu}_0$ unless the breadth of the line is large. Unfortunately it is theoretically unsound as pointed out by Brot [135] amongst others and should not be used if quantitative results are expected. However, because of its simplicity and ease of application, spectroscopists will probably continue to use it to fit all kinds of observed spectral line profiles despite its theoretical drawbacks. The Van Vleck–Weisskopf profile can likewise be criticised but in the particular case of collisional broadening in the gas phase it is much sounder since it derives from a model in which the distribution of energies after collisions is statistically Boltzmannian. Its peak absorption however, just like that for the Lorentz does not occur at $\tilde{\nu} = \tilde{\nu}_0$. Some care might be indicated, therefore, in the determination of line positions when the lines are pressure broadened, but in practice the shifts are found to be very small and they would be revealed only in the most careful work with the most sophisticated equipment. The Van Vleck–Weisskopf line-shape formula is widely used in microwave rotational spectroscopy and since at ambient temperature one then has $h\nu \ll kT$, the full expression can be somewhat simplified with the result

$$\alpha(\tilde{\nu}) = \frac{\pi Nhc\,\tilde{\nu}_0\,\mu^2}{3\varepsilon_0\,(kT)^2} \left[\frac{\tilde{\nu}^2\Delta\tilde{\nu}}{(\Delta\tilde{\nu})^2 + (\tilde{\nu}_0 - \tilde{\nu})^2} + \frac{\tilde{\nu}^2\Delta\tilde{\nu}}{(\Delta\tilde{\nu})^2 + (\tilde{\nu}_0 + \tilde{\nu})^2} \right], \qquad (2.6.7)$$

where N is the total number of molecules present per unit volume and μ is the molecular dipole moment. In deriving this expression, the standard quantum mechanical result has been used:

$$|\mu_{ij}|^2 = \mu^2(J+1)/(2J+1) \text{ for } \Delta J = +1, \qquad (2.6.8)$$

Both the line-shape functions involve two terms, this stemming from the theoretical requirements, on the dielectric response function, that the response at $\tilde{\nu} = \tilde{\nu}_0$ be the same as at $\tilde{\nu} = -\tilde{\nu}_0$. However the term involving $(\tilde{\nu} + \tilde{\nu}_0)^2$ will be slowly varying over the region where the other is strongly peaked and it is common therefore as a purely *practical* matter to ignore it and to write, in the microwave limit,

$$\alpha(\tilde{\nu}) = \frac{\pi Nhc\,\tilde{\nu}_0\mu^2}{3\varepsilon_0(kT)^2} \frac{\tilde{\nu}^2\Delta\tilde{\nu}}{(\Delta\tilde{\nu})^2 + (\tilde{\nu}_0 - \tilde{\nu})^2}. \qquad (2.6.9)$$

This function has its peak at $\tilde{\nu} = \tilde{\nu}_0$ exactly. The pressure-broadening parameter $\Delta\tilde{\nu}$ is proportional to the pressure, and hence to N, so the peak absorption $\alpha_{\max}(\tilde{\nu}_0)$ is paradoxically constant and independent of gas pressure! This strange result is only partly due to the approximations involved in deriving (2.6.9). It need cause infrared spectroscopists, armed only with moderate resolution instruments, no alarm since their resolution interval will usually be considerably larger than the line width and the spectrometer will therefore "see" only the line area and since, this depends on the first power of $\Delta\tilde{\nu}$, it will likewise depend on the first power of the gas pressure. For coherent spectroscopy, however, in which the spectral width can be much less than the line-width it presents a difficulty. It will be seen in the next section how the phenomenon of saturation provides a resolution of the paradox.

If one writes the pressure-broadening parameter Δv in the form

$$\Delta v = \sigma p, \qquad (2.6.10)$$

where p is the pressure, it is found that σ, the self-broadening parameter, lies typically between 1 and 10 MHz per torr. It is therefore possible to delineate the profile of microwave lines with gas pressures of a few torr or so. In this way concentrations can be measured [136] and dipole moments determined. However for the delay-type spectrometers (section 6.6) which most infrared spectroscopists use and which have resolution intervals of 3 GHz or larger, this is not possible and the line will only occupy a small fraction of the scanning area. Under these circumstances the measurement of line intensity by direct means becomes impossible since for path lengths sufficient to produce a measurable absorption all the central part of the line will be essentially black. A way round this difficulty was introduced by Wilson and Wells [55] who developed the technique of foreign gas broadening in which a high (≈ 200 torr) pressure of a foreign transparent gas is introduced. This broadens the line of the absorbing gas and the peak height is no longer invariant. It is thus possible to produce conditions in which the line is made comparable in width to the scanning "slit-width" and of a height which lies within the dynamic range of the spectrometer and meaningful intensities may be deduced. In passing it should be noted that foreign gas broadening parameters are about an order of magnitude smaller than the self-broadening parameters, so a fair pressure of foreign gas is required.

The third cause of finite line widths lies in the random and chaotic motion of the molecules in a gaseous sample. The Doppler effect will shift the observed absorption frequencies according to the resolved component of the molecular velocity along the line of sight and we will therefore get a line profile which reflects the Boltzmann distribution of molecular velocities. It is most important to realise here that absorption within the homogeneous line width at a given frequency is due to a specific set of molecules, all of which have the same resolved velocity component along the beam direction. All other sets have apparent resonant frequencies which are different. We have therefore an *inhomogeneous* line shape. Inhomogeneous line shapes are mostly observed in gas-phase spectroscopy but examples occur for the other phases of matter. A well-known case occurs in solid-state spectroscopy where one has absorbing entities dispersed at random in an otherwise inert matrix. The absorbers will interact with one another and this interaction will induce dipole–dipole broadening. Clearly one again has a large number of possible configurations and the observed line-shape will be the result of the addition of very many independent and much narrower components.

The explicit form for a Doppler broadened line is

$$\alpha = \alpha_{max} \exp \left[\frac{-mc^2}{2kT} \frac{(v_0 - v)^2}{v_0^2} \right], \tag{2.6.11}$$

from which the half-width (FWHM = Full Width at Half Maximum) is

$$\Delta v = 2v_0 \sqrt{[2kT(\ln 2/mc^2)]}, \tag{2.6.12}$$

which can also be written

$$\Delta v = v_0 \, T^{\frac{1}{2}} \, M^{-\frac{1}{2}} \, 7{\cdot}162\,303 \times 10^{-7}, \tag{2.6.13}$$

where M is the relative molecular weight. For room temperature one has

$$\Delta v = v_0 M^{-\frac{1}{2}} \, 1{\cdot}240\,55 \times 10^{-5}. \tag{2.6.13a}$$

Taking a specific example, HF ($M = 20$), and a rotational line at $20\,\text{cm}^{-1}$ ($v_0 = 600\,\text{GHz}$), one has $\Delta v = 1{\cdot}7\,\text{MHz}$, so the Doppler broadening would be negligible unless the pressure was much less than a torr. At microwave frequencies, the effects of Doppler broadening are usually unobservable. In the middle infrared and at all higher frequencies, on the other hand, it becomes the dominant line-shape factor except for those rare cases where high pressures are being used. At 10 μm the Doppler width for a moderately light molecule (say $M = 40$) becomes 60 MHz and this represents a major limitation to high resolution infrared spectroscopy since instrumental line widths less than this are now available. Ways of achieving "sub-Doppler" spectroscopy will be discussed in the next section and again in Chapter 4.

In the submillimetre region, and occasionally in other regions under exceptional circumstances, the situation can be found where the Doppler and the pressure broadened line widths become comparable. In this case the observed line profile will be the convolution of the two—the so-called Voight line-shape. Unfortunately this cannot be expressed in closed form in terms of elementary functions so when it is used it has to be generated in a computer. Fortunately the cost and accessibility of computers are both becoming more reasonable and the Voight profile is finding increasing use.

The discussion of line-broadening so far has been in essentially classical terms but, since the underlying physics is very profound, one would not expect dramatic changes in adopting a quantum formalism. This is indeed found and the equations just given can be used with confidence. The only significant example where quantum effects appear occurs when atoms (especially) or molecules (occasionally) are being studied at long wave-lengths under conditions where the Doppler and pressure broadening are of the same order. Then, and especially if rare gases such as argon are being used to produce the broadening, line-narrowing may occasionally be observed. This phenomenon is called "Dicke-narrowing" after R. H. Dicke who analysed it in detail in 1953 long before there was any real spectroscopic chance of experimental observation for molecules. The basic cause of Dicke-narrowing is that the absorbing entity, atom or molecule, tends to survive in a given quantum state, and thus interact continuously with the probing field, despite the frequent collisions, whilst its velocity undergoes very many random changes. The Doppler shift is then determined not by the free-flight velocity between collisions, which is high, but by the average, or drift, velocity which is small. Under these conditions and provided that the wavelength is large compared with the mean-free-path, so that the electric field spatial variations can be neglected, a reduction of the Doppler contribution to the line width will be observed. The phenomenon is relatively easy to observe with atoms which tend to stay in a given quantum state for a long time, but is much more difficult with molecules because of the rapid rotational interchanges. Molecular examples therefore tend to be observed not in the far-infrared and microwave region but in the mid infrared where, because the Doppler component is larger, the narrowing is more conspicuous. The development of several kinds of very high resolution laser spectroscopy has permitted the observations of Dicke-narrowing for rotational lines in the fundamental vibration bands of the hydrogen halides broadened by argon. Typical experimental conditions would involve a mixture with partial pressures 1 torr of HCl plus 250 torr of A. A good description plus further references has been given by T. Oka in an article (pp. 151–167) in the book "Horizons of Quantum Chemistry", published by D. Reidel in 1980.

2.6.3 *Saturation of absorption lines*

When the radiant field, at the resonant frequency, becomes intense, we are no longer in the weak signal limit and now the spontaneous transitions become negligible compared to the stimulated ones. Under these conditions the rate of upward transitions may approximately equal that of the downward ones and the populations of the upper and lower states become approximately equal. The system will no longer absorb linearly and we have *power saturation*. The effect physically can readily be visualised for since the absorption at line centre will be depressed relative to that in the wings, the line half-width will be observed to increase. The phenomenon can be discussed at various levels and in essence will be taken up again in the next section but, at the simplest, one has the following expression [137] for the power broadened line-profile:

$$\alpha(\tilde{v}) = \frac{2\pi\Delta N\tilde{v}}{3hc\,\tilde{v}_0\varepsilon_0}\,|\mu_{ij}|^2 \left[\frac{\tilde{v}\Delta\tilde{v}}{(\Delta\tilde{v})^2 + (\tilde{v}_0 - \tilde{v})^2 + |\mu_{ij}|^2 I_0/12h^2c^3\varepsilon_0}\right]. \quad (2.6.14)$$

Where ΔN is the *equilibrium* population difference and, as usual, the term involving $(\tilde{v}_0 + \tilde{v})$ has been neglected. I_0 represents the intensity of the monochromatic radiation of frequency v. The concept of monochromatic intensity does, of course, present some difficulties but it can be taken to be the intensity in a bandwidth small compared to Δv. In the microwave limit and for the special cases of the linear or the symmetric rotor (section 5.2.2), equation (2.6.14) can be simplified to read,

$$\alpha(\tilde{v}) = \frac{\pi h N c \tilde{v}_0 \mu^2}{3\varepsilon_0 (kT)^2} \left[\frac{\tilde{v}^2 \Delta \tilde{v}}{(\Delta \tilde{v})^2 + (\tilde{v}_0 - \tilde{v})^2 + I_0 \mu^2 \tilde{v}_0 / 24B(2J+1)h^2 c^3 \varepsilon_0} \right], \quad (2.6.15)$$

where B is the rotational constant (equation 5.2.2) in wavenumber units. From either (2.6.14) or (2.6.15) two important conclusions emerge at once. Firstly, at peak, i.e. when $\tilde{v} = \tilde{v}_0$, the absorption will be less than expected and the width will be greater. Secondly if one imagines $\tilde{v}$ to be fixed at the resonant condition (i.e. $\tilde{v} = \tilde{v}_0$) and $\Delta \tilde{v}$ allowed to approach zero—in other words one is considering ultra-low pressure for the gas—then the denominator in (2.6.15) will approach a constant value but the numerator will vary as $(\Delta \tilde{v})^2$, that is as p^2. This disposes of the paradox mentioned earlier that the absorption at peak is entirely independent of gas pressure and that one would therefore have a finite absorption with no gas at all in the cell! The point where saturation becomes manifest, i.e. the region of the transition from a pressure independent to a pressure dependent α_{max}, is given by the condition

$$I_0 \mu^2 \tilde{v}_0 / 24B(2J+1)h^2 c^3 \varepsilon_0 \approx (\Delta \tilde{v})^2.$$

The probe power necessary to induce saturation effects therefore depends on the square of the pressure and inversely on the square of the dipole moment. In microwave spectroscopy, where $\Delta \tilde{v}$ can be very small because low pressures are used and where I_0 can be large because cavity techniques are used and because the sources, klystrons for example, can be intense, saturation is commonplace and is the usual cause of loss of resolution. In infrared spectroscopy with black-body sources, on the other hand, it hardly ever happens.

When one comes to consider infrared spectroscopy with laser sources, however, the fields may once again be high enough to bring about noticeable saturation. In this case though, one can no longer ignore Doppler broadening and, since the line-profile is then inhomogeneous, the effects of the saturation may not be immediately obvious if the Doppler width is much greater than the homogeneous width. What is happening is that "holes" are being burnt in the population distribution as discussed in section 2.4.2. It follows that if the probing beam is reflected back along its track by means of a plane mirror, then at line centre the same group of molecules will be saturated both ways and there will appear to be a sudden rise in the observed transmission of the cell. This is entirely analogous to the phenomenon which gives Lamb dips in laser emission [69], the only difference being that at the moment we are not considering the absorption cell to be inside the laser cavity. If one has *two* Doppler broadened lines with their centres separated by less than the line width they will not normally be resolved as the laser scans over them, but if the power is high enough, two inverse Lamb dips will appear (Fig. 2.47) and the line centre frequencies may be readily determined. The widths of the inverse Lamb dips are determined by the homogeneous broadening so in principle one can have in this way microwave resolving power transferred to the infrared. In passing it should be mentioned that in multiple-photon-spectroscopy where several photons add together to span a molecular or atomic energy level separation, Lamb dips can be observed for sets of molecules whose velocity components along the tube axis are *not* zero. These have been discussed in a comprehensive review by Jones [138] to which the interested reader is referred.

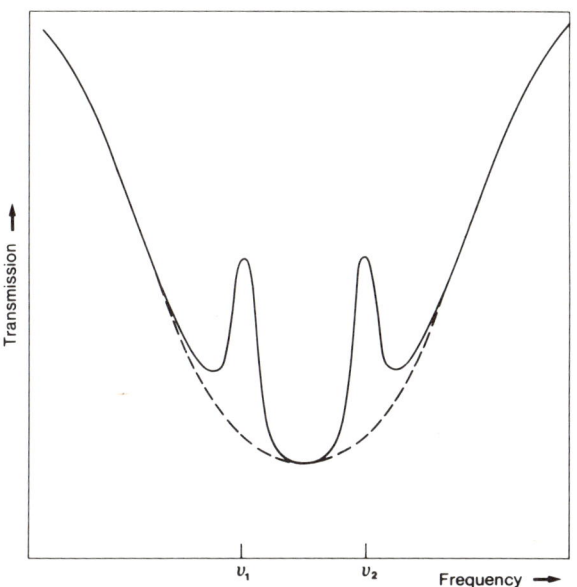

FIG. 2.47. Two equal amplitude gaussian lines separated by an amount equal to their broadening parameter would not ordinarily be resolved (dotted curve), but when the gas is probed with high-power tunable-laser radiation travelling in opposite directions through the gas, two inverse Lamb dips would appear whose positions would reveal clearly the line-centre frequencies v_1 and v_2.

Saturation is primarily of interest in gas-phase spectroscopy but it does occur also in condensed phases if the energy is being pumped in at a faster rate than it can be relaxed away. To describe the situation, a simple phenomenological theory is adequate and one generalises equation (2.6.1a) to read

$$\frac{dI}{dl} = -\alpha I + \beta I^2 + \gamma I^3 + \text{etc.,} \qquad (2.6.16)$$

where $\alpha \gg \beta \gg \gamma$. This equation is readily solved since the variables separate. Thus, taking the first two terms only, one has

$$\frac{I}{1 - (\beta/\alpha)I} = \frac{I_0}{1 - (\beta/\alpha)I_0} \exp(-\alpha l). \qquad (2.6.17)$$

It will be noted that as I_0 increases, I tends to become equal to I_0 and not to depend strongly on l. This argument cannot be taken to the limit of (2.6.17) when the denominators vanish because from physical arguments one would then need to include higher terms from (2.6.16), but it can be shown, nevertheless, that the point is general. One finds therefore, in the modern jargon, "bleaching" of the saturable absorber. The use of saturable absorbers to passively Q-switch lasers was mentioned in section 2.4.2. Dyestuffs are widely used for this purpose in the visible region but usually they tend to be far less absorbing as soon as the infrared is entered but there are some dyes which can be used to Q-switch the 1·06 μm Nd^{3+} laser.

2.6.4 The interaction of molecules with high-power, high-coherence, infrared radiation

When the power of a coherent beam of radiation, resonant with a set of elementary oscillators, is high, new effects are observed. Several of these have formal classical equivalents [139] but one gets a more integrated picture by discussing all the phenomena purely in quantum-mechanical terms. The first point is that it is no longer valid to use the zero-field solutions for the molecular wave functions, since the field is now high enough to perturb the molecules and it must be included, therefore, in the Hamiltonian. There is no possibility, of course, of an analytical solution of the resulting wave equation and one has, as usual, recourse to a perturbation method in which the time-dependent wave functions are developed as linear combinations of the original zero-field functions used as an orthonormal basis set. In many cases one is considering a transition between two fairly isolated levels and then a further approximation is invoked in which one considers *only* those two levels in the expansion and writes for the time dependent wave function

$$\Psi(t) = \Psi_1^0(t) \cos \omega_0 t + i\Psi_2^0(t) \sin \omega_0 t, \qquad (2.6.18)$$

where $\Psi_1^0(t)$ and $\Psi_2^0(t)$ are the "stationary-state" but time-dependent wave functions of the two states involved, thus for example

$$\Psi_1^0(t) = \Psi_1(r) \exp(-i\mathscr{E}_1 t/h). \qquad (2.6.19)$$

The angular frequency ω_0 which appears in (2.6.18) is related to the so-called Rabi frequency,

$$\nu_R = \omega_0/\pi = |\mu_{12}| E_{12}/h, \qquad (2.6.20)$$

where E_{12} is the electromagnetic field strength. Physically what is happening is that the intense electromagnetic field is forcing the molecules to transform from one pure state to the other with a regular periodicity given by half the Rabi frequency [140]. In the absence of any collisional processes, no photons are being absorbed, the system is just nutating between its two extreme positions without remaining in either. One can say that virtual photons absorbed or emitted during the 0–π phase are compensatingly being re-emitted or re-absorbed during the π–2π phase. Furthermore, and most importantly, because the molecules are simultaneously in resonance with the beam, the whole ensemble develops a macroscopic polarisation which nutates synchronously with the field. If the laser beam is not exactly in resonance with the molecules, i.e. if

$$hv \neq \mathscr{E}_2 - \mathscr{E}_1, \tag{2.6.21}$$

then the nutation still occurs with a frequency $(v^2 + \Delta v^2)^{1/2}$, where Δv is the frequency offset but the amplitude of the macroscopic polarisation rapidly falls as Δv increases.

This off-resonant situation has been analysed by several workers to varying degrees of approximation, thus Javan's original treatment [141], essentially to first order, is still widely used to analyse laser pumping and simple double resonance experiments but more sophisticated treatments [142] are necessary to deal with some of the complex triple-resonance experiments now being carried out [143]. These treatments give the amplitude of the interaction cross-sections but the basic physics of the situation can be analysed in much simpler terms. Thus taking the real part of (2.6.18) and substituting form (2.6.19) gives

$$\Psi(t) = \tfrac{1}{2}\Psi(r)\left\{ \exp\left[\frac{-it}{h}(\mathscr{E}_1 - \hbar\omega_0) \right] \right.$$
$$\left. + \exp\left[\frac{-it}{h}(\mathscr{E}_1 + \hbar\omega_0) \right] \right\}, \tag{2.6.22}$$

and the system now has *two* resonant frequencies given by

$$hv = \mathscr{E}_2 - \mathscr{E}_1 \pm \tfrac{1}{2}hv_R, \tag{2.6.23}$$

and is transparent at the original resonant frequency given by (2.6.21). This phenomenon is called *coherence splitting*, the two resonant frequencies being separated by the Rabi frequency v_R. Coherence splitting is not usually directly observable with a single probing beam but in double resonance experiments, and especially microwave-microwave double resonance experiments, it is often very obvious. The simple system analysed as equation (2.6.22) would correspond to a three-level system with a zero dipole moment matrix element connecting $|1>$ and $|3>$. Then pumping strongly the transition $1 \to 2$ and observing the transition $2 \to 3$ with a weak probe will show the doublet structure characteristic of the coherency splitting of state 2. An example taken from the comprehensive review by Baker [144] is given in Fig. 2.48. In the case of infrared–microwave or infrared–infrared double resonance, the heterogeneous Doppler broadening which is usually present complicates matters considerably. It is usually best to do double-resonance experiments in the infrared using a Lamb-dip regime. An exhaustive account of this topic has been given by Jones [138].

In microwave double resonance experiments one has tunable sources (section 2.6) so exact resonance with the transition to be pumped can be ensured but in double resonance experiments using infrared lasers this is seldom possible. One is always dealing, therefore, with off resonant pumping. The Javan theory gives a good account here but the phenomena can also be analysed in terms of a dynamic, high-frequency or AC Stark effect [145]. If the applied laser field is strong enough it can split the nearby non-resonant line

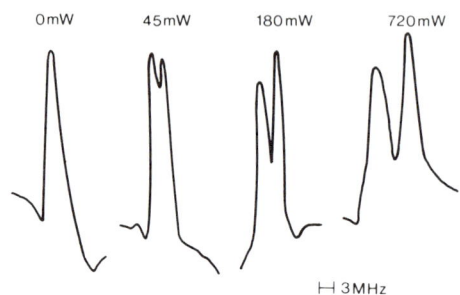

FIG. 2.48. Coherence splitting in the microwave spectrum of the ground torsional state of CH_3NO_2 at increasing pump power. The pumped transition is one $|M| = 1$ Stark component of the J (1 → 2), K (1 → 1) transition at 32·034 GHz whilst the probe is being tracked through the same Stark component of the J (2 → 3), K (1 → 1) transition at 51·979 GHz. (After J. G. Baker [144].)

into its Stark components and a component oscillating back and forth across the laser frequency will lead to observed absorptions of the pump power. At high enough pump powers, very non-linear processes such as stimulated resonant Raman scattering may be observed. These various processes are certainly involved in the operation of some optically pumped pulsed lasers where the frequency mismatch may be as high as 1 GHz.

The theory of the Rabi oscillations was originally worked out for the case of nuclear magnetic resonance or NMR where intense radio frequency tunable power is readily available and the two-level spin system provides an ideal simple case. Even here, however, relaxation effects are encountered and energy is in fact absorbed from the field. Two relaxation times are defined T_1 and T_2 which in the NMR case are the population decay and the ensemble dephasing times. To study these relaxations in NMR spectroscopy requires intense very brief pulses to make all the spins in the system precess coherently and then one can study the time evolution of the system either by following its decay—free induction decay—or else by using rather sophisticated sequences of pulses. This latter approach leads to a large number of very interesting phenomena, spin-echoes, spin-nutation, quantum beats etc. The theoretical formalism for describing all these was developed in a very elegant manner by Bloch [146].

The interest of all this for optical workers is that with the development of intense coherent sources for the optical region it is now possible to produce the optical analogues of all the spin phenomena. The situation is usually rather more complicated in that one seldom has an isolated two-level transition, so population is not conserved and one has more than two relaxation times but, that said, the remaining analogies are perfect. The theoretical basis is the coupling of the Schrodinger–Maxwell–Bloch equations and is associated with many workers starting with Feynman and his colleagues. The most significant theoretical work, however, came from McCall and Hahn [147] and the most elegant experimental work has come from Brewer and his colleagues [148]. The result of the theoretical treatment is that one can formally define a vector $S(t)$ which is a pseudo-spin vector but which, unlike the NMR case, does not have a simple physical interpretation. However, using this vector one can define a rotating coordinate system and apply all the analogies of rotating frame analysis in NMR Pulses applied to the system can then be defined in terms of the phase angles which they are equivalent to in the rotating frame. A π pulse will invert the population whereas a 2π pulse will take the system all the way round and back to where it started from. In quantitative terms, one defines a pulse "area" or integral A given by

$$A(t, z) = \frac{2 \langle \mu \rangle}{h} \int\limits_{-\infty}^{t} E(t', z) \, dt', \qquad (2.6.24)$$

and then carries through the analysis in terms of this dimensionless quantity. If long and therefore weak pulses are considered, then the answers which emerge are identical to the ordinary quantum (or for that matter the simple classical) treatment and the normal weak field dielectrics are recovered. However, if very intense pulses with life times shorter than the relaxation times are considered, very different effects are manifest. The equations cannot be solved absolutely analytically but, with some acceptable approximations and assuming inhomogeneous broadening, one finds,

$$\frac{\partial A}{\partial z} = -\tfrac{1}{2} \alpha \sin A, \qquad (2.6.25)$$

which is the equivalent of Lambert's law under this regime. It follows at once that if A is an integral number of multiples of π there will be no dependence on depth of penetration into the medium. It can further be shown that some special pulse shapes can be propagated unchanged, from which it follows that there is no energy absorption—"Self-induced transparency" has been achieved. If A is not an exact multiple of π, the pulse shape *will* alter so that the pulse can evolve into the nearest even multiple of π. This is the famous "Area Theorem" first derived by McCall and Hahn in 1967 [147]. This phenomenon can be used to make a passive pulse compressor and amplifier. Thus Gibbs and Slusher [149] applied a 3·5 π pulse from a ^{202}Hg II laser to ^{87}Rb atoms placed in a magnetic field and showed both experimentally and theoretically that the pulse was considerably shortened whilst compensatingly being intensified. The

numerical solutions of the coupled field equations show that if one applies a
pulse whose area is greater than π but of arbitrary shape, then, after
propagating a few times the low field optical depth, the pulse envelope will
transform into the type

$$E(t') = \frac{h}{\tau \langle \mu \rangle} \, \text{sech} \left[\frac{t'}{\tau} \right] \qquad (2.6.26)$$

where τ is the original pulse length, i.e. the time required for the pulse to pass a
given point, and t' is the local time, i.e. $t' = t - z/v$. It follows at once by
integrating this expression from $-\infty$ to $+\infty$ that the pulse area is indeed 2π.
Of course one must always bear in mind the fundamental restriction that one is
necessarily considering times short in comparison with the relaxation time, but
since this latter will usually be of the order 10 ns it follows that one would need
to be considering propagation in specimens thicker than 3 m before the
treatment would become inappropriate. For laboratory situations, where the
specimen thickness would be measured in cm, there is seldom any difficulty. A
very interesting result follows, however, since another consequence of the
equations gives the simple relation

$$v = \frac{2c}{2 + \alpha c \tau} \qquad (2.6.27)$$

between the pulse velocity v, the small field absorbtion coefficient α and the
pulse length τ. Since, as mentioned above, $c\tau$ will be of the order of metres and
α will be at least a few cm^{-1}, the pulse velocity will be much reduced. Here one
is usually considering gaseous samples hence the appearance of c for the
continuous-wave propagation velocity. For condensed specimens, the normal
phase velocity can be used to first order (i.e. c/n) to replace c in equation
(2.6.26), but at high field strengths and very short pulse lengths, new
phenomena appear since, by the Kramers–Kronig relations, changes in α will
necessarily imply changes in n.

Experimentally, present day Q-switched lasers are only powerful enough to
generate 2π pulses but the higher members of the family have been investigated
by Lamb [150] who has shown how these 4π, 6π etc. pulses tend to break up
into 2, 3 etc. 2π pulses travelling with different velocities. To give the discussion
some quantitative form it should be noted that for a transition dipole moment
of the order 10^{-30} coulomb metre, the Rabi frequency v_R would be given by

$$v_R \, (\text{MHz}) = 4 \cdot 2 \sqrt{I} \, (\text{W cm}^{-2}), \qquad (2.6.28)$$

so that a powerful laser delivering 57 kW into an area of 1 cm^2 would give a
Rabi frequency of 1 GHz. If, for simplicity, one thinks of an initial pulse which
has a rectangular envelope, then from (2.6.24) it follows that the pulse length τ
would have to be 1 ns to achieve the 2π condition. Lasers with such a
specification can just about be produced. There is no difficulty, of course, in
making even more powerful lasers and of creating picosecond pulses, but for

the present application it is necessary that the laser be tunable so that the resonant condition can be ensured. Nevertheless it can be assumed that tunable high-power picosecond pulse lasers will eventually be available and then some of the fascinating phenomena predicted by Lamb can be searched for experimentally.

The equations which emerge in this area are very simple and don't seem to involve quantum ideas directly, thus equations (2.6.25) and (2.6.27) do not involve h explicitly. Nevertheless it must be stressed that phenomena like self-induced transparency, (short coherent pulses travelling anomalously long distances through resonant absorbers at unexpectedly low velocities) arise entirely from the non-linear quantum theory of dipole oscillators. One can derive apparent classical equivalents: for example since a short pulse must have a spread of frequencies (see Appendix 5) it follows that it cannot be entirely resonant with the absorbers and must therefore show anomalously low absorption, but this is a quite different phenomenon. Nevertheless, the non-linear quantum theory must transform smoothly into the classical theory at low field intensities and this is illustrated nicely by equation (2.6.25) which when A becomes small transforms into Lambert's law since $\sin x \approx x$ for small x.

Chapter 3
The Propagation of Infrared Radiation

3.1 General introduction: basic concepts

For a rigorous treatment of the flow of radiant electromagnetic energy one should ideally set up the relevant Maxwell equations in full vector form and then solve them for the given boundary conditions. Unfortunately, this recipe is impossibly difficult to carry out analytically for all but the simplest systems, only idealised cases such as the diffraction of an infinite extent plane wave by a semi-infinite perfectly conducting plane having in fact been solved. Some progress can be made by abandoning the search for analytical solutions and by going over to numerical solutions in a high-speed digital computer, but this approach is rather more suited to the investigation of fundamental properties of the field itself than it is to discussion of simple practical problems. In these practical cases, one must work at various levels of approximation, but provided one is aware of the limitations of each approximation one can arrive at answers which are correct to any reasonable order.

The simplest approximation is the use of "scalar waves" [60]. This approximation is so commonly made that it tends to be forgotten that it is, in fact, an approximation. The full electromagnetic approach involves two fused and mutually orthogonal vector fields $\mathbf{E}$ and $\mathbf{H}$ and it is the time-variation of these which is sought. A very important quantity is the "Poynting vector",

$$\mathbf{S} = \mathbf{E} \times \mathbf{H} \qquad (3.1.1)$$

with a magnitude giving the instantaneous intensity and a direction which is that of the radiant energy flow. Now at infrared frequencies (several THz) and with conventional detecting systems, the field oscillations are much too rapid to be followed, so the concept of instantaneous intensity is not very meaningful. Also, nearly all the phenomena, absorption for example, which are

190

observed depend only on the electric component. One therefore neglects the vector nature of the field and introduces a purely scalar quantity E which, for the reasons explained in section 1.4.1, is usually written in explicitly complex form as $\hat{E}$. When necessary, the corresponding scalar quantity $\hat{H}$, for the magnetic field, can be calculated from the analogue of (1.2.14), namely

$$\hat{H} = \sqrt{(\varepsilon_r \varepsilon_0 / \mu_r \mu_0)} \, \hat{E}, \tag{3.1.2}$$

where ε_r and μ_r are respectively the relative permittivity and permeability of the medium in which the radiation is propagating. The observable quantity at infrared frequencies is the time-averaged intensity I, which is found from the analogue of equation (1.4.6), namely

$$I = (1/4)[\hat{E}\hat{H}^* + \hat{E}^*\hat{H}]. \tag{3.1.3a}$$

For propagation in free space, one finds, by invoking equation (1.2.11), the result

$$I = (1/2)\varepsilon_0 c E_0^2 \tag{3.1.3b}$$

where E_0 is the amplitude of $\hat{E}$.

3.1.1 Introduction to free-space propagation

The general wave-equation can be found by separating the field variables in the original Maxwell equations. The result [151] is

$$\nabla^2 \hat{E} = \varepsilon_r \varepsilon_0 \mu_r \mu_0 \frac{\partial^2 \hat{E}}{\partial t^2} + \mu_r \mu_0 \sigma \frac{\partial \hat{E}}{\partial t}, \tag{3.1.4}$$

where σ is the conductivity of the medium. The first term on the RHS of equation (3.1.4) represents essentially the displacement current which is responsible for propagation in dielectric media. The second term represents the conduction current which is responsible for propagation in media having a high conductivity, for example metals. It is convenient to discuss these two cases separately since it is rare to find a medium where the two effects are of the same order. Taking therefore a purely dielectric medium (that is $\sigma = 0$) one has, say for propagation in the z-direction in a non-magnetic medium, the simple result

$$\frac{\partial^2 \hat{E}}{\partial z^2} = \frac{\varepsilon_r}{c^2} \frac{\partial^2 \hat{E}}{\partial t^2}. \tag{3.1.5}$$

One looks for solutions of this equation appropriate to the particular situations being considered. If polarisation effects enter, these are tacked on in an *ad hoc* fashion. Various possibilities emerge but the simplest, the *plane wave*,

$$\hat{E} = E_0 \exp[2\pi i \tilde{v} (\sqrt{\varepsilon_r} \, z - ct)], \tag{3.1.6}$$

propagating with velocity $v = c/\sqrt{\varepsilon_r}$, in an unbounded medium, is of very common use. True plane waves and fully unbounded media are, of course, impossible but in practice good approximations can often be obtained and even when they cannot the observed wave can always be regarded as made up from its plane Fourier components.

Another simple and useful case is the spherical wave which can be obtained from (3.1.6) by replacing the cartesian displacement z by the polar distance r. Spherical waves are important from a theoretical point of view since they provide the basis for *Huyghen's construction*. In this, as is well known, all the points lying on an instantaneous wavefront are considered to act as secondary sources giving out spherical waves. Then, by invoking interference between all the members of this infinite set, one can explain non-reverse propagation, diffraction, reflection and in fact all the phenomena of physical optics. A good example, which arises several times in infrared physics, is provided by the treatment of the interaction of a non-absorbing (that is loss-less) component with a beam into which it has been inserted. The answer is derived by imagining the component to contribute additional fields which interact with the original field. Thus on the side remote from the source, one would have

$$\hat{t}\hat{E} = \hat{E} + \hat{s}\hat{E},$$

i.e.

$$\hat{t} = 1 + \hat{s}, \tag{3.1.7}$$

where $\hat{t}$ and $\hat{s}$ are respectively the complex amplitude transmission and scattering coefficients. Similarly on the near side one must have

$$\hat{r} = -\hat{s}. \tag{3.1.8}$$

By definition, the component introduces no loss so by energy conservation it follows that

$$|\hat{t}|^2 + |\hat{r}|^2 = 1 = |\hat{t}|^2 + |\hat{s}|^2. \tag{3.1.9}$$

Equations (3.1.7) and (3.1.9) represent a point which moves on a circle (see, for example, Figs 1.3 and 3.38) in the complex plane with centre at $(1/2, 0)$ and radius $1/2$. The vectors $\hat{t}$ and $\hat{s}$ are perpendicular, that is the fields which they represent are out of phase by $\pi/2$. The exact position of the point on the circle is determined by the nature of the component, by its dimensions and by the wavelength. A particular case, that of scattering by metallic mesh, is discussed later in section 3.10. Huyghen's construction thus gives us a very useful means for arriving at the general features of the electromagnetic behaviour of experimental systems. From an historical viewpoint, it was its successes in this area which finally swayed opinion in favour of the wave theory.

The reason why opinion had to be swayed is that to first order, visible radiation and for that matter near-infrared radiation behaves as though it were made up of rays. The beam from a distant source, or from a near one rendered quasi-parallel by a lens, casts sharp shadows and is capable of giving sharp images. This leads naturally to the commonest of all the approximations, viz. geometrical optics. Here one assumes the beam to undergo rectilinear propagation governed by simple straight-line geometrical constructions and then adds diffraction, interference etc. as first-order perturbations. The reason why this works so well lies in the great mismatch between the wavelength of the radiation (~ 0.5 μm) and the sizes of the various objects, images, stops, optical devices etc. which the beam encounters. Clearly this approach would be totally inappropriate if one were using centimetric or decimetric waves in the laboratory, since the separation of propagation and beam-shaping from interference and diffraction would then be meaningless. The infrared techno-logist thus faces some complications since his region encompasses the entire gamut from that where a full electromagnetic approach is required to that where geometrical optics is entirely adequate. There is also the difficulty that for certain combinations of component sizes and radiation wavelengths the scalar electromagnetic approach, whilst appropriate, would be unusable becaue of mathematical complications and the geometrical optics approach would inevitably fail. In these circumstances, an intermediate approximation, the so-called "beam-optics" becomes relevant. The normal region where this is used is the near-millimetre wave part of the spectrum, that is $3 - 0.3$ mm, but it is rapidly being extended to shorter wavelengths because of the development of powerful laser sources in the mid and near infrared.

Another substantial reason for the delay in accepting the wave theory was the failure to observe the most obvious expectation from such a theory, namely interference. We now understand that the lack of interference effects in conventional optics arises from the incoherent (see section 2.3) nature of the radiation produced by the familiar types of source. On the microscopic scale, this radiation consists of wave trains of variable length which bear only fleeting phase relationships to one another and which are in addition undergoing random sudden jumps in phase. Interference patterns *must* exist when the radiation from two or more independent sources is combined, but these patterns are flickering on the nanosecond time scale and all the eye can see is a steady field with an intensity which is the simple sum of the incident intensities. Stationary interference patterns can be produced, as Young showed in his celebrated experiment, if the radiation from a single source is split into two beams which are recombined after one of them has travelled over a longer path than the other. The basic reason for this is that each wavepacket then interferes with itself and the positions of the features in the interference pattern will not depend on time. The "quality" of the interference, that is the contrast of the fringes, will however depend on the amount of extra travel introduced since the longer this is the greater is the chance of a random phase shift of one partial beam with respect to the other. Using high-grade incoherent sources such as low-pressure single isotope lamps, interference can be observed over path-differences of the order $0.1-1.0$ m. This implies that the "coherence time", i.e. the time over which the beam maintains a defined phase relation, is of the order

10^{-9} s. Microwave engineers call a radiation source of this type a "noise source" because the incoherent addition is the exact analogue of that observed with noise signals where the power is added rather than the amplitude. The usual microwave sources, magnetrons, klystrons, backward-wave oscillators etc. are, by contrast, "coherent sources" and the output from two or more of these which are locked to the same frequency will, when combined, show interference patterns which will be slowly varying and may even be stationary.

Infrared engineers also have both types of source available to them. The broad-band continuous variety, discussed in Chapter 2, are highly incoherent; in fact it is only the finite width of the black-body curve which prevents their being totally incoherent. The laser sources, on the other hand, are highly coherent and, since the coherence times can be of the order of seconds, it has proved possible to observe slowly varying interference effects when the radiation from two or more identical lasers, locked to the same frequency standard, is combined.

In discussing the propagation of an infrared beam, therefore, two points are of crucial importance:

a. The ratio of the wavelength to the size of the objects used to shape the beam.
b. The degree of coherence of the beam.

Geometrical optics plus diffraction corrections would therefore be fine when using a black-body source in the near infrared, whereas it would be totally inappropriate when using a laser source in the near-millimetre wave region.

At very long RF wavelengths, the devices used to launch or receive the electromagnetic waves are arrays of conducting wires usually called antennae. As the wavelength shortens into the microwave region these antennae transform smoothly into metallic parabolic reflectors or "dishes" which focus the radiation either directly onto a single wire antenna plus rectifying diode or else onto a transition to a guided-wave system. This combination in its turn transforms smoothly into the reflecting telescopic receivers used in the near-infrared and visible regions. These telescopes, or infrared/visible lenses for that matter, are nevertheless still antennae and each of them has an *antenna pattern* which describes the field around it. The capabilities of an imaging system are limited in the final analysis by the nature of the antenna pattern in its image plane. At microwave frequencies, for example, imaging is not usually a useful concept because with the moderate sized "dishes" which are used, the lobes of the antenna pattern will be of the same order of size as the detector. One just aims therefore to match the detector to the antenna pattern in order to get the maximum sensitivity. At visible wavelengths, on the other hand, the lobes of the antenna pattern will be minute and one thinks of sharp images which only under very close examination can be seen to be rendered slightly fuzzy by the effects of the antenna pattern. The practice therefore is to treat these effects in terms of the concept of diffraction. Three limiting approximations are involved, depending on the relative sizes of the aperture at which the diffraction is occurring, $2a$, the wavelength λ, and the observing distance d. These are usually combined, as in equation (2.1.6), to give the Fresnel number $N_F = a^2/\lambda d$. When $N_F \gg 1$, we have the "near field" or shadow region where diffraction effects are not apparent. When $N_F \sim 1$, we have the region of Fresnel diffraction whilst for $N_F \ll 1$, we have the far field or Fraunhofer diffraction region. At infrared wavelengths diffraction becomes very significant. Some important examples are discussed in section 3.2.

3.1.2 *Introduction to guided propagation*

The principal alternative to propagation through free-space aided and abetted by beam-shaping elements such as mirrors and lenses is the guiding of the wave by metallic conductors. At radio frequences and below, it is perfectly feasible to propagate the wave along a metallic wire because the high conductivity of the wire will tend to concentrate the field in the neighbourhood of the wire. This simple approach can be improved by surrounding the conductor with a flexible dielectric, such as polyethylene, and then enclosing the composite within an outer metallic conductor either braided or continuous. The result is the co-axial cable which has multifarious uses ranging from the humble down lead from a domestic TV aerial to the very costly intercontinental telephone link. At microwave and higher frequencies, however, this approach starts to run into difficulties caused by the skin effect. The basic problem is that at these super-high and extra-high frequencies the wave tends to become highly localised in the outer "skin" of the conductor and doesn't tend to penetrate the bulk at all. A further consequence is that the ohmic losses per metre (which arise because the metal does not have an infinite conductivity) tend to rise rapidly with frequency. The proper treatment of the response of a metal to an extremely high frequency wave is naturally very complicated but at infrared frequencies and below one can make two reasonably valid approximations which greatly simplify the problem. These are (1) that the displacement term can be neglected in comparison with the conduction term (equivalent roughly speaking to setting $\varepsilon_r = 0$) and (2) that the conductivity does not vary with frequency. This latter will be true provided that the observing frequency is much less than the plasma frequency as it certainly will be for a metal in the infrared since the plasma frequency (from equation 2.2.3) will be of the order 3×10^{15} Hz. With these assumptions equation (3.1.4) for x direction propagation and in a non-magnetic metal becomes

$$\frac{\partial^2 \hat{E}}{\partial x^2} = \mu_0 \sigma \frac{\partial \hat{E}}{\partial t},$$

(3.1.7)

the solution of which is

$$\hat{E} = E_0 \exp\left(-2\pi i v t\right) \exp\left[(i-1)x/d\right],$$

(3.1.8)

where

$$d = (\pi \nu \mu_0 \sigma)^{-\frac{1}{2}}. \tag{3.1.9}$$

This quantity, d, represents two things, firstly it is the "$1/e$" penetration depth of the field into the metal and secondly it is equal to $(\lambda/2\pi)$ where λ is the wavelength *inside* the metal. It is usually referred to as the skin depth. For copper, for example, where $\sigma = 5{\cdot}82 \times 10^7\ \Omega^{-1}\,\mathrm{m}^{-1}$, one would have at a frequency of $\nu = 2{\cdot}998 \times 10^{10}\ \mathrm{Hz}$ (i.e. $\lambda_0 = 1\ \mathrm{cm}$), a skin depth $d = 3{\cdot}81 \times 10^{-7}\ \mathrm{m}$. At millimetre, submillimetre and infrared wavelengths d will be still smaller. At these wavelengths, it is power rather than field strength which is measured and since the former is proportional to the square of the latter (see also equation 3.4.2) one has

$$d_\mathrm{p} = \tfrac{1}{2}d, \tag{3.1.10}$$

where d_p is the power skin depth. From either viewpoint it is immediately clear that guidance by means of conventional conductors is quite impracticable at these extremely high frequencies. Not only would the conducting wire have to have a diameter vanishingly small but propagation in the z-direction would be extremely lossy because, of course, d_p is a measure of the attenuation in this direction as well. The coaxial concept can remain though, provided one is prepared to take the drastic step of omitting the central conductor! One also usually omits the material dielectric as well and thus arrives at a wave guide which is just a hollow metallic pipe of circular or rectangular cross-section. Now since the outer conductor is doing all the guiding it has to be rigid and finished on the inside to an almost optical polish. Essentially what is happening is that the wave is mostly flowing in free space with just occasional reflections at the walls to keep it in line (Fig. 3.1). Ohmic losses only occur at the reflections and the attenuation per metre is much reduced.

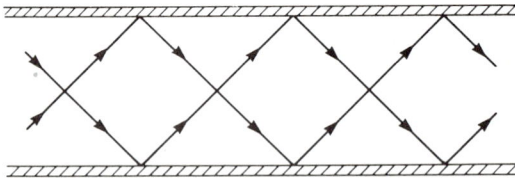

FIG. 3.1. Ray interpretation of propagation in a waveguide.

The full electromagnetic calculation (see Appendix 3 for a summary) of the propagation in a perfectly conducting waveguide shows that this is only possible in a series of modes. For each mode there is a cut-off frequency below which propagation is impossible. The lowest order mode and therefore the longest possible wavelength occurs for λ of the order of the waveguide internal dimension. Therefore as the wave-length shortens the waveguide has to

become smaller. With presently available work-shop techniques, it is impracticable to fabricate a waveguide of adequate quality for λ shorter than a
millimetre or so. In addition, it must be borne in mind that the waveguide will
not have infinitely conducting walls: there will always be the ohmic loss
mentioned above. This is sometimes called the copper loss because waveguide
walls are commonly made of high-purity copper. The copper loss increases
rapidly with frequency as shown in Fig. 3.2.

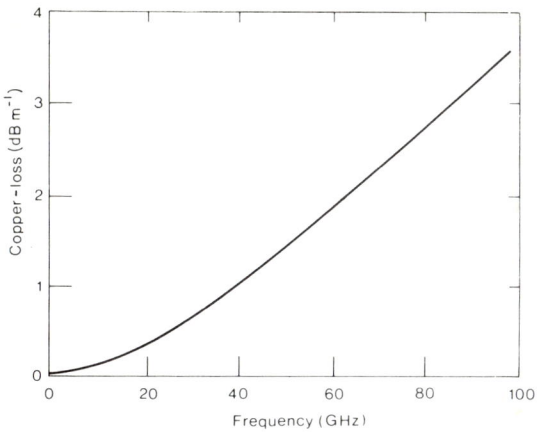

Fig. 3.2. Copper loss as a function of frequency for typical waveguides.

For both of these reasons, fundamental mode propagation in a waveguide is
uncommon at submillimetric and infrared wavelengths.

The options open are to go over to free space propagation with mirrors and
lenses or else to use an over-moded waveguide. This is a pipe much larger in
internal dimension than indicated by the wavelength being propagated.
Several modes can therefore propagate and this can be unwelcome if one of
them is a particularly lossy mode. However from the present viewpoint the
matter of importance is that as more and more modes become available the
propagation starts to look more and more like that of the plane waves of
geometric optics. One can therefore envisage a smooth transition from single
mode propagation, only properly describable in full electromagnetic terms, to
multi-mode propagation, a close approximation to geometrical optics. In
mathematical terms the complete set of modes forms a full basis set and any
wavefront may be synthesised from the correct set of combinations. In the
optical, or multimode limit, one no longer calculates the effect of reflection of a
wavefront in terms of the currents induced. Instead one imagines the metal to
be equivalent to a slab of lossy dielectric with a complex permittivity or
refractive index and calculates the amplitude of the reflected wave from the
components of these complex quantities.

3.2 Elementary treatment of diffraction

As mentioned above, when parallel light is incident on an orifice, and the transmitted light is received on a screen, the pattern on that screen depends on how close the screen is to the orifice. For a narrow slit, as might, for example, be used in an infrared spectrometer, the results are illustrated schematically in Fig. 3.3.

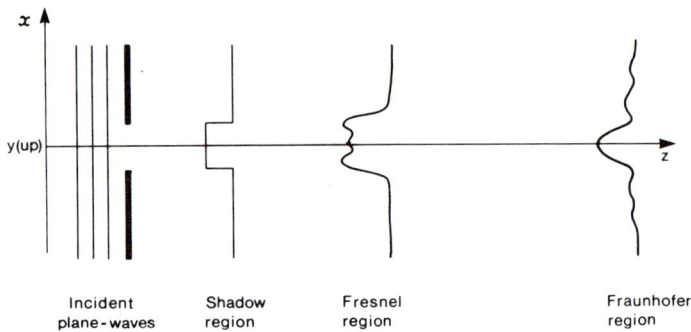

FIG. 3.3. Diffraction of plane waves at a narrow slit.

The displacements along the vertical axis are not to scale

The mathematical analysis of diffraction is, in general, very complicated but Born and Wolf [60] show that within the framework of a scalar wave-optics theory, the complex amplitude of the diffracted beam can be related to that of the incident beam by the relatively simple Fresnel–Kirchhoff equation:

$$\hat{E}(x,y,z) = -(i/2\lambda) \iint\limits_{x',\,y'} \hat{E}(x',y',z') \left[\cos\gamma + \cos\gamma'\right] R^{-1} \exp(2\pi i\tilde{v}R)\,dx'dy'.$$

$$(3.2.1)$$

In this equation the primed coordinates refer to points in the diffracting aperture and the unprimed coordinates to points on the receiving screen. R is the distance between the diffracting point and the observing point. The angles γ and γ' are respectively that between the surface normal at the diffracting point and the direction of observation and that between this normal and the direction of the incident beam. In the particular geometry of Fig. 3.3, γ' is zero.

In the far field ($R \gg a^2/\lambda$), where Fraunhofer diffraction applies, equation (3.2.1) simplifies still further. Firstly, γ will be everywhere small so $\cos\gamma$ can be taken to be unity without much error. Secondly, since the displacements in the x and y directions will be small in comparison with the screen/aperture separation $D (= z' - z)$, one can approximate the exact expression for R, namely

$$R = D[1 + (x' - x)^2/D^2 + (y' - y)^2/D^2]^{\frac{1}{2}}, \qquad (3.2.2)$$

by taking a power series expansion and ignoring higher terms. This gives

$$R = D + (x^2 + y^2)/2D + (x'^2 + y'^2)/2D - (xx' + yy')/D + \ldots \quad (3.2.3)$$

Since diffraction effects are usually only considered for narrow apertures, x' and y' will always be small in comparison with x and y and consequently the third term in (3.2.3) can safely be ignored. With these approximations, one then arrives at the Fraunhofer diffraction equation for a slit, namely

$$\hat{E}(x,y) = \left[-\frac{i}{\lambda D} \exp(2\pi i \tilde{v} D) \exp\left[i\pi \tilde{v} (x^2 + y^2)/D \right] \right]$$

$$\times \iint_{x', y'} \hat{E}(x', y') \exp\left[-2\pi i \tilde{v}(xx' + yy')/D \right] dx' dy' \qquad (3.2.4)$$

The two complex exponential terms before the double integration sign in (3.2.4) are merely phase factors which will cancel identically when $\hat{E}$ is multiplied by its complex conjugate to yield the intensity. The remainder of the RHS of (3.2.4) is of more solid interest since it is, in essence, the two-dimensional Fourier transform of $\hat{E}(x', y')$. This is found to be a general result. The Fraunhofer diffraction pattern is always simply related to the two-dimensional Fourier transform of the illuminated aperture. In the particular case of a uniformly illuminated slit and considering illumination along a line in the plane (i.e. $y = y' = 0$) one has that

$$\hat{E}(x') \sim \prod (x'/2a) \qquad (3.2.5)$$

where $2a$ is the full width of the slit. It follows at once that

$$\hat{E}(x) \sim 2a \text{ sinc } (2\pi \tilde{v} xa/D). \qquad (3.2.6)$$

One thus has, as expected, the familiar Fourier "pair". However, to tidy up the situation, it should be noted that (3.2.4) is not quite the proper Fourier formulation. To achieve this, one should introduce the dimensionless variables $u = (x'/\lambda)$ and $v = (x/D)$ which do form a conjugate pair. This, incidentally, shows that the transform scales linearly so that any change in a, D or λ will merely stretch out or compress the pattern without changing its functional form. In the particular case of the slit, one can say that the far-field intensity diffraction pattern will always be of the sinc^2 form.

Another important example is provided by diffraction at a circular aperture which may, mathematically, be described by the polar rectangle function $\prod(r'/2a)$. Making use of the standard identity (1.6.17) and of the transformations

$$a \rightarrow (a/\lambda) \quad \text{and} \quad q \rightarrow (r/D), \qquad (3.2.7)$$

it follows that the intensity in the diffraction pattern will be given by

$$I(r) = I_{max} [2J_1 (2\pi ar/\lambda D)/(2\pi ar/\lambda D)]^2. \qquad (3.2.8)$$

This is the celebrated Airy formula. The function in square brackets in (3.2.8) is similar in form to the sinc^2 function (Figs 1.4 and 1.5) having a large central maximum together with a series of much weaker side-lobes. The first zero occurs at

$$(2\pi ar/\lambda D) = \pi 1 \cdot 22 \ldots = 3.832 \ldots \qquad (3.2.9)$$

and all the area from $r = 0$ to $r = 3 \cdot 83\overset{.}{2} \lambda D/2\pi a$ is usually called the "Airy disc".

Two further examples are of some interest. These are diffraction of a Gaussian beam (i.e. a beam such as those described in Section 2.1.1., where the radial variation of intensity is of the form $\exp(-kr^2)$) and diffraction by a regular array of slits, that is to say a diffraction grating. The interest lies in the fact that the Gaussian function and the regular array or Dirac comb function (called "Shah" by Bracewell [11]) ⊔⊔⊔(x/a) belong to the exclusive set of functions which are their own Fourier transforms (Appendix 5). Thus a diffracted Gaussian beam will be another Gaussian beam and the diffraction pattern of a diffraction grating will likewise consist of a regular array of identical elements—the orders of the grating. These results have considerable significance since in beam optics (section 3.3) one seeks to have everywhere a Gaussian beam profile and of course diffraction gratings are extensively used to produce infrared spectra and one would wish to avoid difficulties due to overlapping orders.

One would not in practice seek to produce the Fraunhofer patterns in the way indicated in Fig. 3.3 because the required aperture/screen distances become uncomfortably large—usually several tens of metres are required. Instead a lens or mirror is used because with such it is easy to arrange that the diffracting aperture is essentially at infinity. This is shown in Fig. 3.4.

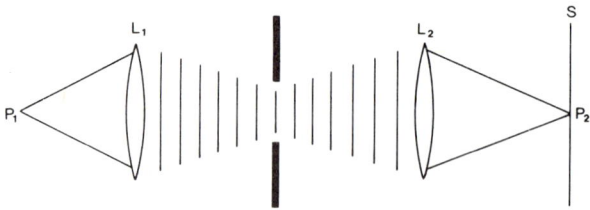

FIG. 3.4. Practical arrangement for producing Fraunhofer diffraction.

The radiation from a small source P_1 is rendered essentially parallel by the lens L_1 and this parallel radiation after passing through the aperture is focussed by the lens L_2 onto the receiving screen S at the point P_2. If an exposed piece of photographic film with density varying with distance from the centre is placed over the aperture, then the Fraunhofer pattern is the square of the Fourier transform of the density variation function. A lens can therefore be regarded as an optical Fourier transform device. It is essential, however, that when used in this way the aperture of the lens be greater than the extent of the beam for otherwise the lens, now acting as an aperture, would superimpose its own diffraction pattern. Analogue Fourier transformers of this type have, in fact, been used, especially in the visible region, and with the availability of continuous-wave gas lasers to provide the necessary parallel light for illumination they are quite attractive. However since by this method the square of the transform is obtained rather than the transform itself, and since digital methods (Appendix 5) using a computer are now so advanced, the lens-analogue method has had, so far, only limited applications in infrared physics and technology.

Fresnel diffraction is considerably more difficult to analyse mathematically [152] because in the intermediate distance regime one can no longer regard x' and y' as being very small in comparison with x and y. One cannot therefore make the crucial assumption that $(x'^2 + y'^2)$ is negligible compared with $(x^2 + y^2)$ and hence arrive at a linear integral transform. It is still permissible, however, to use the truncated power series expansion for R and one is then led to transforms of the type

$$\hat{E}(x) = \int_{-a}^{+a} \hat{E}(x') \exp\left[-i\pi(x'-x)^2/\lambda D\right]dx', \qquad (3.2.10)$$

Introducing now the dimensionless variable w, defined by

$$w = (2/\lambda D)^{\frac{1}{2}}(x'-x), \qquad (3.2.11)$$

enables (3.2.10) to be rewritten in the form

$$\hat{E}(x) \sim E_0 \int_{w(-a)}^{w(+a)} \exp\left(-\tfrac{1}{2}\pi w^2\right)dw, \qquad (3.2.12)$$

for the usual case where there is uniform illumination of the aperture. The variation of intensity in the x direction will be given by

$$I(x) = \tfrac{1}{2}c\,\varepsilon_0\,\hat{E}(x)\hat{E}^*(x) = 2I_0\left[C_F^2 + S_F^2\right], \qquad (3.2.12)$$

where

$$C_F = \int_0^{w(+a)} \cos\left(\tfrac{1}{2}\pi w^2\right)dw \quad \text{and} \quad S_F = \int_0^{w(+a)} \sin\left(\tfrac{1}{2}\pi w^2\right)dw \qquad (3.2.12a)$$

are the Fresnel integrals [152]. These far from elementary functions cannot be expressed in simple closed form for finite limits of integration and resort must be had either to Tables [153, 154] or else to numerical integration. However for integration over the full range, that is for w running from 0 to $+\infty$ (this physically corresponding to the aperture not being there at all), each integral takes on the absolute magnitude $\frac{1}{2}$ to give

$$I(x) = 2I_0(1/4 + 1/4) = I_0 \qquad (3.2.13)$$

which is necessary if energy is to be conserved.

Fresnel diffraction forms a fascinating subject in its own right and it is a very important consideration in the design of the Fabry–Perot resonators used for infrared lasers but, from the infrared optics point of view, the most important application is the Fresnel zone plate, commonly called the Fresnel lens. The simplest form of this has a circular plate made of transparent material and having circles ruled on it with radii proportional to the square-roots of the natural numbers. It can be shown [152] that if the first of the resulting zones has a radius r_0, then at a point on the axis perpendicular to the plate, with a distance

$$f_0 = r_0^2/\lambda, \qquad (3.2.14)$$

the contributions to the wave amplitude produced by the diffraction of an incident plane wave will have alternating phases. If, therefore, alternate zones are covered with an opaque material, then only the reinforcing waves will arrive and the plate will act as a lens of focal length f_0. It will, in fact, have an infinite set of focal points but the principal focus is of the most interest. Such a plate (Fig. 3.5(a)) transmits only half the power incident upon it so better results can be obtained by having a suitable thickness of dielectric (Fig. 3.5(b)) to retard the phase by π in the negative zones. In this version, usually called a "zone-reversal" plate, all the zones act cooperatively at the focus. One can do still better by having a sinusoidal variation of thickness over the zones (Fig. 3.5(c)) but this is very difficult to achieve to sufficient accuracy in a "one-off" laboratory made lens. The importance of the Fresnel lens in infrared technology stems from the lack of completely satisfactory transmissive

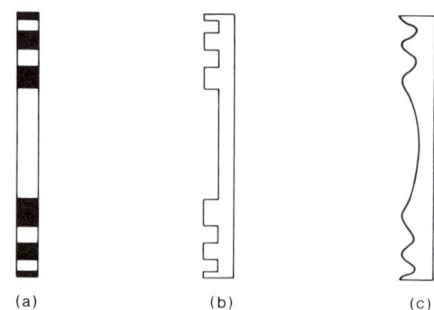

(a)   (b)   (c)

FIG. 3.5. Various forms of the Fresnel zone-plate or lens.

materials for wavelengths longer than a few μm. Technologists working at shorter wavelengths have available to them a most remarkable optical material, namely glass. This is not only rigid, easily worked, easily polished, transparent, resistant to chemical attack and abrasion, and cheap, it is also available with a range of refractive indices and with a range of both positive and negative dispersions. It is, therefore, possible to make large aperture optics essentially free of aberration. In the infrared, and especially at the longer wavelengths, there is no such universal material and the lens maker encounters problems. In particular, since he cannot at the moment construct compensated lenses, he has to rely solely on simple lenses. For such a lens, the focal length will be given by

$$f = R_1 R_2 [(n-1)(R_1 - R_2)]^{-1}, \qquad (3.2.15)$$

where R_1 and R_2 are the radii of curvature,[†] so to avoid highly curved surfaces and the resulting spherical aberration it is necessary that n be as large as possible. Quite apart from the other difficulties, e.g. softness, hygroscopicity and sometimes cost, this condition would be sufficient to rule out the alkali and alkaline earth halides which are, of course, invaluable transmissive media in other connections (section 6.4). Semiconductors (section 5.4) have large refractive indices in the infrared and are popular candidates as lens materials. Germanium ($n = 4$) in particular is in widespread use. The high index has the consequent further advantage that the thickness of the lens is reduced. Thus for a plano-convex lens, the maximum thickness is given by

$$d = D^2/8(n-1)f, \qquad (3.2.16)$$

where D is the diameter of the plano face, so less material is required the larger is n. This reduces cost, but, perhaps even more to the point, it reduces absorptive loss in the lens. Even with highly purified semiconductors, the absorption is not negligible so the thinner the lens the better. The high index unfortunately does mean high reflection losses (see later) but these can to a large extent be mitigated by the use of anti-reflection coatings. For wide-band applications, e.g. thermal imaging (section 7.1), coated germanium lenses are indispensable, but for narrow-band work, e.g. examining scenes illuminated by an infrared laser, the very much simpler and many orders of magnitude cheaper, Fresnel lens made from low-cost, low-index, polymers is very attractive. It is interesting to observe that the severe chromatic aberration of Fresnel lenses which makes them only suitable for virtually monochromatic radiation can be used the other way round and the lens used as a separating filter in much the same way that Rubens used focal isolation to separate the various regions of the far infrared. To be fair to the Fresnel lens, one should point out that the chromatic aberration can, to a certain extent, be compensated for by the use of a combination of zone plates in series. By a

---

†The sign convention adopted here is "direction of propagation positive" and the radii of curvature are measured in the sense "from pole to periphery".

careful choice of the zone radii, it is possible, at least in principle, to bring quite broad-band radiation all to a common focus. Unfortunately this is not a practical proposition in the infrared because the absorptive and reflective losses would soon prove prohibitive.

Before leaving this topic it is worth emphasising that the common practice of mounting a specimen in a converging, or diverging, beam can lead to erroneous results for the absorption coefficient at long wavelengths. At these wavelengths the detector window might well be in the Fresnel diffraction region unless the diameters of the optics and that of the specimen itself were all much greater than the beam width.

The question of diffraction by a *random* array of apertures or protruberances is of considerable practical importance especially when considering large structures such as infrared or millimetre-wave telescopes. It is obviously impossible to make a parabolic dish reflector which might be as much as 50 metres in diameter absolutely true.[†] The surface will wander from the true "figure" and usually the departures will be of random size and will be distributed randomly. The effect will be to degrade the Airy diffraction pattern at the prime focus by reducing the central grand maximum and increasing the subsequent maxima. The reduction of the central grand maximum will be given by the function [155]

$$I_{\text{obs}} = I_0 \exp - \left[\frac{4\pi\sigma}{\lambda} \right]^2, \qquad (3.2.17)$$

where σ is the root mean square variation away from the ideal figure. It will be seen therefore that extraordinarily good finish is required for infrared telescopes. In fact (σ/λ) has to be no more than about 0·02. One cannot pursue the point further without having some knowledge of the properties of the detector. If a coherent detector (see Chapter 4) is being used, it is sensitive to the amplitude of the wave and the full attenuation represented by (3.2.17) will be experienced because none of the negative going amplitude of the next lobes can be allowed to be incident. If an incoherent (that is a power) detector is being used then a lot of the loss due to (3.2.17) can be recovered by allowing the adjacent lobes to enter the detector. Coherent detectors are much more sensitive, so in astronomy they will be preferred wherever they are available— that is principally for millimetre wavelengths and longer. This is the reason behind the apparent paradox that millimetre wave telescopes are often made with a surface finish at least as good as that used for infrared telescopes. Whilst on the subject of coherent detectors it is worth pointing out that, because they detect the *amplitude* of the wave rather than its *intensity*, they can be used to give directly the Fourier transform of a given beam profile. Unfortunately this operation would normally be required for visible or near infrared sources

[†]The main reason is deformation under gravity—if one makes the structure strong enough to resist gravity it will be too heavy to move! The designers get round this by the clever trick of designing a structure which is not right initially but *after* deformation will be!!

and in these spectral regions coherent detectors with reasonable sensitivity are not yet available.

3.3 Beam-wave optics

Beam-wave theory arose originally in the treatment of diffraction at a series of consecutive apertures, the practical spur coming from the scheme to set up long-distance optical communication links in which the natural tendency of the beam to spread out would be corrected by encounters with a series of widely spaced weakly focussing lenses [156]. It was soon realised that this system is exactly analogous to the Fabry–Perot resonator where the long optical path is endlessly folded back upon itself and the two mirrors take the place of the series of lenses [30]. The theory showed that, provided some very reasonable approximations were made, analytical solutions could be reached that displayed a Gaussian dependence of the intensity on the distance away from the beam axis and the beam was therefore essentially confined. The beam could be "focussed" or perhaps better "concentrated" by lenses and mirrors but always retained its Gaussian power distribution so one did not get the highly structured and therefore undesirable Airy patterns in the region surrounding the geometrical focus that one gets when an ordinary plane-parallel beam is focussed by a lens [35]. The theory was originally intended for use at near-infrared wavelengths but it is ideal for dealing with the awkward cases which arise in the millimetre/submillimetre region where the methods of classical optics fail yet the full electromagnetic treatment is impossibly difficult to carry out. It is particularly appropriate at long wavelengths (far-infrared and millimetre waves) where the optical elements in a chain tend to be usually in the near field of their immediate predecessors.

One starts the mathematical treatment by assuming that the time dependence of the field will be the simple sinusoidal oscillations given by

$$E = E_0 \exp\left(-2\pi i v t\right) \tag{3.3.1}$$

and then uses this to transform the basic wave equation (3.1.5) into the form

$$\nabla^2 \hat{E} + k^2 \hat{E} = 0, \tag{3.3.2}$$

that is Helmholtz's equation where k is the magnitude of the wave vector equal to $(2\pi/\lambda)$. One looks for solutions of this equation which represent a beam travelling along the z axis and which have the property that, for very high frequencies, the solution will degenerate into a simple plane wave. The simplest trial solution to satisfy these requirements is

$$\hat{E}(x,y,z) = \psi(x,y,z) \exp\left(-ikz\right), \tag{3.3.3}$$

where ψ is a slowly varying function of z. Substituting (3.3.3) into (3.3.2) gives

$$\frac{\partial^2 \psi}{\partial x^2} + \frac{\partial^2 \psi}{\partial y^2} + \frac{\partial^2 \psi}{\partial z^2} = 2ik \frac{\partial \psi}{\partial z}. \qquad (3.3.4)$$

The next step is to make the paraxial approximation, viz. that the beam is essentially confined to the neighbourhood of the axis and has a negligible curvature with respect to z. This, together with the slow z dependence of ψ, is equivalent to setting $\mathrm{d}^2 \psi / \mathrm{d}z^2$ equal to zero. With this assumption, one arrives at the basic equation

$$\frac{\partial^2 \psi}{\partial x^2} + \frac{\partial^2 \psi}{\partial y^2} = 2ik \frac{\partial \psi}{\partial z}. \qquad (3.3.5)$$

This is basically a wave equation and is, in fact, formally equivalent to the time-dependent Schrodinger equation with z taking the place of time [157]. The solutions therefore involve Hermite Polynomials (section 1.5.3) together with a Gaussian dependence of amplitude on x or y. The totality of solutions forms a complete basis set in terms of which *any* field disposition may be described—thus it includes electromagnetism and geometrical optics as special cases. However the beam-wave approach is only of practical value when the actual field distribution can be described in terms involving merely low-order solutions of (3.3.5). The lowest order solution is the beam wave given by equation (2.1.10) with $F_{\mathrm{mn}}(x, y)$ set equal to the first Hermite polynomial, namely unity. Even so, this expression is rather cumbersome, especially since its parameters w, ϕ and R are all functions of z, but its physical content can nevertheless be expressed quite simply [35, 158]. The results are:

1. The amplitude has a Gaussian dependence on radius with a scale width (i.e. the $1/e$ distance) of w. For small values of z, w is close to its value w_0, the "beam-waist", which it has at the origin. At large values of z, w is directly proportional to z.
2. The phase with respect to a plane-wave reference varies from $-\pi/2$ to $+\pi/2$ as z goes from $-\infty$ to $+\infty$ and the phase varies parabolically with radius.
3. The radius of curvature is infinite at $z = 0$, i.e. we have a plane-wave front, but for large z, r is equal to z.

The practical use of beam-wave theory is also not too difficult. Thus we have already used it to calculate laser "spot sizes" as in equation (2.1.16) and in fact all the calculations of geometrical optics have their beam-wave equivalents. One can, for example, consider the focussing action of a lens as shown in Fig. 3.6. If R_1 and R_2 are the wavefront radii at the lens, then one has, in complete analogy with the ordinary lens equation (equation 3.2.15), the simple relation

$$\frac{1}{f} = \frac{1}{R_2} - \frac{1}{R_1}. \qquad (3.3.6)$$

The wavefront radii can be related to the sizes and positions of the respective beam waists by the use of equation (2.1.12). From this point of view, a lens is a device for converting a Gaussian beam from one waist size and position to another. If an attempt is made to take the algebra further and produce analytical expressions relating the various distances and waist sizes to one another, some severe non-linearities can be found, but these can usually be side-stepped. Thus a useful concept is the waist size at the lens w_L. If one has a waist w at a distance d from the lens, then from equation (2.1.11) one finds

$$2w^2 = w_L^2 \pm \sqrt{[w_L^4 - 4(\lambda d/\pi)^2]}. \tag{3.3.7}$$

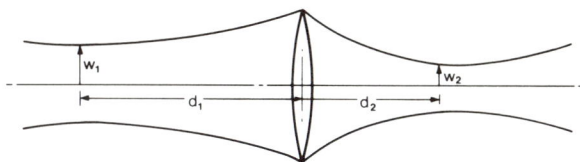

FIG. 3.6. "Focussing" action of a lens on a beam wave.

The beam waist cannot, therefore, be further from the lens than a critical distance $d_c = \pi w_L^2/2\lambda$. The best way to collimate a beam wave is to choose a lens focal length equal to d_c to obtain a total distance of $2d_c$ throughout which the beam waist will be less than w_L. It does follow though that a series of lenses spaced at $2d_c$ and of focal length d_c will serve to confine a beam wave more or less indefinitely.

The complete set of beam modes given by equation (2.1.10) forms, as mentioned above, a basis set in terms of which any arbitrary distribution of amplitude may be expressed, so in the practical case, where one might not have a pure beam mode, all that has been said above will still apply because it will apply to each component mode separately. The principal point in this connection is that the change in phase as one goes through a "focus"—a phenomenon that was rather mysterious in classical optics—is different for odd and even modes. Interference effects are therefore possible when both kinds of mode are present. This is the beam-wave explanation of the complex Airy patterns mentioned above. The phase change emerges clearly from the mathematical analysis but it can also be derived in a more heuristic fashion by noting that the shrinkage of the beam waist as the "focus" is approached is a form of beam containment, so the propagation can be thought of as being in a non-material waveguide which first closes down and then opens up [157]. The phase velocity will then necessarily be a function of z and will vary from mode to mode. This point is taken up again in section 3.6.2. The principal parameter of the beam mode, w_0, is either determined by the geometry of the system itself, as it is for the Fabry–Perot resonator, or else by the geometry of the "feed" system that is the auxiliary optics. The basic point is that the larger the ratio of aperture to wavelength the smaller is w_0.

3.4 Propagation in absorbing media

3.4.1 *The complex refractive index*

As has been stressed earlier, the wavelengths of infrared radiation, though large by visible standards, are still very small in comparison with the sizes of the objects transmitting and shaping the beam and it is usually permissible to treat the flow of energy by simple optical methods. A plane wave propagating in a loss-less medium of refractive index n will have a velocity (c/n), where $n = \varepsilon_r^{1/2}$, and the electric field will be given by equation (3.1.6), the real, space-dependent, part of which is

$$E(z) = E_0 \cos(2\pi n \tilde{v} z). \qquad (3.4.1)$$

In a lossy medium, the power falls as the wave progresses into the medium according to Lambert's law, equation (1.2.16), which may alternatively be written

$$E(z) = E_0 \exp\left(-\tfrac{1}{2}\alpha z\right). \qquad (3.4.2)$$

Combining (3.4.2) with the complex version of (3.4.1), gives

$$\hat{E}(z) = E_0 \exp(2\pi i \hat{n} \tilde{v} z), \qquad (3.4.3)$$

where $\hat{n}$, the complex refractive index, is *defined* by

$$\hat{n} = n + i\alpha/4\pi\tilde{v}. \qquad (3.4.4)$$

In the older literature, this equation is sometimes found written

$$\hat{n} = n(1 - i\kappa) \qquad (3.4.5)$$

in terms of the extinction coefficient κ. In the modern physics literature, it is frequently written

$$\hat{n} = n + ik \qquad (3.4.6)$$

in terms of the absorption index k. This latter is rather unsatisfactory since there is some risk of confusion with the quite unrelated quantity, the wave-vector, also conventionally indicated by the symbol k. However, the use of the absorption index is so well entrenched in the ordinary everyday communication between physicists that to reject it would seem arbitrary and it will, in fact, be used several times later in appropriate contexts. Unfortunately, as mentioned in section 1.4.1, there is no uniform agreement as to the sign to be used for the imaginary component in equations such as (3.4.6), so care is needed when comparing equations taken from different sources. However, with either sign convention it follows from (3.4.4) and (3.4.6) that

$$k = \alpha/4\pi\tilde{v}. \qquad (3.4.6a)$$

The complex refractive index is implicit in equation (3.1.4) and it may therefore also be written in the alternative form

$$\hat{n}^2 = \mu_r \varepsilon_r + i\mu_r \sigma / \varepsilon_0 \omega. \qquad (3.4.7)$$

The conductivity of a medium is thus intimately related to its absorption coefficient. In fact, by identifying the imaginary terms on each side of (3.4.7), one immediately has the result

$$\alpha = 4\pi \tilde{\nu} k = \sigma / nc\varepsilon_0. \qquad (3.4.8)$$

This equation has to be handled with a little care since both n and σ are functions of frequency, but two extreme cases are of interest. Firstly, when σ is very small, as it will be for a dielectric medium, n will vary slowly with frequency and over a narrow range may be taken to be constant. One then has that α and σ are directly proportional. Secondly, in the opposite limit when σ is very large, as it will be for say a metal, one can assume that it is constant at its DC value. Then with $\varepsilon_r \approx 0$, one has $n \approx k$ and the absorption coefficient may be written

$$\alpha = 2\sqrt{(\pi\mu_0 \nu\sigma)}. \qquad (3.4.9)$$

The power skin-depth d_p (see equations 3.1.10 and 3.1.9) follows at once by taking the reciprocal.

3.4.2 Propagation—the dielectric equations

In the far-infrared and submillimetre-wave regions, significant overlap occurs with observations made by radio-frequency dielectric techniques. It is essential, therefore, to be able to relate the optical and dielectric languages so that equivalent results can be compared. The dielectric properties of a medium are completely defined by the complex relative permittivity $\hat{\varepsilon}_r$, which in the positive convention is defined by

$$\hat{\varepsilon}_r = \varepsilon_r' + i\varepsilon_r'', \qquad (3.4.10)$$

where ε_r' is the relative permittivity and ε_r'' is the loss factor. Dielectricians often quote the loss tangent given by

$$\tan \delta = \varepsilon_r'' / \varepsilon_r' \qquad (3.4.11)$$

and since δ is numerically equal to $\tan \delta$ when δ is small, it is not unusual to hear the ratio $\varepsilon_r'' / \varepsilon_r'$ quoted in microradians! The dielectric and optical descriptions of the response of a medium to an electrical stimulation must, of course, be related. This relation, first derived by Maxwell, is given, in essence, by equation (3.4.7), which, for a non-magnetic medium may be written

$$\hat{\varepsilon}_r = (\hat{n})^2. \qquad (3.4.12)$$

Identifying real and imaginary components of this equation gives

$$\varepsilon'_r = n^2 - k^2 = n^2 - (\alpha/4\pi\tilde{v})^2, \tag{3.4.13a}$$

$$\varepsilon''_r = 2nk = n\alpha/2\pi\tilde{v}. \tag{3.4.13b}$$

Thus if n and α are known, ε'_r and ε''_r may be readily calculated. The inverses are, however, rather more complicated, viz.

$$n = \left(\frac{\varepsilon'_r}{2}\right)^{\frac{1}{2}} \left\{ \left[1 + \left(\frac{\varepsilon''_r}{\varepsilon'_r}\right)^2 \right]^{\frac{1}{2}} + 1 \right\}^{\frac{1}{2}}, \tag{3.4.14a}$$

and

$$k = \left(\frac{\varepsilon'_r}{2}\right)^{\frac{1}{2}} \left\{ \left[1 + \left(\frac{\varepsilon''_r}{\varepsilon'_r}\right)^2 \right]^{\frac{1}{2}} - 1 \right\}^{\frac{1}{2}}. \tag{3.4.14b}$$

In most infrared cases, however, one usually has that $\alpha \ll 4\pi n\tilde{v}$, whereupon it is permissible to set ε'_r equal to n^2 without much consequential error. This is tantamount to treating n as a constant.

3.4.3 *The complex propagation coefficient*

For most instances of infrared propagation, the radiation will be essentially incoherent and it is perfectly acceptable to treat it by the methods of geometrical optics. However, with the rapid development of infrared lasers and infrared laser components and with the increasing overlap, at the low frequency end, with microwave techniques, it is becoming frequently necessary to consider the propagation of the electric field rather than the radiant power. In the most general way one can write for propagation in the z direction

$$\hat{E}(z) = \hat{E}(0)\exp[-\hat{\gamma}z] \tag{3.4.15a}$$

where

$$\hat{\gamma} = \alpha_v - i\beta \tag{3.4.15b}$$

is the complex propagation coefficient or factor. By substitution of (3.4.15b) into (3.4.15a) it will be seen that α_v is a voltage attenuation coefficient (equal to $\alpha/2$) whilst β is a phase coefficient (equal to $2\pi\tilde{v}n$).

If the propagation is occurring in a medium in which the separation of boundaries is so large, compared to the wavelength or the beam diameter, that it may be considered unbounded, then $\hat{\gamma}$ may be readily related either to the dielectric parameter $\hat{\varepsilon}_r$ or the optical parameter $\hat{n}$. One has

$$\varepsilon'_r = \left(\frac{\lambda\beta}{2\pi}\right)^2 \left[1 - \left(\frac{\alpha_v}{\beta}\right)^2 \right], \tag{3.4.16a}$$

$$\varepsilon''_r = \left(\frac{\lambda\beta}{2\pi}\right)^2 \frac{2\alpha_v}{\beta}, \tag{3.4.16b}$$

and the converses

$$\alpha_v = \left(\frac{2\pi}{\lambda}\right)\left(\frac{\varepsilon'_r}{2}\right)^{\frac{1}{2}}\left\{\left[1+\left(\frac{\varepsilon''_r}{\varepsilon'_r}\right)^2\right]^{\frac{1}{2}}-1\right\}^{\frac{1}{2}},$$

(3.4.17a)

$$\beta = \left(\frac{2\pi}{\lambda}\right)\left(\frac{\varepsilon'_r}{2}\right)^{\frac{1}{2}}\left\{\left[1+\left(\frac{\varepsilon''_r}{\varepsilon'_r}\right)^2\right]^{\frac{1}{2}}+1\right\}^{\frac{1}{2}},$$

(3.4.17b)

where throughout λ means the *vacuum* wavelength of the radiation.

If the propagation is occurring in a bounded medium then the problem may need an exact, or semi-exact, solution of Maxwell's equations with proper recognition of the boundary conditions. The best-known example occurs for monochromatic radiation propagating in a rectangular section, hollow, perfectly conducting, metallic pipe having dimensions a and b in the x and y directions. In such a *waveguide*, the complex propagation factor is given by

$$\hat{\gamma} = i\left[\left(\frac{2\pi v}{c}\right)^2 \hat{\varepsilon}_r - \left(\frac{m\pi}{a}\right)^2 - \left(\frac{n\pi}{b}\right)^2\right]^{\frac{1}{2}},$$

(3.4.18)

where $\hat{\varepsilon}$ is the complex relative permittivity of the medium (usually air or vacuum) inside the pipe. The propagation is described in terms of *modes* and the integers m and n are labels which designate the modes. Equation 3.4.18 can be rearranged to read

$$\hat{\gamma} = \frac{2\pi i}{\lambda}\left[\hat{\varepsilon} - \left(\frac{\lambda}{\lambda_c}\right)^2\right]^{\frac{1}{2}},$$

(3.4.19a)

where

$$\frac{1}{\lambda_c^2} = \left(\frac{m}{2a}\right)^2 + \left(\frac{n}{2b}\right)^2.$$

(3.4.19b)

Under usual circumstances $\hat{\varepsilon}_r$ will be purely real and equal to unity and it will be seen then that $\hat{\gamma}$ will become purely real and therefore correspond to attenuation without propagation for values of λ greater than λ_c. The critical value λ_c defined by (3.4.19b) is therefore known as the "cut-off" wavelength since the pipe cannot propagate, in that mode, wavelengths longer than λ_c. If $\alpha > b$ and m equals unity and n is zero λ_c becomes the longest wavelength which can be propagated by the waveguide. It will be observed that a guide with ultimate air-filled cut-off wavelength λ_c^0 will be able to propagate longer (free-space) wavelengths up to $(\varepsilon'_r)^{\frac{1}{2}} \lambda_c^0$ when filled with a material having a relative permittivity ε'_r. This is the basis of the "microstrip" technique used for propagating microwaves in circuit boards. Of course the dielectric material should in addition to having a high ε'_r have a low ε''_r or else the gain in being able to propagate the waves in a conveniently sized circuit element will be lost by the absorption in the microstrip. Alumina is very useful for the construction of microstrip since $\varepsilon' = 10$ and ε'' can be as low as 10^{-3}.

When the guide is lossy the components of $\hat{\gamma}$ are given by

$$\alpha = \left(\frac{2\pi}{\lambda}\right)\left(\frac{\varepsilon' - p}{2}\right)^{\frac{1}{2}} \left\{\left[1 + \left(\frac{\varepsilon''_r}{\varepsilon'_r - p}\right)^2\right]^{\frac{1}{2}} - \right\}^{\frac{1}{2}} \qquad (3.4.20a)$$

and

$$\beta = \left(\frac{2\pi}{\lambda}\right)\left(\frac{\varepsilon'_r - p}{2}\right)^{\frac{1}{2}} \left\{\left[1 + \left(\frac{\varepsilon''_r}{\varepsilon'_r - p}\right)^2\right]^{\frac{1}{2}} + 1\right\}^{\frac{1}{2}}, \qquad (3.4.20b)$$

where $p = (\lambda/\lambda_c)^2$. It must be stressed, however, that this simple theory assumes no losses at the bounding surfaces—in electrical language that the conductivity of the metal is infinite, in optical language that the reflectivity is unity. In practice this will not be so, of course, and so-called "copper losses" (Fig. 3.2) will be encountered which will give even an evacuated guide a finite attenuation.

From an infrared point of view one is interested in this theory for the situations where λ is much smaller than a or b. As λ gets smaller and smaller, increasingly higher values of m and n are possible that permit propagation. Eventually, so many modes will propagate that a description of the wave field in mode terms becomes no longer useful and one is back with geometrical optics and the elementary theory of diffraction. In the intermediate state, where a number of modes will propagate but not too many, the guide is said to be "over-moded". One can develop a similar theory for waveguides which have cylindrical section and, apart from the complication that one has degenerate modes, similar conclusions emerge. An overmoded cylindrical waveguide is sometimes called a "light-pipe". Light-pipes are widely used in infrared technology because the manufacture of single-mode waveguides, whose internal dimensions would be very much less than a millimetre, would present nearly insuperable machining problems. Single-mode waveguides have, in fact, been used at frequencies up to 500 GHz but only in very short lengths and even then the losses are daunting. The principal uses which such waveguides have found is as rejection filters for longer wavelength radiation and for coupling incident radiation to a coherent detector (see section 4.6.4). More details of propagation in hollow pipes are given later and the mathematical theory is given at more length in Appendix 3.

3.4.4 *Propagation, the wave vector*

Another very useful quantity is the wave vector **k**. This is introduced by considering a plane wave propagating with constant amplitude and velocity. Such a wave is completely specified by the equation

$$E = E_0 \, (\mathbf{k}\cdot\mathbf{r} - 2\pi v t), \qquad (3.4.21)$$

where **r** is the position vector from the chosen origin to the point of interest. The wave is plane because at any given moment of time corresponding values of E are defined by $\mathbf{k}\cdot\mathbf{r} = $ constant, that is they lie on planes perpendicular to **k**.

The wavevector $\mathbf{k}$ thus points in the direction of propagation and its magnitude $|\mathbf{k}|$ is given by

$$|\mathbf{k}| = 2\pi\lambda^{-1} = 2\pi n\tilde{v} = 2\pi n\lambda_0^{-1} = nk_0. \qquad (3.4.22)$$

Equation 3.4.21 can be cast in complex form as before, viz.

$$\hat{E} = E_0 \exp(-2\pi ivt)\exp(i\mathbf{k}\cdot\mathbf{r}). \qquad (3.4.23)$$

One next proceeds to consider the more general case of a wave propagating in an absorbing medium. In order that the modulus of $\hat{E}$ will decrease as r increases, it is necessary that $\mathbf{k}$ be complex, i.e.

$$\hat{\mathbf{k}} = \mathbf{k}' + i\mathbf{k}'' \qquad (3.4.24)$$

from which it follows that

$$\hat{E} = E_0 \exp(-2\pi ivt)\exp(i\mathbf{k}\cdot\mathbf{r})\exp(-\mathbf{k}''\cdot\mathbf{r}). \qquad (3.4.25)$$

This equation shows that $\mathbf{k}'$ is normal to the surfaces of constant phase whilst $\mathbf{k}''$ is normal to the surfaces of constant amplitude. When $\mathbf{k}'$ and $\mathbf{k}''$ are collinear, the wave is said to be homogeneous and when they are not the wave is said to be inhomogeneous. Inhomogeneous waves are usually only encountered in absorbing media but when one is considering waves incident on the interface at an angle greater than the critical angle (see section 3.7.2) one is led to the concept of an evanescent wave which can be regarded as the extreme case of inhomogeneity with the two vectors perpendicular to one another!

Either by comparing (3.4.25) with (3.4.3) or else by noting that the complex wave vector must obey the equation

$$\hat{\mathbf{k}}\cdot\hat{\mathbf{k}} = k_0^2\hat{n}^2, \qquad (3.4.26)$$

it follows immediately that

$$|\mathbf{k}'|^2 - |\mathbf{k}''|^2 = k_0^2(n^2 - k^2), \qquad (3.4.27\text{a})$$

$$\mathbf{k}'\cdot\mathbf{k}'' = k_0^2 nk. \qquad (3.4.27\text{b})$$

These relations when applied to a lossless medium (for which $k = 0$) permit either a non-complex wave vector specifying a perfectly homogeneous wave or else an infinite number of pairs of $\mathbf{k}'$ and $\mathbf{k}''$ perpendicular to one another and therefore specifying inhomogeneous waves. For a lossy medium there is again an infinite number of solutions amongst which there is a unique homogeneous solution with $\mathbf{k}'$ parallel to $\mathbf{k}''$. However even for this solution there is no guarantee that $\mathbf{E}$ will be perpendicular to $\mathbf{H}$. The whole matter of polarisation in a lossy medium is certainly very complicated. An illuminating discussion can be found in the standard work of Stone [133]. One thing which does emerge very clearly from equations (3.4.27) is the confusion introduced by using the same symbol k for two very different quantities.

3.4.5 *The Kramers–Kronig dispersion relations*

The combination of α and n (or α_v and β) into a single complex quantity is not just a mathematical artifice—there are profound physical reasons which compel this approach. These stem from the Principle of Causality, an absolute law of physics which can be crudely stated "causes must always come before their effects". This Principle obliges α and n (or any other conjugate pair) to be functions of one another, so that if one is known over the whole frequency range, the other can be calculated exactly at any chosen frequency. This result can be derived by a "thought experiment" which goes as follows [159]. Imagine that a brief impulse (Fig. 3.7) is applied to a system which has an input (to which the impulse is applied) and an output (where the effects are monitored). Now the input pulse (of the form $I = 0$ for $t < t_0$, $I = 0$ for $t > t_0 + \Delta t$) can be resolved into its Fourier components which stretch from $t = -\infty$ to $t = +\infty$ and the output will be the sum of these after modification by the system. Imagine next that the system has a narrow absorption band lying between the frequencies v_0 and $v_0 + \Delta v$: the Fourier components lying within the absorption band will be removed or strongly attenuated which is equivalent to saying that their complements, entire or in part, will be *added* to

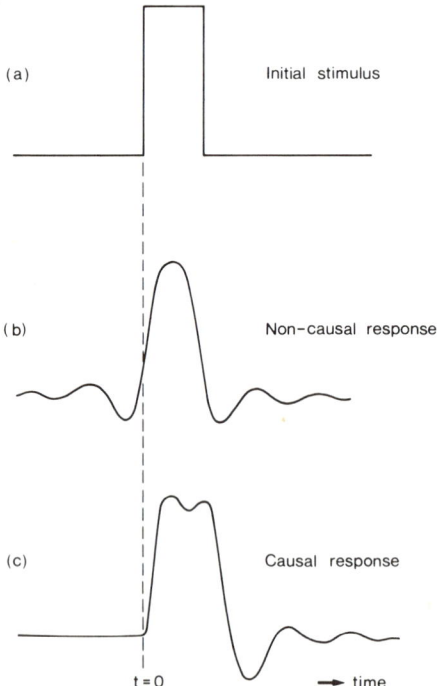

FIG. 3.7. A brief pulse applied to a selectively absorbing system at time zero might be expected to give the output shown in (b) but the principle of causality demands that all response be confined to times greater than zero and the observed response looks like (c).

the output. The principle of causality states that no effect is possible at times preceding the application of its cause. It follows therefore that, in this case, no output may be observed before $t = t_0$. Yet since the sum of all the Fourier components must be zero before $t = t_0$, their sum minus the absorbed components cannot be zero. This means that ripples will be detectable at the output at times before $t = t_0$ in anticipation of the impulse which has not yet been applied! This is obviously physical nonsense (though not mathematical nonsense) violating the principle of causality. It follows that if the principle of causality (an arbitrary boundary condition from a mathematical viewpoint) is to be universal, the Fourier components *outside* the region of the absorption band must be altered in some way such that the sum of all the components should once more be identically zero for all times prior to t_0. By definition their amplitudes are not to be affected, so all that can change is their phases β and the change of phase will be frequency dependent; in other words there must always be dispersion associated with absorption. This result is quite general; it applies for example to electrical networks, to stress and strain in mechanics and to the propagation of acoustic waves, but most importantly in the present context to the propagation of electromagnetic waves through absorbing materials. In all these cases, we are considering the response of a system to an external stimulus and this response is described by a complex response function. Usually the further assumption is made that the response is linear, i.e. the response equation involves only the first power of the stimulus. One may write therefore as the general equation

$$\hat{R}(t) = \hat{X}(v)\hat{S}(t), \tag{3.4.28}$$

where $\hat{R}$ is the response, $\hat{S}$ the stimulus and $\hat{X}$ the response function ($\hat{X} = X' + iX''$). The Principle of Causality then requires the real and imaginary parts of $\hat{X}$ to be related to one another. R and S may be vectors and $\hat{X}$ is then a tensor but the only difference is that it is now equivalent components of X' and X'', which are necessarily related to one another.

To clothe the bare ideas of the "thought experiment" in mathematical dress[†] it is very useful to introduce the concept of complex frequency. Physically, frequency must be real and positive, but if one imposes this restriction too early in the mathematical process, the treatment becomes awkward. From a mathematical point of view, time and frequency are Fourier conjugates and, since the former runs from $-\infty$ to $+\infty$, it is useful to have the symmetry that the latter should run between the same limits. Furthermore having taken this step one might as well go the whole hog to complex frequency, for then the very powerful theorems which arise in the theory of functions of a complex variable are available to handle the mathematical difficulties. The notions of the "thought experiment" can then be translated as follows. The principle of causality introduces an arbitrary asymmetry into the time variable and it must

---

[†]For an alternative treatment see Appendix 1.

therefore introduce an equally arbitrary asymmetry into the frequency variable. This asymmetry takes the form that all the poles of the complex response function (i.e. values of $\hat{v}$ where $\hat{X}(\hat{v})$ is infinite) lie in the lower[†] half of the complex plane. $\hat{X}(\hat{v})$ is therefore analytic throughout the upper half and it can be shown [160], that for any function with this property, the real and imaginary parts are related by

$$\hat{X}(\hat{v}) = \frac{i}{\pi} P \int_{-\infty}^{+\infty} \frac{\hat{X}(\hat{v}')}{\hat{v}' - \hat{v}} d\hat{v}', \tag{3.4.29}$$

where P signifies the Cauchy principal value, that is the singularity at $v' = v$ is treated by assuming that

$$\int_{-\infty}^{+\infty} \frac{\hat{X}(v')}{v' - v} dv' = \underset{\delta \to 0}{Lt} \left[\int_{\infty}^{v'-\delta} \frac{\hat{X}(v')}{v' - v} dv' + \int_{v'+\delta}^{+\infty} \frac{\hat{X}(v')}{v' - v} dv' \right]. \tag{3.4.30}$$

The path of integration in (3.4.29) is along the real axis in the complex plane, so both v' and v in equations (3.4.29) and (3.4.30) are real.

Identifying the real and imaginary parts of (3.4.29) one now has

$$X'(v) = +\frac{1}{\pi} P \int_{-\infty}^{+\infty} \frac{X''(v)}{v' - v} dv', \tag{3.4.31a}$$

and

$$X''(v) = -\frac{1}{\pi} P \int_{-\infty}^{+\infty} \frac{X'(v)}{v' - v} dv'. \tag{3.4.31b}$$

These equations and their analogues for the various explicit response functions are known as the Kramers–Kronig dispersion relations. They arise purely because of the form of $\hat{X}(\hat{v})$ and, since this has its origin in the principle of causality, it follows that equations of the form (3.4.31) will be general and will apply regardless of any particular choice of a system of mechanics. This is one reason why the quantum mechanical theory of infrared propagation differs only qualitatively from the classical Lorentz theory [133]. Equations (3.4.31), however, still involve negative frequencies and, whilst this is no handicap in mathematical treatments of explicit functions, it is an impossible hurdle if experimental values of one part of $\hat{X}(v)$ are to be transformed to give

---

[†] This particular asymmetry comes from the choice of a positive sign for the imaginary component of $\hat{X}$. Sometimes the opposite convention is encountered and then the poles of $\hat{X}$ must lie in the *upper* half of the complex plane.

the other. We therefore require modifications of equations (3.4.31) to apply for the case where v is not only real but also positive. To do this, it is necessary to derive relations which give $X'(-v)$ and $X''(-v)$ when $X'(v)$ and $X''(v)$ are known. The first step is to define a Fourier transform of $S(t)$, thus

$$\hat{S}(v) = \int_{-\infty}^{+\infty} \hat{S}(t)\exp(2\pi ivt)\,dt. \qquad (3.4.32)$$

Strictly, in this equation, different symbols should be used since $\hat{S}(v)$ and $\hat{S}(t)$ have different dimensions, but the convention to imply the difference by explicit quotation of the argument, is so well established that there is now little point in flouting it. Using (3.4.32), one may now write an expression for $\hat{R}(t)$ thus:

$$\hat{R}(t) = \int_{-\infty}^{+\infty} \hat{X}(v)\,\hat{S}(v)\exp(-2\pi ivt)\,dv. \qquad (3.4.33)$$

It should be noted in passing that $\hat{S}(v)$ will be symmetrical about the ordinate axis, i.e. $\hat{S}(-v) = \hat{S}(v)$. One next considers the application of a purely real stimulus and makes the physical requirement that the response also be purely real. The imaginary part of (3.4.33) is

$$\text{Im}\left[\hat{R}(t)\right] = -i\int_{-\infty}^{+\infty} S(v)[X'\sin 2\pi vt + X''\cos 2\pi vt]\,dv. \qquad (3.4.34)$$

If this is to be zero, then from the symmetry properties of $S(v)$ it follows that $X'(v)$ will likewise be symmetrical about the ordinate axis whereas X'' will be antisymmetric; in other words

$$X'(-v) = X'(+v), \qquad (3.4.35a)$$

$$X''(-v) = -X''(+v). \qquad (3.4.35b)$$

The final step is to write, say for X'',

$$X''(v) - X''(-v) = \frac{1}{\pi}P\int_{-\infty}^{+\infty} \frac{X'(v')}{v'-v}\,dv + \frac{1}{\pi}P\int_{-\infty}^{+\infty} \frac{X'(v)}{v'+v}\,dv' \qquad (3.4.36)$$

and, applying (3.4.35),

$$2X''(v) = -\frac{1}{\pi}P\int_{-\infty}^{+\infty} X'(v')\left[\frac{1}{v'-v} - \frac{1}{v'+v}\right]dv' \qquad (3.4.37)$$

and therefore

$$X''(v) = -\frac{v}{\pi}P\int\limits_{-\infty}^{+\infty}\frac{X'(v')}{(v')^2 - v^2}\,dv'. \tag{3.4.38}$$

This function, unlike (3.4.31), involves an integrand which is symmetrical about $v' = 0$ and therefore one may write:

$$X''(v) = -\frac{2v}{\pi}P\int\limits_{0}^{\infty}\frac{X'(v')}{(v')^2 - v^2}\,dv'. \tag{3.4.39}$$

The corresponding expression for the real part of $\hat{X}$ is

$$X'(v) = +\frac{2}{\pi}P\int\limits_{0}^{\infty}\frac{v'X''(v')}{(v')^2 - v^2}\,dv'. \tag{3.4.40}$$

The use of equations (3.4.39) and (3.4.40) was once widespread in experimental physics, since it used to be much easier to determine one component of a complex response function over a wide range of frequency than it was to measure the other at a series of spot frequencies. New techniques, such as dispersive Fourier transform spectrometry (section 4.3.4) in which both components are simultaneously determined are, however, now providing a serious challenge especially in the long wavelength infrared but there are still situations where the Kramers–Kronig approach is useful. Refractive index is a good case in point. The analogue of (3.4.40) for the response function $n = n + i\alpha/4\pi\tilde{v}$, is [161]

$$n(v) - n(\infty) = \frac{c}{2\pi^2}P\int\limits_{0}^{\infty}\frac{\alpha(v')}{(v')^2 - v^2}\,dv'. \tag{3.4.41}$$

The quantity $n(\infty)$ has been introduced here so that the LHS of (3.4.41), like the RHS, will be zero when v is infinite. Equation (3.4.41) is extremely useful in experimental infrared spectroscopy, for it enables the dispersion to be calculated once the absorption spectrum is known. Dispersion measurements, by the classical methods, are never easy and for heavily absorbing specimens, where the wave may well be essentially extinguished within a wavelength, often impossible. Yet if reasonable α values are known good dispersion data can be computed. It is very important to note that from the form of (3.4.41), the dispersion arises from all regions where $\alpha(v')$ is finite and not just from the regions where $\alpha(v')$ is high. It is for this reason that, even if peak α values cannot be reliably determined, good n values may nevertheless result.

    The inverse case, i.e. calculating α, given data on the refractive index over a wide frequency range via the analogue of equation (3.4.41),

$$\alpha(v) = -\frac{8v^2}{c} \int_0^\infty \frac{[n(v') - n(\infty)]}{(v')^2 - v^2} dv', \qquad (3.4.42)$$

is very uncommon, but calculating integrated strength from refraction data is frequently resorted to. If one has an isolated resonance as shown in Fig. 3.8, then there will be a "step" in refractive index in going over the band. If $n(v_-)$ is the refractive index at a frequency in the low-frequency plateau and $n(v_+)$ that at a frequency in the high-frequency plateau, then one may write (via 3.4.41)

$$\Delta n = n(v_-) - n(v_+) = \frac{c}{2\pi^2} \int_0^\infty \alpha(v) f_v \, dv, \qquad (3.4.43a)$$

where

$$f_v = (v^2 - v_-^2)^{-1} + (v_+^2 - v^2)^{-1}. \qquad (3.4.43b)$$

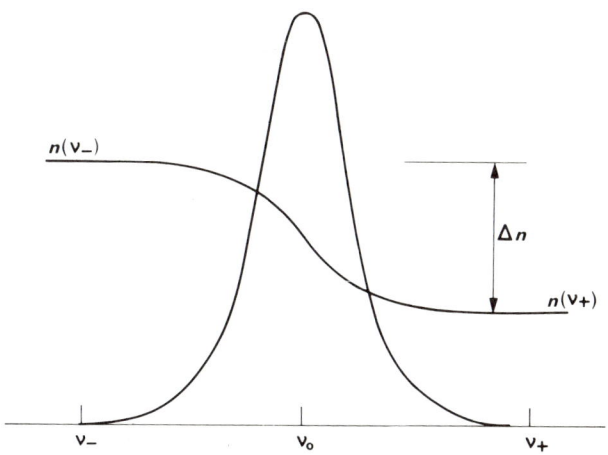

FIG. 3.8. An isolated resonance at frequency v_0 will have associated with it a sigmoid variation of refractive index whose slope is a maximum at $v = v_0$. The "step" in refractive index Δn, is a measure of the strength of the resonance.

The approximation is now made that one can truncate the integral by calculating it only over that narrow range, which includes most of the band, that is say over a frequency range several times the half-width. With this approximation f_v becomes essentially constant and can without serious error be written

$$f_v' = (v_0^2 - v_-^2)^{-1} + (v_+^2 - v_0^2)^{-1} \qquad (3.4.44)$$

and it follows that

$$\Delta n = \frac{c}{2\pi^2} f'_v \int_0^\infty \alpha(v)\,dv.$$
(3.4.45)

Thus, from two measurements of n, the integrated intensity may be inferred. The quantity f'_v is not very sensitive to uncertainties in v_0 (the band-centre frequency), so one may accurately deduce integrated intensities of bands so intense that their peaks are inaccessible.

There are, however, circumstances where the absorption is so intense and covers so wide a spectral range that even the methods outlined above fail. The experimentalist is not necessarily beaten even in these cases and he can still deduce values for n and α, if he can measure the amplitude and phase shift of a ray reflected from the very lossy material. The complex reflectivity (see next section) can be written

$$\hat{r} = \rho^{1/2} \exp(i\theta),$$
(3.4.46)

where ρ is the power reflectivity and θ is the phase change on reflection. The optical properties of the medium can be extracted from $\hat{r}$ regardless of the angle of incidence, but, for all except normal incidence, this is a difficult procedure complicated by a dependence on the degree of polarization of the incident beam. Experiments, therefore, are usually done at as close an approach to normal incidence as is possible. Under this restriction, the complex reflectivity may be written

$$\hat{r} = (1 - \hat{n})(1 + \hat{n})^{-1},$$
(3.4.47)

from which it follows that

$$n = (1 - \rho)(1 + \rho + 2\rho^{\frac{1}{2}}\cos\theta)$$
(3.4.48a)

and

$$\alpha = \frac{4\pi v}{c}\left(\frac{-2\rho^{\frac{1}{2}}\sin\theta}{1 + \rho + 2\rho^{\frac{1}{2}}\cos\theta}\right).$$
(3.4.48b)

The minus sign in this equation nicely compensates for the negative sign of $\sin\theta$ (which is a consequence of θ being constrained to be equal to or greater than π radians) and α is positive as, of course, it must be.

The measurement of ρ is a fairly straightforward procedure, but the measurement of θ is extremely difficult. The method essentially is to use a Michelson interferometer which, as will be shown in Chapter 4, can be used to measure changes of phase occurring in one of its arms. If one of the mirrors of the interferometer is replaced by the substance under investigation, in such a way that there is exact spatial substitution of the mirror's surface by the front surface of the substance, then any phase changes measured will arise entirely from the changes of phase on reflection. The difficulties arise from the tolerances expected in the substitution. These have to be of the order $10^{-2}\lambda$ and therefore, even in the far infrared, it is necessary to replace one surface by another to an accuracy of 1 μm. It is for this reason that the measurements so far done have been restricted to far infrared wavelengths. Bell [162], Russell and Bell [163], Gast and Genzel [164], Parker et al. [165], Birch et al. [166] and Staal and Eldridge [167] have managed to overcome some of the difficulties and have reported absorption and dispersion data for highly absorbing solids, obtained from complex reflectivity measurements. Studies on highly absorbing liquids, such as water, began with the work of Chamberlain [168] and continued with that of Afsar and his colleagues [169]. This work is particularly important because the determination of the optical parameters promises to give insight into the microdynamics of this most important liquid.

The direct method just mentioned is very attractive and it will doubtless continue to be applied in the future to specialised systems, especially heavily absorbing liquids at mid- to far-infrared wavelengths, but there will continue to be specimens and wavelength regions for which it may prove very difficult or even impossible to measure θ to the required accuracy. Recourse must then be had to the less attractive Kramers–Kronig methods. The mathematics is now much more complicated because, unlike the previous cases, one is not measuring one component of a complex response function and inferring the other, one is measuring one combination (the modulus) and trying to infer another (the phase). The relation between ρ and θ, namely [170]

$$\theta(v) = \frac{1}{\pi} \int_0^\infty \left[\frac{d}{dv'}(\ln\rho)^{\frac{1}{2}} \right] \left[\ln\left| \frac{v+v'}{v-v'} \right| \right] dv', \qquad (3.4.49)$$

is therefore rather complicated and the transformation is really only practicable in a computer. A major difficulty arises because of the truncation of the integral. This is a problem in all cases but particularly so for reflection work and here some form of analytical continuation is necessary and has to be assumed. There is clearly a wide range of choice for the analytical continuation and the uncertainties which may thereby be introduced should be borne in mind when assessing quantities derived via this technique. If the continuations seem natural or even inescapable the calculations can be accepted with confidence, but if they seem arbitrary or dictated by mathematical convenience rather more caution is advised. Some idea of the power of the method is shown in Fig. 3.9 which gives dielectric data for the ferroelectric material thiourea derived from far infrared power relection measurements made on a single crystal [171].

3.4.6 *Propagation in dispersive media*

It has been remarked several times in what has gone before that mono-chromatic radiation traversing a lossless medium of relative permittivity ε_r, propagates with a velocity

$$v_p = c/(\varepsilon_r)^{1/2} = c/n \tag{3.4.50}$$

where $n = \varepsilon_r^{1/2}$ is the refractive index. The velocity v_p is usually referred to as the phase velocity since it is the rate at which co-phasal surfaces propagate. This relation, however, should be regarded more as a definition than as a statement of physically verifiable fact since the endless cosine wave which corresponds to the monochromatic input will of necessity lack identifiable features and there will therefore be no way of measuring its rate of propagation. This difficulty would be immediately resolved if one could assume that n was constant over a reasonably wide frequency band since one

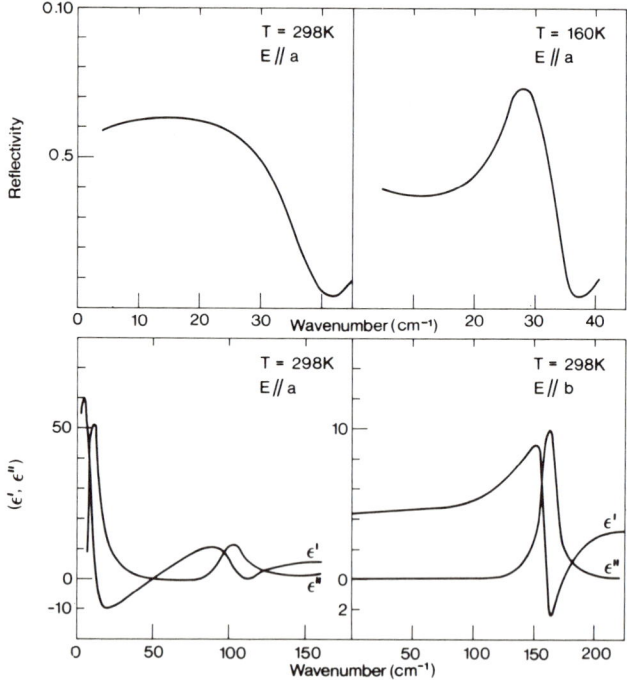

Fig. 3.9. Power reflectivity and the derived dielectric functions for the ferro-electric crystal thiourea $CS(NH_2)_2$. The upper insets show the reflectivity for $E//a$ above and below the phase transition at 202 K. The lower insets show ε' and ε'' for $E//a$ and for $E//b$ at room temperature. That for $E//b$ is the best example known so far of perfect classical oscillator behaviour in the internal mode spectrum of a molecular crystal. The original data come from the work of Fleming *et al.* [171].

could then imagine that a pulse of radiation was propagating and that this pulse had an identifiable grand maximum. All that would be required would be for the spectrum of the pulse to lie wholly within the spectral band over which n was constant to ensure that the pulse propagated without any change in its shape. The velocity of the pulse could now be readily determined by time-of-flight methods. Unfortunately, for this recipe, all real media are found to be more or less dispersive and we have to face up to the fact that we have no physically exact way of defining the velocity of propagation.

This problem is actually quite old and was encountered early on by lens makers who were trying to combat the effects of chromatic aberration in their optical systems. The terms they introduced, *normal* when the dispersion was in the sense of n increasing with frequency and *anomalous* when it was in the reverse sense, are used to this day. Most optical glasses have n increasing as one goes from the red to the blue, this being a Kramers–Kronig consequence of the existence of intense absorption bands in the ultraviolet. Hence the choice of the epithet "normal" for this sort of behaviour. These terms have been transferred to the infrared but, at these longer wavelengths, anomalous behaviour is not so rare. Thus near the centre of a strong isolated resonance (Fig. 3.8), there will always be a region of anomalous dispersion. Dispersion can be quite intense in the infrared and it has numerous consequences for infrared technology. It sets, for example, a fundamental limit to the rate at which information can be sent down a fibre-optic communication cable (see later). This is because the shape of an input pulse will steadily alter with distance of propagation. This pulse-shape change may eventually become so large that the engineer is forced to adopt band-width restrictions in order to prevent the messages becoming garbled. Band-width restriction, of course, implies a slowing down of the rate of information flow.

Although it is thus true in the strictest sense that the concept of rate of propagation though a dispersive medium may be meaningless or at least not useful, optical physicists nevertheless try to find some definition which will work in practice. One can begin by taking the endless wave given previously as equation (3.4.21) and, considering this as a plane wave propagating in the z direction, write it in the form

$$E(z, t) = E_0 \cos(2\pi v t - kz).$$ (3.4.51)

To take one step nearer reality one next imagines that a second wave is present whose frequency is $(v + \delta v)$ and whose wave vector is $(k + \delta k)$. This wave will be written

$$E(z,t) = E_0 \cos[2\pi(v + \delta v)t - (k + \delta k)z].$$ (3.4.52)

These two waves combine to give the resultant

$$E(z, t) = 2E_0 \cos\left[2\pi\left(v + \frac{\delta v}{2}\right)t - \left(k + \frac{\delta k}{2}\right)z\right]\cos\left(\pi\delta v t - \frac{\delta k}{2}z\right).$$ (3.4.53)

If one can assume a narrow spread of frequency and only moderate dispersion so that $\delta v \ll v$ and $\delta k \ll k$, then equation (3.4.53) can be approximated to read

$$E(z, t) = 2E_0 \cos(2\pi v t - kz) \cos\left(\pi \delta v t - \frac{\delta k}{2}z\right). \qquad (3.4.54)$$

This represents a wave with "beats", i.e. there is a relatively high frequency wave, modulated at much lower frequency. Individual crests propagate at the phase velocity given previously (3.4.50) since

$$v_p = 2\pi v/k = c/n, \qquad (3.4.55)$$

but, as mentioned earlier, this quantity may not be physically meaningful. It certainly would not be in a region of strong anomalous dispersion where n may be less than unity and v_p would exceed c. This of course violates the postulates of the theory of relativity which forbid velocities greater than that of light in free spece for any wave capable of carrying a signal or transferring energy.

Since we are interested only in waves which do both we enquire whether there is any quantity in (3.4.54) which is related to the velocity of energy propagation. We therefore fix our attention not on the individual crests but on the momentarily greatest crest. This is found when the modulating function in (3.4.54) is unity, that is when

$$\pi \delta v t - \frac{\delta k}{2}z = 2m. \qquad (3.4.56)$$

Differentiation of this gives

$$\delta k dz = 2\pi \delta v dt, \qquad (3.4.57a)$$

that is

$$\frac{dz}{dt} = 2\pi \frac{\delta v}{\delta k} = \frac{\delta \omega}{\delta k}. \qquad (3.4.57b)$$

The momentarily greatest crest is therefore propagating at a velocity given by (3.4.57b). Since δv and δk are small quantities one may write

$$v_g = \frac{dz}{dt} = 2\pi \frac{dv}{dk} = \frac{d\omega}{dk}, \qquad (3.4.58)$$

where v_g, being the rate of propagation of the momentarily greatest crest, will be also the rate of propagation of a wave group, provided the dispersion is not too great. Using equation (3.4.22), it readily follows that this group velocity v_g is given by

$$v_g = v_p\left[1 + \frac{v}{n}\left(\frac{dn}{dv}\right)\right]^{-1} \approx v_p\left[1 - \frac{v}{n}\left(\frac{dn}{dv}\right)\right]. \qquad (3.4.59)$$

Three important conclusions can be drawn from this equation. Firstly when (dn/dv) is zero, i.e. when there is no dispersion, the group velocity equals the phase velocity. Secondly, when the dispersion is weak and normal, the group velocity is a good measure of the rate of propagation of the disturbance and hence of the energy flow. Thirdly, in regions of strong dispersion or in regions of anomalous dispersion, v_g may become formally negative or else may exceed c and it is then no longer a useful concept. In either case one is back to the problem we started with of having no unequivocal way of defining the rate of propagation of the electromagnetic disturbance through the dispersive medium.

Dispersion in a medium is often demonstrated pictorially by means of a dispersion diagram in which one plots circular frequency ω as ordinate against the magnitude of the wave vector k as abscissa. A representative example is shown in Fig. 3.10. The phase velocity at a point P is given by the slope of OP whilst the group velocity is given by the slope of the tangent at P (i.e. O'P). When the plot of ω versus k is a straight line through the origin, there is no dispersion; when the plot is curved, there is dispersion, normal when the slope of O'P is less than that of OP and anomalous when it is greater.

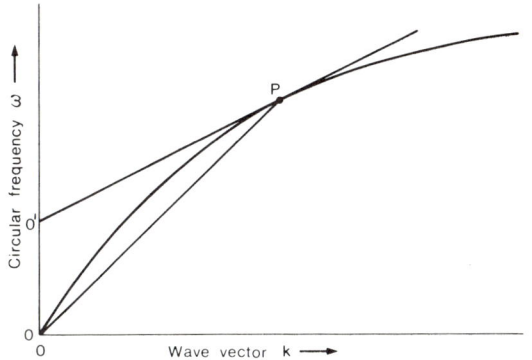

FIG. 3.10. A schematic dispersion diagram showing a possible dispersive variation of ω with k.

Dispersion diagrams are very valuable in both the classical and quantum mechanical treatment of many diverse systems. Thus they are used in treating electromagnetic propagation, acoustic wave propagation and the vibrations of crystals. They are particularly valuable in quantum mechanics because frequency is equivalent to energy and wave vector to momentum and both axes, therefore, correspond to conserved quantities.

3.4.7 Propagation in non-linear media

In sections 2.6.3 and 2.6.4, we saw how the absorption coefficient could depend on the intensity and how this could give rise to phenomena such as saturation.

The refractive index will also, in general, depend on the intensity because of the necessary (causality) connection between the two. Two important cases arise: firstly where the non-linear response is due to population shifts and secondly where it is due to a basic non-linearity of a constitutive property of the medium. In the first case, usually observed with gaseous specimens, the depression of the absorption maximum is matched by a corresponding reduction of the amplitude of the associated dispersion as shown in Fig. 3.11.

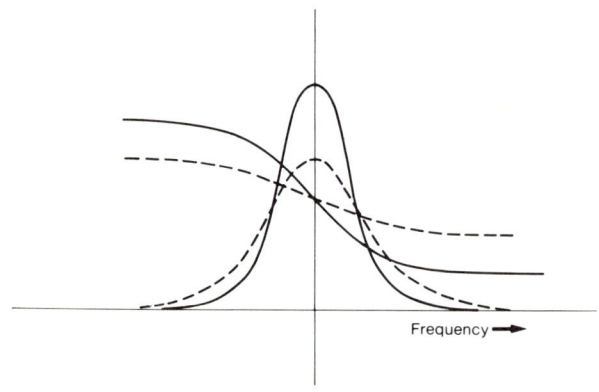

FIG. 3.11. Absorption and dispersion due to an isolated resonance for the cases of normal observation (solid lines) and power-broadened observation (broken lines).

The effects of the non-linearity are very local and there is no effect where the gas does not absorb. In the second case, there can be a power dependence of the refractive index even at frequencies where the specimen does not absorb significantly or where it shows no power dependence of the absorption. This type of behaviour is found most easily in condensed phases. One may write

$$n_v(I) = n_v(0) + \left(\frac{\partial n}{\partial I}\right)I + \tfrac{1}{2}\left(\frac{\partial^2 n}{\partial I^2}\right)I^2 + \text{etc.} \tag{3.4.60}$$

for the refractive index at a frequency v and at a beam intensity I. The coefficients appearing in this equation could, in principle, be calculated using *ab initio* quantum mechanical methods but such calculations are too difficult to contemplate for the far from ideal systems used in practical science and technology so the usual practice is to determine them experimentally.

The literature associated with non-linear optics is complicated by the almost universal use of the cgs/esu system of units and by the use of similar symbols for quantities which differ dimensionally. Thus equation (3.4.60) is often given in the form

$$n = n_0 + n_2 \langle E^2 \rangle + n_4 \langle E^4 \rangle + \text{etc} \ldots = n_0 + \gamma I + \ldots \tag{3.4.60a}$$

where n_2 is given in esu units whilst γ is given in practical units! However, where numerical results are quoted in this book, they will be in SI units to maintain continuity with the rest of the text.

The intensity dependence of the refractive index is usually only observable with laser radiation and then only when either the wavefront is not uniformly intense or else when there is interference taking place. The commonest example of the former arises when there are Gaussian or beam modes propagating. Because of the form of the beam modes (see section 3.3), there is a radial variation of intensity and in a non-linear medium there will inevitably be some distortion of the wavefront. Assuming the beam to have the normal Gaussian profile, two simple results emerge depending on the sign of $(\partial n/\partial I)$, that is of n_2 or γ. If this non-linear parameter is positive, then the central part of the beam will be slowed down relative to the extremities and the beam will be converged just as though it had passed through a convex lens—one thus has *self-focussing*. If on the other hand $(\partial n/\partial I)$ is negative, then the central part will be speeded up relative to the extremities and one will have *self-defocussing*. These two cases are illustrated schematically in Fig. 3.12. The alteration to the wavefront brought about by the non-linearities is often a major nuisance in high-power laser design (where the effect is to cause mode-scrambling) and in the design of optical systems to be used with high-power lasers where inadvertent focussing effects can lead to unexpected damage problems.

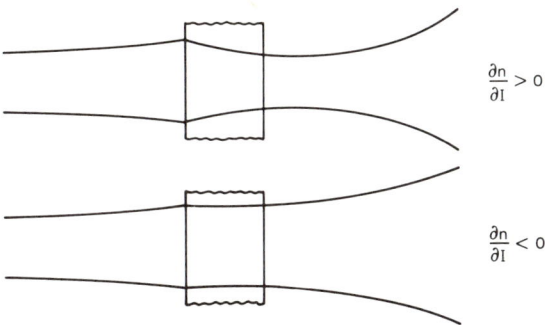

FIG. 3.12 Self-focussing (upper inset) and self-defocussing (lower inset) in a non-linear medium.

It was shown earlier that, for electromagnetic propagation at optical frequencies through linear media, diffraction effects can be treated quite adequately by a linearised analytical theory based on wave-optics. This approach is, however, quite inadequate for non-linear media where something rather closer to electromagnetism is required. Fortunately, for most cases of non-linear propagation, three conditions are satisfied which permit a reasonably tractable treatment. These are:

a. The beam is parallel and closely confined to the z axis so the propagation is paraxial.

b. The envelope function, that is ψ of section 3.3, is a slow function of z.

c. The non-linear coefficients, in the expansion of the permittivity as a function of field strength, are much smaller than the linear terms.

The formalism which led to the beam-wave equations is therefore valid and we may take equation (3.3.5) as it stands and just make an appropriate non-linear addition. The result is

$$\frac{\partial^2 \hat{A}}{\partial x^2} + \frac{\partial^2 \hat{A}}{\partial y^2} = 2ikn_0 \frac{\partial \hat{A}}{\partial z} - \beta \hat{A}, \qquad (3.4.61)$$

where n_0 is the linear refractive index, $\hat{A}$ is the analytical signal which has been used as a more general replacement for ψ and β is a propagative factor including both linear and non-linear terms. Its explicit form is

$$\beta = (3/4)k^2 \varepsilon_0 \varepsilon_2 |\hat{A}|^2 + 2ikn_0 \alpha_0, \qquad (3.4.62)$$

where, as usual, α_0 is the small signal gain of the medium and ε_2 is the quadratic non-linearity defined by

$$\varepsilon_r(E) = \varepsilon_r(0) + \varepsilon_2 |\hat{A}|^2. \qquad (3.4.63)$$

It is thus closely related to n_2 and its sign determines whether the medium will be self-focussing or self-defocussing. In the esu system it follows from (3.4.12) and (3.4.60a) that ε_2 is simply equal to $2n_0n_2$. Equation (3.4.61) reduces, as it must, to (3.3.5) for propagation in vacuum but, unlike (3.3.5), it does not in general have analytical solutions. Computer programs for solving partial differential equations are, however, well advanced and they can be applied to the analysis of the propagation in any particular high-power system. The answers which emerge are quantitatively invaluable to the system design engineer but their qualitative content can be understood within a much simpler heuristic approach. Here one notes that for ε_2 small the beam profile will not be much affected as it propagates and the energy extraction will be that predicted by linear theory, whereas for ε_2 large the central parts of the beam will be saturated and the outer parts, that is the diffraction lobes, will thus be relatively enhanced. Energy extraction from the system is then entirely diffraction dominated. This is an unsatisfactory situation so the engineers try to find ways of reducing ε_2. For the solid-state laser systems such as Nd/glass this can be done by a careful choice of the glass composition so that components which would confer a positive ε_2 are balanced by those which would confer a negative value. In this way non-linear diffraction effects can be minimised.

The effect of an intensity dependent refractive index on interference phenomena is a more recent field of study. The usual experimental arrangement consists of a fixed separation Fabry–Perot etalon made from a plane-parallel plate of a suitable semiconductor [36]. Depending on the laser wavelength, the interferometer will either be at resonance, at anti-resonance, or else somewhere in between. If one plots a diagram of power-out versus power-in, the result would be expected to be a straight line lying somewhere between the two extremes of Fig. 3.13. If, however, the etalon material is non-linear, then the behaviour would be expected to oscillate between the two extremes as the changing refractive index brings the interferometer firstly into resonance and then into anti-resonance. The sense of the oscillation will be determined by the sign of $(\partial n/\partial I)$. It is found experimentally that semiconductors such as indium antimonide and gallium arsenide show large values of $(\partial n/\partial I)$ for wavelengths near their band edges and behaviour similar to that shown in Fig. 3.13 may be easily demonstrated using the powerful radiation from a CO or CO_2 laser [172, 173].

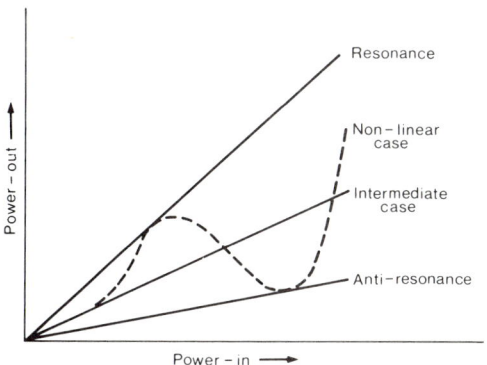

FIG. 3.13. Linear (solid lines) and non-linear (broken line) response of a Fabry–Perot etalon.

The actual behaviour observed is, however, even more dramatic than might be expected from the analysis so far. To see why this might be so one imagines that one has a low-power beam at a frequency which is out of resonance with the etalon. As the power is increased the etalon starts to be pulled into resonance by the intensity inside it, but, of course, as it approaches resonance its transmissivity increases and so, therefore, does the internal intensity. One clearly has a potentially run-away situation and in principle can get sudden catastrophic switching from a low transmission state to a high transmission state and vice versa. The graph of power-out versus power-in is not gently varying: it is more like a staircase! This phenomenon is called *optical bistability* [174]. Smith and his colleagues at Heriot Watt University have investigated optical bistability in indium antimonide using the radiation from a CO laser which happens to lie near the band edge (1900 cm^{-1}) for very low temperature (~ 5 K) observation. They not only observed the "staircase"

[173], they also observed modulation of a weak "signal" beam transferred to a strong pump beam [175, 176]. They suggest that this phenomenon could lead to optical devices similar to transistors and have even gone so far as to coin a suitable word for the device—the "transphasor". The consequences of this sort of device for the development of true (i.e. "all optics, no wires") integrated optical circuitry can hardly be exaggerated.

3.5 Propagation at a discontinuity of refractive index—the Fresnel equations

When a plane wave encounters a discontinuity of refractive index, part of the intensity is transmitted and part is reflected. The distribution of the incident energy into the two resulting beams depends on the magnitude of the discontinuity, on the state of the polarisation of the incident beam and on the angle of incidence. The explicit relations can be derived from an analysis of the boundary conditions which prevail at the discontinuity. One first considers the wavevector of the incident beam and its relationship to that of the transmitted beam. The treatment can be general, that is the incident beam can be propagating in a lossy medium and after crossing the boundary continue in a second lossy medium; however for nearly all the situations considered in infrared physics, one or the other medium is virtually transparent, so one need only consider this case. One thinks, therefore, of a wave with wave vector $\mathbf{k}_i = n_1 \mathbf{k}_0$, incident at an angle θ on a medium with a complex refractive index $\hat{n}_2 = n_2 - i\alpha_2/4\pi\tilde{\nu}$. The situation is shown schematically in Fig. 3.14. The wave vector of the transmitted beam is complex and for all angles of incidence except zero the transmitted wave is inhomogeneous. The

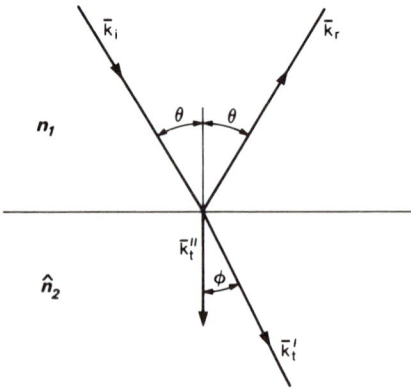

Fig. 3.14. A wave travelling in a transparent medium of refractive index n_1 is incident upon an absorbing medium of index n_2. The incident and reflected wave vectors are both pure real and are equal in magnitude. The transmitted wave vector is complex with components k'_t and k''_t.

boundary conditions which must be satisfied are that (1) the tangential component of wave vector must not change across the boundary and (2) the complex wave vector of the transmitted beam must obey equation (3.4.27). These two conditions lead to the equations

$$n_1 k_0 \sin \theta = k_t' \sin \phi,$$
$$k_0[(n_2 - i\alpha_2/4\pi\tilde{v})^2 - n_1^2 \sin^2 \theta]^{1/2} = k_t' \cos \phi + ik_t''. \tag{3.5.1}$$

When α_2 is zero, that is both media are transparent, these two equations reduce to the well-known relationship

$$\sin \theta/\sin \phi = n_2/n_1, \tag{3.5.2}$$

which is known as Snell's Law. By applying similar boundary conditions to the electric and magnetic vectors of the wave, the Fresnel equations

$$\hat{r}_{//} = \frac{\hat{n}_2 \cos \theta - n_1 \cos \phi}{\hat{n}_2 \cos \theta + n_1 \cos \phi} \qquad \hat{t}_{//} = \frac{2n_1 \cos \theta}{\hat{n}_2 \cos \theta + n_1 \cos \phi}$$

$$\hat{r}_{\perp} = \frac{n_1 \cos \theta - n_2 \cos \phi}{n_1 \cos \theta + \hat{n}_2 \cos \phi} \qquad \hat{t}_{\perp} = \frac{2n_1 \cos \theta}{n_1 \cos \theta + \hat{n}_2 \cos \phi} \tag{3.5.3}$$

result. In these equations, r and t are the *amplitude* reflection and transmission coefficients. It should be noted carefully that they are functions *only* of the optical changes *at the interface*. To derive the optical behaviour of a real component involves the consideration of at least *two* interfaces and, of course, of what happens to the beam in traversing the distance between them. This will be discussed in some detail later. For the special case of normal incidence, $\theta = \phi = $ zero and there is no longer any defined plane of incidence. The moduli of $r_{//}$ and $r_{\perp}$ become equal as do those of $t_{//}$ and $t_{\perp}$. One then has the relations

$$r = |\hat{r}| = \left| \frac{\hat{n}_2 - n_1}{\hat{n}_2 + n_1} \right| \quad \text{and} \quad t = |\hat{t}| = \left[\frac{2n_1}{n_1 + \hat{n}_2} \right]. \tag{3.5.4}$$

The first of these was quoted earlier in connection with reflection spectroscopy. It is worth noting that although (3.5.3) gives the *magnitudes* of $r_{//}$ and $r_{\perp}$ equal when $\theta = \phi = $ zero, it gives them opposite *signs*. This difference has no physical significance, it is purely conventional, merely "reflecting" the fact that a mirror apparently interchanges left and right whilst leaving up and down unaffected. This point will be taken up again later.

When both media are transparent, or nearly so, it is useful to combine Snell's law with the Fresnel equations to produce the alternative forms

$$r_{//} = \frac{\tan(\theta - \phi)}{\tan(\theta + \phi)}, \qquad t_{//} = \frac{2\sin \phi \cos \theta}{\sin(\theta + \phi)\cos(\theta - \phi)},$$

$$r_{\perp} = -\frac{\sin(\theta - \phi)}{\sin(\theta + \phi)}, \qquad t_{\perp} = \frac{2\sin \phi \cos \theta}{\sin(\theta + \phi)}. \tag{3.5.5}$$

A remarkable consequence of the first of these relations is that when $\theta + \phi = \pi/2$, the denominator is infinite and r, therefore zero. This means that if an unpolarized beam of parallel radiation is incident on a dielectric surface, with this special angle of incidence, the reflected beam will be 100% plane polarized. The special angle of incidence θ_p is sometimes called the polarizing angle or more commonly the Brewster angle, after Brewster who first noticed the phenomenon early in the nineteenth century. From Snell's law it follows that the polarizing angle is given by

$$\theta_p = \tan^{-1}(n_2/n_1). \tag{3.5.6}$$

The division of *power* at the interface can readily be derived by invoking equation (1.4.11), together with Lambert's law equation (2.2.66). One then has

$$\rho = rr^* = |\hat{r}|^2, \quad \tau = \left(\frac{n_2}{n_1}\right)\left(\frac{\cos\phi}{\cos\theta}\right)|\hat{t}|^2. \tag{3.5.7}$$

It can be verified immediately, by substituting from equation (3.5.3) that

$$\rho + \tau = 1 \tag{3.5.8}$$

and energy is conserved, at the interface, as, of course, it must. The consequences of equation (3.5.7) for the case of an interface between a lossless dielectric with $n = 1\cdot5$ and the vacuum are illustrated in Fig. 3.15. Because of

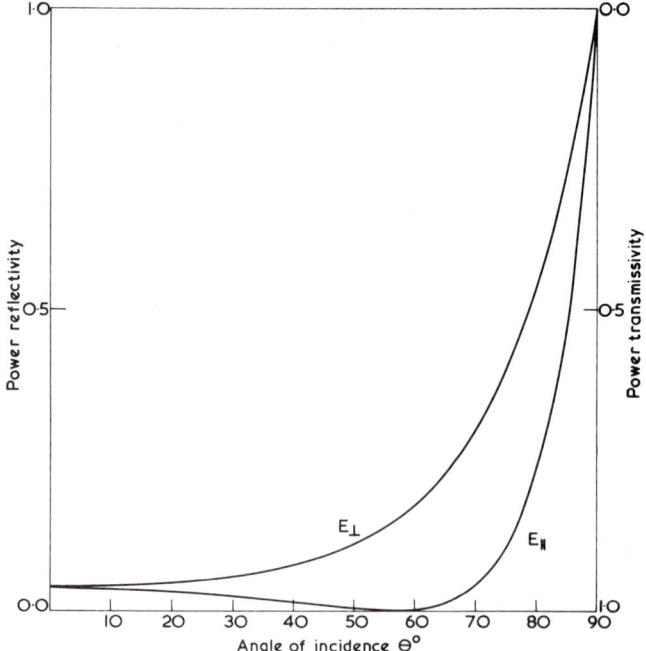

FIG. 3.15. Reflectivity and transmissivity at a dielectric interface.

(3.5.8), the same curves with merely a suitable relabelling of the ordinate axis can be used to display both ρ and τ.

The Fresnel equations have many applications in infrared physics but one of the most important is in the analysis of the interaction of a beam with a thin plane-parallel dielectric sheet. Plane parallel specimens are encountered in such diverse components as cell windows, filters and beam dividers. In all these cases, the beam is travelling in air or vacuo before it strikes the specimen and n_1 may therefore be set equal to unity and the subscript dropped from n_2. In general, the plane parallel specimen will be absorbing and the angle of refraction is therefore formally complex. Now it is possible to carry through the analysis with a complex angle of refraction but, since there is no simple visualisation of the meaning of this concept, it is more usual to separate propagation and absorption. One considers (Fig. 3.16) a narrow pencil of monochromatic radiation incident at an angle θ.

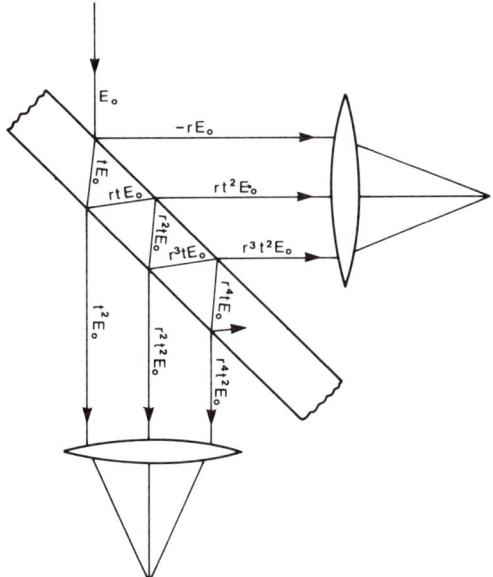

FIG. 3.16. Generation of multiple beams in a thin film. In the case of a lossless film, the sums of the squares of all the amplitudes will equal E^2 since energy must be conserved.

The beam is divided into a series of reflected beams and a series of transmitted beams; each series is recombined by a lens and the problem is to calculate the intensities at the focal points of the two lenses. For simplicity one can assume a linearly polarized beam, either parallel or perpendicular, so that one has only to consider an amplitude reflection coefficient r and an amplitude transmission coefficient t. The explicit values of r and t can be evaluated using the Fresnel equations but for the moment they will be treated as parameters. The last remaining point before the analysis continues is to note that for all the dielectric films considered in infrared systems the absorption is in reality very slight, that is $4\pi\bar{v}n \gg \alpha$, and therefore the phase shifts, in the reflected beams, which arise from the absorption can be ignored. With this assumption—equivalent to settng n_2 real in equation (3.5.3), all the reflection and transmission coefficients are real and the phase angle must be either zero or π, for these are the only values which make $\exp(-i\phi)$ real. It is, in fact, readily shown that ϕ is zero for all cases except for a reflection from and into a less dense medium at its interface with a more dense medium. Thus the first reflected ray suffers a phase change of π but all the rest of the beams only have phase changes by virtue of the distance travelled. Introducing now the delay parameter δ given by

$$\delta = 4\pi n d\,\bar{v} \cos\phi, \tag{3.5.9}$$

where d is the thickness of the plate and $\bar{v}$ the wavenumber, one can write down explicit relations for the electric field vectors in the two focal regions. The reflected beams, for example, contribute

$$\hat{E}_r(t) = \hat{r}E_0\big[-\cos 2\pi vt + t^2 \exp(-\alpha l)\cos(2\pi vt + \delta) \\ + t^2 r^2 \exp(-2\alpha l)\cos(2\pi vt + 2\delta) + \ldots\big] \tag{3.5.10}$$

where l is the slant thickness, i.e. $d(\cos\phi)^{-1}$ and where, because the beams are all in the same medium and are parallel to one another, $t^2 = 1 - r^2$. The resulting intensity is found by summing the series and taking the time average of the square modulus.

This procedure can be very tiresome and it is usual therefore to introduce powerful techniques which bypass much of the lengthy algebra. These involve the idea of the complex amplitude of a wave as developed in section 1.4. Using this approach, equation (3.5.10) may be written

$$\hat{R} = \hat{r}\big[-1 + t^2 a^2 \exp(i\delta) + t^2 r^2 a^4 \exp(2i\delta) + t^2 r^4 a^6 \exp(3i\delta) + \ldots\big]. \tag{3.5.11}$$

In this equation, R is the amplitude reflection coefficient of the *specimen* and a^2 has been written as a shorthand for $\exp(-\alpha l)$. The *amplitude* attenuation per pass is then a since α is, of course, a *power* absorption coefficient. The series (3.5.11) is readily summed, making use of $t^2 = 1 - r^2$, to give

$$\hat{R} = -\hat{r}\left[\frac{1 - a^2 \exp(i\delta)}{1 - r^2 a^2 \exp(i\delta)}\right], \tag{3.5.12}$$

which can be written

$$\hat{R} = \left[\frac{r^2(1 + a^4 - 2a^2 \cos\delta)}{1 - 2r^2 a^2 \cos\delta + r^4 a^4}\right]^{\frac{1}{2}} \exp i \arctan\left[\frac{-a^2 \sin\delta(1 - r^2)}{1 + r^2 a^4 - a^2 \cos\delta(1 + r^2)}\right]. \tag{3.5.12a}$$

The equivalent expressions for the transmitted beam are

$$\hat{T} = \frac{(1-r^2)a}{1-r^2\,a^2\exp{(i\delta)}}$$

(3.5.13)

and

$$\hat{T} = \left[\frac{(1-r^2)^2\,a^2}{1-2r^2\,a^2\cos\delta + r^4\,a^4}\right]^{\frac{1}{2}}\exp{i}\arctan\left[\frac{r^2\,a^2\sin\delta}{1-r^2\,a^2\cos\delta}\right],$$

(3.5.14)

where $\hat{T}$ is the amplitude transmission coefficient of the specimen.

Equations (3.5.12a) and (3.5.14) refer to different origins of phase, the latter being relative to that of the first emerging beam. To put them relative to a common origin, the quantity $i\,\delta/2$ should be added to the argument in equation (3.5.14). In the lossless case, it is found, as expected, that

$$|\hat{R}|^2 + |\hat{T}|^2 = 1.$$

(3.5.15)

3.5.1 *Normal incidence—alternative treatment*

Snell's law becomes indeterminate at normal incidence and this indeterminacy carries over to equations (3.5.5). There is no real difficulty here since one can always go back to the basic equations to derive the essential results, but it is nevertheless worth considering the normal incidence case from a rather different point of view since in this way a simple treatment emerges which can give some new insights into the underlying physics.

The boundary conditions adopted are (1) continuity of electric field across an interface, and (2) conservation of energy. We imagine an incident wave of amplitude $\hat{E}_i$ dividing into a transmitted wave of amplitude $\hat{t}\hat{E}_i$ and a reflected wave of amplitude $\hat{r}\hat{E}_i$. This is illustrated schematically in Fig. 3.17. Applying the boundary conditions at the first interface and assuming for the moment that n_2 is real gives

$$1+r = t,$$

(3.5.16)

$$n_1 = n_1 r^2 + n_2 t^2,$$

(3.5.17)

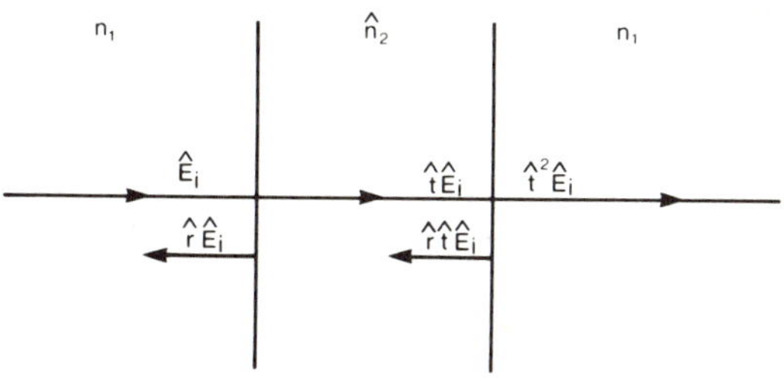

FIG. 3.17. Division of amplitude at two interfaces for the case of normal incidence. The reflected rays are shown vertically displaced for clarity. The index n_2 may be complex but n_1 is assumed to be real.

from which it follows that

$$r = \frac{n_1 - n_2}{n_1 + n_2} \quad \text{and} \quad t = \frac{2n_1}{n_1 + n_2}. \tag{3.5.18}$$

The solution for r given above agrees with that for $r_\perp$ given by equation (3.5.3) when θ is set equal to zero but it does not agree with $r_{//}$ under the same conditions. This is a well-known paradox, $r_{//} \to -r_\perp$ when $\theta \to 0$, despite the obvious physical requirement that $r_{//}$ should be identical with $r_\perp$ in the normal incidence case where the parallel and perpendicular labels lose their significance. The paradox arises from the formal definition of the sense of the field amplitudes, for, as shown in Fig. 3.18, the in-plane amplitude is formally reversed on reflection whereas the perpendicular one is not.

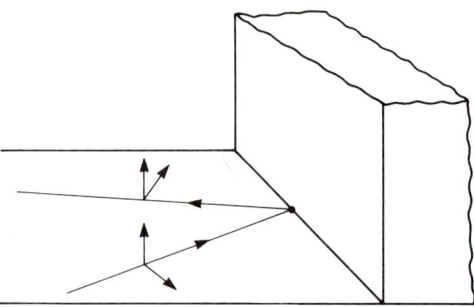

Fɪɢ. 3.18. Formal definition of the senses of the perpendicular and parallel components of the field before and after reflection at an interface.

Equations (3.5.18), because of the way they were derived, have no conventional content and give immediately the main physical result that if $n_1 > n_2$ there is no phase change on reflection, whereas if $n_1 < n_2$ there is a phase change by π. It should be noted here that, if one considers n_2 to vary from values greater than n_1 to values less, there is no difficulty with a discontinuity in the phase since the amplitude is zero at the point of switch-over. Another interesting point is that when crossing a boundary from a dense to a less dense medium the amplitude may *increase*! This does not, however, violate energy conservation since the intensity in a medium depends on the refractive index according to

$$I = 1/2 \, cn\varepsilon_0 |E|^2, \tag{3.5.19}$$

and the increase in amplitude just compensates for the fall in refractive index.

In the more general case where medium 2 has a complex refractive index one can derive the optical properties by the usual means of merely putting in the complex quantity wherever its real analogue occurs in the treatment of the non-absorbing case. With this recipe, one has

$$\hat{r} = \frac{n_1 - \hat{n}_2}{n_1 + \hat{n}_2} \quad \text{and} \quad \hat{t} = \frac{2n_1}{n_1 + \hat{n}_2}, \tag{3.5.20}$$

which, on rationalization, becomes

$$\hat{r} = \frac{n_1^2 - n_2^2 - k^2 - 2in_1 k}{(n_1 + n_2)^2 + k^2} \quad \text{and} \quad \hat{t} = \frac{2n_1[(n_1 + n_2) - ik]}{(n_1 + n_2)^2 + k^2}. \tag{3.5.21}$$

These equations show that for both n and k ($= \alpha/4\pi\bar{\nu}$) large, the reflectivity can approach unity, a point obviously germane to the properties of metals, but for many infrared situations we are interested in the properties of systems for which $4\pi\bar{\nu}n \gg \alpha$ and then it is permissible to approximate (3.5.21) thus

$$\hat{r} = \left(\frac{n_1 - n_2}{n_1 + n_2}\right) - i\left(\frac{n_1 \alpha}{2\pi\bar{\nu}(n_1 + n_2)^2}\right) \quad \text{and}$$

$$\hat{t} = \frac{2n_1}{(n_1 + n_2)} - i\left(\frac{n_1 \alpha}{2\pi\bar{\nu}(n_1 + n_2)^2}\right). \tag{3.5.22}$$

This shows that, as the absorption coefficient rises from zero, the complex number $\hat{r}$ moves into the fourth quadrant for $n_1 > n_2$ and into the third quadrant for $n_1 < n_2$. The shifts are always, of course, miniscule as was mentioned earlier and their determination is one of the greater difficulties in reflection spectroscopy.

The power reflection and transmission, at normal incidence, follow immediately from (3.5.20). Thus

$$\rho = \hat{r}\hat{r}* = \frac{(n_1 - n_2)^2 + k^2}{(n_1 + n_2)^2 + k^2}, \tag{3.5.23a}$$

$$. \tau = \hat{t}\hat{t}* = \frac{4n_1^2}{(n_1 + n_2)^2 + k^2}. \tag{3.5.23b}$$

The division of intensity *at the interface* therefore satisfies energy conservation since

$$\rho + (n_2/n_1)\tau = 1. \tag{3.5.24}$$

The reflected and transmitted beams from the sample, as a whole, will not satisfy (3.5.24) since some of the energy will be absorbed within it.

3.5.2 Channel or Edser–Butler fringes

From inspection of equation (3.5.14), it will be seen that the transmissivity of a plane parallel plate will be expected to vary as the probing frequency is varied because of the frequency dependence of δ. For a non-absorbing plate (i.e. $a = 1$) the transmitted intensity will be given by

$$\tau = \hat{T}\hat{T}* = \frac{(1 - r^2)^2}{1 - 2r^2 \cos\delta + r^4} \tag{3.5.25}$$

and will show maxima when $\cos \delta = 1$ and minima when $\cos \delta = -1$. Between these limits τ will vary in a smoothly oscillatory fashion as a function of δ. The plate therefore behaves as a low finesse Fabry–Perot interferometer (cf. equation 2.1.23a) and since the oscillations have their origin in multiple-beam interference, they are usually called "fringes" and in particular "channel fringes" or "Edser–Butler" fringes. They are familiar occurrences to anyone who has operated any form of infrared spectrometer for any length of time. The form of the fringes for the case of a silicon ($n = 3.4$, $r^2 = 0.3$) plate of thickness 0.1 mm is illustrated in Fig. 3.19. In practical spectroscopy, the contrast of the fringes is often degraded because of non-parallelism or non-homogeneity of the specimen or else by insufficient resolution of the spectrometer.

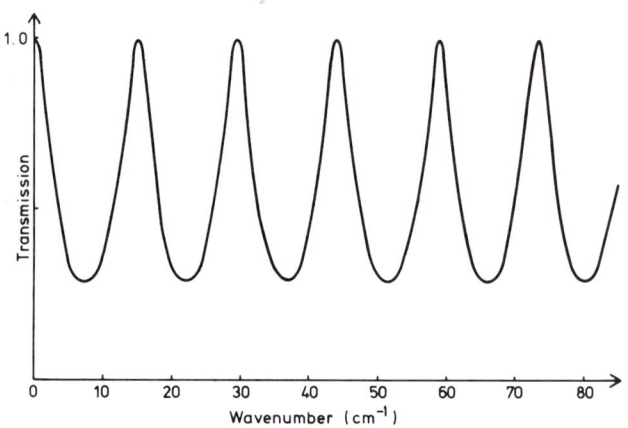

FIG. 3.19. Far-infrared transmission spectrum of a plane-parallel silicon wafer of thickness 0.1 mm. The curve shows the theoretically expected variation; practical spectra might show reduced contrast because of optical imperfections.

Channel fringes may, in principle, be observed in any spectral region but they are most commonly encountered in the infrared. The reason for this is that, if they are to be resolved, the instrumental resolution has to be better than $(2nd \cos \phi)^{-1}$. At normal incidence, for $n = 1.5$ (a typical value) and for an instrumental resolution of 0.5 cm^{-1}, d has thus to be no more than 7 mm. Now a plate of this thickness (or less) can readily be made with its two faces plane parallel to one another, over a reasonable area, to within an infrared wavelength but it is much more difficult to make such an "optical flat" for the visible region and one would never occur fortuitously. Infrared windows are often good enough in surface parallelism to show troublesome channel fringes especially at the long-wave end where the wavelengths are approaching 1 mm. Normal workshop practice on far-infrared components commonly produces superb "optical flats"! There are some aspects of infrared technology where this is an advantage, reduced costs of manufacture for example, but from the spectroscopist's viewpoint it is a most unwelcome bonus since the resulting high contrast channel fringes can mask genuine weak absorption features or else be mistaken for them. The only sure way of getting rid of the fringes is to mutilate the offending component either by "wedging" it or else by roughening its surface. The fringes are particularly troublesome when they arise in the specimen since it is usually undesirable to mutilate the specimen, especially if quantitative spectroscopy is being attempted. When the spectrum under study contains only broad features or else is made up of sharp features widely separated, one can usually live with the channel fringes since they can easily be identified because of their regularity. If, however, the spectrum consists of a rich array of sharp lines, the touchstone or regularity is not available and a channel fringe can easily be mistaken for a genuine absorption feature. Methods for dealing with this problem will be discussed later. It should be noted, however, here, that in the transmission case unlike the reflection case, the *position* of the channel fringes does not depend on a, that is on the absorption coefficient.

Channel fringes, as remarked above, are usually an annoying hazard, but they can be turned to advantage as a means for determining the refractive index of a plane parallel transparent plate. There are two methods available; the first is used with a fixed frequency source, for example an infrared laser [177], whilst the second is used with a broad-band source [178]. In the first method, the specimen is rotated about an axis in its body and the observed transmission will then show maxima and minima as its optical thickness varies. From the angular positions of these turning points, the refractive index can be inferred. In the second method, the plate is mounted normal to the beam and the frequency of the beam is scanned. Again maxima and minima will be observed and from their frequency positions, via equation (3.5.25), the refractive index can be calculated. The value so derived will be of necessity an average value since observations have to be made over a finite range of wave number. The significance of the refractive index, determined in this way, depends on the strength of the dispersion over the wave number range in question. If it is negligible then the average value will equal the constant value and be an excellent and meaningful quantity. If it is strong, as manifest by obvious irregularity of the positions of the channel fringes, then the average will not be physically meaningful but that is not to say that it will not be valuable. Modern methods, which give the refractive index as a continuous function of wave number, have virtually replaced the channel fringe method in practice, but since the method can be inverted, i.e. knowing n infer the plate thickness, it does form the basis of a valuable diagnostic method both in industry and in the research laboratory. Industrially, the main application is to the automatic sensing of film thickness in production plant, but it can also be used for the non-destructive measurement of the thicknesses of the coatings on finished objects [179]. These coatings are usually very thin (a few micrometres) so near-infrared wavelengths are indicated for the probing beam. In the research laboratory, the main application is to the quantitative spectroscopy of liquids. The problem here is that one has to first assemble an empty cell and then measure the cell spacing—i.e. the thickness of the cell void which is to be filled with the liquid. This measurement is naturally not easy since one cannot assume that the cell spacers will retain their nominal values after they have been subjected to the compressive forces unavoidably involved in the assembly of a liquid-tight cell. However, if the empty cell is placed in the sample beam of a spectrometer, channel fringes will usually be discernible and, from their spacing, the cell void thickness may be inferred.

This leads us on to the vexed question of how best to determine the quantitative absorption spectrum of a liquid. The usual procedure, at least in principle, is to determine the transmission spectra of the cell first empty $[\tau_e(v)]$ and then filled $[\tau_f(v)]$ with the liquid: one then assumes that the absorption coefficient is correctly given by

$$\alpha(v) = -[l]^{-1} \ln[\tau_f(v)/\tau_e(v)]. \tag{3.5.26}$$

This procedure, though commonplace, is open to much criticism. The liquid has to be confined between two transparent windows and even if it can be assumed that the windows are thick enough or else have been wedged sufficiently for no channel fringes due to them to be observable, some errors remain. Thus, since the refractive index of the liquid is different from that of air or vacuum, the channel fringes of the filled void will be differently spaced from those of the empty void and, on division, perfect cancellation will not be obtained. In addition, because of the change in the magnitude of the discontinuity of refractive index at the interfaces, the amplitude of the fringes will alter, leading again to failure of cancellation. These effects can lead to the "observation" of spurious features in the spectrum and the determination of incorrect intensities. Sometimes, and especially if the absorption spectrum is diffuse, the cell thickness can be chosen so that the channel fringes are not resolved and no artefacts result, but the "absolute" intensities will not be correct unless some attempt is made to calculate a correction. More usually, the spectrum is structured and then it becomes impossible to satisfy the twin requirements of delineating the peaks of the absorption lines and avoiding channel fringes in their wings. There are several ways that one might think of to side-step some of these difficulties—one might for example think of compensating for the change in period of the channel fringes by altering appropriately the thickness of the cell for the empty run—but then one would still have to deal with the intensity problem by calculation. Fortunately there is a relatively simple procedure which can virtually eliminate all the difficulties— at least in favourable cases. This is to record spectra of the cell filled with liquid at two different thicknesses. If these are l_1 and l_2, then

$$\alpha = \frac{1}{l_1 - l_2} \ln (\tau_2/\tau_1), \tag{3.5.27}$$

provided that *both* l_1 and l_2 are large enough to prevent the channel fringes being resolved. The method can be refined to reduce the effects of random errors (section 4.2) by measuring at several nominal thicknesses and plotting a graph of ln τ against l. The slope of this linear plot will be α as will be seen from equation (3.5.27). A calibrated variable path cell is very useful for this kind of experiment. For the case of slowly varying absorption, and especially for slowly varying low absorption, this two or more thickness method works very well. That it is necessary can readily be shown by considering the case of a highly transperent liquid—cyclohexane in the far infrared for example. In the orthodox technique, the observed transmission will be given by

$$\tau_{\text{obs}} = |\hat{T}_{\text{f}}|^2/|\hat{T}_{\text{e}}|^2 = (1-r_{\text{f}})^2 a^2/(1-r_{\text{e}}^2)^2, \tag{3.5.28}$$

where the subscripts f and e refer to full and empty respectively and as usual it is assumed that the channel fringes are not resolved. Now for cyclohexane in the far infrared, a^2 for a 5 mm cell thickness will be in excess of $0·9$ and since there is a very good match of refractive index between the liquid and the polymeric windows used to confine it the quantity $(1 - r_{\text{f}}^2)^2$ will be roughly $1·15$ and the RHS of (3.5.28) will be greater than unity! The two-thickness method eliminates the reflection effects but even so it is clear that one of the thicknesses will have to be in excess of 5 mm if meaningful measurements are to be made.

The technique is also necessary at higher frequencies for determining integrated band absorption, i.e. line strengths. This is because the cell spacing will have to be small to accommodate the peak of the line but then uncertainties will be introduced into the "wings" which contribute a significant amount to the line strength. The origin of the effect can be seen by rewriting (3.5.28) in the form

$$\alpha_{\text{obs}} = \alpha_{\text{true}} - 2[l^{-1}] \ln[(1-r_{\text{f}}^2)/(1-r_{\text{e}}^2)]. \tag{3.5.29}$$

The error is thus larger, the smaller is l. Under these circumstances the percentage error in α increases as α becomes smaller—for example in the wings of a line. The two-thickness method is therefore essential if meaningful results are to be obtained. An ingenious solution to some of these problems has been suggested by Kilp [180] who has devised what is essentially a fixed spacing variable path-length cell. Basically he uses a fixed cell defined by two thick windows but the space between the windows is divided into two chambers by a transparent septum. The idea is to have the first chamber filled with the lossy liquid under investigation and the rear chamber filled with a transparent liquid of hopefully the same refractive index. If the refractive index of the septum also matches that of the two liquids, one will have the situation that the multiple beam interference will be independent of septum position but the absorption will not. A graph of ln τ against first chamber thickness will therefore be a straight line with a slope that immediately gives a true value of α. If there are slight mismatches of index, the plot will still be sensibly linear but with superimposed small amplitude ripples. Even so, a good estimate for α will be obtained.

The two-thickness method can readily be applied to solid specimens but if one is only interested in the spectrum over a narrow region an alternative technique can be invoked. This is to use two identical pieces arranged in series in the beam but with one normal and the other inclined at an angle of incidence θ_i. The two sets of channel fringes will have different periodicities and it can be arranged by suitable choice of θ_i that they are out of phase and therefore cancel over a significant wavelength interval surrounding the chosen wavelength. By the use of this technique, the spectrum can be investigated sequentially with a view to identifying weak absorption bands which would normally be lost in the channel fringes.

3.6 The propagation of infrared radiation in guiding structures

As has been mentioned earlier, the wavelengths of infrared radiation are too short for effective propagation in low-order-mode waveguides. However confinement of the beam by devices other than the classical optical devices, lenses and mirrors, is nevertheless fairly commonplace. These transmission guides can either be regarded as over-moded waveguides, when looked at from the microwave standpoint, or as "light-pipes" when looked at from the optical viewpoint. The choice of device depends crucially on whether one is dealing with coherent or incoherent radiation. The latter because of its nature can be made at best only quasi-parallel rather than strictly parallel and any device which depends on non-spread of the beam will be ineffective. For coherent radiation, on the other hand, where beam spread can be made negligible, there is available the whole gamut of fibre optics, graded index waveguides, integrated optics etc. which have been developed in recent years for communication systems but which can just as well be used for spectroscopic purposes. A good review of this field together with numerous references has been given by Kompfner [156].

3.6.1 *The light-cone or cone-channel condenser*

Perhaps the only guiding device used widely for incoherent infrared radiation is the "light-cone" which was called more correctly a "cone-channel condenser" by its inventor Williamson [181]. This is illustrated in Fig. 3.20 which also indicates Williamson's geometrical construction for designing the condenser. This construction shows that the axis and the dimensions of the condenser (radius of entrance aperture = s, radius of exit aperture = c, length of condenser = x) define a polygon and images of the polygon can be used to tesselate the surrounding plane. A ray such as 1 with a projection which fails to intersect the polygon will eventually be reflected back out of the condenser whereas one such as 2 with a projection which does meet the polygon must emerge finally from the narrow end. The number of reflections which a ray will undergo in traversing the cone follows immediately from Williamson's construction. This number is simply the number of intersections which the ray makes with the cone wall and its images before reaching the polygon. The condenser is used, in practice, with a field lens which forms an image of the source in the entrance aperture. The construction for finding the maximum permissible divergence angle $2v$ is shown in the lower inset of Fig. 3.20. The circle which circumscribes the polygon is called the reference circle by Williamson and the extreme rays, when projected, are tangential to this circle. From the geometry of Fig. 3.20(b), it follows that

$$\frac{c}{s} = \frac{s \cos v + x \sin v}{x + s \cos v}.$$

$$(3.6.1)$$

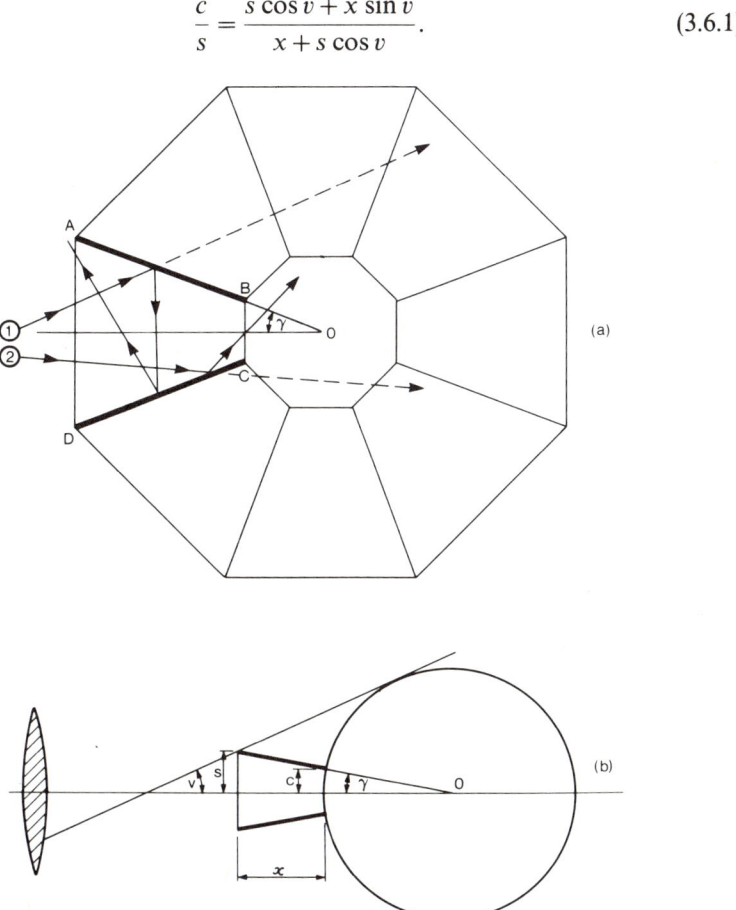

FIG. 3.20. Williamson's [181] geometrical constructions for the design of a cone-channel condenser.

The maximum reduction of image is therefore sin v which is in agreement with the sine condition of standard imaging optics. Williamson also gives some approximate relations from which c/s can be usefully calculated. These arise because practical light-cones will be long and slender and the half-cone angle γ will therefore be small. Williamson goes on to show that the f-number of an ideal system will be 0·5. This assumes, of course that the reflectivity of the inside surface of the cone is unity and since this will not be so and since many reflections may be involved, a more realistic f-number will be closer to unity. In practice, and especially with long cones, it may be worthwhile to consider gold-plating the inside surface. Williamson's original construction applied only to rays propagating in a plane containing the cone axis but several authors [182] have shown that it may be extended, relatively easily, to the skew-ray case.

The cone-channel condenser has the property that it scrambles the image so it is most often used immediately before the detector when the requirement for image preservation no longer holds. In fact the narrow end of the condenser (i.e. c) is usually chosen to match the detector window or else, in cryogenically cooled detectors, the size of the detecting element. The scrambling has the advantage that the detector entrance aperture is more evenly illuminated and the detector positioning becomes less critical. However, since the condenser is a means of increasing the effective solid angle of radiation collection, its use in an interferometric system will reduce the ultimate resolution. In most cases this is not a serious drawback since the spectrometer is either unable to or else not intended to reach its theoretical resolution limit and the use of a cone-channel condenser can bring a welcome increase in signal-to-noise ratio. This is especially true for two-beam interferometers used to obtain moderate resolution spectra of heavily absorbing or otherwise difficult specimens.

Propagation over long distances (> 10 m) becomes difficult with light-cones because of the wall losses but, since the whole pipe is readily evacuated, the cone does present a reasonably attractive propostion for guiding far-infrared radiation since the losses in the pipe will be less than those which the beam would have experienced in travelling an equivalent distance through the atmosphere. The reason for the high loss is, as mentioned above, that the loss due to imperfect reflection is multiplied up by the large number of reflections which occur along the length of the pipe. Interestingly this effect can be used deliberately in the design of a black-body cavity. If the reflectivity is made very low and if the cone is continued virtually to its apex, then, although every ray which enters must re-emerge, it will do so at negligible intensity. In other words one has made a black body. This type of black body is of considerable importance in fundamental standards work. Historically it proved very valuable to the pioneer spectroscopists in the form of the "Wood's Horn" which was used to facilitate the observation of weak luminescence (such as fluorescence, Raman scattering etc.) excited by intense primary radiation.

3.6.2 *Propagation in cylindrical hollow waveguides*

It was remarked earlier that fundamental mode propagation of infrared radiation in hollow waveguides is very difficult because the internal diameter of the guide has to be of the same order as the wavelength and the internal surface has to be polished to a high finish and maintained to a high figure. Also, since the copper (or for that matter gold) loss increases rapidly with frequency (Fig. 3.2), the attenuation per unit length soon becomes prohibitive. One can, to a certain extent, get round some of the mechanical difficulties by using a pipe with an internal diameter much greater than the wavelength. One then has an "overmoded" waveguide which can transmit very many modes of the fundamental frequency. This is commonly done with incoherent radiation when, with suitable fore-optics, one has essentially a weakly guiding structure or "light pipe". Light pipes of cylindrical or toroidal cross-section are quite lossy, but since they can be used to guide radiation round corners and since they can be evacuated they are quite popular in far-infrared laboratories to transfer radiation over moderate distances of the order of tens of metres [183]. It turns out, however, that when pipes of this nature are used with coherent radiation they can transmit power over very much longer distances because there exist special very low loss modes of propagation. Experiments have so far been confined to the high microwave region ($\sim 70 - 140$ GHz), but since these have confirmed the basic theory [184] there is no doubt that propagation in this manner is possible at infrared frequencies as well.

The solutions of the electromagnetic problem of propagation in a pipe can vary from the relatively easy to the formidably difficult but since they all have features in common it is instructive to consider a simple example. We take the case of propagation in a cylindrical pipe and take as our starting point Helmholtz's equation (3.3.2). The next step is to express ∇ in terms of the more appropriate cylindrical coordinates (r, ϕ, z) as shown in section 1.7 and we arrive at

$$\frac{\partial^2 E}{\partial r^2} + \frac{1}{r}\frac{\partial E}{\partial r} + \frac{1}{r^2}\frac{\partial^2 E}{\partial \phi^2} + \frac{\partial^2 E}{\partial z^2} + k^2 E = 0. \tag{3.6.2}$$

This equation has an infinity of solutions but one special class of solution is of particular interest. This is those functions for which the variables become separate and we may write

$$E(r, \phi, z) = R(r)\Phi(\phi)Z(z). \tag{3.6.3}$$

Substituting (3.6.3) into (3.6.2) gives the result

$$\frac{1}{R}\frac{d^2 R}{dr^2} + \frac{1}{Rr}\frac{dR}{dr} + \frac{1}{r^2}\frac{1}{\Phi}\frac{d^2\Phi}{d\phi^2} + k^2 = -\frac{1}{Z}\frac{d^2 Z}{dz^2}. \tag{3.6.4}$$

The RHS of this equation is a function of z alone whilst the LHS is a function of r and ϕ alone. The only way this can be true for arbitrary values of the variables is if each is equal to the same constant. This constant is called the constant of separation and can be defined in any way but for the present case it is convenient to designate it γ^2 and we arrive at the separate equations

$$\frac{1}{R}\frac{d^2R}{dr^2} + \frac{1}{Rr}\frac{dR}{dr} + \frac{1}{r^2}\frac{1}{\Phi}\frac{d^2\Phi}{d\phi^2} + k^2 = \gamma^2, \tag{3.6.5a}$$

$$-\frac{1}{Z}\frac{d^2Z}{dZ^2} = \gamma^2. \tag{3.6.5b}$$

The second of these has the solution $Z = Z_0 \exp(i\gamma z)$ and describes the propagation in the direction of the tube axis. The value of γ is chosen to match the wavelength in the pipe. The first equation can be multiplied through by r^2 and separated, as before, with a second constant of separation, say m^2 and we obtain

$$r^2\frac{d^2R}{dr^2} + r\frac{dR}{dr} + \left[(k^2 - \gamma^2)r^2 - m^2\right]R = 0 \tag{3.6.6a}$$

$$-\frac{1}{\Phi}\frac{d^2\Phi}{d\phi^2} = m^2 \tag{3.6.6b}$$

As before, the second of these two equations has a simple analytical solution either $\Phi = \Phi_0 \exp(im\phi)$ or else

$$\Phi = A \cos(m\phi) + B \sin(m\phi) \tag{3.6.7}$$

as appropriate and shows that the field amplitude undergoes m changes of sign as ϕ goes through its full range of $0 - 2\pi$ radians. The first equation (3.6.6a) is not however elementary and in fact is a form of Bessel's equation (section 1.5). The solution is

$$R(r) = J_m\left[(k^2 - \gamma^2)^{\frac{1}{2}}r\right], \tag{3.6.8}$$

neglecting the arbitrary multiplicative constant. To finish off the treatment, one puts in the boundary condition that $R(r)$ must be zero at the tube walls, i.e. that

$$(k^2 - \gamma^2)^{\frac{1}{2}} = U_{mn}/a, \tag{3.6.9}$$

where U_{mn} is the nth zero of the Bessel function of order m. The final solution is therefore

$$E_{mn}(r, \phi, z) = J_m(U_{mn}r/a)(A \cos \gamma z + B \sin \gamma z) \exp(im\phi) \exp(i\omega t) \tag{3.6.10}$$

where A and B are arbitrary constants to be decided by the intensity and phase of the chosen wave and the explicit time dependence has been added for completeness. In the electromagnetic case that we are considering here, the remaining boundary conditions compel m to be integral as well but, apart from this restriction, any choice of m or n is possible and these give rise to a doubly infinite family of modes. If we consider the usual case where the medium within the pipe (air or vacuum) is loss-less, than we would have $k = \omega/c$ and $\gamma = 2\pi/\lambda_g$, where λ_g is the wavelength in the guide. It follows from (3.6.9) that

$$\lambda_g^2 = \lambda^2 [1 - (\lambda/\lambda_c)^2]^{-1} \qquad (3.6.11)$$

where

$$\lambda_c = 2\pi a/U_{mn} \qquad (3.6.12)$$

and λ is the free space wavelength. This is a very interesting result, for, since λ_g^2 must be positive, it follows that free space wavelengths greater than λ_c, the so-called cut-off wavelength, cannot be admitted to and propagated down the pipe in that mode. What happens is that with λ greater than λ_c, γ becomes purely imaginary and the propagating function $\exp(i\gamma z)$ becomes non-oscillatory and purely attenuating taking on the form $\exp(-|\gamma|z)$. The field therefore can penetrate into the pipe but is very rapidly attenuated with respect to distance into it. The lowest frequency which can propagate in the pipe is obtained from (3.6.9) by setting $\gamma = 0$ and choosing U_{01}/a for the RHS. Considering now frequencies increasing past this value, one will eventually encounter the cut-off frequency of the next mode up and so on. The gives rise to the idea of a mode spectrum and to the concept, touched upon earlier, of the simultaneous existence of very many modes when optical frequencies are propagated along pipes with $a \gg \lambda$.

This simple example contains all the features of the more exact and general treatments and brings out the physics of the situation clearly; however a rather more detailed analysis is required to derive the more recondite results such as the variation of loss with frequency in a given mode and in particular to deal with the case where the refractive index of the pipe material is not real and infinite but complex and finite. But the basic results which emerge can be quoted quite simply and these show that there exist modes which show decreasing loss as the frequency rises. This is naturally a very interesting result for infrared and millimetre-wave practitioners. The modes in question are the TE_{0n} (sometimes called the H_{0n}) for metallic pipe and the hybrid EH_{11} mode for dielectric pipe with $n > 2$. The mode which has been discussed above is technically a TM (transverse magnetic) mode since we have taken the radial field, E_r, to vanish at the tube walls. TE (transverse electric) modes have the circumferential field E, vanish at the boundaries and since E is proportional to dH_r/dr, one is looking for a solution equivalent to (3.6.10) but invoking the magnetic field and consequently the derivatives of the Bessel functions. These derivatives can be defined interms of the Bessel functions themselves by means of the identity

$$J_n'(x) = nx^{-1} J_n(x) - J_{n+1}(x). \qquad (3.6.13)$$

For the particular case of the very important TE_{01} mode, we would have $J_0'(r) = -J_1(r)$ and can therefore write for the circumferential field

$$E(r, \phi, z, t,) = J_1(3{\cdot}8317r/a)(A\cos\gamma z + B\sin\gamma z)\exp(i\omega t), \qquad (3.6.14)$$

where $3·8317$ is the first zero of $J_1(x)$. The intensity corresponding to this expression is shown schematically in Fig. 3.21. It is easy to see from this Figure why the epithet "smoke ring" has been used for this mode. The components of the complex propagation coefficient $\hat{\gamma}$ under the assumption that $\lambda \gg a$ have been derived by Marcatili and Schmeltzer [185] They are

$$\beta_{mn} = \frac{2\pi}{\lambda}\left[1 - \frac{1}{2}\left(\frac{U_{mn}\lambda}{2\pi a}\right)^2\right] \tag{3.6.15a}$$

$$\alpha_{mn} = \left(\frac{U_{mn}}{2\pi}\right)^2 \frac{\lambda^2}{a^3} \, \mathrm{Re} \begin{cases} (\hat{n}-1)^{-1/2} & \text{for } TE_{0n} \text{ modes} \tag{3.6.15b} \\ 1/2\,(\hat{n}^2+1)(\hat{n}^2-1)^{-1/2} & \text{for } EH_{mn} \text{ modes} \tag{3.6.15c} \end{cases}$$

Equation (3.6.15a), in fact, follows immediately from (3.6.9).

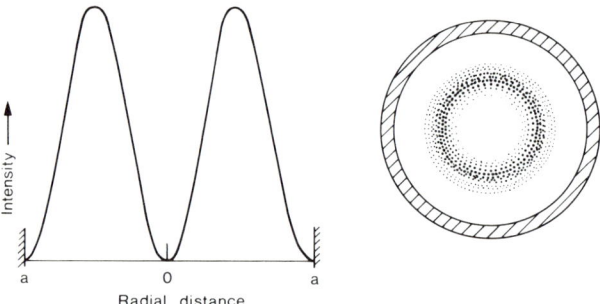

FIG. 3.21. Intensity distribution as a function of radial distance in a wave-guide propagating the TE_{01} mode. At the left is shown the analytical form and at the right a schematic representation where regions of high field strength are heavily shaded.

Communication engineers are faced with an ever increasing demand for bandwidth as new forms of communication, television, computer/computer links etc. compete for what is available with the established forms, telephony for example. The exploitation of the virtually unused high microwave and millimetre bands therefore seems inevitable and one obvious way of doing this would be to use cylindrical waveguides operating in the TE_{01} mode. This is an attractive proposition because between 30 and 300 GHz there is nearly ten times as much information carrying potential as there is in all the lower frequencies put together. Additionally, waveguide communication systems are almost completely free from interference and any form of covert surveillance. Several countries have on-going research programmes to investigate the feasibility of laying TE_{01} waveguides between their major cities and to use them for intercity telephony. The operating frequencies being considered in the first case would be between 70 and 140 GHz. The main difficulties with this system are two-fold. Firstly the TE_{01} mode is very unstable with respect to decay into other much more heavily attenuated modes. To overcome this difficulty the guide has to be made as a helix with insulation between the turns. There is thus little resistance to the circumferential currents associated with the TE mode but very high resistance to the axial currents associated with the other modes. The second difficulty is that the low-loss modes are very intolerant of any bends in the guide. The TE_{01} mode is more tolerant than the EH_{11} mode but, even so, a bend having a radius of curvature as much as 48 m will double the loss per unit length. This is why the waveguides will only be used for the intercity links. Within the city, where sharp bends are inevitable, other means for transmitting the signal will have to be used. These two difficulties reduce the performance of the guides from their ideal figures but nevertheless the values which have been observed, of the order 2 dB/km, are sufficiently low for a viable system. Unfortunately for the waveguide system there is now severe and growing competition from fibre-optic systems working in the near infrared (see next section). This kind of system can be much cheaper for intercity telephony because the small highly flexible guides can be inserted into already existing conduits and repairs, which unfortunately are frequently necessary due to the inadvertent attention of mechanical diggers, are not too expensive. TE propagation of infrared radiation in cylindrical waveguides has not so far been attempted. This is because the technology of coherent operations with infrared radiation is still in its infancy and there would be considerable difficulties with launching the mode. In addition, the only reason at the moment for wanting to transmit infrared radiation over long distances would be for communication purposes and here one has the options not only of fibre optics but also of the lens systems mentioned in section 3.3. However, it is clear that high power submillimetre systems are on the horizon and, for these, material absorption problems rule out the fibre optic option and atmospheric absorption makes the use of an evacuated pipe essential. For this kind of application TE mode propagation in a pipe may well be an attractive option.

It is interesting here to consider the operation of the waveguide from an optical viewpoint. One thinks of wavelets being guided down the pipe by multiple reflections at its inner surface. A ray making an angle α to the axis of the pipe will travel an axial distance 2a cot α between reflections but at each reflection it will suffer some measure of loss. For the metallic pipe the loss is due to imperfect reflection, whereas for the dielectric pipe the losses are due to refraction into the dielectric material. In both cases the loss per reflection can be calculated knowing the complex refractive index and this is the origin of equations (3.6.15b). The loss per reflection also depends on the angle of incidence via the Fresnel relations (3.5.3), so for dielectric reflection high angles of incidence will be preferred (see Fig. 3.15). This means that after traversal of a long length of dielectric pipe the radiation will be being propagated at near grazing incidence. In an ordinary dielectric pipe n will be very much greater than k and the imaginary component of n will have little effect on the propagation. For a metallic pipe on the other hand this situation, depending on frequency, may be reversed. At microwave frequencies k will usually be larger than n though both are very large for a typical metal such as copper. That is to say, the optical properties of a metal, at microwave frequencies, are dominated by its conductivity (see equation 3.4.7):

$$\sigma = 2\pi\varepsilon_0 \varepsilon'' v = 4\pi\varepsilon_0 nkv. \tag{3.6.16}$$

At near infrared frequencies, on the other hand, the stronger dispersion to which k is subject will have reversed the roles and metals in this region of the spectrum can be considered to be dielectrics with very large refractive indices. The modulus of n will be high if either n, k or both are large and metals have high reflectivities throughout the microwave and infrared regions; in fact it is not until the ultraviolet that one encounters examples of metals with reflectivities less than 90%. Because the reflectivity is high and not very sensitive to angle of incidence the metal pipe propagating the TE_{01} mode is more tolerant of bends than is a dielectric pipe propagating the EH_{11} mode. To take the discussion further, it is necessary to consider the polarization of the propagating wavelets. The TE_{01} mode has the electric vector entirely perpendicular to the plane of incidence: the situation is therefore simple, the larger is n the larger is the reflectivity. The EH_{11} mode, on the other hand, has a component of the electric vector in the plane of incidence and phenomena of the kind that give rise to the Brewster angle (see Fig. 3.15) are encountered. By differentiation of equation (3.6.15c), it is found that the losses are minimal when $n = \sqrt{3}$. In the optical region, it is not easy to achieve this optimal value since most convenient dielectrics, glass for example, have refractive indices of the order of 1·5. In the middle and far infrared it is much easier since the strong dispersion associated with the reststrahlen process often permits accurate matching. However, even with the non-ideal value of 1·5, EH_{11} propagation at optical frequencies in dielectric pipe can be achieved with power losses of only 0.426×10^{-3} Np m^{-1}, that is 1·85 dB km^{-1} (equations 3.6.15 and 1.2.27).

It has already been remarked that the EH_{11} mode is extremely intolerant of bends in the guide. Marcatili and Schmeltzer calculate that a radius of curvature as gentle as 10 km will double the losses! Clearly this mode is not a practical proposition for long distance communications since gravity alone could induce larger deformations in the guide. Marcatili and Schmeltzer point out, however, that the mode is highly relevant to the operation of lasers since the laser tubes are at most only a few metres long and there is no difficulty in keeping them perfectly straight. They therefore proposed the development of a new kind of resonant cavity in which two spherical mirrors were to be used to image the radiation emerging from a length of dielectric waveguide back onto itself. The arrangement is shown schematically in Fig. 3.22.

This hybrid arrangement is now known as a waveguide resonator and lasers based upon it are usually called *waveguide lasers*. The prime reason for developing this configuration was the observation that, with the product of gas pressure and tube radius constant, the gain per unit length of helium/neon lasers is inversely proportional to the tube radius. The waveguide resonator permits the use of capillary tubing and hence can, in principle, give very high gain. Marcatili and Schmeltzer were thus mostly concerned with the use of their ideas in the construction of lasers operating in the visible and near-infrared regions, but strangely the first experimental evidence of waveguiding

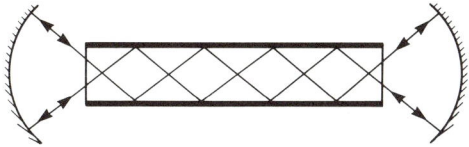

FIG. 3.22. Matching of a cylindrical waveguide to a Fabry–Perot interferometer.

in a laser cavity came from Kneubuhl and his colleagues in Switzerland [186] who had made an ICN laser operating at 773·5 µm! Steffen and Kneubuhl later made a considerable contribution to the basic theory [187, 188]. Since then waveguide lasers based on the He/Ne [189], CO_2 [190], CO [191] and HCN [192] systems have been developed and shown to possess all the properties predicted by the theory.

Waveguide lasers are now very popular throughout the visible and infrared regions so it is worth examining their properties in some detail. Firstly it has to be admitted that the original motivation was based on at best a half-truth since scaling laws have been subsequently established which show that the gain and efficiency of waveguide lasers are independent of tube diameter, so the performance available is no better than that of a conventional laser. The real advantages of the waveguide format are:

1. The laser size and bulk can be drastically reduced, an important point for commercial applications.

2. Discharge or electron beam excited lasers can be operated at much higher pressures. The homogeneous line width thus increases and the laser has a larger tuning range. This is a particularly important point for the CO_2 lasers used as the pumps for far-infrared optically pumped lasers. Current waveguide CO_2 lasers have tuning ranges in excess of 1 GHz so it is easy to get a good match with the pumped transition.

3. Doppler-broadened laser lines can show increased gain and it thus becomes possible to make short low to medium power lasers. This is an important point for the lasers used in field work and in military applications.

4. The reduced diameter of the laser tube increases the area to volume ratio and thus improves the efficiency of the cooling system. This is a valuable gain for those lasers where there is a "thermal bottleneck", i.e. a build-up of lower state (but still highly excited) molecules or atoms.

5. There can be a much improved matching of the laser excitation region with the mode volume.

6. Finally, but far from least importantly, these lasers have excellent mode characteristics and can give out a close approximation to a pure TEM_{00} Gaussian beam.

The operational principles of waveguide lasers have been described in detail in a number of excellent reviews [193]. The basic point is that the laser has to work on a mode pattern which is such that the electric field, after one round trip, repeats itself, apart from a scaling factor, both in amplitude and phase. One thus has waveguide resonator modes which are, in general, different from those of the waveguide itself. There are several ways of deriving the form of the resonator modes. Thus iterative methods in the style of Fox and Li [30] might suggest themselves but an attractive alternative is to express the resonator modes as a power series in terms of the waveguide modes used as the basis set.

Glass or fused silica is widely used for visible and near-infrared lasers, but it is not the best material for mid-infrared lasers, because the refractive index is relatively low and the absorption index is quite high. The most promising material, especially for the very important CO_2 laser, is beryllium oxide which, because of the low mass of its constituent ions, has very high phonon frequencies. Thus Loh [194] gives for the two values of the transverse optic mode ($E//c$ and $E \perp c$) the frequencies 20·4 THz (680 cm^{-1}) and 21·7 THz (724 cm^{-1}) respectively. This strongly ionic crystal, which can also be made in the form of a sintered glass, is therefore very highly reflecting in the 9·5 − 10·6 μm region where the laser operates.† Beryllia is also very resistant to thermal shock and to plasma erosion, has a high thermal conductivity and furthermore has a low coefficient of expansion, so, on these important counts, the material is very attractive for the tubes of high-power CO_2 lasers. In other respects, however, it is less satisfactory. Thus, like all beryllium compounds, it is intensely toxic and special handling facilities are required in the optical workshop where it is to be cut and polished. It is also a difficult material for optical working. It is likely therefore that fused silica will continue as a strong contender for at least the foreseeable future.

The optically pumped lasers, that is those which involve stimulated fluorescence under the influence of the radiation from a powerful pump laser, offer excellent opportunities for the use of waveguide techniques [195]. These lasers do not have to sustain a high voltage discharge and the confining pipe can be made of metal. This type of laser is discussed in more detail in Chapters 2 and 6; all that needs be said here is that hole-coupling is an effective way of getting the pump and the resulting infrared radiation into and out of the cavity respectively and that the narrow diameter good thermally conducting metal pipe is very effective in increasing the otherwise slow rate of vibrational relaxation of the lasing lower level.

---

† By a remarkable coincidence, the refractive index of BeO in the visible region is almost exactly the ideal value of $\sqrt{3}$.

3.6.3 *Propagation in hollow periodic structures*

The use of periodic "slow-wave" structures to introduce feedback and therefore amplification in electron beam tubes was mentioned briefly in section 2.5.2. The same principle can be applied to laser construction where instead of having a smooth waveguide one can have a corrugated one [196]. In such a laser, the feedback does not take place discontinuously at the surface of the cavity mirrors, as it does in a conventional laser, but continuously at each of the corrugations. For this reason such lasers are called distributed feedback lasers. At infrared wavelengths, where these lasers operate, an optical rather than an electromagnetic formalism is appropriate so one thinks in terms of backward Bragg diffraction at the periodic structure and sees that the tube will show gain at a wavelength which satisfies the Bragg condition. A DFB laser does not need any mirrors and this can be an advantage in two connections; firstly in the design of high-power lasers where damage problems are avoided and, secondly, in the design of optically pumped lasers where one does not then have to satisfy the requirements of having mirrors transparent at the pump wavelength yet nearly perfectly reflecting at the lasing wavelength. So far, however, only one DFB gas laser has been made to work. Kneubuhl and Affolter [197] have succeeded in getting emission at 496 μm from an optically pumped CH_3F laser. They used a rectangular metallic waveguide with mechanically milled corrugations. The period was 248 μm (i.e. $\lambda/2$) and the amplitude of the corrugations was 124 μm (i.e. $\lambda/4$). With a long laser waveguide, the reinforcing effect of the large number of Bragg reflections gives a very sharply peaked spectral response and because of this it is possible to select longitudinal modes even when the gain curve is very broad. One does not need to have metallic corrugations, however; any periodic variation of the optical constants will do. Thus periodic variations of refractive index work well [198]. In this connection one immediately sees that the free-electron laser discussed in section 2.5.4.2 is really a form of DFB laser. The question of whether these devices are lasers or electron tubes therefore seems settled—they are both!

3.6.4 *Propagation in solid dielectric guides*

3.6.4.1 *General introduction*
In general, dielectric media have permittivities greater than that of free space so an electromagnetic wave incident upon a dielectric medium will tend to become concentrated within the medium. Also, once propagating inside the dielectric, the wave will tend to remain confined because of the interface effects described in section 3.5. Dielectrics are therefore usable as effective guides for electromagnetic radiation. When properly designated it is even possible to retain the image-forming property and the use of fibre-optic telescopes to see inside inaccessible objects—for example human beings!—is becoming quite

widespread. At infrared and millimetre wavelengths, there are three basic types
in current use. Firstly there is the conventional extended dielectric guide,
usually made from a polyolefine such as polyethylene, and which has circular
or rectangular cross-section. Guides of this type are mostly used at very long
wavelengths ($\lambda > 1000$ μm). Secondly there is the "core-and-cladding" or
"optical fibre" type of guide which is normally only used in the very near
infrared ($\lambda < 2$ μm). Thirdly there is the slab waveguide which can be used
anywhere in the infrared but which is mostly used at the shorter wavelengths
for the fabrication of diode lasers and integrated optics devices. These three
types are illustrated schematically in Fig. 3.23.

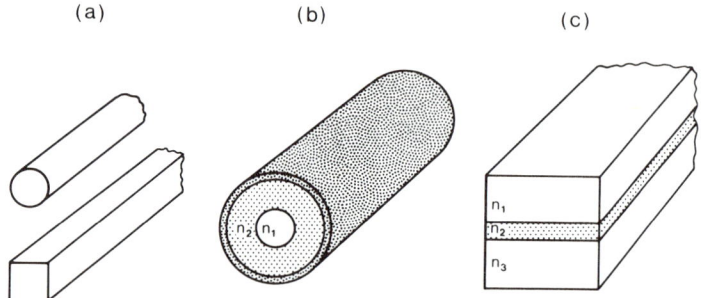

FIG. 3.23. Various types of dielectric waveguide. In (a) are shown two types of solid-rod
guide; in (b) is shown the core and cladding type of guide and in (c) the slab waveguide.
Types (a) and (b) are conceptually alike but differ in practice because of the
considerably greater difference of index for the former as compared with the latter.

Simple dielectric waveguides are often used for propagating microwaves
since they are easy to cut to shape and, as will be shown later, are quite tolerant
of bends. They are therefore quick and convenient things to use for
experimental work when compared with metallic guides but they do lose
radiation into space and a human passing near them can couple power out of
them! Nowadays there is much concern over the hazards of exposure to
microwaves [199], so dielectric guides in the open laboratory are rather less
popular than they once were. They still remain viable candidates for coupling
the high-frequency microwave (94–140 GHz) equipment used as parts of a
communication link. Indeed it was suggested that dielectric guides be used for
the final stages of intercity trunk waveguide microwave communication
systems but now that the trunk waveguides themselves have been abandoned
this use has gone as well. Any imperfection in the guide, such as a change in
transverse dimension, will cause power to be radiated and this is a nuisance
since it is not possible to make absolutely perfect guides. However this loss
mechanism can be deliberately invoked to make an effective antenna which
launches any desired form of wave field. Thus by having a guide ending in a
cone one can launch an almost spherical wavefront. This type of antenna has

been used in concert with a 100 GHz Impatt oscillator (section 6.5) to make such things as Doppler burglar alarms. The propagation in a cylindrical guide can be analysed both from a full electromagnetic approach [200] or else from a simplified scalar-wave approach. This latter leads to an equation of the Helmholtz type and the solutions are akin to those derived for the hollow pipe in section 3.6.2. These solutions are in the form of modes but the nature of the modes is more complex because one doesn't have the finite boundary conditions that one has for a metal pipe. In fact there are three types of modes [201].

a. Guided modes which have $E(r)$ zero at $r = \infty$ and which therefore do not lose energy by radiation.

b. Radiation modes which are not guided but which are simultaneously solutions of the wave equation with the same boundary conditions.

c. "Leaky" modes which are strongly coupled to the rod but which do lose energy by radiation. $E(r)$ is not therefore formally zero at $r = \infty$.

There are in addition tangential modes which essentially arise from the circular cross-section acting as a resonator. These so-called "whispering gallery" modes are very interesting in some laser connections but since they do not correspond to propagation in the z-direction they are of only passing interest in the present analysis.

The guided modes form a discrete set which can be labelled in the usual way with a radial and an azimuthal mode number, but the radiation modes form an infinite continuous set. Both kinds are necessary in order that any arbitrary function can be expanded in terms of the modal functions as a basis set and in the present context the radiation modes describe the diffraction of the beam when it encounters the entrance aperture of the guide. The "leaky" modes are connected with cut-off phenomena and will be discussed again later under this heading. The quantitative expressions for the modes are rather more complicated than those for the metallic pipe because the field does not vanish at $r = a$, the guide radius, but basically the form is the same. The field *inside* the guide is determined by normal Bessel functions $J_m (U_{mn} r/a)$ whilst that *outside* the guide is determined by Hankel functions. The exact functions have been listed by several authors [200, 201, 202]. A very important parameter which emerges from these relations is the dimensionless normalised frequency F given by

$$F = 2\pi\tilde{v}\,[n^2 - 1]^{1/2}\,a. \qquad (3.6.17)$$

The larger is F, i.e. the larger is $\tilde{v}$ for a fixed a or a for a fixed $\tilde{v}$, the more the power is confined to the guide; conversely the smaller is F the more the power is flowing in the free space surrounding the guide. For any given mode, there will in general be a cut-off value of F, given in the usual way by the roots of the Bessel functions, but for F values greater than this cut-off value, the ratio of power incident on the guide to power staying in the guide rapidly approaches the value of unity appropriate to geometrical optics [203]. The analytical transition from wave-optics to geometrical optics has been described by Arnaud [158].

In the real world one needs to consider the effects of bends in the guide and dielectric loss in the guide material. A bend must lead to radiation loss because at a certain distance out from the bent guide the field will be constrained to move at a speed greater than that of light if it is not to break up. This is impossible. Therefore the field does break up and to provide a solution of Maxwell's equations an outward flow of energy is needed. In other words the guided mode breaks up into leaky modes. Dielectric rods are nevertheless relatively tolerant of bends. Thus Neumann and Rudolph [204] show that the loss in a 90° bend is a simple power function of the ratio (R/λ_0) where R is the bend radius and λ_0 the free space wavelength. The results indicate that losses less than 0·1 dB would be experienced if (R/λ_0) were 300 or greater. This makes dielectric rods attractive for guiding millimetre waves since bends of about 30 cm are quite feasible. They would be even more attractive for guiding submillimetre waves were it not that absorption in the polymer makes them prohibitively lossy. Some measure of compromise can be sought, when faced with a lossy guide, by appropriate choice of the parameter a. If this is too big, the wave will be mostly in the guide and the loss will be high: if it is too small the wave will be only weakly guided and the slightest imperfection or bend will lead to heavy loss. An intermediate value will therefore lead to an optimally low loss. Even so, in the 100 GHz band, the best that can be done is about 20 dB km^{-1} and, good though this sounds, it pales into insignificance when compared with the 0·2 dB km^{-1} available from third-generation optical fibres (see next section).

The cut-off phenomena in circular section guides have their basic origins in the roundness of the surface [158]. They are therefore fundamentally different from slab waveguides (see later) where cut-off is due to critical angle effects. The fundamental E_{00} (or HE$_{11}$ in the electromagnetic terminology) mode does not have a mathematically exact cut-off condition so it can, in principle, propagate an arbitrarily low frequency wave but in practice for $F < 0.5$ the bending and imperfection losses become prohibitive. The next mode, E_{10}, has a cut-off frequency $F = 2.405 \ldots$, determined by the first root of J_0. It follows that for F values between 0·5 and 2·405, the guide will sustain only a single mode. This is sometimes called the monomode condition. In loss contexts it is sometimes important to note that "leaky" modes may nevertheless be capable of propagating over considerable distances with quite acceptable loss figures.

The simple dielectric guide may be limited in its applications but its close relative, the optical fibre, seems to be one of the most promising technological innovations of recent times with apparently limitless scope for bringing about a true communications revolution [205]. Basically the low-loss and enormous bandwidth capability of the optical fibre more than compensate for the extra (quantum) noise present at optical frequencies and one has the remarkable situation that communication systems which had slowly clawed their way up to higher and higher radio frequencies suddenly leaped over the entire mid infrared to land very much on their feet in the near infrared. It will not be long before *all* trunk communication systems will be working in the near infrared using optical fibre links. The optical fibre (Fig. 3.23) has an inner core of refractive index n_1, a cladding of index n_2 (with, of course, $n_1 > n_2$) and a totally absorbing outer cover. The fibre can be made of plastic, but much more usually glass and because of this n_1 and n_2 will be very similar. The change of index at the discontinuity is therefore small and for this reason optical fibres are sometimes said to be "weakly guiding" devices. This phrase has to be taken, however, in a technical sense since radiation once launched properly into the fibre remains in it indefinitely subject only to attenuation by absorption. To analyse this propagation, at near-infrared wavelengths, it is more appropriate to use ray optics and then one interprets the guiding to arise from total internal reflection at the index discontinuity. A ray propagating down the fibre will therefore follow a sort of spiky helicoidal path. The analogue of the normalised frequency is now

$$F = 2\pi\tilde{\nu}[n_1^2 - n_2^2]^{1/2}a. \qquad (3.6.18)$$

So in mode language one sees that (a/λ) can be quite large, since $(n_1 - n_2)$ is small, and yet the fibre can propagate only the fundamental mode. This is an important point since it means that monomode fibres working near 1 µm, although very fine, will nevertheless be large enough to see and handle. The fundamental mode corresponds to propagation at grazing incidence so the total infrared reflection condition will always be satisfied (section 3.7) but for all other modes one has, in ray language, a finite angle of incidence and this angle varies with wavelength. It follows that optical fibres will show cut-off phenomena. Beyond cut-off the rays penetrate the interface and energy is lost to the cladding. There is no discontinuity in behaviour here, however. For wavelengths shorter than the cut-off wavelength, there will exist a field in the cladding but this field will be evanescent (section 3.7) and will propagate with wave vector strictly parallel to the surface. At the point of cut-off the phase velocity will become equal to c/n_2 but beyond cut-off it would be formally required to exceed this value which would of course be impossible. What in fact happens is that the wave vector becomes inclined to the surface at a finite angle such that its projection in the direction of propagation is c/n_2 but it then has of necessity a finite projection in the perpendicular direction and energy is propagated away from the guide. In this connection one should consider the effect of the absorbing outer wrapper since this will have a perturbing effect on the evanescent wave propagation. However it can be shown [206] that losses due to this effect rapidly become negligible, in any mode, provided the cladding is at least a few wavelengths thick, as soon as the normalised frequency F becomes significantly larger than the cut-off normalised frequency F_c. The analysis mentioned earlier of bend-losses applies just as well to the fibre optic. The smaller change of index at the boundary tends to make the guide more sensitive to bends but this effect is completely cancelled by that due to the much smaller wavelength. In fact bends with radii of 1 metre or so have no perceptible effect. As an example British Telecom have demonstrated transmission with high efficiency through 49 km of fibre optic cable wound onto a drum of diameter 1 m!!

Absorption in the fibre material is the limiting factor on the use of fibre optics and it is this which has dictated the use of the near infrared. Attempts have been made to make fibre optic guides for the mid infrared from glasses of the arsenic/sulphur system but these have not been very successful. Plastics can be used as fibre optic guides in the near infrared and in fact Texas Instruments market [207] a complete range based on the Du Pont material PIFAX PIR 140, but the high losses—600 dB km^{-1}—limit this type of guide to short-range links, for example connecting computers together. It was the development of very low loss glasses which transformed the situation. The use of glass fibre optic guides for long-distance communications will be discussed in more detail in the next section.

Slab guides which have three plane-parallel regions with indices $n_1 > n_2 \geqslant n_3$ are relatively easy to analyse [158, 201], using only elementary mathematics. The confinement depends entirely on critical angle phenomena. One can conceptually think of two types of guide, symmetric when $n_2 = n_3$ and asymmetric when this is not so. It is also useful to consider weakly guiding systems for which $n_1 \approx n_2 = n_3$ and strongly guiding ones for which there are substantial differences of index. The modes of the guide can be divided into TE modes and TM modes. TE modes have the electric field rigorously transverse to the direction of propagation, roughly speaking they correspond to perpendicular polarisation in the sense of Fig. 3.18. TM modes have the magnetic field transverse and correspond therefore to the parallel polarisation of Fig. 3.18. The symmetric guide, like the simple dielectric rod, has the feature that its lowest mode is not cut-off for arbitrarily low frequency, but this behaviour does not hold if there is *any* asymmetry. For this reason this mode also tends to be cut-off in practice. One might at first think that there would be a continuum of modes since any ray launched into the guide such that its angle of incidence was greater than the critical angle would propagate indefinitely. However, a ray is not a very physical sort of quantity; it is really an abstraction, the normal to a plane wavefront, and what one really needs to analyse is the progress of a wavefront through the guide. The condition to be satisfied is that the phase shifts of multiply reflected wavefronts differ from that of the unreflected wavefront by multiples of 2π. In other words one is thinking of the guide as a multiple reflection resonator and for a given guide thickness, wavelength and set of refractive indices, there will be only a discrete set of angles which give the resonance condition Marcuse [201] gives for the symmetric guide the appropriate eigenvalue equation

TE modes $\qquad \tan(\kappa\delta) = \kappa(\gamma + \delta)/(\kappa^2 - \gamma\delta),$ $\qquad\qquad$ (3.6.19)

where

$$\kappa = 2\pi\tilde{v}n_2 \sin\theta, \quad \gamma = [(n_1^2 - n_2^2)4\pi^2\tilde{v}^2 - \kappa^2]^{1/2}$$

and $\qquad\qquad \delta = [(n_1^2 - n_3^2)4\pi^2\tilde{v}^2 - \kappa^2]^{1/2}.$

This equation can be solved graphically to give the allowed values of θ, the angle of incidence at the intersection. Slab guides will be discussed again in connection with integrated optics (section 3.6.4.3) and diode lasers (section 6.5).

3.6.4.2 Optical-fibre systems for telecommunications

Before analysing modern optical fibre communications [208, 209, 210, 211] in detail, it is instructive to consider the conventional systems which they are supplementing, or in some cases completely replacing. The basic arrangement is shown very schematically in Fig. 3.24.

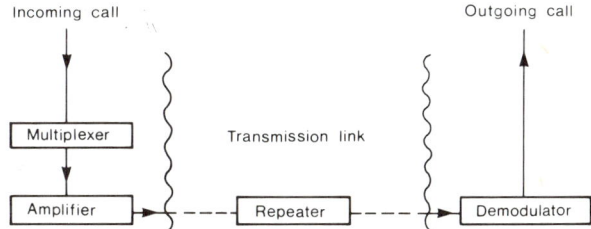

FIG. 3.24. Schematic of a long-distance telephony system.

Long-distance communication until very recently involved solely telephony and even today when digitally encoded colour television programmes are sent down communication links and when computers are starting to "talk" meaningfully to one another over distances of many kilometres, telephony still remains the principal use for a telecommunication system. In this application each signal has a very limited bandwidth ($< 10\,\text{kHz}$) whereas the overall system may have a considerable bandwidth ($\sim 10\,\text{MHz}$) so multiplex techniques naturally suggested themselves in order to make optimum use of the link. The multiplexer divides the total bandwidth into a number of "slices" or channels and generates a carrier frequency at the centre of each channel. Each incoming call is assigned a channel and modulates the carrier for that channel. The division of the frequency space is a compromise of economics (as many channels as possible per link) and quality of reception which demands that each carrier and its associated sidebands are separated sufficiently from one another for cross-talk to be reduced to an acceptable minimum. In practical systems this may still mean that any given link is transmitting up to a thousand different telephone calls at the same time! The multiplexed signal after suitable broad band amplification then passes to the crucial element of the system—the transmission link. The original transmission links were just twisted wire-pairs but these soon came to be replaced by coaxial cables which are capable of working at higher frequencies and therefore of carrying more channels. This pressure on bandwidth has even led to the use of microwave links through free-space or in waveguides and it is the fundamental reason for the development of fibre optics. The coaxial cable, however, has proved surprisingly resilient and nearly all land telephony at the moment and all undersea telephone links are based on coaxial technology. It is a familiar story of an established technology improving itself much more than anyone would have thought possible when it encounters a vigorous challenge from a new and radically different technology. Nevertheless the optical fibre systems must win. It is not just a matter of bandwidth, it is a matter also of attenuation in the link. All transmission links must of necessity have a finite loss (resistive heating in the case of wire or coaxial links) and this loss will have the twin effect of not only attenuating the signal but also of introducing noise. For acceptable SNRs over long distance links, repeaters are essential. These are active (battery powered) devices that receive the signal, amplify it, remove the noise and then retransmit it. The use of repeaters is necessary but it is a most unwelcome necessity since with conventional systems the repeaters have to be spaced at intervals of only a few kilometres and that means that they will sometimes have to be located in remote terrain either buried underground or else mounted high on pylons. Repeaters occasionally fail, though to be fair, modern solid-state ones are very reliable, but they can easily be damaged by either natural phenomena (lightning, flooding etc.) or else by inadvertent human intervention (mechanical diggers). Also they have to have their batteries serviced from time to time. Digging up (or in the case of links under the sea, fishing up) a repeater is not only a costly operation, but the loss of revenue whilst the link is out of commission can be very substantial. With fibre optics, repeater spacings of better than 10 km have already been achieved. The second-generation systems will work with a spacing of 50 km or more and already research is going on into systems which will enable a submerged transatlantic link to work without repeaters at all!

The elimination of noise by the repeater might at first sight be thought magical but the trick is done by the use of a very clever dodge, pulse-code modulation or PCM for short. Basically what happens is that a complex waveform (a voice conversation for example) is converted into a stream of rectangular pulses before transmission and reconverted back to a recognisable audio signal on reception. The frequency and amplitude information in the original is not lost provided the waveform is sampled frequently enough (section 1.6.2) so the reconstruction on reception can be perfect. Each pulse is in fact encoded in binary digital form (that is in "bits") and the flow of information consists of a stream of regularly spaced pulses each of which is either a 1 or a 0. Now, provided the accrued additive noise does not exceed one bit, the incoming noisy pulses can be used to trigger a noise-free generator which will thus provide an exact copy of the original signal and the noise has been eliminated without in any way affecting the signal!! The operation is illustrated in Fig. 3.25.

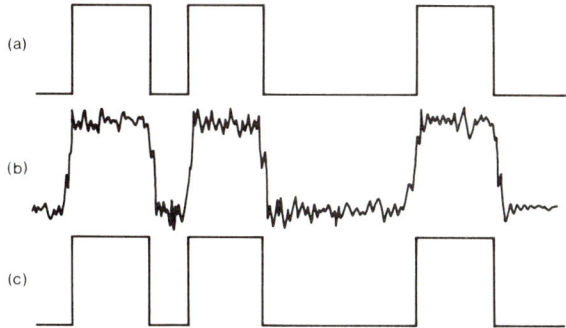

FIG. 3.25. Noise suppression in a digital communication link. The number eleven (1101) as transmitted (a), as received (b) and as processed (c).

The information carrying capacity of a link is determined by how many pulses, that is how many bits, it can transmit per second. There are two factors involved, the bandwidth and the system signal-to-noise figure. The bandwidth has to be, at an absolute minimum, equal to the number of bits per second and obviously, if possible, should be considerably more. Current coaxial systems can transmit at the rate of 50 Megabits per second, but there are immediate plans for intercity telephony operating at 140 Megabits per second. This is an uncomfortably high rate for even the latest coaxial technology and if higher rates are needed (as indeed they are) it is going to be necessary to use much higher frequency systems, microwave, millimetre-wave and optical. Microwave links operating on S or X band can operate through the atmosphere and the tall towers carrying a large array of dishes are rapidly becoming familiar landmarks in developed countries. Atmospheric absorption, however, rules out free space propagation for the higher frequency bands. Millimetre links, if they are to be used, therefore require the use of waveguides either evacuated or else filled with dry nitrogen. At optical frequencies one can again transmit through the atmosphere if one so wishes since intrinsic absorption need not set a limit, but scatter due to water droplets and dust haze does set such limits. Even on a perfect summer's day the scatter attenuation is 2 dB km^{-1} whilst on a normal day it would be 10 dB km^{-1} and with moderate fog it would be more than 100 dB km^{-1}. The optical systems too will therefore have to work in waveguide if they are to be absolutely reliable and independent of the weather. The system signal-to-noise, the second factor, is a rather subtle concept in a PCM repeated system but it makes itself manifest by occasional errors, that is a 1 changed into a zero or vice versa. Error rate is a crucial parameter and as a high-quality system it must be kept below 10^{-9}, that is only one error in a thousand million.

The costs of a telecommunication link arise initially in the installation and recurrently in the subsequent maintenance. It is very expensive to lay conduits under the ground to carry telephony signals and despite every care these are constantly being damaged and having to be repaired. It was mostly the enormous cost of laying the proposed underground waveguides that persuaded the British Post Office to abandon the scheme for TE_{01} intercity telecommunications. Likewise the Bell Company abandoned their plans for a beam-wave system using a chain of beam-shaping lenses even though the problem of reflection/absorption losses at the lenses had been solved by the introduction of gas lenses [156]. The rival optical fibre systems which were first suggested by STL(UK) in 1966, have, on the other hand, many attractions:

a. They are flexible, easily stored and installed.
b. They can be mechanically very strong—in fact almost as strong as metallic wires.
c. They require no additional ducting, a fibre-guide can be installed in the same ducts that are used for conventional cables.
d. They are cheap to produce (1–10 p per metre).
e. They have very low losses so repeaters are only required at spacings of 5–10 km in the first generation and at 20–40 km in the second. .
f. They are immune to electrical noise and are very hard to tap.
g. Properly developed and engineered fibre systems can operate at very high bit rates, maybe as much as 1 Gigabit per second.

The choice of the near infrared for optical fibre communications is dictated by several factors. Thus:

a. The intrinsic loss of glass is a minimum in this region and this minimum is accompanied by a minimum of dispersion.
b. The LED and diode laser technology necessary for the manufacture of the launching sources is most developed in this region.
c. The desirability of using a room-temperature detector requires that a frequency be chosen high enough to make thermal noise negligible. The most obvious choice for the detector would be an avalanche or a PIN diode based on either the well-established silicon technology or else on the emerging gallium arsenide technology. This again means operation in the 1–2 μm region.

The reasons for the adoption of glass for the transmission medium have been touched on above but the main ones are:

a. Glass readily forms fibres of indefinite length and of arbitrary diameter.
b. Glass, being non-crystalline and non-stoichiometric, is readily "doped" with inorganic ions in order to change its optical properties.

c. The various doped types are mechanically and thermally compatible so the cladding will bond to the core without difficulty.

Very fine glass fibres are also very flexible, as mentioned above; in fact a monomode fibre can readily be bent with a radius of curvature of 1 mm without being damaged. However all these excellent properties would count for nothing unless it was possible to achieve adequately low loss, that is attenuation figures of less than 10 dB km^{-1}. Ordinary glasses as made by conventional processes have near-infrared losses of about 20 dB km^{-1} in their most transparent regions and are therefore virtually useless for even moderate distance communication. A typical spectrum is shown in Fig. 3.26. The overall loss arises from several causes but some important ones are:

a. The long-wave "tails" of the visible and ultraviolet absorption bands (d → d transitions) of residual transition metal ions, principally Fe^{2+} and Fe^{3+}.
b. Near-infrared "forbidden" bands of these ions, thus Fe^{2+} has a peak at 1·2 μm and Cu^{2+} at 0·8 μm.
c. Overtone bands of H_2O molecules chemisorbed into the silica lattice, of hydroxyl groups chemically bonded into the structure or else of OH^- ions trapped in the glass. Thus OH^- has absorption peaks at 0·725, 0·875 and 0·950 μm.

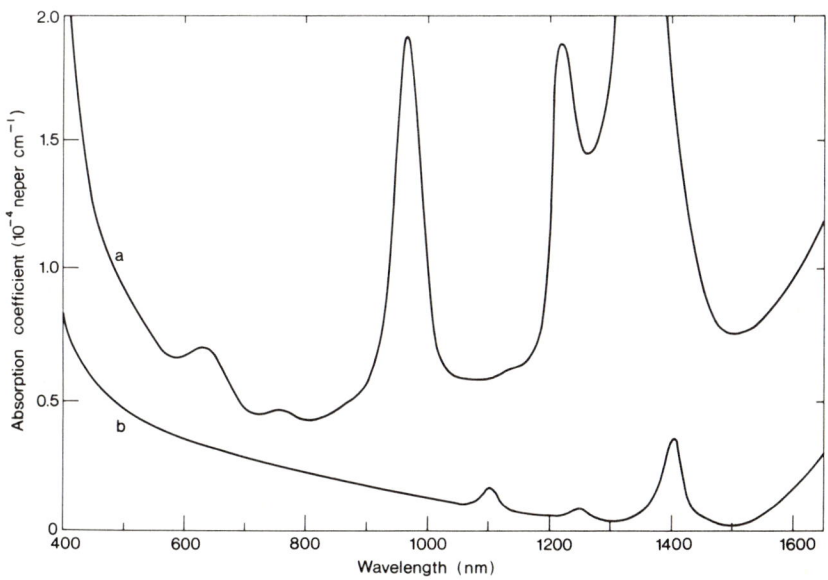

FIG. 3.26. Absorption coefficient of (a) normal good quality glass and (b) very carefully purified borosilicate glass suitable for optical fibres.

d. Scattering, both of the Tyndall and Rayleigh varieties. These arise respectively from large-scale and small-scale inhomogeneities in the glass. The Rayleigh scattering can be thought to be due to the amorphous nature of the glass which has its high temperature thermal fluctuations of density "frozen in".

e. Absorption due to the overtones and combinations of the fundamental vibration frequencies of the amorphous silica structure.

Only the last is something fundamental and the losses due to the rest can be reduced if the technologist is prepared to work hard enough. The first thing to do is to get rid of inadvertent impurities and the obvious way of doing this is to make the glass from vapour phase components. The developed form of this concept is usually known as the Modified Chemical Vapour Deposition (or MCVD) technique. It was pioneered by Bell Laboratories and by Corning Glass Works in the late 60s but is now used worldwide. There are two basic approaches to actual MCVD manufacture, depending on whether the deposition takes place on the inside or the outside of a forming tube. The typical fibre compositions aimed at use claddings of silica doped with boron or fluorine and core materials also of silica but doped with germanium, boron and fluorine in various proportions. The technique is to pass a stream of dry oxygen through liquid $SiCl_4$, $GeCl_4$ and BCl_3, combine the resultants in the correct proportions by means of valves or "mass flow controllers" and then pass them into a very hot tube heated by a burner. A glass "soot" is formed on the inside of the silica tube and when this inside coating is thick enough the whole tube is collapsed, by increasing the heat, into a solid rod about 1 cm in diameter and this is then drawn out into a narrow fibre. The latest (1979) state of this technology has been patented by Le Sergent and Liegois [212]. The outside deposition technique just reverses everything but is less popular because it is the core rather than the cladding for which the low-loss is important. One can therefore start with the best silica tube one can find for the inside process, and devote all one's efforts to getting the unwanted impurities in the core down as much as possible. The principle variant techniques are the axial deposition process [213] in which the vapours are deposited on the top of an already existing silica rod, and the various double crucible methods [214]. This latter starts with the relevant oxides (very carefully purified) and melts them in a silica crucible. After treatment the glasses are removed in cone-form from the melt, loaded into a platinum double crucible, heated in a furnace and then converted into fibres by flowing through the bottom nozzle of the crucible. Double-crucible fibre was developed by British Telecom but is now being made by several manufacturers, e.g. STL(UK), GEC and Plessey.

Using these techniques, fibres were produced a few years ago that had the remarkably (for then!) low loss of 20 dB km^{-1}. This was made up of a 13 dB km^{-1} absorptive loss and a 7 dB km^{-1} scatter loss. Since then the losses have dropped to a typical figure of around $2 \, dB \, km^{-1}$ for conventional multimode fibre and even to values as low as $0.35 \, dB \, km^{-1}$ for carefully fabricated monomode fibre. These values are adequate, as mentioned above, for very long repeater spacing telecommunication links. In the future, new glasses, e.g. fluoride glasses, will be introduced [215] and then by increasing the operating wavelength to 4 μm losses as low as $10^{-3} \, dB \, km^{-1}$ could be possible [216]. With such a cable one could go half-way round the world without needing a repeater!

Up till now, however, it has not been the losses in the glass fibre which have dictated the operating wavelength, rather it has been the state of diode laser and LED technology. The GaAs diode, variously doped, tends to radiate somewhere in the band 830–860 nm and diode detectors made from the same material respond optimally in the same region. This is why first-generation optical fibre systems work near 850 nm. The second generation will almost certainly work at 1·3 or 1·55 μm where glass is more transparent and has less dispersion. New types of laser are required and the British Telecom Research Laboratory (BTRL) [217] amongst others has developed double heterostructure GaAlAs lasers for these longer wavelengths. Both diodes and lasers are required since there are uses for both. The two types do not differ much in construction; the lasers give more power (1–3 mW) and the output radiation is more nearly monochromatic but on the other hand they cost more, are stigmatic, non-linear, non-Lambertian and less reliable. These drawbacks are not so troublesome in a monomode system and in particular the non-linearity does not matter so much in PCM operation. Originally the lasers had to be cooled to low temperatures because of the high threshold currents which were involved, but with the introduction of the double heterostructure monolithic stripe geometry (section 6.2.5) this is no longer necessary and the lasers can operate at well above room temperature. Diodes give an incoherent Lambertian emission with a considerable spread in wavelength. In fact the spectral bandwidth is sufficient to limit the use of diodes to the lower reaches of the bits per second spectrum. On the other hand the diodes, being linear in their response, can be readily modulated. Both diodes and lasers are fabricated with a "pigtail" which enables them to be connected immediately to a fibre optic cable. Early diodes and lasers had very short operating lifetimes but currently these have increased to 10^{6} h for diodes and 10^{5} h for lasers and are still slowly increasing. The improved reliability of diodes and lasers has been an important factor affecting the cost of fibre optic systems since every repeater needs a source within it and, from what has been said before, repeater failure, especially under the sea, can be calamitous. But it is not only the sources which are undergoing rapid development, new types of receiver and processing electronics all working on the same "chip" are already operative and these have performed well at data rates of 320 Megabits per second. The detector for this system [218, 219] is a PIN–FET hybrid optical receiver based on $In_{x}Ga_{1-x}As/InP$. The laser source and the detector can each be coupled to the fibre by means of a microhemispherical lens made from a high-index glass and already it seems that systems based on these plus a monomode fibre, will permit data rates of 1 Gigabit per second over distances of the order of 50 km. It seems that the simpler systems based on LED sources and multimode cable will be used for long-distance communication and for medium data-rates. The avalanche diode which has the merit of internal gain seems, however, to be becoming obsolete because of the development of the much more trouble free PIN–FET combination, but avalanche diodes based on $In_{0.53}Ga_{0.47}As/InP$ are still under intense development for the longer wavelength bands [220]. Power levels in optical fibre systems are commonly quoted in dBm, that is decibels relative to 1 mW. So a system which was delivering 200 photons per pulse in a 100 Megabits per second system operating at 900 nm would represent an average power level of -57 dBm and a peak power level of -54 dBm. The avalanche diodes and PIN–FET detectors need at least 200 photons per pulse to give an error rate of 10^{-9} or better [209] so insertion losses of 57 dBm are the maximum that can be endured without the need for a repeater.

Apart from absorption, the other limitation on the performance of an optical fibre is set by dispersion. In optical fibre connections this can have two meanings: firstly material dispersion, that is a variation of refractive index with frequency and secondly mode dispersion where each mode actually propagating does so at its own characteristic mode velocity.

$$v_{nm} = v\lambda_{nm}. \tag{3.6.20}$$

In either case one can see that the set of sharp pulses (the "bits") initially launched will start to deform more and more with distance along the fibre as their Fourier components get more and more out of step. Eventually, as the spreading pulses start to overlap, the error rate will start to rise alarmingly and the link can no longer be used at that data rate. Obviously by reducing the data rate one can bring the system back into commission. Any link can therefore be characterised by a bandwidth times distance product and this product is determined essentially by the two types of dispersion. Typical values (in units of $MHz\,km^{-1}$) are ~ 100 for short-distance fibres and more than 1500 for long-distance highgrade fibres [221]. Two ways of dealing with dispersion suggest themselves: firstly to work at a wavelength where material dispersion is a minimum (for glass 1·3 μm) and secondly to use a monomode fibre. This latter solution is, in fact, adopted where high data rate transmission over long distances is required, but monomode fibres are costly and delicate and other solutions are sought for the more "bread-and-butter" applications. Principal amongst these is the use of "graded-index-fibre" in which the core refractive index varies smoothly with radial distance. In this way one can arrange for the ray to be travelling faster the further it is away from the axis and thus compensate for its having to travel further. The refractive index variation usually adopted is the so-called "alpha" profile

$$n(r) = \{n_m^2 - [n_m^2 - n_c^2] (r/r_g)^\alpha\}^{1/2}, \quad r < a,$$

$$n(r) = n(r_g), \quad r > a, \tag{3.6.21}$$

where n_m is the peak value of the index at the centre of the fibre, n_c its value in the cladding and r_g is the radius of the guide. This particular form is adopted because (a) the wave equation can be solved exactly with this particular choice [222] and (b) guides can actually be manufactured with this particular variation. Step index guide, where there is a discontinuity of index (Fig. 3.27(a)) is given by $\alpha = \infty$ but the solution of the wave equation shows, as expected, that this choice gives high dispersion. The minimum of mode dispersion occurs for α close to 2; this is the so-called "parabolic" profile; it is illustrated in Fig. 3.27(c). Graded index guide with this profile and operating at 1·3 μm is capable of a band-width distance product of 1000 and so can compete quite well with the expensive single-mode guide. It is also possible with this type of guide to increase the information carrying potential of the fibre even more by the use of wavelength multiplexing. In this one has several sources operating at several different wavelengths which are all coupled into a common fibre. At the far end the various wavelength channels (each of course carrying a large number of frequency multiplexed channels) are separated by a polychromator.

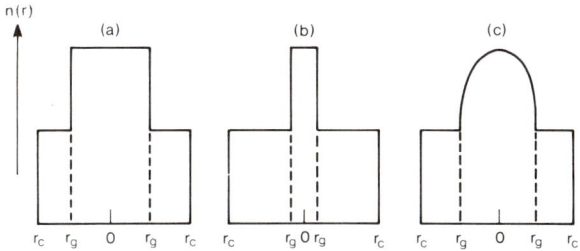

FIG. 3.27. Various types of refractive index profile used in optical fibres. They are (a) step index, (b) monomode and (c) parabolic profile.

In the early days of glass fibre optics there were difficulties with connectors since glass is a hard substance which cannot easily be cut and the fibres were very fine. It was this point which suggested the use of thicker polymeric cables which are easily joined. However, intensive research has come up with several practical forms of connector and there are even connectors which permit broken cable to be repaired in the field. A schematic diagram of a simple but effective type of connector (BTRLs prototype) is shown in Fig. 3.28. Commercial ruggedised versions of this "expanded-beam" connector are now available from several sources, e.g. STL [223]. The optical connection can be either across an air filled gap, when there is the inevitable reflection loss, or else across a thin film of index matching jelly. In the latter case reflection loss is minimised but there may be dust difficulties.

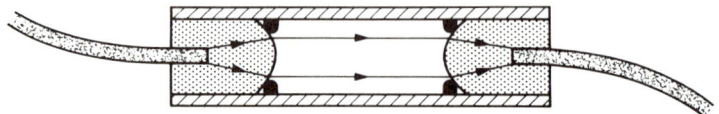

FIG. 3.28. Schematic of the "expanded-beam" type of fibre optic coupler.

The fibre optic cables actually used will usually carry very many separate guides. BTRL have described a modular optical fibre cable which has a hexagonal close-packed (or "honeycomb") cross-section. At the centre are seven oriented polypropylene tubes each of which can loosely house up to three fibres. These tubes are surrounded by twelve strength-giving members of similar size and made of polypropylene-coated glass fibre bundles. The hexagonal structure is held in place by an overlay of polyethylene ter-ephthalate tape and finally protected by a high-density polyethylene sheath [224]. Other designs, popular in the USA, have a central metallic wire which both gives strength and serves for passing normal electrical signals (and battery charging voltages) down the cable. The tests which have been so far carried out using these cables [225] have been most impressive. It is clearly only a matter of time before *all* long-distance land telecommunications willl be by optical fibre. The recent successful tests of underwater links imply that eventually this is going to be true of *all* telecommunications. Midwinter's book [209] gives a very clear and up-to-date account of the present situation.

3.6.5 *Integrated optics*

One of the most remarkable developments in technology, in recent times, has been the introduction of integrated circuits (ICs) and large-scale integration (LSI) in which very many circuits can be assembled on a single silicon chip or wafer. Naturally there have been attempts to mimic this development at optical frequencies and these have led to the concept of integrated optics. The basic idea is to have laser beams propagating in very small transverse dimension waveguides formed within very low loss glass slabs. The waveguides are created within the bulk material (though actually very near to its surface) by irradiation with high-energy particles (protons or Li^7 nuclei for example) and this gives regions in which the refractive index may be as much as 1 % higher than that of the surrounding material. If the guide so formed has a square section, then, for fundamental mode propagation, the side of the square must be $4.9 \, \lambda/n_1$, where n_1 is the guide index and n_2 (the outside index) is 1 % less. With GaAs laser radiation ($\lambda = 0.85$ μm) this gives a dimension of 2·8 μm so one is talking about very small scale fabrication and the techniques of photolithography, developed for the electronics industry, are highly appropriate. Using suitable masks, it is possible to lay down several guides in one block of glass and to have these bend so as to approach one another and then diverge away again. In this way one can make the analogue of a microwave directional coupler, as shown in Fig. 3.29. Bends inevitably lead to loss of power by radiation but the losses of a beam confined by total internal reflection need not be high and coupling (essentially by tapping the evanescent waves) can be very effective even for bend radii of less than 1 mm. The laser beam can be coupled into the guide using a prism (as in Fig. 3.29) or else using a diffraction grating etched onto the surface of the substrate. The grating method is elegant and naturally adapted to the integrated optics approach, but both methods are comparable in their performances, transferring about 70% of the incident power into the guide. Modulators can be made by depositing thin metallic films, to serve as electrodes, and fabricating the guide from a suitable electro-optic material. Only low driving voltages are required since the plates will only be a few micrometres apart. Resonant circuits can be made by terminating a section of guide with grating reflectors. Active circuits too can be constructed because the power densities in the guides are quite high. If the guide is fabricated from a non-linear material it is possible in this way to make a mixer. Devices of this type are obviously very important for the development of the topic but the fabrication technology at the moment is very expensive and it requires highly skilled operatives, so mass production of integrated optics components is still some way off. Nevertheless, what has been achieved so far (see Miller [226] for a very clear review) is impressive and it has established a firm place for integrated optics in electro-optical technology [227].

(a) (b)

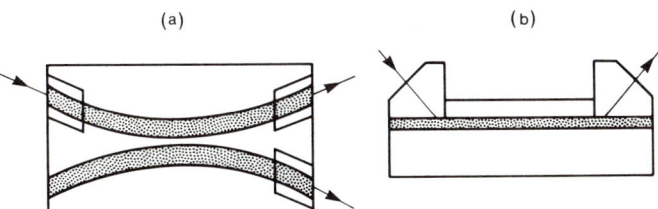

FIG. 3.29. A schematic in (a) plan and (b) elevation of a prism-coupled integrated-optics directional coupler.

An exciting recent development in this field has come from the attempts to fuse together the approaches of integrated electronics and of integrated optics. This work has sprung from the need to have absolutely reliable and absolutely stable optical repeaters for underground (and especially undersea) fibre-optic telecommunication systems. The idea, in a nutshell, would be to develop a "chip" which would have merely a fibre-input, a fibre output and two DC electrical connections! Such a unit could be assembled on an InP substrate with In-Ga-As-P quaternary layers laid down through suitable masks by epitaxy. The variations of refractive index so produced could provide wave-guidance but in addition the layer structure could be used to provide laser/modulator or detector/amplifier combinations in-line in the "wave-guide". The incoming near-infrared optical signal would be detected, the resulting electrical signal digitally cleaned up (see Fig. 3.25) and then used in its turn to modulate the radiation from a launching laser feeding the output fibre. The future possibilities of such integrated optoelectronic components seem almost limitless.

3.7 Reflection spectroscopy

3.7.1 *Introduction*

The conventional way of recording an absorption spectrum is to determine the transmission spectrum of the specimen and then to correct this for the losses due to single and multiple reflections. Sometimes, however, this procedure is either very difficult or actually impossible and the spectroscopist then must have recourse to determining the spectrum from reflection measurements. This is possible because the Fresnel equations (Appendix 2) hold just as well when either n_1, n_2 or both are complex and it follows therefore that the power reflectivity, at any given frequency, will carry information about both n and k. The problem is to disentangle the information and thereby to deduce the separate values of n and k. We have already discussed one method of doing this, namely the application of the Kramers–Kronig equations. Clearly another would be to take measurements of the reflectivity at two widely different angles and then to use the two resulting simultaneous equations to evaluate n and k directly. Unfortunately it turns out that this method is limited to the study of strongly absorbing media and is therefore of little use for the investigation of the organic compounds and mixtures which make up the bulk of the routine spectroscopists work. To show this one need only consider the variation of ρ with k for a fixed n and angle of incidence. For convenience normal incidence will be chosen and n will be given the value 1.5. One therefore has from (3.5.23a):

$$\rho = \frac{(n-1)^2 + k^2}{(n+1)^2 + k^2} = \frac{0.25 + k^2}{6.25 + k^2}. \tag{3.7.1}$$

This function is plotted in Figure 3.30. It will be immediately seen that ρ is virtually independent of k for small k and only becomes sensibly dependent when k is of the same order as $(n-1)$. Now k values of the order 0.2 correspond to intense absorption (equation 3.4.6a) at infrared frequencies and it follows that analysis of the reflectivity of organic substances (for which k would usually be less than 0·01) will give no precision whatever in the determination of k.

The occasions when one might wish to investigate the spectrum of a specimen by reflection methods even when it is relatively transparent arise nearly always when the specimen is in the form of a thin film. Often the film is not free standing but can only be made as a surface layer on some substrate. A good practical example is found in the study of the active entities on the

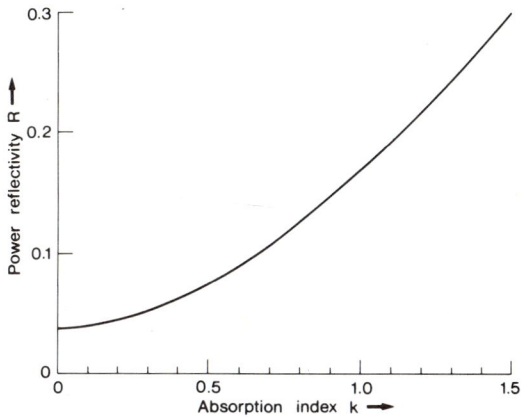

FIG. 3.30. Power reflectivity at normal incidence as a function of absorption index.

surfaces of heterogeneous catalysts. It might be thought that there would be little problem since one could send the probing beam *through* the film, have it reflected at the substrate interface and then detect it after it has come back through the film. However, on deeper analysis, it will be realised that the ingoing and outgoing beams will interfere with one another leading to a standing wave pattern. If the substrate is metallic there will be a node in the standing wave pattern at the metallic surface and interaction of the beam with the film will be slight. Even with non-metallic substrates there are many difficulties with this method and the number of practical cases where it may be used are few and far between.

The urgent need of practical infrared spectroscopists for a means of recording the spectrum of this kind of specimen led several investigators, principally Harrick [228] and Fahrenfort [229], to develop a radically new way of obtaining an infrared spectrum by reflection techniques. This technique, usually called attenuated total reflection (ATR for short), will be described in some detail later but first it will be necessary to return to the topic of propagation at a discontinuity of refractive index in the special case of incidence from a more dense onto a less dense medium.

3.7.2 *Theory of internal reflection*

3.7.2.1 Case 1: *both media transparent*

The treatment given so far of refraction/reflection at an interface has been general but, whenever an illustration was attempted, it was assumed implicitly that one was considering a ray travelling in a less dense medium (usually air or vacuum) and encountering a more dense medium at an interface. When the situation is reversed, some remarkable phenomena can result. Thus if one writes Snell's law in the form

$$\sin \theta = n_{12} \sin \phi, \tag{3.7.2}$$

where $n_{12} (= n_2/n_1)$ is the relative refractive index, it will be seen that when n_{12} is less than unity there exists a maximum value of θ for which ϕ can still be a real angle. This maximum value given by

$$\theta_c = \sin^{-1} n_{12} \tag{3.7.3}$$

is known as the critical angle. For values of θ greater than θ_c, $\sin \phi$ is real (but greater than unity) whereas $\cos \phi$ is pure imaginary and given by

$$\cos \phi = \pm \frac{i}{n_{12}} \sqrt{(\sin^2 \theta - n_{12}^2)}. \tag{3.7.4}$$

If one substitutes the above value for $\cos \phi$, the Fresnel reflection coefficients take on the form

$$\hat{r} = \frac{a - ib}{a + ib} \tag{3.7.5}$$

since $\cos \phi$ is pure imaginary. Now, as is well known, any complex number of the above form has unit modulus (since its complex conjugate equals its reciprocal) and it follows therefore that the power reflection coefficient is unity, that is the reflection at the interface is total.

Total reflection is a useful phenomenon throughout the whole spectral range. It is even available at X-ray wavelengths though there, because material refractive indices are usually less than unity, it is the atmosphere or vacuum which is the denser medium and one has formally total *external* reflection rather than the more usual total *internal* reflection. However, it is at much longer wavelengths that total reflection reaches its maximum utility for there it can be used to make polarisers, variable couplers, filters etc. and to provide an invaluable technique for determining the infrared spectra of surfaces. All of these applications depend on the remarkable phenomena which occur at the interface during total reflection and to elucidate these, one again uses the concept of a complex wave. If one writes

$$\hat{E} = \hat{E}_0 \exp \left[-2\pi i (vt + \mathbf{k} \cdot \mathbf{r}) \right] \tag{3.7.6}$$

for the general complex wave, then, for the transmitted wave, one would have

$$\hat{E}_t = \hat{E}_{t0} \exp\left\{ -2\pi i \left[vt - \frac{n_2 v}{c}(x \sin \phi + z \cos \phi) \right] \right\}, \qquad (3.7.7)$$

which by (3.7.2) and (3.7.4) becomes

$$\hat{E}_t = \hat{E}_{t0} \exp\left[-2\pi i v \left(t - \frac{n_1 x \sin \theta}{c} \right) \right] \exp\left(-\frac{2\pi n_1 vz}{c} \sqrt{(\sin^2 \theta - n_{12}^2)} \right).$$
$$(3.7.8)$$

This is a strange sort of wave, for whilst it is propagated along x it is not propagated at all along z. All that happens is that the electric field amplitude falls extremely rapidly with depth of penetration into the second medium. Equation (3.7.8) can be interpreted as follows: when θ_i approaches θ_c, the effective wavelength in the z direction (i.e. $\lambda_0/n_2 \cos \phi$) approaches infinity. When θ_i equals θ_c the electric field in the second medium is uniform everywhere, that is the wavelength *is* infinite. When θ_i exceeds θ_c the electric field in the second medium is not time dependent, i.e. it does not propagate, but simply falls exponentially. In this light the behaviour is continuous through the critical angle, the electric field passes from exponential behaviour with an imaginary argument to exponential behaviour with a real argument via the condition where E is uniform everywhere in the second medium. Since all the incident energy is reflected, there can be no net flow of energy into the second medium but this does not imply that the Poynting vector is zero. In fact it is everywhere finite but its time average is zero, a fact which has profound consequences for practical infrared spectroscopy since it follows that energy flows into and then out of the second medium. This peculiar situation is usually described by saying that there is an evanescent wave present in the second medium. The presence of this evanescent wave has been more or less confirmed by a series of experiments which involve bringing up such things as razor blade edges and fluorescent materials close to the interface. One has to say "more or less confirmed" of these experiments; because of its nature, it is impossible to detect the evanescent wave without distorting the conditions producing it. In the final analysis one has to rely on the basic electromagnetic theory being correct and regard these experiments as merely signifying that there is no obvious disagreement.

3.7.2.2 *Second medium absorbing—Attenuated Total Reflection or ATR*

In many real cases, the second medium is not perfectly transparent; $\cos \phi$ is now complex instead of pure imaginary and there will be a propagating but rapidly damped wave in the second medium. Alternatively, one can still think of an evanescent wave but as it flows back and forth through the absorbing medium it loses energy. This leads to an observable loss and the reflection is no longer total—we have the phenomenon of *attenuated total reflection*.

The discussion of the phenomena taking place during attenuated total reflection (ATR for short) follows quite naturally from the ideas already developed; all that is necessary is to replace the refractive indices, which appear in the Fresnel equations, by their complex analogues. The basic theory was first developed for dealing with reflection from metals and the notion is sometimes abroad that these mathematical techniques have something to do with free electron absorption. This is not so. It can be shown quite rigorously that the complex forms of the Fresnel equations apply strictly whenever there is a discontinuity between two media, one or both of which is or are absorbing. Metals are a special case of this, but the treatment applies just as well to non-conducting, but absorbing, dielectrics. In nearly all infrared cases only one of the media has appreciable absorption; the other can therefore be taken as perfectly transparent. We can thus write Snell's law in the form

$$\sin \theta = \hat{n} \sin \hat{\phi} \qquad (3.7.9)$$

where the suffices 1, 2 have been dropped,

and hence

$$\cos \hat{\phi} = \frac{1}{\hat{n}} \sqrt{(\hat{n}^2 - \sin^2 \theta)}. \qquad (3.7.10)$$

Multiplying top and bottom by $\hat{n}^*$ and expanding terms gives

$$\cos \hat{\phi} = \frac{1}{n^2 + k^2} \sqrt{[(n^2 + k^2)^2 - \sin^2 \theta (n^2 + k^2) - 2ink \sin^2 \theta]} \qquad (3.7.11)$$

which, making use of the standard identity (section 1.4.2),

$$(a + ib)^{1/2} = \left[\frac{a + (a^2 + b^2)^{1/2}}{2} \right]^{1/2} + i \left[\frac{-a + (a^2 + b^2)^{1/2}}{2} \right] \qquad (3.7.12a)$$

$$= (a^2 + b^2)^{1/4} \exp i \tan^{-1} \left[\frac{-a + (a^2 + b^2)^{1/2}}{a + (a^2 + b^2)^{1/2}} \right]^{1/2}, \qquad (3.7.12b)$$

may be written in analytical form. However in this form the equation is rather complicated and the equations resulting from substituting it into the Fresnel equations are even more so and may really only be handled using a computer. This tends to prevent one gaining an insight into what is happening, but by making use, once again, of the "infrared" assumption, namely that $k \ll n$, the equations may be considerably simplified. If one makes also the contingent approximation that the angle of incidence is that value which would be the critical angle for a lossless interface then one has

$$\cos \hat{\phi} = \left(\frac{k}{n} \right)^{1/2} (1 - i). \qquad (3.7.13)$$

The parallel reflection coefficient now becomes

$$\hat{r}'' = \hat{E}_r'' / \hat{E}_i'' = \frac{1 - \cos \hat{\phi} / (\cos \theta \, \hat{n})}{1 + \cos \hat{\phi} / (\cos \theta \, \hat{n})} \qquad (3.7.14)$$

which, again making use of the approximations consequent upon $n \gg k$, becomes

$$\hat{r}'' = \frac{\{1 - [(n+k)/(n^2 \cos \theta)] (k/n)^{1/2}\} + i[(n-k)/(n^2 \cos \theta)] (k/n)^{1/2}}{\{1 + [(n+k)/(n^2 \cos \theta)] (k/n)^{1/2}\} - i[(n-k)/(n^2 \cos \theta)] (k/n)^{1/2}}.$$

(3.7.15)

The modulus of $\hat{r}''$ determines the reflected power; it is in fact the power reflection coefficient and from the relation $\hat{r}\hat{r}^* = \rho$ one has

$$\begin{aligned}
|\hat{r}''|^2 = \rho'' &= \frac{\{1 - [1/(n \cos \theta)](k/n)^{\frac{1}{2}}\}^2 + \{[1/(n^2 \cos^2 \theta)](k/n)^{\frac{1}{2}}\}^2}{\{1 + [1/(n \cos \theta)](k/n)^{\frac{1}{2}}\}^2 + \{[1/(n \cos \theta)](k/n)^{\frac{1}{2}}\}^2} \\
&= \frac{1 - [2/(n \cos \theta)](k/n)^{\frac{1}{2}} + 2\{[1/(n \cos \theta)](k/n)^{\frac{1}{2}}\}^2}{1 + [2/(n \cos \theta)](k/n)^{\frac{1}{2}} + 2\{[1/(n \cos \theta)](k/n)^{\frac{1}{2}}\}^2}.
\end{aligned}$$

(3.7.16)

When (k/n) is so small that the square terms can be neglected in equation (3.7.16), the expression reduces to

$$1 - \rho'' \approx \frac{4}{n \cos \theta} \left(\frac{k}{n}\right)^{\frac{1}{2}}.$$

(3.7.17)

A typical value for n might be 0.707 which conveniently makes $\cos \theta = 0.707$ as well and with this substitution one has

$$1 - \rho'' \approx 8 \left(\frac{k}{n}\right)^{1/2}$$

(3.7.18)

Therefore if $k/n = 10^{-4}$, $(1 - \rho'')$, i.e. the fractional change in reflected power, would be 8%. This illustrates the great sensitivity of the ATR technique. Similar conclusions apply to the perpendicular power reflection coefficient except that this quantity is not as sensitive to k as is ρ''.

3.7.3 Practical implementation of ATR spectroscopy

We have just seen that when infrared spectroscopy is being applied to the study of thin films, ATR provides a very useful way of enhancing the sensitivity. Still greater enhancement is available if the beam can be arranged to undergo further internal reflections. One then gets the very widely applied technique of Multiple Internal Reflection (or MIR for short) spectroscopy. A common way of achieving this is shown in Fig. 3.31.

MIR spectroscopy is much used in industry [230]. Thus in the polymer industry where thin films and sheets are produced it is very useful, and in the packaging industry where it is used to characterise thin coats applied to a wide range of materials it is invaluable. However, it is just as useful in almost any research investigation where one has thin films or thin layers adsorbed on a substrate and it is becoming widely applied throughout the fields of surface chemistry and physics. In favourable cases, and especially when MIR is combined with Fourier transform spectrometry, it may even be used to study monolayers. An example of a more conventional application—the study of thin polymeric films—is shown in Fig. 3.32. In this Figure, one immediately obvious advantage of MIR spectroscopy, the absence of channel fringes, is well illustrated.

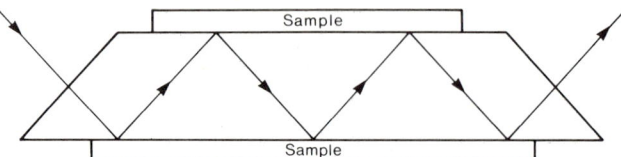

FIG. 3.31. Use of a high refractive index prism to achieve multiple internal reflection. In the prisms (or ATR plates) actually used it is possible to realise the equivalent of up to 25 reflections.

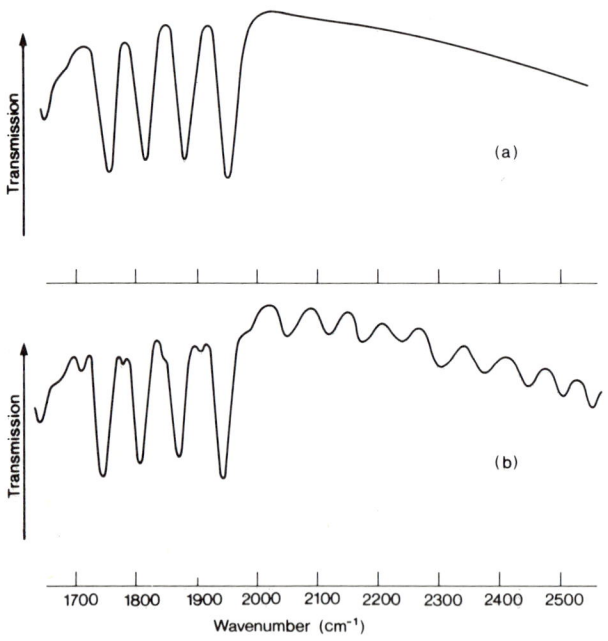

FIG. 3.32. Part of the mid-infrared spectrum of a polystyrene film (0·05 mm thick) obtained by (a) MIR spectrometry and (b) by conventional transmission spectrometry. The confusing channel fringes so conspicuous in (b) are completely absent from (a).

MIR spectrometry is a most valuable technique, but its use does introduce some fairly stringent experimental requirements which are either absent or else only marginally present in conventional spectrometry. Thus, because the beam travels a long way on its zig-zag path through the prism, the prism material has to be very transparent and that usually means very expensive as well. Secondly, in order to get undistorted spectra, it is necessary that the refractive index of the prism material be as high as possible. The reason for this is that the sample will have a refractive index that varies with frequency and, if the oscillations of sample refractive index take its value close to that of the prism, the penetration depth will rapidly increase and enhanced attenuation of the beam will be observed. This anomalous absorption, almost entirely dispersive in origin, gives rise to very distorted line shapes [231]. One can actually take advantage of this phenomenon and by the deliberate use of a low index prism measure the complex permittivity of a thin film [232], but most spectroscopists are interested solely in the absorption spectrum and are therefore keen to avoid all distortion and that implies the use of a high index prism. The available choices are rather limited. One has KRS-5 ($n = 2.4$), germanium ($n = 4.0$) and zinc selenide ($n = 2.4$) which can be used over wide spectral ranges and silicon ($n = 3.0$) which is usable over a much narrower range. Germanium is attractive because of its high index but it tends to be brittle and, when solid specimens have to be clamped to it under pressure, it is liable to fracture. On the other hand it is excellent for the study of liquids. KRS-5 is a soft material free from fracture problems and virtually transparent throughout all of the conventional infrared. Its drawbacks are that it is attacked by aqueous solutions and that its dust is toxic. Recently zinc selenide has become available in a very pure form and monocrystalline slabs several cm in diameter can be purchased at a not unreasonable price. This material is almost ideal: it is insoluble, temperature insensitive and transparent out to 18 μm. It will probably become the most widely used prism material as the older types are gradually replaced.

Two practical aspects of ATR spectrometry which merit discussion are the choice of the angle of incidence and the effects of beam divergence. The larger the angle of incidence, the less the chance of line-shape distortion due to dispersion, since one is further from the critical angle. However the larger the angle of incidence, the fewer are the number of reflections in a given length of prism, so some degree of compromise is necessary. The usual compromise is to decide on 45 degrees as the preferred angle of incidence though some commercial equipment does permit operation at 30 and 60 degrees. For a refractive index of 2.4, the critical angle is 24.6 degrees, so even for a choice of 30 degrees there would still be a reasonable safety margin. Beam divergence has the effect that the various elementary pencils into which one might imagine the beam to be resolved undergo different numbers of reflections. The observed intensities will therefore vary from instrument to instrument and, if one is using ATR as a diagnostic tool, some caution is required. The usual way of dealing with this problem and the related one that the intensities in ATR, though very similar to those observed in conventional operation, are not exactly the same, is to assemble a library of reference spectra appropriate to a given type of prism and a given type of spectrometer.

3.8 Propagation in stratified media

A common occurrence in infrared technology is to have the beam propagating through a series of plane-parallel dielectric slabs. This situation might arise, for example, in such disparate applications as the use of a windowed cell in spectroscopy and the use of a multilayer interference filter for atmospheric monitoring. Each slab will generate its own set of channel fringes and, if there is no simple relation between the optical thicknesses of the slabs, a complicated variation of transmission with wavelength will result. This is the normal case for the spectroscopic cell and it is the cause of the difficulties mentioned in section 3.5.2. For many other cases, however, the situation is much simpler because the optical thicknesses of the slabs are related to one another and to the wavlength. Examples of devices of this nature are provided by the antireflection coating and by the interference filter. These will now be described in outline.

3.8.1 *Multilayer infrared interference filters*

The simplest type of multilayer interference filter is the anti-reflection coating [233] shown in Fig. 3.33. One has a layer, thickness d_1, of a transparent material whose refractive index is n_1 deposited on a very much thicker substrate of refractive index n_2.

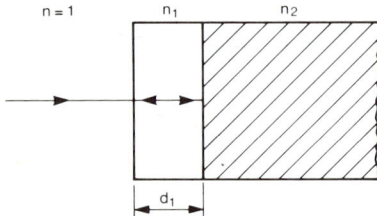

FIG. 3.33. Optical arrangement of the anti-reflection coating.

The beam, incident from air or vacuum $(n = 1)$, will undergo multiple internal reflections as described in section 3.5, but the analysis is now slightly more complicated because there are two reflection coefficients, r_1 at the first interface and r_2 at the second. For an anti-reflection coating, n_2 will be greater than n_1 so, in adding up the multiple beams, the phase change of π on reflection at the second interface has to be taken into account. The overall amplitude reflection coefficient is, therefore,

$$\hat{r} = -r_1 - t^2 r_2 \exp(i\delta) + t^2 r_1 r_2^2 \exp(2i\delta) - t^2 r_1^2 r_2^3 \exp(3i\delta) + \text{etc.}$$

$$\hat{r} = -\left(\frac{r_1 + r_2 \exp(i\delta)}{1 + r_1 r_2 \exp(i\delta)} \right). \tag{3.8.1}$$

The reflection coefficient will be zero if and only if $\exp(i\delta) = -1$ and $r_1 = r_2$. The first of these conditions demands that the optical thickness be a quarter wave length so such coatings are often referred to as quarter-wave layers or as quarter-wave plates. The second condition, from equation (3.5.4), is equivalent to writing

$$\frac{n_1 - 1}{n_2 + 1} = \frac{n_2 - n_1}{n_2 + n_1}, \tag{3.8.2a}$$

from which it follows that

$$n_1 = \sqrt{n_2}. \tag{3.8.2b}$$

This square-root condition is the fundamental requirement to be met in the construction of an antireflecting layer. In the visible region, where refractive indices tend to cluster about the value 1·5, it is very difficult to satisfy since homogeneous media with refractive indices of 1·22 are not available. The best that can be done is to use materials such as MgF_2 ($n = 1·35$) for the 'blooming" of lenses etc. and one has to be content with merely reducing the reflection loss rather than with eliminating it altogether. In the infrared, on the other hand, where there is a much wider spread of indices, it is far easier to get near perfect coatings and this is very convenient because the high index of germanium, for example, leads to unacceptably high losses with uncoated optics. A concise review of the possible materials for this application has been given by Evans *et al.* [234]. These authors also point out that by using two layers, one having an index $n = n_2^{2/3}$ and the other having an index $n = n_2^{1/3}$, acceptable antireflecting properties can be achieved over a wide spectral band. As an example, with thallium iodide and caesium bromide coatings on germanium, it is possible to have transmissions greater than 50% over most of the infrared.

The next type of filter involves combinations of quarter-wave layers and half-wave layers and is, in essence, a way of achieving a high finesse Fabry–Perot interferometer without having to have heavily absorbing metallic reflectors. The filter consists of alternating quarter-wave thicknesses of high (H) and low (L) index materials and the simplest type is symmetrical about a central half-wave layer. The convention used to describe such a filter is to write them say as

$$H\ L\ H\ H\ L\ H$$

or $\qquad\qquad H\ L\ H\ L\ L\ H\ L\ H \qquad$ etc.

where each symbol represents a quarter-wave thickness. The calculation of the transmission of such an ensemble at an arbitrary wavelength is very difficult and is best tackled by matrix methods, but fortunately near the peak transmission wavelengths given by

$$\lambda = \lambda_{\max}/m, \tag{3.8.3}$$

where $\lambda_{\max}$ is four times the optical thickness of the layers, the behaviour can be accurately modelled in terms of an effective reflectance R_e for the "reflectors" of the central half-wave Fabry–Perot resonator. This quantity can then be inserted in equations (2.1.23) to give the observed performance of the filter. R_e itself can be calculated by summarizing the consecutive effects of each layer. One needs to consider two cases: (a) where the central half-wave is LL and (b) where it is HH. In the first case there will be an even number of quarter-waves each side of the half-wave whilst in the second case there will be an odd number. Houghton and Smith [235] give expressions for calculating R_e in both cases. Thus for case (a), one has

$$R_e = \left(\frac{n_H^{2m} - n_s n_L^{2m}}{n_H^{2m} - n_s n_L^{2m}} \right)^2. \tag{3.8.4}$$

It follows from this and for the equivalent expression for case (b), that the larger the difference of index, the fewer will be the number of layers required to give a specified performance. The filter is constructed by depositing the layers on a suitable substrate, index n_s, and is usually completed by the deposition of a final single antireflecting layer. The ensemble may contain a suitable absorptive blocking filter to isolate the transmission maximum of interest or if convenient this may be used somewhere else in the system.

More complex interference filters can be constructed. Thus, by making two normal filters back to back, one has a double half-wave filter (DHWF) and the process can be continued onwards to almost any degree of complexity. The analysis of the optical properties of these multiple half-wave filters is very complicated indeed: some details have been given by Seeley [236]. The performance of infrared interference filters is ultimately determined by absorption in the dielectric materials used. This limits the number of layers which can be used and also determines the maximum wavelength at which dielectric filters can be employed. This is about 40 μm where restrahlen absorption becomes prohibitively large. Filters for the region beyond 40 μm can, however, be constructed using the metallic grids and meshes described in section 3.10.

3.9 Propagation in heterogeneous media

In all that we have discussed so far it has been implied that the medium supporting the propagation of the beam is perfectly homogeneous. In the limit this is not possible, of course, since ultimately all media are heterogeneous, but for the majority of infrared cases the heterogeneities are much smaller than the wavelength and the medium can be taken, for all intents and purposes, as perfectly homogeneous. In the rare cases where this is not acceptable, the theory has to be modified to take account of scattering of radiation out of the beam direction. It is convenient to consider the three cases where we have the diameters of the scattering particles (a) much less than the wavelength, (b) of the same order as the wavelength and (c) much larger than the wavelength. Case (a), as mentioned above, is usually not observable because it is always very weak, the scattering here being due to the Rayleigh and Raman effects. Rayleigh scattering arises solely from *random* fluctuations of the refractive index. These induce random fluctuations of the phases of the secondary Huyghen's wavelets and therefore imperfect cancellation of energy propagation in directions away from the beam. The fluctuations of index can arise from the random distribution of atoms in space (entropy fluctuations) or else from the incoherent vibrational motion of the atoms about their equilibrium positions (thermal fluctuations). The Raman effect arises essentially from inelastic collisions of photons with atoms or molecules. It is usually completely incoherent and the scattered photons are either shifted down (Stokes scattering) or else up (Anti-Stokes scattering) in frequency by an amount equivalent to an internal energy change of the scattering atom or molecule. Raman scattering is usually many orders of magnitude weaker than Rayleigh scattering which is itself very weak, so under normal circumstances it can be ignored. At high power levels, however, the normal incoherent form can be replaced by a coherent form which is far more intense. This stimulated Raman scattering is best discussed in the context of non-linear optics rather than in that of scattering theory so it will be postponed to chapter 6. Both Raman and Rayleigh scattering depend on the fourth power of the frequency so they tend to be more obvious at visible and ultraviolet wavelengths (cf. the well-known phenomena such as the blueness of the sky and the converse, the reddish colours of sunset) rather than in the infrared, but now, with the development of very low loss cables for fibre optic applications, it is a measurable source of loss when integrated over distances of the order of kilometres.

Type (b) scattering is often called Tyndall scattering after John Tyndall, a nineteenth century British physicist who carried out several classic investigations of the scattering of sunlight. One of these, still carried out today as a laboratory demonstration experiment, is the blowing of smoke into the otherwise nearly invisible beam of sunlight traversing a darkened room. Tyndall showed that there was a wavelength dependence (of the order of λ^{-2}) to this particulate scattering and this dependence is usually obvious in the smoke-blowing experiment since the scattered light is markedly blue. In type (c) scattering there is no wavelength dependence and the phenomenon is really a branch of diffuse reflectance. Cases (a) and (c) are rather easy to analyse theoretically but the intermediate case (b) is very difficult indeed. Most attempts rest on the work of G. Mie [237] in the early years of this century who applied rigorous diffraction theory and managed to get an analytical solution for the simple case of the dielectric sphere. The importance of his work was that it could be applied quantitatively to an ensemble of such spheres randomly spaced, provided the mean separation was considerably greater than the wavelength. This is the usual situation for the propagation of an infrared beam through fog and Mie theory is frequently used for the analysis of atmospheric transmission experiments in the infrared.

The important parameter, then, in scattering is the ratio of the diameter of the scattering entity, a, to the wavelength λ. A schematic plot of how the scattering efficiency might be expected to vary with (a/λ) is shown in Fig. 3.34. This Figure shows well that, for wavelengths larger than the particle size, the scattering rapidly falls off. Now the average size of particles making up atmospheric fogs and hazes is about 1 μm so the use of infrared to "see"

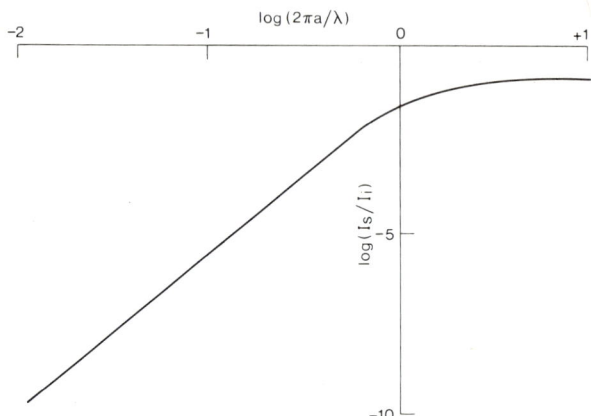

Fig. 3.34. Variation of scattering efficiency at 90° to the beam direction as a function of the ratio of particle diameter to the wavelength. The slope, magnitude 4, characteristic of Rayleigh scattering is well demonstrated for values of $2\pi a/\lambda$ less than unity.

through fog, alluded to above, is an attractive proposition. The principle users of infrared for this purpose are the military mainly in the connection of the "heat-seeking" missiles which are commonly used in modern aerial warfare. These missiles can find their targets—the hot jet engines of high-speed military aircraft—even through dense natural clouds. Counter measures are naturally resorted to, but, whilst these can be relatively effective for slow moving objects such as tanks, there are considerable difficulties with rapidly moving objects such as aircraft. The principle countermeasure is the use of smokescreens made up of particles having diameters in excess of 10 μm. In fact with a judicious choice of smokescreen aided and abetted by atmospheric absorption it is possible to virtually black-out the entire infrared. Faced with this situation the military technologists have switched a lot of their effort to the millimetre-wave band where atmospheric transmission is once again reasonable and where smokescreens are impracticable. The obvious countermeasure to a conventional millimetre-wave radar system is to use highly reflecting but militarily worthless decoys—an up-to-date version of "window" or "chaff". To be effective therefore the millimetre system has to be an "imaging" radar so that the human operator can steer the missile to its true target. Millimetre wavelengths are sufficiently short for quite good pictures to be built up, but on the other hand they are sufficiently long for broad-band antireflecting coatings to be applied by relatively low technology methods, spraying for example. It is the usual story of measure followed by countermeasure but nevertheless the 1980s will certainly see widespread use of the millimetre waveband for military radar systems.

From the point of view of the receiving system, particulate scattering has two effects: firstly attenuation of the beam and secondly loss of angular correlation. The attenuation obeys Lambert's law quite closely but the value of the apparent "absorption" coefficient depends on the solid angle of the receiving optics. Thus one can partly compensate for the scatter loss from a quasi-parallel beam by using an oversized primary collecting mirror. It follows that if the infrared beam is being used solely to carry time-varying information and is not being used for imaging, slight haze or fog need not be a major handicap. If, however, good imaging is required, the loss of angular correlation becomes a major consideration and haze or fog can therefore interfere seriously with near-infrared systems, but from the arguments developed above it will be seen that little interference will be expected with mid-infrared systems. Stillwell [238], for example, using a 10 µm thermographic camera, has reported clear observation from Brentford of Battersea Power Station six miles away through a fairly dense London fog!

3.9.1 The Christiansen effect

In most discussions of the propagation of infrared radiation through heterogeneous media, one can assume the refractive indices of the components to be constant and the only effect that a change in wavelength has is to change the transmission because of the change in (a/λ). There is, however, an important exception which arises when one or both constituents of the heterogeneous medium displays strong dispersion. It is then possible that, as the wavelength changes, the varying refractive indices of the two components will suddenly coincide and the scatter loss will vanish. This phenomenon is usually known as the Christiansen effect. If one is trying to study the infrared absorption spectrum of the heterogeneous medium, this rapid, non-absorptive change in transmission can be seriously misleading. Rapid changes in n are associated with rapid changes in α (see section 3.4.5), so the spurious consequences of the Christiansen effect are mostly manifest as distortions of the line profiles [239]. This sometimes takes the form of the breaking up of a genuinely single line into two apparent "components". The Christiansen effect is a hazard because in practical infrared spectroscopy it is not always possible to have a specimen, normally solid, in the form of dilute solution in a suitable solvent or else in the vapour phase. Often one has no choice but to use either nujol mulls or pressed discs and both of these are heterogeneous. Care must therefore be taken and if one suspects that a certain feature is a spurious artefact it is advisable to repeat the experiment with the solid specimen ground to a different particle size. Alternatively, one can use a different support medium. As an example, potassium bromide is widely used as a dispersing medium for pressed discs, but it would be wise to check a suspect run by using, say, potassium iodide or even caesium iodide.

The Christiansen effect, whilst ordinarily a nuisance, can nevertheless be turned to advantage in the manufacture of narrow-band transmission filters. These are readily constructed using an ordinary disc press, so they can be much cheaper than the competing interference filters since costly fabrication to fine tolerances is not required.

3.10 The interaction of infrared radiation with metallic grids and meshes

At wavelengths beyond about 15 µm, materials absorption problems connected with the reststrahlen process and its analogues (section 5.3.2) make the construction of conventional optics difficult, but fortunately a new type of device, based on metallic grids and meshes, then becomes available to adequately fill the gap. The metallic grid, as its name suggests, consists of a set of metallic wires, tightly stretched to make them accurately parallel and spaced at a distance greater than but still of the same order as the wire diameter. The arrangement is shown schematically in Fig. 3.35.

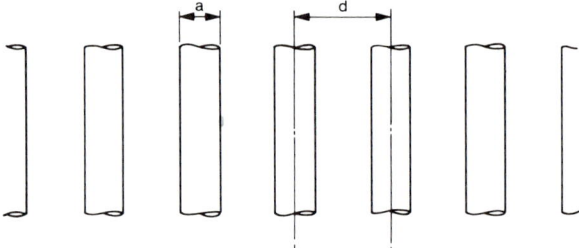

FIG. 3.35. Schematic of the construction of a free-standing fine-wire grid.

Most modern work on the properties of grids stems from Costley and his colleagues [240], but mention should be made of the similar work of Auton [241] and of the earlier theoretical work of Larsen [242]. The mesh is a sort of two-dimensional version of the grid. There are two possible types, the inductive and the capacitative, as illustrated in Fig. 3.36. These have complementary properties. For ease of manufacture, the elements of the meshes usually have square or rectangular sections whereas the elements of the grids, being made of wire, usually have circular cross section. These differences alter the boundary conditions, but since a full rigorous electromagnetic calculation is impossible for either case, it is usual to blur over the variations and to discuss both grids and meshes within a common formalism.

The closest approach to a rigorous electromagnetic calculation of the properties of grids and meshes has come from Beunen *et al.* [243] who take as their starting point the Green's function method introduced by Ham and Segall [244] for dealing with the electromagnetic properties of periodic structures. This theory has been used very successfully for calculating the

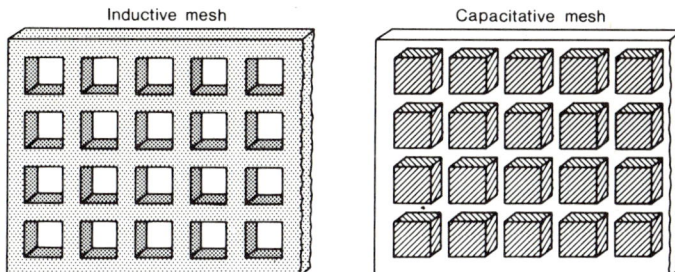

FIG. 3.36. The two types of mesh used for infrared work. The squares are of side d and are separated by rectangular strips of thickness $2a$. It is assumed that the metal foil is thin on the scale of the wavelength.

band-structures of crystalline semiconductors (section 6.5), but it is just as applicable to the case of regularly spaced scatterers. The basic differential equation is transformed into an integral equation with a Green's function kernel and this equation is solved under the given periodic boundary conditions. The Green's function itself has to be evaluated by summing an infinite series. This summation can be speeded up considerably by a method described by Ham and Segall. The solution of the integral equation then proceeds in terms of the appropriate cylindrical partial waves and the final result is expressed in terms of the coefficients of the Green's function series and the logarithmic derivatives of the partial waves at the surfaces of the scatterers. The theory does take account of resistive losses in the metallic parts of the structure, but the finite conductivity is found to make only a very small difference at long (i.e. $\lambda > d$) wavelengths. This calculation sounds very complicated and indeed it is, but, oddly enough, once a computer has been programmed to carry it out, the actual time of computation is found to be relatively short. Some results quoted by Mok *et al.* [245] for the case of the grid are shown in Fig. 3.37.

The agreement with experiment is quite good, especially the observation of maxima for integral values of (d/λ). Mok *et al.* attribute the slight discrepancies to a residual lack of regularity in the wire spacing.

The basic physical ideas behind this approach can be understood in qualitative terms where one notes that for $\lambda \ll d$, the grid looks like a diffraction grating whereas in the opposite limit, i.e. $\lambda \gg d$, it looks like a set of tall waveguides laid side by side. The similarities between diffraction gratings and waveguides are not usually stressed but they are nevertheless real. Thus the modes of a waveguide and the orders of a grating are formally equivalent. At low frequencies, a waveguide may be either cut-off or else be transmitting only the lowest order mode: similarly a grating would be transmitting only the zero'th order. As the frequency rises, the waveguide becomes able to sustain

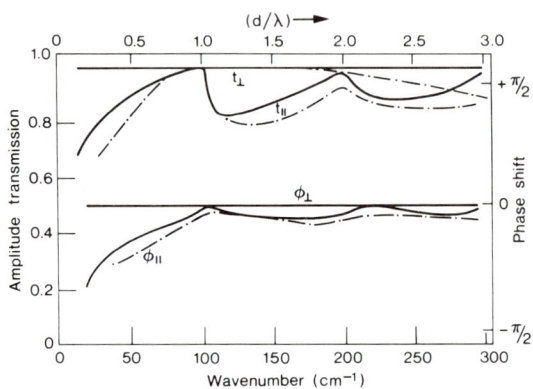

FIG. 3.37. The calculated (—) amplitude ($t_{\shortparallel}$ and $t_{\perp}$) and phase ($\phi_{\shortparallel}$ and $\phi_{\perp}$) transmission coefficients for a wire grid wound from 5 μm diameter tungsten wire with $d = 100$ μm. For comparison, some measured [245] results (-·-·-) for a grid with that specification are shown.

further modes and the grating, correspondingly, is able to channel power into initially its first order and then progressively into the higher orders. A grid, therefore, will, in the long-wave limit, be cut off for waves polarised parallel to the wires but will transmit waves polarised perpendicular to the wires. Consequently, when natural radiation, of wavelength much greater than d, is incident normally on the grid, 50% will be reflected with perfect parallel polarisation whilst 50% will be transmitted with perfect perpendicular polarisation. This result emerges also very clearly from the quantitative data given in Fig. 3.37 but the unexpected result that $t_{\shortparallel} = t_{\perp}$ when (d/λ) is integral could not be deduced from the heuristic approach. This strange property of the grid seems to be due to a coherence effect which occurs when $a \ll \lambda$. This effect would only be noticed at the longer wavelengths since, at the shorter, one would usually have a, d, and λ of comparable size. Nevertheless one would expect that, in all applications, the performance of the grid would fall off as the wavelength shortens towards d. Looking at the grid now as a diffraction grating, one has the result mentioned above, that when $\lambda > d$ only the zero'th order of diffraction is possible and there is thus no separation of the wavelengths. This is a most important property since, combined with the excellent polarising properties, it makes the grids near ideal candidates for the beamdividers in polarising interferometers (section 6.7.3). Costley and his colleagues

[243], amongst others [246], have investigated the properties of the Martin and Puplett [247] polarising interferometer when equipped with grids as the beam-dividers and find that the performance is excellent. Surprisingly though, they find that the interferometer still works well at wavelengths shorter than $2d$ where, from the arguments given above, one would expect the polarising properties to begin to fail. It seems that, at these shorter wavelengths, the interferometer is working in a conventional Michelson mode and one has an unexpected but, of course, very welcome addition to the working range. This extension is particularly felicitous because the very fine grids which one might expect to have to use would have to be fabricated on a substrate and this could introduce problems due to absorption, either discrete or continuous, in the substrate medium. Apart from their use as beam-dividers, the grids can also be used as mirrors for polarised light and of course as polarisers and analysers for the far infrared.

Costley and his co-workers [248] have described an ingenious method for the manufacture of the grids, based on the use of a modified coil-winder. The grids are made in pairs by placing two beam-divider frames, of the type used in the cube interferometer, on a former which is rotatable between centres. The coil-winder gearing is set for the desired value of d, the leading end of the wire is taped to the frame and the coil-winder started. A winding speed of about 4 rev/min is used with a wire tension only just less than the breaking tension. For 10 μm tungsten wire this would be about 20 g. When the winding is complete, the wires are fixed permanently to the frame with a suitable epoxy glue and, when this has set, the two frames are parted from one another. Grids made in this way, using both 10 and 5 μm wire, and with appropriate spacings ranging from 100 μm down to 12·5 μm, are now available commercially from the Specac company. Walker et al. [249] have shown that this type of grid can be used very effectively to make the "mirrors" of a scanning submillimetre Fabry–Perot interferometer.

The properties of the two-dimensional structures, the inductive and capacitative meshes, could also be analysed along the lines adopted by Beunen et al. [243] but, since the mesh elements are usually square, polarisation effects do not enter and it is simpler to adopt an analytical formalism, introduced by Ulrich [250, 251], which views the structure in terms of transmission-line theory. One begins by invoking the Huyghen's principle ideas sketched out in section 3.1.3, then defines the amplitude transmission, reflection and scattering coefficients to be $\hat{t}_i$, $\hat{r}_i$ and $\hat{s}_i$ and $\hat{t}_c$, $\hat{r}_c$ and $\hat{s}_c$ respectively, where the subscripts refer to the inductive (i) and capacitative (c) cases, and thus derives the result

$$\hat{t}_i = 1 + \hat{s}_i, \quad \hat{t}_c = 1 + \hat{s}_c, \quad \hat{r}_i = -\hat{s}_i, \quad \hat{r}_c = -\hat{s}_c. \tag{3.10.1}$$

By complementarity, an electromagnetic form of Babinet's principle, it follows that

$$\hat{t}_i = -\hat{s}_c \quad \text{and} \quad \hat{t}_c = -\hat{s}_i \tag{3.10.2}$$

and, by energy conservation, it follows that

$$|\hat{t}_i|^2 + |\hat{s}_i|^2 = |\hat{t}_c|^2 + |\hat{s}_c|^2 = 1. \tag{3.10.3}$$

These equations serve to define all of the coefficients as soon as one of them is known. The analytical solutions are given by

$$
\begin{aligned}
\hat{t}_i &= \;\;\;(1/2)[1 - \cos 2\phi - i \sin 2\phi] \\
\hat{s}_i &= -(1/2)[1 + \cos 2\phi + i \sin 2\phi] \\
\hat{t}_c &= \;\;\;(1/2)[1 + \cos 2\phi + i \sin 2\phi] \\
\hat{s}_c &= -(1/2)[1 - \cos 2\phi - i \sin 2\phi]
\end{aligned}
\tag{3.10.4}
$$

These can be plotted as vectors in the complex (Argand) plane, as described earlier. The result is shown in Fig. 3.38.

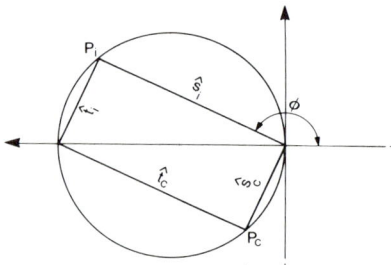

FIG. 3.38. Complex plane representation of the transmitted and scattered amplitudes for a mesh.

Clearly, from the form of equations (3.10.4), the four vectors define a rectangle in the complex plane.

The actual calculation of any one of these amplitudes can be done to various degrees of sophistication, as mentioned above, but for the case of practical importance, i.e. $\lambda \gg d$, where diffraction does not arise and where finite conductivity effects can be ignored, Ulrich shows that the mesh can be thought of as a characteristic impedance

$$Z_0 = [2 \ln \operatorname{cosec}(a\pi/2d)], \tag{3.10.5}$$

placed across a transmission line. The mesh with holes corresponds to a pure inductance of magnitude $L_0 = Z_0/\omega_0$, whilst that with the complementary structure corresponds to a pure capacitance of magnitude $C_0 = \omega_0/Z_0$. This is the origin of the names used for these devices. In passing it should be noted that a grid behaves as a capacitance for radiation polarised perpendicular to the wires and as an inductance for radiation polarised perpendicular in the other sense. The dimensionless normalised "resonant" frequency ω_0 is expected to be unity, but small departures (see later) are found in practice. The presence of the impedance across the "line" causes a division of the incident wave into a reflected and a transmitted wave and this is an exact model of the division of the radiant wave at the real mesh. The merit of the approach is that one may use simple AC theory to calculate the amplitudes of the reflected and transmitted beams. Ulrich gives for the inductive mesh the result

$$\tau_i = |\hat{t}_i|^2 = \omega^2 L_0^2/(1 + \omega^2 L_0^2),$$

i.e.

$$\tau_i = (\omega/\omega_0)^2 Z_0^2/[1 + (\omega/\omega_0)^2 Z_0^2], \qquad (3.10.6)$$

where $\omega = d/\lambda$. The transmission is thus formally zero at DC but it rises as the wavelength shortens. The complementary capacitative mesh is consequently highly transmissive at long wavelengths but its transmission falls as the wavelength shortens. For this reason it can be used as a long-wave passing filter.

Equation (3.10.6) and its equivalents can be used to give an excellent account of the action of the mesh provided a/d is less than 0·2 and ω is less than 0·8, but serious discrepancies occur for higher values. In particular these equations do not predict the resonance at $\omega = \omega_0$. To produce this necessary result, the single reactive element is combined with another of the opposite type to make a resonant circuit across the line. Ohmic losses in the metal of the mesh can then be coped with by including a dimensionless series resistor in the modelling network. The transmission equation for the inductive mesh then becomes

$$\tau_i = |\hat{t}_i|^2 = [R_z^2 + \Omega^2 Z_0^2]/[(1 + R_z)^2 + \Omega^2 Z^2], \qquad (3.10.7)$$

where Ω is the so-called "generalised frequency":

$$\Omega = \omega\omega_0/(\omega^2 - \omega_0^2). \qquad (3.10.8)$$

Equation (3.10.7) gives a very good account of the properties of real meshes. Thus the transmission reaches unity at the resonance and approaches low values for low frequencies. Some typical results are shown in Fig. 3.39.

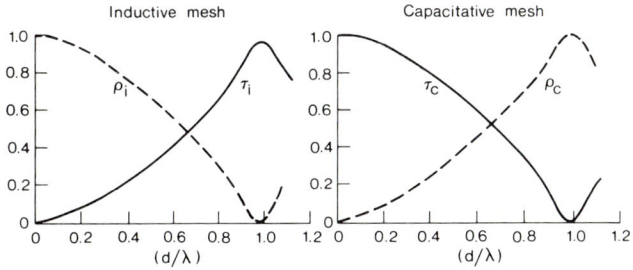

FIG. 3.39. Transmission and reflection characteristics of the two types of mesh.

The dimensionless ohmic resistance correction R_z can be regarded purely as a parameter to be determined by experiment, but Ulrich shows that one can estimate it in terms of the bulk conductivity of the metal and of the fraction of the area of the grid which can conduct current. The result is

$$R_z = (4\pi\varepsilon_0 c/\lambda\sigma)^{1/2}\eta/2, \qquad (3.10.9)$$

where η, the obscuration factor, is $d/2a$ for the inductive grid and $1 - (2a/d)$ for the capacitative one. However, from what was said above it will be realised that the effects of the ohmic losses are usually very small. They would only be of real significance in multimesh filter assembilies where they would lead to reduced peak transmission and lower contrast. In general, as will be seen from equation (3.10.9), R_z will vary only very slowly with frequency across the frequency range where these meshes would be used. The resonant frequency of the meshes, as mentioned earlier, does not occur exactly at $\omega = 1$. From experiment, Ulrich suggests the empirical relation

$$\omega_{obs} = 1 - 0{\cdot}27(a/d). \qquad (3.10.11)$$

For the meshes readily available, a/d is about $0{\cdot}15$ so the resonant frequency will be displaced downwards by about 4%. Buckbee Mears of St Paul Minnesota manufacture a range of meshes some with even smaller values of a/d.

Meshes have been used to make Fabry–Perot interferometers for the far infrared and especially to make far infrared interference filters [252]. In both applications it is essential that the meshes be flat so it is usual to stretch them in a tensioner. The treatment of these multimesh devices is carried out by an obvious extension of the method sketched out above in which one has a series of reactive elements placed across the line. The resonances thus become sharper and one then has a series of regularly spaced narrow-band transmission regions. The situation is almost identical to that of the dielectric interference filters discussed earlier. Present technology is capable of fabricating meshes which will work up to roughly a few hundred cm^{-1} and the interference filters made using them take over most felicitously from the conventional type which become limited by material absorption problems as one attempts to extend their range of utility into the far infrared. In recent times very sophisticated micro-fabrication techniques have been introduced for the manufacture of components for the microelectronics industry and one can imagine these being used to manufacture meshes which could be used throughout the infrared. This underlines a central theme of this book, that the artificial division that used to separate the low-frequency region where electromagnetism regined supreme from the high-frequency region where geometrical optics ruled is rapidly becoming a thing of the past.

3.11 Infrared imaging systems—the optical transfer function

Imaging systems, such as cameras, projectors, microscopes, telescopes etc. working in the visible region represent probably the most important branch of physical optics and the need for a quantitative measure of their performance was apparent almost from the beginning. Early limitations were set almost exclusively by aberrations, so the traditional measure which emerged was the resolving power, that is the maximum number of lines per unit distance in the object plane which could still be distinguished in the image plane. As lens systems improved and each type of aberration was overcome, imaging performance approached closer and closer to the limitations set by diffraction and quantitative measures of the performance were suggested based on Rayleigh's criterion, namely that two points could be said to be resolved when the grand maximum of the Frauenhofer diffraction pattern of one lay on the first zero of that due to the other. This concept is still widely used to define the limiting resolution of spectrometers (section 4.3.2) and indeed of imaging systems, but it is open to the objection that it is only valid for completely incoherent illumination and, whilst this might be the case for a spectrometer equipped with a black-body source, it would not usually be the case for a microscope working at the limits of optical technology where one would be trying to resolve detail at the 1 μm level. Here one has to take into account that image quality depends on the coherence of the illumination and also on the nature of the object being imaged. The first attempts to take these two effects into consideration came from Abbe at the turn of the century but it was not until fifty years later when physicists were paying much more attention to the coherence of optical fields that these ideas were developed into a comprehensive theory of imaging quality. The technical need that spurred this work on was the demand for an assessment of the performance of the high-aperture, multi-element, highly corrected zoom lenses used in cine photography and television. The approach adopted has much in common with transmission line-theory and hence with linear response theory. It leads to a quantity, the *Optical Transfer Function* (or OTF for short), which describes exactly all the image transmitting qualities of the overall system. The OTF can, in some rare circumstances, be complex, in which case its modulus, the *Modulation Transfer Function* (or MTF), can be used to describe just the imaging qualities. Infrared imaging systems have always lagged behind their visible region counterparts but modern versions are starting to approach performance levels where the OTF (or MTF) of infrared imaging systems is now becoming available and it is therefore worthwhile sketching out the fundamental ideas of the OTF approach [253].

One starts by considering an object plane which has a brightness variation given by the function $B(x, y)$. This is imaged by the optical system onto an image plane which occurs further along the z axis. Actually one can work in one dimension without much loss of generality so one considers an object defined by the brightness function $B(y)$. The crucial point of the OTF approach is to resolve $B(y)$ into its Fourier components over the whole range of spatial frequencies (i.e. wavenumbers) k_y. The optical system is assumed to be linear so each Fourier component will be transmitted undistorted but the attenuation and phaseshift (if any) will in general vary with k_y. The image is reconstructed from the recombination of the Fourier components and it will, therefore, to some degree, be degraded in quality when compared with the original object. That this is inevitable can be shown at once by noting that there is a limit to the range of k_y since Fourier components lying beyond the limit will be diffracted out of the system and not contribute to the image. This is, of course, merely stating the Rayleigh point in a different language—large values of k_y correspond to small separations in the object plane. The resolution, i.e. the permitted range of k_y, depends on the wavelength λ and on the numerical aperture N, so it is convenient to work in terms of the normalised spatial frequency

$$s = k_y / Nk \qquad (3.11.1)$$

where, as before, $k = 2\pi / \lambda$. For completely coherent illumination s is restricted to the range 0–1 but for incoherent illumination it will usually be even less. The resolution of $B(y)$ into its Fourier components is given by

$$B(s) = \int_{-\infty}^{+\infty} B(y) \exp(isy)\,\mathrm{d}y \qquad (3.11.2)$$

and the OTF, $D(s)$, is then defined to be ratio of the contrast of a Fourier component of spatial frequency s in the image plane to that of the same frequency component in the object plane. However, the OTF is not usually measured with a continuous object since a Fourier transformation would be required. Rather a well ruled grating or grill is used so that B(s) is only finite for a few discrete values of s. The complete function can then be obtained by using several such gratings. The results are quite interesting. Thus an aberration free system illuminated with perfectly coherent light gives an OTF

$$D(s) = \prod\left(\frac{s}{2}\right), \quad s > 0, \qquad (3.11.3)$$

whereas the same system irradiated with incoherent light gives a function closer to the triangle function $\Lambda(s)$, $s > 0$. Coherent light can therefore reveal fine detail better than can incoherent light. Another interesting result which emerges from OTF studies is that certain objects can be imaged well even by systems which contain aberrations. The condition is that they contain only a narrow band of spatial frequencies. When this is so, a good image can be obtained merely by adjusting the focus. Aberrations in, or loss of quality of, the imaging system show up well on the OTF and an important use of automatic

OTF testing equipment is the monitoring of the performance of lenses, those used in TV for example, which have to sustain rather rough usage. The OTF is hence a very valuable measure of the performance of an imaging system but it does have the drawback, which the admittance of a transmission line does not, that it is a property of the overall system and cannot be easily calculated in terms of the OTFs of the constitutent parts. More sophisticated approaches in which each optical element has associated with it a matrix defining its properties and in which the overall result is obtained merely by multiplying all the defining matrices together have been explored for the visible region but so far do not seem to have been used for infrared imaging systems.

Chapter 4
Infrared Radiometry and Infrared Spectrometry

4.1 Infrared measurements

4.1.1 *Introduction—the infrared measuring system*

Infrared measuring systems do not differ in any fundamental way from the equivalent systems used in the visible and ultraviolet regions but the variations, though small, are nevertheless significant since they reflect the basic differences between the infrared and the shorter wavelength optical regions. These differences have been touched on several times in what has gone before. A typical infrared experiment might be laid out as illustrated schematically in Fig. 4.1.

Radiation from the source passes first to a modulator where either its amplitude or its phase is made to vary periodically in time. The modulated radiation then passes into the experiment area. Here it is processed in some way and also interacts with the specimen under study. There are very many forms of processing to which the beam might be subject. It might, for example, be passed through a monochromator to spatially separate its spectral components or it might be passed through an interferometer or interact with a non-linear crystal or with a point contact diode. The specimen is usually just that, a piece of solid or else some liquid or gas confined in a suitable cell, but it may be a plasma or, by an extension of the basic concept, be another beam of electromagnetic radiation. The beam suitably modified by its processing and by its interaction with the specimen is then incident on the detector. This may be just a simple detecting element or else may be a "head"—that it a combination of an elementary detector with some form of preliminary electronic processing but for the moment one can think of it as an entity which produces an electrical signal whose amplitude is hopefully linearly proportional to the incident intensity. The signal from the detector is amplified (if necessary) up to the level required by the *Phase-Sensitive Demodulator* (or PSD) and the final output is an analogue (that is a slowly varying "DC") signal which can either be used as it stands or else may be digitised and then processed further in a computer.

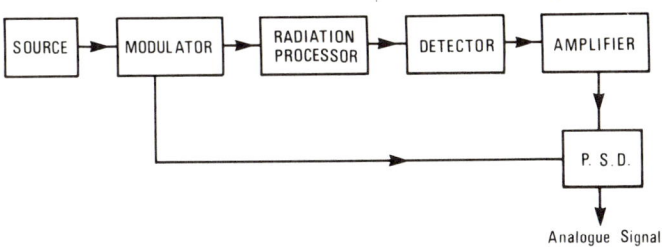

Fig. 4.1. Schematic lay-out of an infrared measurement experiment.

The beam is modulated because (a) it is much easier to process AC signals than DC ones, (b) the problems of drift are largely eliminated, and (c) it is much easier to suppress "noise" down to a tolerable level when using an AC system. These points will be explained in more detail later. The frequency adopted for the modulation should ideally be as high as possible but not be so high that the detector is unable to respond. In practice values as low as 16 Hz are necessary with thermal detectors and values as high as several kHz may be used with photodetectors. Some sources are inherently modulated, pulsed lasers for example, but for the common black-body sources and for continuous-wave lasers one must use an external modulator. Electro-optical modulators can be used in the near infrared but the shortage of suitable materials makes them unavailable at longer wavelengths. In the majority of cases therefore the practice is to use a rotating blade chopper. The principal exceptions occur in *Fourier Transform Spectrometry* (FTS) where one has available the alternative options of modulating the phase by "jittering" one of the interferometer mirrors or else of achieving AC operation without explicit modulation by the use of rapid-scan techniques. When a mechanical chopper is being used for visible or ultraviolet work, its exact location is not critical since, as explained in Chapter 2, the background is everywhere dark with only the source bright. In the infrared, however, where not only is the background glowing faintly but the specimen itself may be producing significant amounts of radiation, the positioning of the chopper is very important. To make sure that only the source is seen by the detector, the modulator has to be placed as close to the source as possible. In this way the unmodulated radiation from say a hot specimen will be ignored by the recording system. However, it is nevertheless still arriving at the detector and the experimentalist must be careful that this real if "invisible" radiation is not driving the detector into a non-linear portion of its characteristic. Another hazard is that this unwanted radiation can either transmit noise or else induce it in the detector leading to a deterioration of the observed signal-to-noise-ratio (SNR). The remedy is to place the specimen as far from the detector as can be arranged so that the detector window will intercept the minimum solid angle. In mid-infrared grating spectrometers it is usual, for these reasons, to place the specimen *before* the monochromator. In this way one not only obtains a reduction because of the solid angle effect but one also achieves an enormous attenuation of the specimen radiation at the slits of the monochromator and by dispersion within it. When using an interferometer these latter two benefits are not available, which incidentally is one of the very few disadvantages of FTS compared with grating spectroscopy. The best location for the specimen in FTS depends on the spectral region and on the specimen temperature. In the far infrared where power levels are always very low there is little risk of non-linear operation so the specimen can be placed in the "beam-divider-detector" arm (see Fig. 4.15) rather than in the "source-beam-divider" arm where it might be uncomfortably close to the very hot mercury arc lamp. In the mid infrared where sources are less hostile and usually somewhat cooler it is permissible to place the specimen in the source arm. This would be essential if the specimen were hot, a molten inorganic salt for example. The compensatory other side of the interferometric coin is that one has, for the specimen, two further possible locations which have no analogues in grating spectroscopy. These positions are in the two "partial-beam" or "active" arms. When the specimen is placed in one of these arms, one is doing Dispersive *Fourier Transform Spectrometry* (or DFTS) a topic which is discussed in more detail later. In the particular case of DFTS with phase modulation [254] there is no possibility of modulated radiation from the specimen reaching the detector but if one is using amplitude modulation, there is a risk of this happening, especially in the far infrared. If it does occur, to a significant extent, one immediate effect is that power transmissivities calculated by ratioing a spectrum against that of the background may be erroneous.

After traversing the specimen and leaving the experiment area, the radiation is received by the detector which produces as its output an AC signal whose frequency is that of the modulation and whose envelope contains all the experimental information. To obtain solely the envelope requires that the output be demodulated. One could do this very simply by passing the output voltage through a half-, or full-wave rectifier but much better practice is to use a PSD. The operation of the PSD will be explained later but in simple terms it is a linear synchronous switch ending in an RC smoothing network. The RC network has a minor role to perform, that of short-circuiting the modulation frequency, but its major role is to reduce the noise excursions down to acceptable levels. The final output will therefore be a slowly varying DC (or analogue) signal with satisfactory SNR and the information which the experimenter seeks is either immediately available or will be after some mathematical processing of the signal. This in outline is the operation of an infrared measuring system but the various components making it up will merit study in some detail and this will be carried out in the remainder of this Chapter.

4.1.2 Modulation of radiant flux in infrared instrumentation

The beam traversing an infrared experimental set-up is modulated, as remarked earlier, to ensure three main advantages: these may be summarised as

   a. the use of AC rather than DC amplifiers in the chain following the detector;
   b. the suppression of reciprocal frequency noise;
   c. the availability of a wide-range of detectors most of which are AC coupled.

AC electronic systems are cheaper, easier to use and much more free from drift than are the corresponding DC systems and they can readily be made to have a narrow band frequency response so that they transmit only at the modulation frequency. These points, when combined with the advantages of AC coupled detectors, alone present an overwhelming argument in favour of modulation but the advantages of noise suppression can in practice be just as considerable. Most solid-state electronics show a random noise characteristic which is not "flat" or "white" but which follows rather the law [255, 256]

$$P_N \sim v^{-\alpha}, \ 0 < \alpha \leqslant 1. \tag{4.1.1}$$

This inverse power law presumably cannot continue down to zero frequency since otherwise the total power would be infinite but it has nevertheless been found to hold down to frequencies as small as 10^{-4} Hz with no sign of an incipient failure. In the time domain, one is talking therefore of hours, i.e. the sort of times involved in even the longest infrared experiments and one can presume (4.1.1) to hold exactly. The drawback to unmodulated, that is DC, observation is now immediately obvious since the signal will be located in the frequency band where the noise is strongest, whereas if modulation is used one is in essence shifting the experiment to a frequency region where the noise is less. This is the justification for the remark made earlier that one always chooses the maximum permissible modulation frequency.

4.1.2.1 Amplitude modulation
The commonest form of modulator is a rotating blade chopper, the principle of which is illustrated in Fig. 4.2.

As the chopper rotates, the detector will "see" alternately the source and the surface of the chopper blades. In the mid and near infrared the amount of thermal radiation emitted by the chopper blades is negligible in comparison with that arriving from the source, so it is usual to blacken the chopper blades to prevent stray radiation, from unrelated external sources, being reflected back along the path towards the detector. In the far infrared, on the other hand, "black" chopper blades would be almost one third as bright as the source, so it is desirable to use brightly polished metallic blades in an attempt to reduce the emissivity. The output from the detector will be a sort of squared-off saw-tooth whose minimum corresponds to the radiation from the chopper blades and whose maximum corresponds to the radiation from the source. The chopper is nearly always made with a symmetrical disposition of the blades so that it will be balanced and run evenly. The number of blades together with the rotation speed determines the modulation frequency.

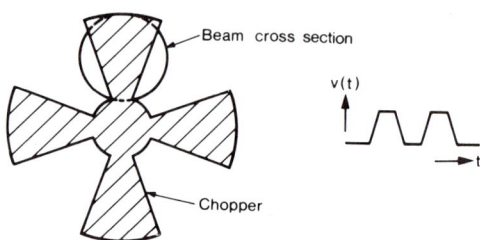

FIG. 4.2. Mode of operation of the rotating-blade chopper. In the right inset is shown the corresponding voltage output from the detector.

Modulation confers great advantages, but it must be realised that the flow of energy to the detector is being interrupted and this must inevitably lead to a loss of information and therefore to some deterioration in the SNR. It is, however, possible, with some experimental ingenuity, to avoid the information loss altogether. One situation where this is possible is in some forms of double-beam spectroscopy where the chopper blades can be fashioned as mirrors and the interrupted beam can be deflected and used for the blank channel. The amplitude of the saw-tooth from the detector will then be a measure of the absorption in the specimen channel.

The amplitude modulated radiation arriving at the detector has a periodic envelope, but in general this periodic waveform cannot meaningfully be Fourier analysed since some of the hypothetical components into which we might mathematically resolve it could be negative and this would be physically meaningless. The detector however transforms the periodic intensity wave-form into a (hopefully!) identical periodic *voltage* waveform and we need have no inhibitions about resolving this into its Fourier components. Indeed discussion of what happens to the signal in the subsequent electronic chain, which will contain elements such as tuned amplifiers, PSDs etc. responding only at a fixed frequency, naturally demands such an approach. Having carried out the resolution, one needs some criterion for the efficiency of the modulation, that is of how effectively the chopper channels signal into the chosen response frequency which naturally will either be the modulation frequency or one of its harmonics. Two criteria suggest themselves: (a) the relative magnitudes of the respective voltages and (b) the relative magnitudes of the corresponding powers. These two sets of quantities are necessarily related by Parseval's theorem (a close relative of Rayleigh's theorem (section 1.6.1) which states that if a periodic function $f(t)$ with period $1/v$ is resolved into a Fourier series

$$f(t) = \sum_{0}^{\infty} a_n \cos 2\pi n v t \qquad (4.1.2)$$

then

$$\int_{0}^{1/v} |f(t)|^2 \, dt = a_0^2 + (1/2) \sum_{1}^{\infty} a_n^2. \qquad (4.1.3)$$

As an illustration, the square wave function which has the value $+1/2$ from 0 to $\pi/2$, $-1/2$ from $\pi/2$ to $3\pi/2$ and back to $+1/2$ from $3\pi/2$ to 2π (see Fig. 4.3) can be resolved into the series

$$f(t) = (2/\pi)[\cos t - (1/3)\cos 3t + (1/5)\cos 5t - \ldots]. \qquad (4.1.4)$$

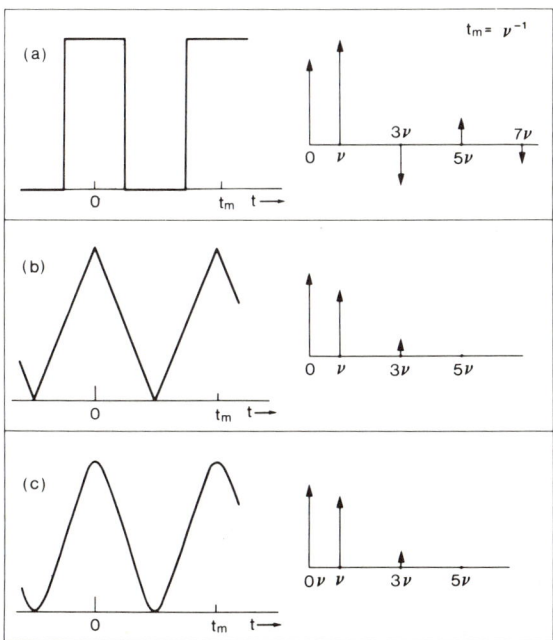

FIG. 4.3. Some periodic modulation waveforms and their Fourier decompositions.

The square of $f(t)$ has the constant value $1/4$, so over the repeat interval 0 to 1 its integral is obviously $1/4$. The sum of the squares of the coefficients inside the square bracket on the RHS is

$$1 + (1/9) + (1/25) + (1/47) + \ldots = \pi^2/8.$$

The RHS of (4.1.3), for this example, therefore takes on the value $(1/2)(2/\pi)^2$ $(\pi^2/8)$, i.e. $1/4$, and is therefore equal to the LHS. The theorem is thus demonstrated.

This result can be looked at in a more physical way in which one imagines an amplifier having unit gain in very narrow bands centred on the frequencies v, $2v$ etc. followed by a set of filters terminating in identical resistors. If each filter passed only its own frequency (nv) then the total power dissipated in the infinite set of resistors would be the same as if the original voltage were applied to any one of the resistors. One can therefore talk meaningfully of the power "channelled" by the modulation into a given harmonic. In this light pure cosinusoidal modulation

$$V(t) = (1/2) V_0 (1 + \cos 2\pi vt) \tag{4.1.5}$$

channels only one third of the available power into the (only) harmonic and not the one half which is the result in voltage terms. Square-wave modulation (see Fig. 4.3(a)) in which we would have

$$V(t)/V_0 = (1/2) + (2/\pi)[\cos 2\pi vt - (1/3)\cos 6\pi vt + \ldots] \tag{4.1.6}$$

is not analysable in an obvious fashion in terms of the voltages because of the presence of the negative contributions, but in power terms it produces the components

$$1 = (1/2) + (4/\pi^2) + (4/9\pi^2) + \ldots$$

The AC fundamental therefore carries 40% of the total power!

Square-wave modulation is thus very efficient, but it is almost impossible to achieve in practice using only mechanical choppers and so other modulation profiles have to be considered. A considerable amount of attention has in fact been given to the question of the design of the blades of the chopper so that, in combination with a particular shape of source, they give an approximation to a desired modulation envelope. If the source is long and thin, as it would indeed be in a grating or prism instrument, then a straight-sided blade with an edge that moved steadily from the top to the bottom of the source (or an image of it) and with a thickness exactly equal to the length of the source (or image as the case might be) would give an excellent approximation to triangular modulation. The replicated triangle function (see Fig. 4.3(b)) has the Fourier representation

$$f(t) = (1/2) + (4/\pi^2)[\cos 2\pi vt + (1/9)\cos 6\pi vt + \ldots], \tag{4.1.7}$$

and this would produce components, in power terms,

$$1 = (3/4) + (24/\pi^4) + (24/81\pi^4).$$

The modulation is therefore relatively inefficient with 75 % of the available power being wasted in the DC term (as compared with the inevitable 50 %) but, since the higher harmonics are rapidly attenuated, nearly 25 % of the power goes into the desired first harmonic and in practice one gets reasonably good performance.

Interferometric spectrometers usually feature a round, or approximately round, source to give a good match to the cylindrical symmetry which the instrument possesses and to ensure thereby the full throughput advantage which this kind of instrument enjoys. If one is using a blade chopper, close to the source, then the modulating function over the half-period $0 < t < T/2$ is given by

$$f(t) = (1/\pi [\cos^{-1}[(4t/t_m) - 1] - [(4t/t_m) - 1] \sqrt{[2(4t/t_m) - (4t/t_m)^2]}$$
(4.1.8)

with suitable reversals and repetitions at later intervals. This function is shown in Fig. 4.3(c). It will be seen that the curve is a good approximation to the triangular modulating function and its Fourier decomposition is very similar. Sometimes amplitude modulation is achieved in an interferometer by introducing a vane chopper into a part of the instrument before the beam divider, where there is a quasi-parallel beam (see Fig. 6.20). This is often resorted to when one is using a solid-state detector for then it is permissible to modulate at a relatively high frequency (800 Hz for example). Looked at from the viewpoint of the detector the rotating circular vane looks like an ellipse with one axis fixed at a value say $2a$ and the other varying periodically from 0 to $2a$. Since the area of an ellipse is ab where a and b are the semimajor and semiminor axes respectively this type of chopper gives nearly perfect triangular modulation.

The conclusion therefore on the question of amplitude modulation is that the only kind readily available is triangular modulation and that this channels only 25 % of the available power into the desired first harmonic. Nevertheless the benefits of selective AC amplification and phase synchronous rectification are so great that they more than compensate for the inefficiency of the modulation.

4.1.2.2 Phase modulation

In phase modulation [257], which is only available with two-beam inter-ferometers, one of the mirrors is "jittered" with an amplitude a at a frequency v_m and the periodic motion of this mirror provides the modulation. To analyse this form of modulation one considers the spectroscopically simplest system, namely irradiation of the interferometer with monochromatic radiation of wavenumber $\tilde{v}$ (see Fig. 4.4). If the displacement of the moving mirror from the zero-path-difference position introduces a path difference x, then the intensity at the detector when the jittering mirror is not moving will be

$$I(x) = \tfrac{1}{2}I_0(1 + \cos 2\pi\tilde{v}x).$$
(4.1.9)

Switching on the mirror will produce a time-dependent intensity

$$I(x,t) = \tfrac{1}{2} I_0 \left[1 + \cos 2\pi\tilde{v}\,(x + a\sin 2\pi v_m t) \right] \qquad (4.1.10)$$

that is

$$
\begin{aligned}
I(x,t) = {}& \tfrac{1}{2} I_0 + \tfrac{1}{2} I_0 \cos 2\pi\tilde{v}x \cos (2\pi\tilde{v}a \sin 2\pi v_m t) \\
& - \tfrac{1}{2} I_0 \sin 2\pi\tilde{v}x \sin (2\pi\tilde{v}a \sin 2\pi v_m t)
\end{aligned}
\qquad (4.1.11)
$$

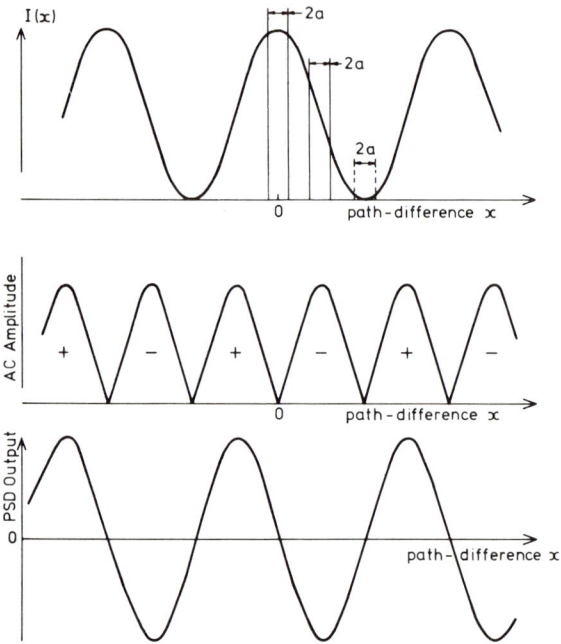

FIG. 4.4. The mechanism of phase modulation. The variations of intensity (upper inset) with, for example, monochromatic irradiation, are converted into a varying AC voltage out of the detector (middle inset) by the oscillation of the mirror. The AC voltage has either a zero (positive) or π (negative) phase shift with respect to the reference. After passing through the PSD, the output is the voltage interferogram shown in the lower inset which oscillates about a true zero.

The Fourier resolution of the time-dependent parts of (4.1.11) is not elementary and the coefficients involve Bessel functions. Explicitly one has

$$
\begin{aligned}
\cos (2\pi\tilde{v}a \sin 2\pi v_m t) = {}& J_0 (2\pi\tilde{v}a) + 2J_2 (2\pi\tilde{v}a)\cos 4\pi v_m t \\
& + 2J_4 (2\pi\tilde{v}a)\cos 8\pi v_m t + \dots
\end{aligned}
\qquad (4.1.12a)
$$

and

$$
\sin (2\pi\tilde{v}a \sin 2\pi v_m t) = 2J_1 (2\pi\tilde{v}a)\sin 2\pi v_m t - 2J_3 (2\pi\tilde{v}a)\sin 6\pi v_m t + \dots
$$
$$(4.1.12b)$$

It will be seen from these equations that the operation of the modulation is rather subtle. At a *fixed* value of x the varying signal at the detector is produced by the varying interference dividing the total intensity between the beam going to the detector and that returning to the source, in a time-varying, but periodic, manner. When x is changing (at a rate much less than the modulation frequency), this process is still going on but additionally the detector signal is being divided between the even and odd harmonics of the modulation frequency in an x-dependent manner. Thus at zero-path-difference there is no power in the odd harmonics and if we were observing an odd channel a true zero of signal voltage would be obtained. A definition of modulation efficiency is therefore not so straightforward as the corresponding amplitude modulation definition, especially since the modulation is wave number dependent.

In normal operation one will be using an amplifier/PSD combination locked to the modulation frequency and to maximise power in this channel one will set $J_1(2\pi\tilde{v}a)$ to its maximum value, namely $0\cdot58187$ which occurs for $2\pi\tilde{v}a = 1\cdot8411$. The other Bessel functions then take on the values shown in Table 4.1.

n	$J_n(1\cdot8411)$	$J_n(3\cdot05)$
0	0·31608	− 0·27654
1	0·58187	+ 0·32019
2	0·31601	+ 0·48650
3	0·10470	+ 0·31784
4	0·02519	+ 0·13876
5	0·00478	+ 0·04612
6	0·00075	+ 0·01244
7	0·00011	+ 0·00057
8	0·00001	+ 0·00057

When $\sin 2\pi\tilde{v}x$ is unity the maximum power goes into the odd terms (zero into the even terms) and then, since

$$1 - \sin(2\pi\tilde{v}a \sin 2\pi v_{\mathrm{m}} t) = 1 - 2J_1(2\pi\tilde{v}a)\sin 2\pi v_{\mathrm{m}} t$$
$$+ 2J_3(2\pi\tilde{v}a)\sin 6\pi v_{\mathrm{m}} t \text{ etc.,} \qquad (4.1.13)$$

it follows at once that the relative powers are in the proportions

$$1, 0\cdot677\,15, 0\cdot021\,92, 0\cdot000\,04 \text{ etc.}$$

and the modulation is very efficient with nearly 40% of the power going into the desired channel. The modulation is, however, wave number dependent, reaching its maximum value when, as remarked above, $2\pi\tilde{v}a = 1\cdot8411$, that is when $\tilde{v} = 0\cdot2930\,a^{-1}$. It is zero at the trivial case of $\tilde{v} = 0$ but also when $2\pi\tilde{v}a = 3\cdot8317$, that is when $\tilde{v} = 0\cdot6098\,a^{-1}$.

The rectified signal emerging from the phase-synchronous rectifier oscillates about a true zero and since this holds for any value of $\tilde{v}$ and since, by the principle of superposition, one can resolve any observed interferometric curve into pure cosinusoidal components, it follows that all interferograms observed in this way will have this property. Now it is a great advantage for the interferogram to oscillate about an average value of zero: not only does this make much more efficient use of the limited dynamic range of the recording system but there is also a very useful suppression of noise due to source fluctuations. Looked at in a simple-minded way, this latter arises from the fact that when there is no fringe to be recorded, i.e. when the interferogram ordinate is not varying (within the digital discrimination), then a sudden change of source output will lead to *no* recorded effect. One can only "see" source noise when there is a signal to be observed and not all the time as in conventional amplitude modulation. The reason why the interferogram has an average value of zero lies in the nature of phase modulation and in the use of phase-synchronous rectifiers. In AM, the phase of the modulated signal is constant and, with the phase of the rectifier set to agree with that of the signal, the output of the rectifier will be the familiar full-wave rectification pattern, with a steady positive DC signal whose magnitude is proportional to the original radiant intensity. The interferogram voltage is therefore always positive oscillating between zero and I_0. In the broad-band case, each cosinusoidal resolved component will always be positive and the whole interferogram will be everywhere positive having a maximum value of I_0 (at $x = 0$) and an asymptotic value of $I_0/2$ (as $x \to \infty$). In first harmonic observation PM, the signal phase changes discontinuously (from 0 to π) as the cosinusoidal interferogram intensity passes through zero. So when the intensity is decreasing we will get, say, a negative voltage out of the PSD but when it is increasing we will get a positive voltage. It is not difficult to see therefore that over a whole fringe the average value will be zero. In the light of this one can venture a more subtle interpretation of the source noise suppression and say that at values of x where the interferogram function is virtually constant source fluctuations *slow* in comparison to the modulation frequency will be suppressed, those *fast* in comparison will get through but, since one will have additional RC smoothing and since one *must* make sure that the digital system cannot respond at the modulation frequency, the effect will be negligible.

From the above it will be realised that the use of PM with first harmonic detection results in an analogue record in which each cosinusoidal element has been replaced by an equivalent sinusoidal element with an amplitude that has been modified by the inclusion of the wavenumber dependent factor $J_1(2\pi\bar{\nu}a)$. If a is very small, so that $\bar{\nu}_{\max}$, the wavenumber where $J_1(2\pi\bar{\nu}a)$ is maximal lies outside the observed spectral range, then $J_1(2\pi\bar{\nu}a)$ can be roughly approximated by half its argument, namely $\pi\bar{\nu}a$, and one has the situation that the derivative of the original interferogram, which will be composed of Fourier components of the form $2\pi\bar{\nu} \sin 2\pi\bar{\nu}x$, will be very closely similar to the PM interferogram. This is why people often loosely speak of the PM interferogram as the "first derivative of the AM interferogram". This terminology is, however, rather misleading for one will always choose $\tilde{\nu}_{\max}$ in practice to lie in the observed region and then there is only a very rough coincidence of the derivative of the AM with the PM interferogram. An example of a PM interferogram and the corresponding spectrum are shown in Fig. 4.5.

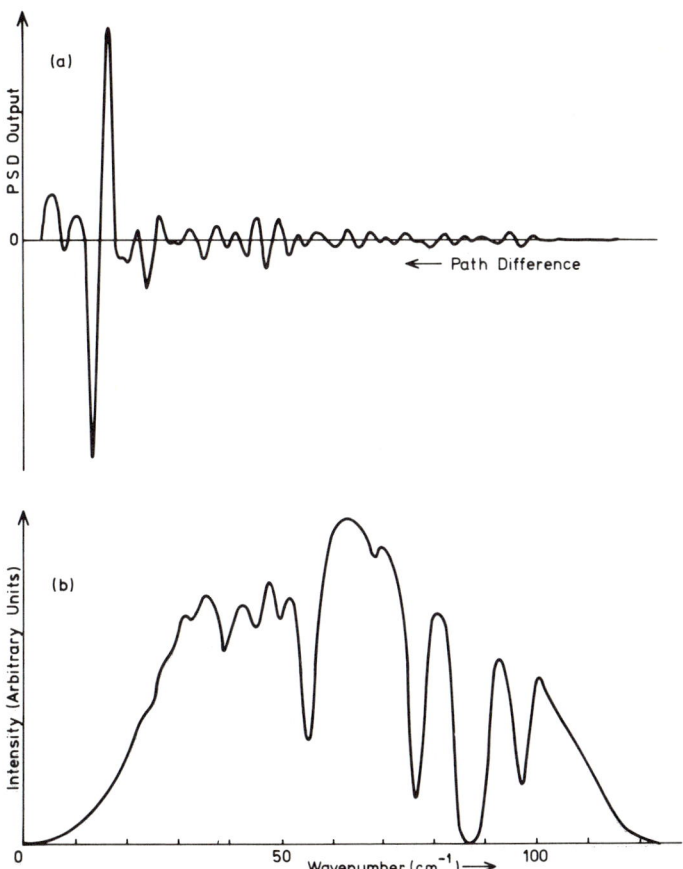

FIG. 4.5. An example of a phase-modulated interferogram (upper inset) and the corresponding spectrum (lower inset). The specimen was a hydrate of potassium fluoride dispersed in polytetrafluorethylene (courtesy of Dr Elisabeth Nicol).

One is not restricted, of course, to observing at the modulation frequency provided one has a detector with an amply short response time. The second-order Bessel function $J_2(2\pi\tilde{\nu}a)$ has its first maximum (see Fig. 1.6) when $2\pi\tilde{\nu}a = 3.05$. At this value, the other Bessel functions take on the values shown in Table 4.1 and it can readily be calculated that observing at twice the modulation frequency gives a modulation efficiency of 46%. The interferogram is now symmetrical once more but it still oscillates about a true zero. The transform will give the original spectrum modified by the modulation factor $J_2(2\pi\tilde{v}a)$. From this it follows that in principle a value of a could be chosen which was optimal for some experimental reason and then spectroscopy could be carried out in the region where the response was falling off with fundamental modulation frequency observation, one could switch to observation at the first harmonic. One arrives therefore at the final subtlety of phase modulation: as the radiation frequency increases, the total power is redistributed amongst the harmonics of the modulation frequency in a complex way. So with PM one not only obtains improved signal-to-noise characteristics, one achieves an additional degree of experimental freedom.

4.1.3 *Phase-sensitive demodulation of the detector output signal*

The modulated signal emerging from the detector will usually not be large enough to meet the requirements of the subsequent analogue/digital analysis system, so some amplification will be required. Typically, the detector output will be a few millivolts whilst the analysis system will work at the volt level. Because one is working with an AC system, the choice exists of using either a broad-band (i.e. video) amplifier or else of using a narrow-band tuned amplifier. The broad-band amplifier has the advantages of not being sensitive to shifts of the modulation frequency and of not having any frequency dependent phase shifts (another illustration of the Kramers–Kronig relations!). The narrow-band amplifier has the advantage of suppressing a lot of the noise before the demodulating stage is reached. This could be an important consideration if very large noise excursions were present which might drive the PSD into a non-linear region. This point is, however, the only real reason for preferring the use of a narrow-band amplifier since, provided the system is linear, all the noise suppression can be achieved just as well in the RC filter alone. The broad-band amplifier has the experimental convenience that the modulation frequency can be changed without having to change the electronics. If, nevertheless, a narrow-band amplifier is to be used, then it is desirable that it should have the "flat-topped" sort of response as shown in Fig. 4.6.

When it comes to the process of demodulation, there are several reasons for preferring a PSD to a simple diode network [258]. Chief amongst these is that the PSD, unlike the network, is a linear device which does not harmonically mix the signal with the noise. Another benefit springing from the same source is that there is no rectified noise in its output to serve as a misleading DC offset on the measured signal. To achieve this linear operation, the PSD needs two inputs, one being the signal to be demodulated, the other being a preferably noise-free reference signal at the same frequency. This latter can be provided either by a small alternator locked on to the shaft of the motor driving the chopper, or else by the use of a lamp/photodiode combination mounted one each side of the chopper. The reference signal, whatever its waveform (though it is almost always close to sinusoidal) is converted into a square-wave with clearly defined zero crossings. The signal to be demodulated is multiplied by the now square-wave reference in an appropriate circuit and the output is smoothed by an RC filter. The multiplication by the square-wave is exactly

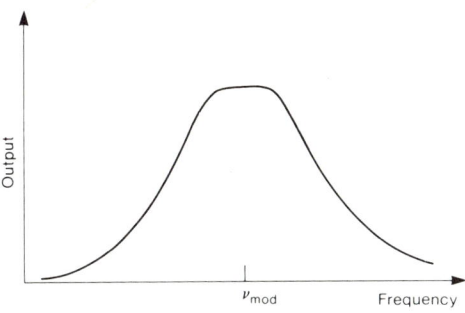

FIG. 4.6. Desirable response function for a band-limited preamplifier.

equivalent to the opening and reversing of a *linear* gate and a signal wave of exactly the same frequency will give a DC output. The value of the DC output depends on the relative phase of the two waveforms and to optimise the output it is necessary to have a means of adjusting this phase. In early systems, this had to be done by tedious manual adjustments of the lamp/photodiode positions with respect to the chopper blades, but in modern systems the phase adjustment is provided by a rotary device on the front panel. The principle of the PSD is shown schematically in Fig. 4.7. Conceptually, one can split the square-wave reference into its Fourier components which, as is well known, take the form already indicated in equation (4.1.4). The PSD will therefore pass the odd harmonics of the modulation frequency and this too indicates the desirability of some form of bandwidth limitation prior to the PSD. It goes without saying that the modulation frequency should *never* be chosen near an odd subharmonic of the mains frequency.

A PSD has the great advantage over all other demodulating systems in that it remains locked to the signal automatically no matter how the signal frequency varies. The PSD therefore permits stable AC operation of the infrared system—a benefit that it would be difficult to exaggerate. The whole system of amplifier (tuned or untuned), PSD and RC filter is often, for this reason, known as a "lock-in" amplifier/demodulator or more simply just as a "lock-in" [258, 259]. The performance of modern lock-ins is very impressive and it is quite normal for them to be able to recover signals buried more than 40 dB down in the noise [260]. The operation of phase-synchronous demodulation does itself suppress some of the noise since noise power near the signal frequency tends to be spread out into more remote regions and similar effects arise because of the randomly fluctuating phase of the noise, but, nevertheless, it is the final RC filter which suppresses the great bulk of the noise. However, to understand how it does this and to appreciate many further aspects of the infrared system does require some familiarity with the elementary theory of random fluctuations, so the systematic development will be interrupted at this point in order that the relevant points of this theory may be outlined.

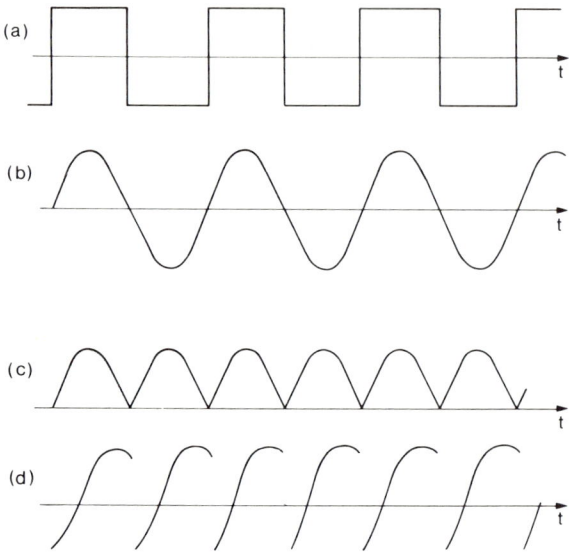

FIG. 4.7. The operation of a *Phase Synchronous Rectifier* (PSR). The reference waveform is converted to a square-wave (a). This is multiplied by the signal waveform (b) to give the output waveform (c). This has a DC component—the desired signal—and an AC component which can be attenuated by an RC filter. The RC filter will also attenuate noise accompanying the signal. If there is a phase mismatch between the signal and the reference (d), the DC signal will be reduced.

4.2 Measurement uncertainties—random variables and the elementary theory of noise

4.2.1 *Introduction*

When making any physical measurement, it cannot be assumed that the readings obtained will be accurate to the ultimate limit set by the sensitivity of the measuring equipment. For example unknown factors may be present which make all the values consistently either too high or too low and the readings will therefore be subject to *systematic error*. Errors of this kind are very insidious since, by definition, they do not affect the precision of the measurement. What they do is affect the accuracy by setting an ultimate limit which is independent of the observer's care and diligence. Much of the effort in making the most refined measurements therefore consists in tracking down and eliminating the various sources of systematic error. Provided one has achieved this and provided the measurement equipment is sensitive enough, the limitation on accuracy will then be set by the residual errors which will now be purely random in character.

Random errors are unavoidable but they have the property that provided one is prepared to work hard enough for long enough one can reduce their effects below any arbitrarily chosen level. Some of their properties can be illustrated by considering a simple case, say the measurement of a voltage not thought to be varying in time but which is in fact fluctuating due to the presence of random error. If M estimates were made of this voltage then each estimate could be written

$$V_m = V_t + \varepsilon_m, \tag{4.2.1}$$

where V_t is the "true" value and ε_m is the random error. What one needs to do now is to consider how one can arrive at a reliable estimate of V_t. Clearly several things could be done. One could, for example, carry out N simultaneous experiments and arrive at N records of the fluctuations over the experimental period. Such a set of records is said to form an *ensemble*. Alternatively one could carry out a single experiment but lasting N times as long. If it is found that the fluctuations over the ensemble at a fixed observing time are statistically equivalent to the fluctuations along any arbitrarily chosen member of the ensemble, then the random process characterised by the ε_m is said to be *ergodic*. If in addition any stretch of the $N \times M$ long record is equivalent statistically to any other stretch of equal length then the random process is said to be *stationary*. Statistical analyses of random processes [261] are not restricted to the stationary ergodic variety but there is no gainsaying the fact that the statistical properties of stationary ergodic processes are very much easier to appreciate. Fortunately nearly all random processes encountered in infrared science and technology are of this type. In particular the voltage measurement under discussion would be expected to be governed by stationary ergodic statistics and these enable one to write at once

$$V_t = \operatorname*{Lt}_{M \to \infty} \bar{V}_M, \tag{4.2.2}$$

where $\bar{V}_M$ is the average defined by

$$\bar{V}_M = (1/M) \sum_1^M V_m. \tag{4.2.3}$$

Further analytical expressions for the statistical parameters will be given later. All that will be pointed out here is that the statistical analysis does permit us to arrive (via equation 4.2.2) at a meaningful definition of what we mean by "true" value. The basic assumption made throughout the analysis of random or *stochastic* processes is that they can be characterised by statistical properties and equation (4.2.2) is a good example of this.

The time scale of the measurements, implied in what has gone so far, is of the order of hundreds per day. One could regard these data points, so laboriously collected, as a time sequence of "signals" arriving at the rather low frequency of 10^{-4} Hz. Modern experimental physics, however, often involves measurements equivalent to data streams arriving at audio or even radio frequencies. The analysis of the random errors is not now fundamentally any different since all that is involved is a change of time scale, but because the experimental set-up is so different it becomes useful to look at random fluctuation theory in a different light and to use different concepts and language in its analysis. These concepts arose first during the middle years of this century and especially under the imperative of global war when it became vitally necessary to ensure the optimum information flow capability of telephone and telegraph links [262]. The carrier frequencies involved in such links may be anywhere from kHz to MHz but the message being transmitted nearly always lies in a band (50–10 000 Hz) appropriate to reception by the human ear. The random fluctuations are thus audible as noise and the word "noise" came rapidly into general usage to describe random fluctuations in any signal transmitting system even if there is no possibility of the fluctuations ever being heard. Perhaps if television systems had been invented first we might all be referring to random fluctuations as "snow" for the effect to the eye when a TV set is picking up a very "noisy" video signal is of the picture being confused by an apparent snow storm. There is thus no basic difference between random measurement errors and "noise"—all that is involved is a very different time scale. Telecommunications were, however, so important, technically, commercially and politically that the mathematicians interested in the subject were led to expand and develop the basic subject of random fluctuations and this led to one of the more elegant branches of modern mathematics—information theory. The central topic of this theory is the determination of how fast information can be sent down a noisy link. One aspect of the theory has already been mentioned in section 3.6.3—the use of pulse code modulation in telephony to achieve essentially noise-free transmission. Another example well known to Hi-Fi fans is the use of the Dolby system [263] to reduce the noise in magnetic tape recording. Information theoreticians showed that for any system there was an irreducible noise level but they also showed that for most practical systems the performance came nowhere near this limit. This is well known to experimental infrared workers who commonly have amplifiers and other electronic processors which have noise figures several orders of magnitude down on the noise accompanying the actual signal they are trying to process. In such a circumstance, the important quantity is the *signal-to-noise ratio* (or S/N) since this will be an invariant of a broad-band low-noise processing chain. For the ultra low frequency physical measurements mentioned earlier, the S/N will usually be high (> 100), since the experimentalist would hardly think the experiment worthwhile doing otherwise, but for telephony under difficult circumstances it may be much smaller or may even be less than unity. Nevertheless the telephone operator is required to get some sort of sense out of the message and in equivalent circumstances an infrared physicist will want to derive some sort of a conclusion when all he has available is a very noisy signal with the S/N close to unity. It is here that information theory comes into its own. The basic points are twofold: (a) the signal occupies only a finite bandwidth so one gains by suppressing all other frequencies where by definition there is only noise; and (b) by repeating the same signal over and over again one gets an inexorable improvement of S/N since the identical signals will add coherently whilst the noise, being random and uncorrelated, will only add incoherently. Thus, provided one is prepared to restrict signal bandwidth, rate of signal information transfer or both, one can eventually transmit a meaningful message no matter how low is the S/N. This is a most important conclusion for modern experimental physics where, as hinted above, low values of S/N are unfortunately only too common and because of this the language and concepts of information theory have tended to become laboratory commonplaces. Thus one often hears of people "digging out signals buried in the noise" or else referring to some minor trifle as "lost in the noise". But it is worthwhile pointing out once again that there is no conceptual difference between these procedures and the older ones aimed at improving experimental precision. The real gain with information theory is that it provides us with the correct formulation of the principles appropriate to the very changed circumstances. Having made this point we now go on to consider the theory in detail, making a rather arbitrary division into "low" and "high" frequency concepts.

4.2.2 The statistical theory of measurement uncertainties and fluctuations

The inevitable fluctuations present in the repeated measurement of a notionally constant quantity V_t, form, as mentioned above, a zero-mean set defined by

$$\varepsilon_m = V_m - \overline{V}_M. \tag{4.2.4}$$

The statistical properties of this set are very important since from them one might hope to extract a numerical estimate of the probable error bounds when one quotes $\overline{V}_M$ as the most likely estimate of V_t. Whenever one examines such a set, three properties are usually immediately obvious [264]:

a. small fluctuations are much more common than large ones;
b. very large fluctuations are exceedingly rare;
c. positive and negative fluctuations of the same absolute magnitude tend to occur with equal probabilities.

One can understand the size properties by noting that the ε_m often are, and always could be, random voltages, so the average of their squares, being a mean power, cannot exceed the limits set by thermodynamics. Put crudely if large voltage fluctuations were to occur they could be harnessed to extract usable energy from a system in equilibrium. The symmetry property is rather more subtle. The majority of fluctuations which occur in physical systems do tend to be symmetric, but even if they are not there exist powerful mathematical constraints which tend to force observably symmetrical behaviour even if the fundamental fluctuations are not symmetric. However, before going into this mathematical aspect in detail it is first necessary to adopt a suitable numerical definition for the degree of spread of the ε_m about their zero mean, that is for the uncertainty of the determination of V_t. The point just made about power restrictions suggests that a good measure of the uncertainty would be given by the mean square error

$$\overline{\varepsilon_M^2} = (1/M) \sum_1^M \varepsilon_m^2. \tag{4.2.5}$$

The positive square root of this quantity would then be the number quoted as the likely error, i.e. one might give the result of the experiment in the form

$$V_t = \overline{V}_M \pm (\overline{\varepsilon_M^2})^{1/2}. \tag{4.2.6}$$

However, for reasons buried rather deeply in the theory of random processes, statisticians prefer to use the slightly different quantities

$$\sigma_M^2 = \frac{1}{M-1} \sum_1^M \varepsilon_M^2 \tag{4.2.7}$$

which they call the *variance* and its positive square root σ_M which they call the *standard deviation*. This variation of usage can, in principle, be confusing but fortunately in nearly all experimental situations M will be large and the actual difference therefore small. The results of experiments can thus be quoted, analogously to (4.2.6), in the form

$$V_t = V_m \pm \sigma_M, \tag{4.2.8}$$
$$V_t = V_m \pm 2\sigma_M, \tag{4.2.9}$$

etc. depending on the *degree of certainty* expected. By this phrase one means that $V_M \pm n\sigma_M$ defines a band of values and the probability that V_t actually lies within the band increases with n. The value of statistical analysis is that it shows us that the probability can reach high values for quite small values of n. To show why this is so requires some assessment of the distribution of the ε_m and one therefore introduces a *distribution function* $N(\varepsilon)\,d\varepsilon$ which represents the number of fluctuations having values between ε and $\varepsilon + d\varepsilon$. For mathematical convenience these distribution functions are nearly always normalised so that the integral over the full range of ε will be unity. There are, in principle, a very large number [265] of possible distribution functions but in very many cases the one actually found is

$$N(\varepsilon)d\varepsilon = \left[\frac{1}{2\pi\sigma_M^2}\right]^{1/2} \exp\left[-\varepsilon^2/2\sigma_M^2\right] d\varepsilon. \tag{4.2.10}$$

This distribution is, in fact, so common that it is usual to refer to it just as the "normal curve of error". The only other distribution commonly encountered in infrared physics is the *Poisson distribution*. This arises when one is counting events which are statistically independent, that is they are uncorrelated. If, in a reasonably long observing time, one might expect to count an average of $\langle n \rangle$ events, then the probability of actually counting n events is given by

$$P(n) = \frac{\langle n \rangle^n}{n!} \exp\left[-\langle n \rangle\right]. \tag{4.2.11}$$

For small values of n this is obviously not symmetrical about its mean $\langle n \rangle$ but for larger values of n it becomes more and more symmetrical and eventually for large values of n it becomes indistinguishable from the normal type distribution:

$$P(n) = \left[\frac{1}{2\pi\langle n \rangle}\right]^{1/2} \exp\left[-\frac{(n-\langle n \rangle)^2}{2\langle n \rangle}\right]. \tag{4.2.12}$$

By comparing this with (4.2.10) it follows that for a Poisson distribution the variance is equal to the average number, i.e.

$$\overline{[n-\langle n \rangle]^2} = \langle n \rangle. \tag{4.2.13}$$

This distribution is particularly important since it gives to good accuracy the statistics of photon-counting in the near-infrared and optical regions.

The transition from (4.2.11) to (4.2.12) illustrates well the point made earlier about the mathematical constraints towards symmetry. But these constraints are more specific than that: they constrain distributions to converge on the normal distribution function. This is a consequence of the *central limit theorem* which in general form states [264]: "The folding together of a large number of distributions of independent size and shape will tend to the normal distribution provided that the individual variances are finite and that the variances are all of the same order of magnitude, i.e. no one distribution dominates the others." In more specific terms the theorem indicates that physical distribution functions will tend to the normal one as some defining parameter approaches a limiting value. Thus, for the Poisson distribution, $P(n)$ tends to the normal form as $\langle n \rangle \to \infty$. One difficulty with the normal curve is that it does not have an elementary analytical integral and this makes it that much harder to work out the probability that the true value lies within a given band of values. Fortunately for experimental scientists and engineers, the integral was evaluated numerically many years ago and in tabulated form is readily available [266]. It is usual, in these tabulations, to introduce the dimensionless variable

$$x = \varepsilon/\sigma_M.$$ (4.2.14)

So that the normal curve takes on the form

$$\phi(x) = \frac{1}{(2\pi)^{1/2}} \exp\left[-\frac{x^2}{2}\right].$$ (4.2.15)

The integral of this function is usually referred to as the error function, $\mathrm{Erf}(t/\sqrt{2})$; it likewise involves a dimensionless argument. The error function

$$\mathrm{Erf}(t/\sqrt{2}) = \frac{1}{(2\pi)^{1/2}} \int_{-t}^{+t} \exp\left[-x^2/2\right] \, \mathrm{d}x$$ (4.2.16)

and the corresponding normal curve of error are illustrated in Fig. 4.8. From the error function curve it will be seen that "one sigma" error limits correspond to only a 68 % confidence level whereas a "two sigma" quote would correspond

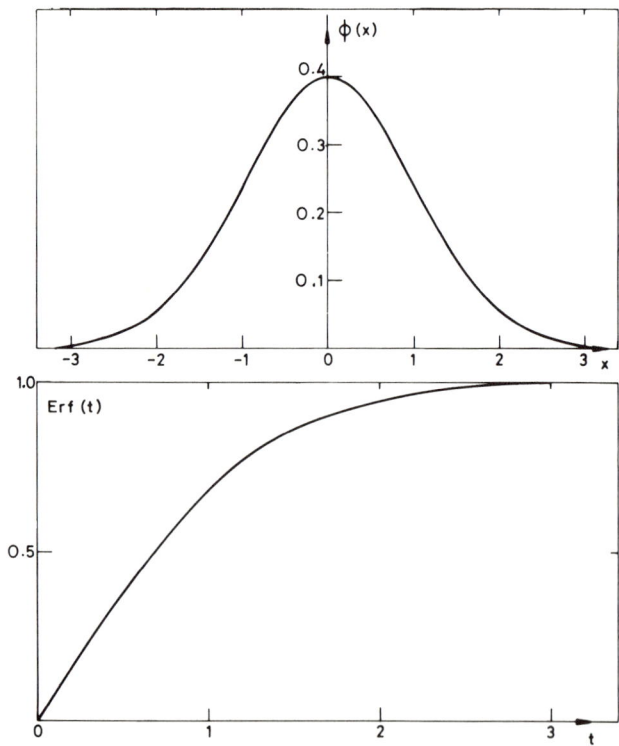

FIG. 4.8. The normal curve of error (upper inset) and its integral, the error function (lower inset). The abscissae axes are labelled in terms of the dimensionless variables discussed in the text, i.e. they are in "sigma" units.

to a 95% confidence level. In practical work one always seeks a balance between wide error bounds which guarantee virtual certainty at the price of low precision and narrow bounds which do the opposite. The two-sigma error bound seems to be a good practical compromise.

4.2.3 *Noise in detectors and electronic processing chains*

4.2.3.1 *Case 1. Signal constant*

As mentioned earlier, there is no conceptual difference between measurement uncertainties and electronic noise; they both belong to the general class of random fluctuations but because the time scales involved are so different it is useful to discuss the two phenomena within rather different formalisms and in particular to use many of the ideas of information theory to analyse noise effects. The most important tool imported from information theory is the use of the Fourier transformation which enables one to look at the phenomena in terms of the two conjugate domains, time and frequency. This is nowadays almost a reflex response of the theoretician—so deeply have the ideas of information theory suffused modern optics [267].

If one has a stationary ergodic noise record then the Fourier transform of a particular stretch will give a noise voltage spectrum. This will also be a "random-jitter" sort of record, but it will have an envelope and if many records are taken and the average calculated the envelope will be more distinct. This envelope will be taken to define the noise voltage spectrum. Its square is the noise power spectrum. Various kinds of noise power spectra are encountered in practice: thus one might have "white" noise in which there is no frequency dependence of the noise power or, as another example, one might have reciprocal frequency noise (often called "one-over-f" noise) in which the noise power varies inversely with the frequency. These two examples would have to be regarded as ideals since neither could occur in practice because in both cases the integral of the noise power spectrum diverges and the total noise power would therefore be infinite! However, over quite wide frequency ranges, noise power spectra are often encountered which do approximate quite closely to these ideals. Thus the Johnson noise is essentially white and has a power in a bandwidth Δv given by

$$P_N^J = 4kT\Delta v. \qquad (4.2.17)$$

This noise arises from very similar processes (spontaneous oscillation of charges, dipoles etc.) to those responsible for black-body radiation and equation 4.2.17 is therefore only valid in the Rayleigh–Jeans limit (see section 2.2.3) but, since the frequencies where it will be tested lie at radio frequencies or below, it will never fail in practice since sufficiently low temperatures will not be reached. It is interesting in passing to note that this restriction removes the divergence difficulty and, furthermore, that within its realm of applicability equation (4.2.17) provides a basis for a practical thermometer—the so-called "noise thermometer". As a second example the noise from solid-state devices, such as transistors, the so-called "flicker" noise, is often a close match to the $1/v$ type (or the more general $1/v^\alpha$ type where $0 < \alpha \leqslant 1$) even down to very low frequencies. This dependence, as mentioned earlier, must ultimately fail since otherwise the noise power would be infinite but over the range of practical electronics it can be assumed to hold exactly [256].

Back in the time domain, one can define a root mean square noise voltage by the analogue of (4.2.5) and then one could give a numerical value for the S/N in the form

$$S/N = \bar{V}/2\bar{\varepsilon}, \tag{4.2.18}$$

where $\bar{V}$ is the average signal voltage and $\bar{\varepsilon}$ is the root mean square noise voltage.

The idea of averaging observations made laboriously by hand, discussed earlier, to get a steadily improving estimate of a notionally constant quantity can be applied just as well to noisy signals, but obviously the averaging has to be done automatically. Electronic averagers which take samples of the input signal at fixed increments of time, add them all up and divide by the number of samples are commonplace tools of modern experimental physicists. If one sends a noisy signal into such an averager, its running contents

$$V_M = \left(\frac{1}{M}\right) \sum_1^M V_m, \tag{4.2.19}$$

will, for all the reasons discussed earlier, converge inevitably on V_t. It is nevertheless worth examining this process from a slightly different point of view. For a stationary record, the constant signal will build up in an adding machine monotonically whilst the random fluctuations of zero mean and symmetrical distribution will tend to cancel one another. Provided the noise fluctuations are truly random, then the total noise power accumulated will simply be M times the average power: in other words

$$\varepsilon_M^2 = M (\bar{\varepsilon})^2. \tag{4.2.20}$$

This is a very general result with repercussions throughout much of modern physics where, of course, statistical concepts are much to the forefront. A familiar illustration is that the total power produced by an array of incoherent lamps is simply equal to the sum of their individual powers. If the lamps have any degree of coherence this simple result will not hold and, likewise, if there is significant correlation of the noise signals, equation (4.2.20) will not hold exactly, but under most circumstances it can be assumed to be true without serious error. It follows that the quantities in the averager before the division by M takes places are:

accumulated signal $M\bar{V}_M = M\bar{V}$, \hfill (4.2.21)

accumulated noise $\quad \varepsilon_M = M^{\frac{1}{2}}\bar{\varepsilon}.$ \hfill (4.2.22)

The averaged signal-to-noise ratio will then be

$$(S/N)_{ave} = M\bar{V}/M^{1/2}\,\bar{\varepsilon} = M^{1/2}\,(S/N)_{original}. \qquad (4.2.23)$$

The effect of the averager is therefore to improve the signal-to-noise ratio by a factor equal to the square root of the number of readings being averaged. Thus if one wants to improve the S/N tenfold, one must be prepared to make 100 observations. This argument is a little imprecise since, of course, the S/N is not defined for one observation; however the way to tighten it up immediately suggests itself. One would say that if one had observed a noisy signal for a sufficient time to define a viable S/N, then the S/N available from a stretch P times longer would be $\sqrt{P}$ times larger.

4.2.3.2 *Case 2. Signal varying*

It has just been shown that when the signal is constant, the (S/N) can be made to reach arbitrarily high values provided one is prepared to wait long enough. However, for much experimental science this is a purely academic result since the signal will be varying in time and the real question to be asked is how should one arrange an experiment to give the optimum S/N when the signal is known to be varying slowly with time. The important questions now are whether the signal is periodic or not, whether it is band-limited and whether the noise-power spectrum is white or not. If the signal is aperiodic, as for example it would be in the scanning of a spectrum in the normal fashion, one will need to restrain the response bandwidth to just fit the signal bandwidth, one will use modulation to shift the operation to a region where the noise power is lower and in addition one will use noise-suppression at the end of the processing chain. When all this is done the desired S/N will determine the maximum permissible scan speed since the noise suppression will establish a fundamental response time τ and the scanning will have to be slow enough for the maximum resulting audio frequency to be much less than τ^{-1}. If the signal is periodic an averager of the kind just mentioned can be used but it should have a large number of channels. One then takes samples of the noisy signal at fixed values of the scanning abscissa variable and these are averaged for as long as necessary to achieve the desired S/N. This, in bold outline, is what is done; the details follow immediately in section 4.2.4 and are taken up again later in section 4.3.5 where the spectroscopic consequences of using noise suppression are discussed.

4.2.4 *The suppression of electronic noise by averagers and RC networks*

The operation of signal averagers in the time domain has been outlined above but considerably more insight into their functioning can be gained by considering the operation of noise reduction in the frequency domain. The basic formalism introduces a response function $R(t)$ (section 3.4.5) to describe the result of applying a delta function pulse to the noise reducer at time zero. The noise reducer, being a device with an input and an output, has a transfer function $\hat{T}(v)$ which describes its frequency response both in amplitude and phase. The transfer function is the complex Fourier transform of the time domain response function, $R(t)$. From its definition, the transfer function is a causal function whose real and imaginary components are Hilbert transforms of one another (see section 3.4.5 and Appendix 1). The signal averager has a response function $R(t)$ which has the constant value t_m^{-1} up to the integrating time t_m and is zero for all longer times. This is shown in Fig. 4.9. The transfer function is therefore

$$\hat{T}(v) = \frac{1}{t_m} \int_0^{t_m} \exp(2\pi i v t)\,dt = \frac{\exp(2\pi i v t_m) - 1}{2\pi i v t_m} \tag{4.2.24}$$

whose square modulus (see Fig. 4.9) is

$$|T(v)|^2 = (\sin \pi v t_m / \pi v t_m)^2 \tag{4.2.25}$$

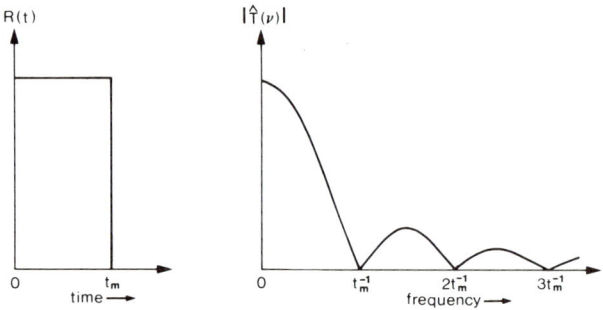

FIG. 4.9. Response and transfer functions for the signal averager.

It follows that the averager is a filter whose spectral response is zero for $v = n/t_m$, where n is an integer, and whose power response falls as v^{-2}. To obtain the variance of the noise one merely multiplies the noise power at a given frequency by the square modulus of the transfer function and integrates. However, it is often convenient to include negative frequencies in this integration (to preserve the symmetry of the underlying Fourier transformation) and then

$$\varepsilon^2 = \int_{-\infty}^{+\infty} P_N^e(v)|\hat{T}(v)|^2\,dv, \tag{4.2.26}$$

where $P_N^e(v)$ is the even part of the noise power spectrum. In the particular case of white noise, one can take P_N^e outside the integral and one is left with

$$\varepsilon^2 = P_N^e A, \qquad (4.2.27)$$

where

$$A = \int_{-\infty}^{+\infty} |\hat{T}(v)|^2 \, dv \qquad (4.2.28)$$

is the "area" of the filter transfer function. This argument has the slight difficulty that "white" noise implies infinite power but this can be overcome by merely requiring $P_N^e(v)$ to be constant for all values of v where $|\hat{T}(v)|$ is sensibly non-zero. With this amendment, it can be said that noise suppressing devices "work" by reducing the "area" of the transfer function and thus reducing ε^2. Applying this to the particular case of the averager, one has

$$\varepsilon^2 = P_N^e \int_{-\infty}^{+\infty} \left(\frac{\sin \pi v t_m}{\pi v t_m}\right)^2 dv = (P_N^e / \pi t_m) \int_{-\infty}^{+\infty} \left(\frac{\sin u}{u}\right)^2 du. \quad (4.2.29)$$

The integral on the right-hand side (obtained by the substitution $\pi v t_m = u$) is far from elementary, over arbitrary limits, but its value over the range $-\infty$ to $+\infty$ is readily obtained from Rayleigh's theorem since $(\sin u/u)^2$ is the Fourier transform of the triangle function. One has therefore

$$\int_{-\infty}^{+\infty} \left(\frac{\sin u}{u}\right)^2 du = \pi \qquad (4.2.30a)$$

and hence

$$\varepsilon^2 = P_N^e / t_m. \qquad (4.2.30b)$$

The variance is thus inversely proportional to the integrating time. The signal, by definition, is constant so one has as before the result that

$$S/N \sim (t_m)^{\frac{1}{2}}. \qquad (4.2.31)$$

The other main type of noise suppressor is the RC filter. In the time domain, one readily sees that this works as a consequence of the "damping" effect it has on rapidly changing wave forms. Thus if we have the simple circuit shown in Fig. 4.10 and apply an instantaneous DC voltage V_0 to the terminals, then the differential equation for the charge on C is

$$V_0 = (dQ/dt)R + Q/C, \qquad (4.2.32)$$

the solution of which is

$$Q = CV_0 [1 - \exp(-t/CR)]. \qquad (4.2.33)$$

The voltage across C is therefore

$$V = V_0[1 - \exp(-t/C)] \tag{4.2.34}$$

and the system responds sluggishly to the rapid change at $t = 0$ as shown in Fig. 4.11. The product RC has the dimensions of time and so it is usually called the time constant and given the symbol τ. One can see intuitively that the larger is τ, the longer will a voltage have to be maintained across the input terminals before it appears substantially across the output. The more rapidly changing is a signal, the more it will be attenuated by the network. To put these ideas into quantitative form requires the choice of an analytical expression for the variation of the input voltage with time. In many practical cases, the variation is sinusoidal, but even if it is not one may use Fourier's theorem to decompose the periodic waveform into its pure sinusoidal components so one can, without loss of generality, assume a sinusoidal input

$$V = V_0 \sin(\omega_0 t). \tag{4.2.35}$$

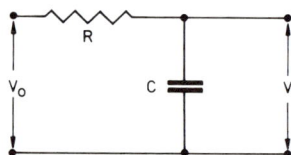

Fig. 4.10. Simple RC smoothing network.

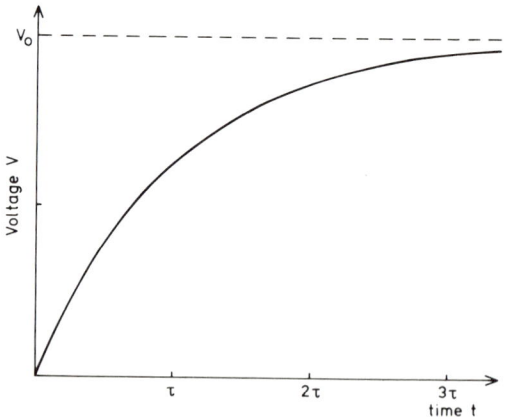

Fig. 4.11. Response of an RC network to the application of an instantaneous voltage at $t = 0$.

Substituting this expression for V_0 in (4.2.32) and considering the long-time or steady-state situation when starting transients will have died down leads to the solution

$$V = V_0[1 + (\omega_0 CR)^2]^{-\frac{1}{2}} \sin(\omega_0 t - \phi) \tag{4.2.36}$$

where $\tan^{-1}\phi = \omega_0 CR = \omega_0\tau$. The same result can be obtained by means of the complex amplitude theory described in section 1.4. One regards the capacitor as a reactance with impedance

$$\hat{Z} = (i\omega_0 C)^{-1} \qquad (4.2.37)$$

and then calculates the total impedance of the two elements in series. One now has a complex voltage divider whose input voltage will be

$$V = iV_0 \exp(-i\omega_0 t) \qquad (4.2.38)$$

(the analogue of 4.2.35) and whose output voltage will be

$$V = iV_0[1 + (\omega_0 CR)^2]^{-\frac{1}{2}} \exp[-i(\omega_0 t - \phi)]. \qquad (4.2.39)$$

Taking the real part of (4.2.39) gives (4.2.36) as before.

The third way of analysing the RC filter is in the frequency domain in terms of its response and transfer functions. In the case of the RC filter, the response function is

$$R(t) = \tau^{-1} \exp(-t/\tau), \qquad (4.2.40)$$

and the transfer function is hence

$$\hat{T}(\omega) = \int_0^\infty R(t) \exp(i\omega t)\, dt \qquad (4.2.41a)$$

$$= (1 - i\omega t)^{-1} \qquad (4.2.41b)$$

$$= (1 + \omega^2 \tau^2)^{-\frac{1}{2}} \exp(i\phi), \qquad (4.2.41c)$$

which is equivalent to equation (4.2.39) above. From any of these approaches one gets an immediate qualitative understanding of how the RC filter works. It suppresses noise by progressively attenuating the high-frequency components. Carrying through the analogue of (4.2.29) with (4.2.41c) as the transfer function gives, for the case of white noise, the result

$$\varepsilon^2 = P_e^N / \tau \qquad (4.2.42)$$

entirely analogous to (4.2.30b). The signal emerging from the filter will be proportional to the modulus of the transfer function and it follows therefore that the signal-to-noise ratio will be given by

$$S/N \sim |\hat{T}(\omega)|/\tau^{-\frac{1}{2}} \sim \tau^{\frac{1}{2}}/(1 + \omega_0^2 \tau^2)^{\frac{1}{2}}. \qquad (4.2.43)$$

This function is plotted in Fig. 4.12. It will be seen that there exists an optimum value of τ and that beyond this value the signal is being attenuated more than the noise.

In this "white noise" situation, AC operation with a PSR gives merely stability and convenience of operation, the RC network does all the noise suppression that can be done. One would get the same answer if one used a broadband, i.e. video, system and merely terminated this with an RC network.

Physically, the content of equation (4.2.43) is that one must not reduce the bandwidth (i.e. τ^{-1}) to the extent where it is less than the frequency range covered by the signal and merely shifting both along the frequency axis makes no difference since the noise, by assumption, is independent of frequency. If

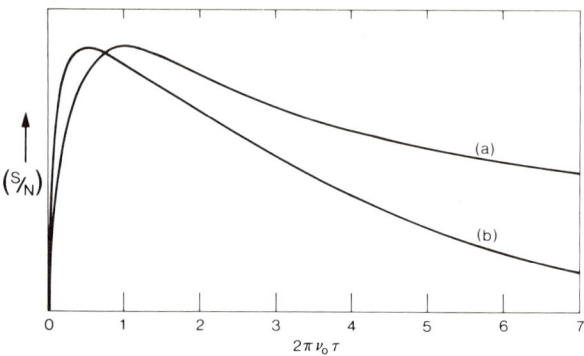

Fig. 4.12. Variation of S/N for a monochromatic signal accompanied by (a) white noise and (b) reciprocal frequency noise, in both cases after passage through an RC filter of time constant τ.

one is dealing with reciprocal frequency noise, however, very different considerations apply. In this case, the bulk of the noise power lies below the signal frequencies and, although there is a small improvement in S/N as τ increases from zero, one soon gets into the regime where the signal is being attenuated whilst the noise is hardly being affected. A video system with an RC smoother would now be a poor buy compared with a modulator/PSR system working at as high a frequency as the detector can handle. The PSR system essentially transfers the shifted signal back down again into the low audio range where the signal was originally but in the transfer it is only the high-frequency noise, i.e. that near the modulation frequency, which is transferred and, by definition, this will be small compared to the total noise power. Indeed, at the modulation frequency, $1/\nu$ noise in a limited band looks quasi-white so the RC network will give some additional noise suppression. Reciprocal frequency (that is "flicker") noise is the chief form of noise in modern solid-state devices and this, together with the insensitivity to "drifts" is one of the more important points in justifying the universality of modulation plus phase-sensitive rectification in infrared systems. It must be realised, however, that all noise suppression depends on bandwidth reduction and this necessarily makes the recording system sluggish. The record produced by this system will not, therefore, be a true one unless some precautions are taken. This topic will be taken up again in section 4.3.5.

4.2.5 *Noise in the electromagnetic field*

In the quantum mechanical approach to the electromagnetic field, one imagines the field to be confined within, a cavity which, as detailed in section 2.1, will have a discrete set of modes. Each mode is then regarded as an harmonic oscillator and has the appropriate (equation 1.5.27) set of energy levels. One continues the treatment by considering the cavity to expand without limit but always retaining the quantisation. One can, in this formalism, therefore talk meaningfully of the quantisation of the free electromagnetic field. The most probable state of the field has any given mode unexcited—that is it is in the vacuum state. If, however, a mode *is* excited, we say that a photon is present. The photons are thus the particles, of spin unity, which are produced by the quantisation of the field. The unit spin is the quantum analogue of the classical result that the electromagnetic field is a vector field. Any mode of the field may have any arbitrary degree of excitation, just as an harmonic oscillator may be excited to an arbitrary degree. Physically this means that the photons do not obey the Pauli exclusion principle which applies rigorously to the more "material" particles—the fermions of half-integral spin—such as electrons and protons. Photons are in fact "bosons" and as such obey the Bose–Einstein statistics:

$$N(\mathscr{E}) = N(0) \left[\exp(\mathscr{E}/kT) - 1\right]^{-1}. \tag{4.2.44}$$

If a mode has a frequency v, then at a temperature T it will contain an average number of photons $\langle n \rangle$ given by

$$\langle n \rangle = \left[\exp(hv/kT) - 1\right]^{-1}. \tag{4.2.45}$$

It will be recalled that this factor, with this meaning, appeared in the derivation of the Planck radiation formula in section 2.2.3. The average energy of a mode is therefore [268]

$$\langle \mathscr{E} \rangle = hv\left[(\exp(hv/kT) - 1)^{-1} + \tfrac{1}{2}\right]. \tag{4.2.46}$$

The extra term in the square brackets represents the zero-point energy of the mode. It was not introduced in section 2.2.3 because the zero-point energy can never be extracted from an oscillator and cannot therefore contribute to the steady radiation flux. It can, however contribute to the *fluctuations* of the flux because of the operation of the uncertainty principle. There is an obvious difficulty at this point in that one is invoking an unbounded field and one will therefore have an infinity of modes each with a finite zero-point energy. The resolution of this paradox via the trick of renormalisation (tantamount to merely subtracting away the awkward infinity!) leads on to quantum electrodynamics, one of the most spectacularly successful of the branches of modern theoretical physics [269]. This apart though, the fluctuations are very real. They can, for example, be invoked to explain the phenomenon of spontaneous emission and to provide the initial "noise" signal which is necessary to trigger a laser into oscillation.

To derive the magnitude of the fluctuations, one needs a relation between the thermal noise power in the mode, the average energy and the bandwidth. Marcuse [268] shows that this takes on the simple form

$$P_N = \langle \mathscr{E} \rangle \Delta v, \qquad (4.2.47)$$

so the thermal noise power is given by

$$P_N = hv\Delta v \{ [\exp(hv/kT) - 1]^{-1} + \tfrac{1}{2} \}. \qquad (4.2.48)$$

This power is composed of two parts. The first is the classical thermal noise power which, being virtually incoherent, represents the black-body radiation in the mode. Its fundamental origin is in the wave-aspect of the field so it is sometimes called "wave-noise". The second term, which arises entirely from the quantum aspects of the field, defines the *quantum noise*, sometimes called *shot noise*. The alternative usage can, however, be a little misleading since shot noise is really a property of electron streams and although the variations of such a stream in a photomultiplier *can* mirror the photon fluctuations of the field, this need not necessarily be so.

The noise power given by (4.2.48) has two interesting limiting cases. Firstly, when $hv \ll kT$, i.e. in the Rayleigh–Jeans regime, it reduces to

$$P_N \approx kT\Delta v. \qquad (4.2.49)$$

This is the thermal noise which radio engineers and radio astronomers have to contend with. The noise power is evidently independent of frequency, i.e. it is "white". Noise of this type was first encountered in resistors and other electronic components working at finite temperatures. It is almost universally called *Johnson noise* after one of the early investigators. Nyquist analysed the particular case of noise power extraction from a resistor regarded as an antenna in thermal equilibrium inside a cavity [270] and showed that, for this configuration, there should be an extra factor of 4. This result has already been quoted in equation (4.2.17). In the opposite limit, i.e. when $hv \gg kT$, the thermal noise becomes negligible and one now has noise which is frequency dependent. It is easy to see the physical origin of this frequency dependence of quantum noise. At a constant power level, the number of photons arriving per second is inversely proportional to the frequency, so the fluctuation due to the arrival or non-arrival of a photon becomes larger and larger as the frequency rises. The transition from thermal noise being dominant to quantum noise being dominant occurs at $hv = kT$, that is when the temperature equals the equivalent temperature. Equation (4.2.48) incidentally shows that there is no frequency for a free-space communication channel which is devoid of noise. At very high frequencies, the thermal noise will fall off but the quantum noise will then take over.

The electromagnetic field is always noisy therefore but the noise is only obvious when one is using a detector sensitive enough to respond to the fluctuations. The most important case arises with detectors capable of photon counting, that is photomultipliers and the more sensitive photoconductors. It can be shown [268] that for both broad-band incoherent radiation and narrow-band coherent radiation, the photon counting statistics obey the Poisson distribution. From equations (4.2.13) and (4.2.45) it then follows that the variance of the photon arrival fluctuations will be given by

$$\sigma_n^2 = \overline{(n - \langle n \rangle)^2} = \langle n \rangle [1 + (\exp(h\nu/kT) - 1)^{-1}]. \qquad (4.2.50)$$

The first term on the RHS of (4.2.50) is the shot (or more properly quantum) noise and corresponds to the fluctuations expected in the arrival of completely independent particles. The second term, the wave noise, shows from its frequency dependence that the arrival cannot be totally independent for arbitrarily small counting intervals. In other words there must exist intensity correlations in the field. Photon counting systems are only available for the optical and near-infrared regions so even for hot sources one will always have $h\nu \gg kT$ and so equation (4.2.50) may be written in the approximate form

$$\sigma_n^2 = \langle n \rangle [1 + \exp(-h\nu/kT)]. \qquad (4.2.51)$$

The second term represents the intensity correlations or *excess fluctuations*. Their existence forms the basis of the intensity interferometry [271, 65] technique introduced by Twiss and Hanbury-Brown for measuring stellar diameters.

The widespread use of lasers has led to considerable interest in the question of laser noise. This, as mentioned above, follows the Poisson distribution but what has emerged from the further studies [268] is the intriguing conclusion that a photon counter can tell the difference between true laser radiation and very narrow-band radiation derived either by filtering from a broad-band original or else naturally from an intrinsically narrow transition. The narrow-band incoherent radiation obeys the Bose–Einstein statistics in which the probability of a count of n when the expected average is $\langle n \rangle$, is [268]

$$p(n) = (1 + \langle n \rangle)^{-1} (1 + (\langle n \rangle)^{-1})^{-n}. \qquad (4.2.52)$$

This has the usual Bose–Einstein feature of predicting a maximum probability for $n = 0$! The Poisson distribution on the other hand (equation 4.2.11) always predicts a maximum probability for n values close to $\langle n \rangle$. Photon counting can therefore tell you whether a laser is in fact working above threshold or not. This point is a most interesting one since it means that the commonly held idea of a close relationship between the coherence and the inverse of the line width is not always necessarily true. Thus relatively broad-band devices such as the free-electron laser described in section 2.5.4.2 can be highly coherent. With coherent sources it is useful to think in terms of a Bose condensation in which all the laser photons have collapsed into a single mode whose degeneracy can be enormous. The counting statistics will therefore be very different. It is not difficult to see why this should be so. Perfectly coherent radiation corresponds to a perfect eternal classical sine wave but since the field variables must be governed by the uncertainty principle, it follows that this perfection of one conjugate variable will be accompanied by total uncertainty in the other. The modes of the field and hence the photons occupying them are thus totally uncorrelated and the photons will then obey the Poisson statistics.

Photon counting using laser sources has become an important diagnostic tool in several branches of physics and physical chemistry since, from the statistics of the scattered light, much can be inferred about the nature of the scatterers [272]. Commercial instruments based on this principle are now being manufactured and widely used.

4.2.6 *Correlation of noise signals*

The concept of photon correlation takes us on quite naturally to consider the question of noise correlation in general. Obviously all noise signals must be correlated to a greater or less extent since, as mentioned earlier, truly random uncorrelated or "white" noise is physically impossible because it would correspond to the transmission of infinite power. All real signals must be spectrally band limited and therefore more or less correlated. This is sometimes expressed in terms of the same colour metaphor by saying that physical noise must be "pink". If one has a stationary ergodic noise record described by the stochastic process $E(t)$, then one can measure the correlation by means of the auto-correlation function

$$F(\tau) = \operatorname*{Lt}_{t_m \to \infty} \left[t_m^{-1} \int_0^{t_m} E(t)\, E(t+\tau)\, dt \right], \qquad (4.2.53)$$

or in terms of the auto-correlation coefficient

$$\gamma(\tau) = F(\tau)/F(0). \qquad (4.2.54)$$

The noise power spectrum is then proportional to the Fourier transform of $F(\tau)$, i.e.

$$S(v) \sim \int_{-\infty}^{+\infty} F(\tau) \exp(2\pi i v \tau))\, d\tau. \qquad (4.2.55)$$

In the absence of correlation, $F(\tau)$ will be a delta spike and $S(v)$ will be correspondingly constant. Conversely, in the presence of total correlation, $F(\tau)$ will be constant and $S(v)$ will be a delta spike. It will be seen from this that the concepts of correlation and coherence are closely related. To show this connection more clearly, one imagines that a Michelson interferometer is being illuminated with a stationary ergodic field characterised at any moment of time by the stochastic variable $E(t)$, then, with the interferometer in perfect alignment, i.e. with the image points from the two arms exactly superposed in the detector plane, one would have that the field at the detector would be given by the simple addition of $E(t)$ and $E(t+\tau)$ where τ is the time delay of one arm with respect to the other. The overall intensity recorded by the detector can be derived by taking ensemble averages over the totality of the field of view or in more precise terms, by taking averages over tiny areas, each a space coherence length wide. The result is

$$I_{\text{obs}} = \tfrac{1}{2}\,\varepsilon_0 c \; \underset{t_m \to \infty}{\text{Lt}} \left[\frac{1}{t_m} \int_0^{t_m} \overline{(E(t)+E(t+\tau))^2}\,dt \right], \qquad (4.2.56a)$$

i.e.

$$I_{\text{obs}} = (\varepsilon_0 c/2t_m) \; \underset{t_m \to \infty}{\text{Lt}} \left[\int_0^{t_m} \overline{(E(t)^2 + E(t+\tau)^2)}\,dt + 2 \int_0^{t_m} \overline{E(t)\,E(t+\tau)}\,dt \right].$$
$$(4.2.56b)$$

In other words

$$I_{\text{obs}} = \tfrac{1}{2} I_0\,(1 + \gamma(\tau)). \qquad (4.2.56c)$$

Thus the oscillating part of the interference record, that is the interferogram, is directly proportional to $\gamma(\tau)$ and one can say that to within a common factor

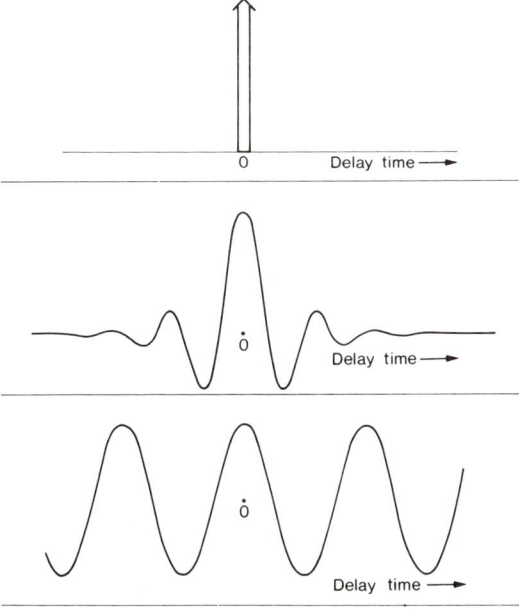

FIG. 4.13. Interferograms corresponding to incoherent (upper inset), partially coherent (middle inset) and fully coherent (lower inset) irradiation of a Michelson interferometer.

the interferogram *is* the autocorrelation coefficient of the incident field. If one now looks at the same phenomena from the point of view of coherence theory, one will recall, as explained in section 2.3, that the Michelson interferometer can be regarded as a time-coherence meter. If the incident field has a coherence time τ_c then it may, for convenience of thought, be decomposed into random Gaussian noise pulses of width τ_c. As the interferometer delay time increases from $\tau = 0$ towards $\tau = \tau_c$, the overlap between the two sub-pulses, into which an incident pulse is divided by the beam splitter, gets less and less upon recombination. For delay times much longer than τ_c, interference no longer occurs because the sub-pulse from the shorter arm will have been absorbed by the detector before its partner has arrived. Thus the interferogram is a rapidly damped function of τ and its time constant is a measure of τ_c. One then has immediately that the autocorrelation coefficient $\gamma (\tau)$ is a direct measure of the time-coherence of the field. If one has a spectrum in which the power does not vary with the frequency, that is the spectroscopic equivalent of incoherent "white" noise, then the interferogram will be a delta spike at the origin. As the spectral bandwidth decreases, one starts to get observable, albeit rapidly damped, oscillation near the origin and, finally, when the line width becomes negligible, and one hence has a completely coherent field, the interferogram will have become an endless cosine wave. This is illustrated in Fig. 4.13. In all cases, it follows from inverting (4.2.55) that the spectrum can be obtained by Fourier transforming the interferogram. This is the basis of Fourier transform spectrometry which has been mentioned earlier and which will be described in some more detail in the next section.

4.3 Principles of infrared spectrometry

4.3.1 *Direct or frequency scanning systems*

One of the commoner operations in infrared science and technology is the determination of the absorption spectrum of a specimen, that is how its absorption coefficient varies with frequency. The conceptually simplest way of doing this is to take a tunable source of quasi-monochromatic radiation, to pass its radiation through the specimen and then to measure the transmitted power by means of a suitable detector. For the reasons discussed earlier the beam would be modulated either by a chopper or else by pulsing the source. This simple system would certainly reveal all the detail in the spectrum but would not in general give good radiometric precision because both the source output and the detector sensitivity would be expected to vary as the operating frequency is changed. The system can therefore be much improved by splitting the incoming beam into a probe beam which goes through the specimen and a monitor beam which passes through an equivalent air or vacuum path. If these are separately detected, the ratio of the detector outputs will give the sample transmission corrected for most of the gross variations. One says most, because if one has two separate detectors there is no way of ensuring that the variations of their sensitivities with frequency will be identical. A very clever way of getting round this difficulty is to use a single detector, as shown in Fig. 4.14. The driving unit sweeps the frequency of the source and simultaneously controls the abscissa position of the pen on the X/Y recorder. The output beam from the source is split by means of a reflecting chopper into two beams, each modulated, but with the phase of the modulation differing by π. After passage through the specimen and reference channels respectively, the two beams are recombined at the detector. The detector output, at the modulation frequency, will then be a minimum when the absorption in the specimen is zero and a maximum when the absorption is total. The rectified output of the PSD can thus be fed to the ordinate amplifier of the X/Y recorder, whose gain and offset are carefully adjusted so that the full range covers 0 to 100% absorption. The X/Y recorder can be replaced by a storage oscilloscope for normal work, or else by a high-speed oscilloscope if rapid-scan spectroscopy is required (see Fig. 7.5). The only remaining drawback to this arrangement is the absence of an easy frequency (that is x-axis) calibration. The system is formally equivalent to a microwave spectrometer, but the usual way of calibrating a microwave spectrometer, via a chain of harmonic multiplication down to a standard frequency, is not readily available in the infrared when the source power is low. The present practice is therefore either to calibrate with a standard substance whose spectrum is known accurately or else to use a high "Q" Fabry–Perot etalon to put frequency markers on the trace.

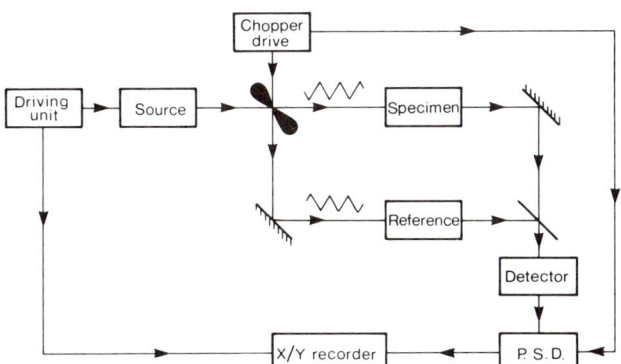

FIG. 4.14. Schematic layout for a frequency-swept spectrometer.

At the moment, there is not exactly a surfeit of suitable sources for swept-frequency infrared spectroscopy, but there are *some* and the number is likely to increase. Colles and Pidgeon have given a comprehensive account of the present position [273]. Diode lasers are very good, but their tuning range is extremely narrow ($\sim 1\,\mathrm{cm}^{-1}$); spin-flip Raman lasers have a much wider tuning range but do have, at the moment, severe mode-pulling and hopping problems which limit their usefulness. Colour centre lasers show considerable promise, but so far they are limited in their spectral range and their long-term performance is not established. Parametric oscillators, whilst broadly tunable, are less attractive than the other sources in this connection because of the relatively large line width and because of their cost and difficulty of operation. So far only the diodes have been used extensively, but the spin-flip and colour centre lasers have been used for some specialised applications. The diodes have proved particularly useful in molecule-specific pollution monitoring (section 7.2.1.2), whilst the spin-flip laser has been used for broad band scans and for characterising narrow-band mid-infrared interference filters. The colour centre lasers which work in the near infrared have some important applications, via the overtones, in pollution monitoring.

It should perhaps also be mentioned in this context that instead of using a coherent source and a broad band detector, one could in principle use a broad band source and a coherent detector. This technique is not so far available over wide spectral regions because of the lack of suitable detectors, but a close relative, heterodyne spectroscopy, has some applications in very high resolution spectroscopy.

4.3.2 *Indirect or delay-type infrared spectrometers*

The direct frequency-scanning infrared spectrometers are an exciting development and with their aid many difficult experiments have become possible, but until they are capable of scanning the whole infrared and until their cost is drastically reduced, they will not pose a serious threat to the traditional methods of infrared spectroscopy. These methods use a broad band, incoherent, black-body source which, of course, has no frequency coding, so one needs an extra piece of optical equipment, e.g. a prism, grating or interferometer to impress this information on the beam. All these frequency encoding devices operate by virtue of the *delay principle*. The essential point of this principle is that the incoming beam of radiation is split into two or more beams, which are subjected to progressive time delays. The beams are recombined in the image plane where consequently an interference pattern is produced. Each sub-beam is oscillating at frequencies of the order of THz, so the slow detectors used for most normal infrared work cannot hope to respond, but the interference leads to a *stationary* pattern extended in *space* and from this spatial distribution the spectral information can be inferred. Thus, although by the act of detection all the frequency information is inexorably lost, this information can nevertheless be recovered either by scanning the detector over the interference pattern or vice versa. This is the delay principle which, as just mentioned, is fundamental to the operation of all indirect spectrometers. Historically all spectrometers were of the delay type;

only in the last fifty years have we had the direct type available, and people are so familiar with, say, a prism spectrograph that it requires a distinct mental effort to appreciate that the indirect approach is not exactly the most obvious way of doing spectroscopy. Microwave spectroscopists who have nearly always used direct methods see things in a very different light, and this has been one of the factors in erecting the highly undesirable and completely unnatural barrier between microwave and infrared spectroscopy.

The number of sub-beams involved depends on the type of instrument. Thus, a prism corresponds formally to an infinite number of partial beams, the grating and Fabry–Perot interferometer to a finite (if large) number, whilst the Michelson interferometer has just two partial beams. Because of the simplicity of this latter system, it provides an easy illustration of the operation of the delay principle, but it must be stressed again that the principle is general. For convenience of reference the basic optical arrangement of the two-beam interferometer is reproduced again in the upper inset of Fig. 4.15. In the lower inset is shown the equivalent (if entirely hypothetical!) plane-parallel optical interferometer. It will be seen from this that the beam-divider, introduced by Michelson, is an exceedingly ingenious way of getting round the practical problem that one mirror cannot pass through another. The material mirrors

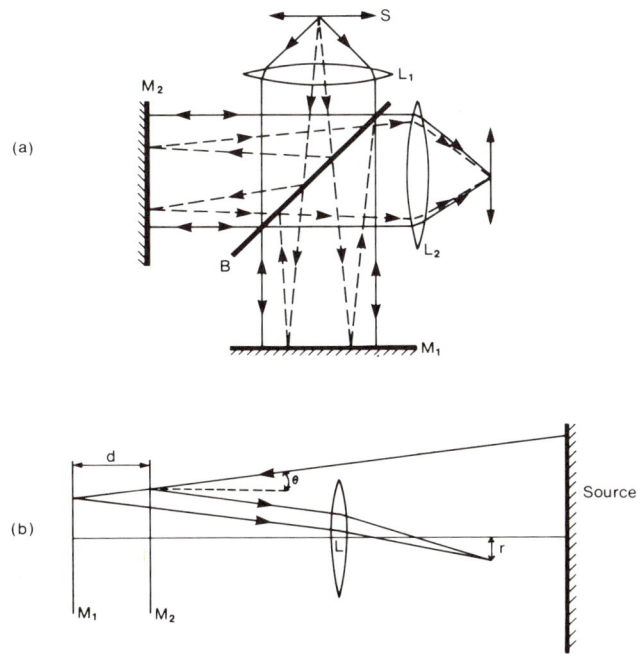

FIG. 4.15. Schematic arrangement of the Michelson interferometer in (a) its collimated (solid-lines) and divergent (dotted lines) versions. In (b) the equivalent plane-parallel interferometer is shown.

are replaced by their optical images and these *can* pass through one another without any difficulty.

The incoming beam from the source S, is rendered quasi-parallel by a collimator in the Twyman–Green version of the interferometer, or else is simply allowed to diverge in the classical Michelson form. In either case the beam is split into two partial beams by the beam-divider B and each of these beams travels to a mirror, M_1, M_2. The beams return from their respective mirrors and are recombined at the beam-divider. Part of the flux then proceeds to a condenser, which produces an image of the source in the focal plane. We are interested in the case where the two mirrors are so adjusted that the two images of the source are made exactly coincident, for then radiation from an arbitrarily small region of the source, and therefore of arbitrarily high spatial coherence, will be recombined with itself. Interference will thus occur, leading to the formation of a stationary pattern of rings for the interference fringes. If the incident field is monochromatic then one gets a set of rings for all values of the path difference, whereas if it is broad band, the pattern of rings is only detectable in the vicinity of zero-path difference and one gets the well-known and rather beautiful phenomenon of "white-light fringes". The ring pattern is a consequence of the angular dependence of the path-difference. Obviously in the case of the divergent version of the instrument, there will be a range of angle of incidence on the mirrors and this remains true—but to a lesser extent—when one uses collimation. A ray making an angle of incidence θ on one of the mirrors (Fig. 4.15(b)) will suffer a path-difference delay or advance of $2d \cos \theta$ compared with its companion which has been reflected, at the same angle, of course, at the other mirror. Each value of the angle then corresponds to a fixed radial distance in the image plane so, at values of r which correspond to the usual constructive interference condition

$$m\lambda = 2d \cos \theta, \qquad (4.3.1)$$

there will be a bright fringe. Michelson introduced his interferometer originally just for work in the visible region using the eye as detector [64] and for measurements confined solely to near-monochromatic sources. For the particular geometry of his instrument (Fig. 4.15) one has that the bright fringes are located at radial distances

$$r = f \sqrt{[(2d/m\lambda)^2 - 1]}, \qquad (4.3.2)$$

and thus by scanning a travelling microscope over the ring pattern one could readily determine λ. The technique may also be used to investigate fine structure in a line since with a suitably large setting of d and at a large enough value of r the ring patterns arising from the close components will get out of step and the loss of fringe "visibility" will reveal the fine structure.

This method of scanning over the image plane is, however, of little value in the infrared for two principal reasons. Firstly nearly all infrared sources that one would be interested in studying spectroscopically are broad band and the non-linearity of equation (4.3.2) does not permit any easy way of recovering the spectral information by means of a convenient integral transform applied to $I(r)$. Secondly it is not really practicable to scan infrared detectors, especially those operating at cryogenic temperatures, in a lateral direction. For these reasons, all infrared two-beam interferometry is done in an alternative mode, also invented and used extensively by Michelson, in which the detector is *fixed* at the centre of the ring pattern and the path-difference is scanned by translating one of the mirrors. The signal which is received and recorded, in this regime, is proportional to the integral of the radiant intensity over the detector entrance aperture. As usual one may consider monochromatic illumination since the more general broad-band case can be subsequently derived by means of the principle of linear superposition, that is Fourier's integral theorem. The power received in an annulus corresponding to angles lying between θ and $\theta + d\theta$ will then be [7]

$$dP = (P_0/\Omega_F) (1 + \cos (2\pi \tilde{v} x \cos \theta)) d\Omega \tag{4.3.3}$$

where Ω is the solid angle, given by

$$\Omega = \pi r^2/f^2 \approx 2\pi (1 - \cos \theta) \tag{4.3.4}$$

and Ω_F is that corresponding to full aperture. The path-difference x is equal of course to $2d$. Making use of (4.3.4) one can write for the total power passing through the aperture,

$$P = \int_{\Omega_F} dF = (P/\Omega_F) \int_{\Omega = 0}^{\Omega = \Omega_F} [1 + \cos 2\pi \tilde{v} x (1 - \Omega/2\pi)] d\Omega, \tag{4.3.5}$$

that is

$$P = P_0 [1 + \text{sinc} (x \tilde{v} \Omega_F/2\pi) \cos 2\pi \tilde{v} x (1 - \Omega_F/4\pi)]. \tag{4.3.6}$$

The interferogram function is therefore, as expected, oscillatory, but there are two unexpected complications. Firstly, the amplitude of the oscillations is damped by the premultiplying sinc function; in other words the interferogram suffers instrumental tapering. Secondly, the period of the oscillations is not given by the true wavenumber $\tilde{v}_0$, but by

$$\tilde{v}_c = \tilde{v}_0 (1 - \Omega_F/4\pi). \tag{4.3.7}$$

The frequency scale in the calculated spectrum is therefore contracted. This latter point is usually only of importance in high-resolution spectroscopy where corrections [274] have to be applied to derive true wavenumbers. It is not easy to determine the exact value of Ω_F in a real interferometer, so these corrections are normally derived empirically by measuring the spectrum of a gas such as CO [275] whose high-J line positions can be accurately predicted from extrapolating the microwave data.

The damping of the interferogram, unavoidably present when, as it must, the detector has a finite aperture, arises physically from the presence of more than one ring within the detector aperture. At small values of x only the zeroth order ring enters the detector but as x increases the effects of the next ring out, the first-order ring, begin to make themselves felt, and at a value of x equal to the so-called *extinction distance*

$$x_e = 2\pi/\Omega_F \tilde{v}_0 \qquad (4.3.8)$$

the first-order ring just enters the rim of the detector aperture. This value of x corresponds to the first zero of the sinc function in (4.3.6). The envelope of the modulated interferogram therefore falls to zero at the extinction distance. Because of the damping, it follows that even were one able to observe the interferogram out to infinite values of the path-difference, one would still not calculate the correct line-shape. In fact instead of the delta spike corresponding to the monochromatic illumination, one would get an apparatus function consisting of two rectangle functions [7] of width

$$\Delta\tilde{v}_c = (\Omega_F \tilde{v}_0/2\pi) \qquad (4.3.9a)$$

located respectively at

$$\tilde{v}_c = \pm (1 - \Omega_F/4\pi)\tilde{v}_0. \qquad (4.3.9b)$$

This result follows by taking the Fourier transform of the varying part of (4.3.6) or else almost by inspection by noting that the tapering function sinc $(x\tilde{v}\Omega_F/2\pi)$ has the Fourier transform $\Pi(2\pi\tilde{v}/\tilde{v}_0\Omega_F)$ and then applying the convolution theorem. This result is, however, almost of merely academic interest since one would hardly ever, in real interferometry, have sufficient mirror travel available to be able to approach the extinction distance. In the usual regimen, one can say that to a very fair approximation the interferogram is purely cosinusoidal; in other words we have derived the same result that was obtained in the previous section by coherence and correlation arguments. Necessarily now, the range of observation is restricted and one derives the spectrum from truncated Fourier transforms of the type

$$S(\tilde{v}) = \int_0^{x_{max}} P_0 \cos(2\pi\tilde{v}_0 x) \cos(2\pi\tilde{v}x) \, dx, \qquad (4.3.10)$$

as discussed extensively in section 1.6. Either by evaluating (4.3.10) or else by again invoking the convolution theorem, one sees that the line-shape or apparatus function will be given by sinc functions of the form

$$S(\tilde{v}) \sim P_0 x_{max} (\text{sinc} (\tilde{v}_0 + \tilde{v})x_{max} + \text{sinc} (\tilde{v}_0 - \tilde{v})x_{max}) \qquad (4.3.11)$$

as shown in Fig. 4.16. Under normal circumstances, the contribution from the "image" non-physical feature at $\tilde{v} = -\tilde{v}_0$ can be neglected and one can say that the spectral window or apparatus function is the simple expression

$$S(\tilde{v}) \sim P_0 x_{max} \, \text{sinc} (\tilde{v}_0 - \tilde{v})x_{max}. \qquad (4.3.12)$$

Before continuing this analysis of one particular infrared spectrometer, it is worthwhile considering the general points which have emerged. Firstly, the finite line width which is observed even with infinite path difference available is a consequence of the need in real spectroscopy to use finite apertures. For the

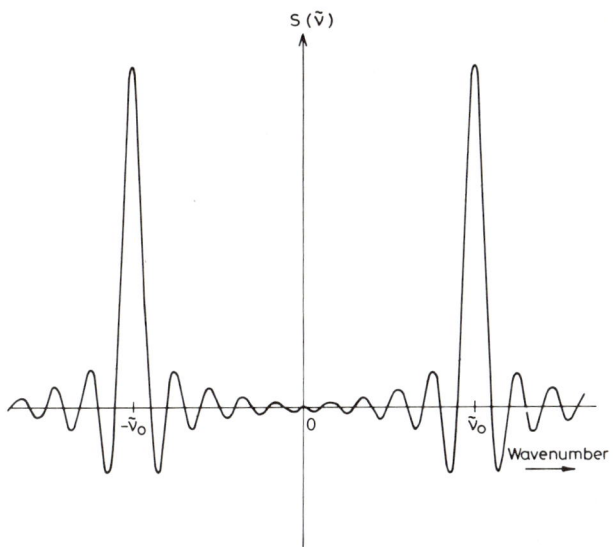

FIG. 4.16. Spectral window or apparatus fuction corresponding to truncated observation of a monochromatic source with a two-beam interferometer.

grating or prism instruments, this would mean the need to use finite slit widths more commonly than the need to use a finite detector aperture but the point remains. If an instrument is to give a finite signal-to-noise ratio it must have a finite limiting stop and will therefore have an apparatus function which has a finite width. This limiting aperture broadening should properly be derived within the framework of a linearised diffraction theory but in practice the simpler theory outlined above gives the same answers. In the case of a dispersive instrument, this diffraction approach leads, as mentioned earlier, to an apparatus function of the sinc^2 form. If it were applied to the interferometer, it would lead to an apparatus function involving combinations of Airy functions, but as just mentioned this has never proved necessary in practice. The second point is that it is very rare to approach resolving powers so high that the fundamental aperture broadening provides the limitation. Nearly always it is the finite size of the optics or the finite amount of mirror travel which provides the practical limitation.

These points made, we return to the analysis of the interferometer because it is the simplest system both from an optical and a mathematical point of view and because its various stages of operation are actually, as opposed to conceptually, separate. It might be thought that the interferometer is fundamentally different from the other instruments of this class because it gives its primary data in sampled rather than continuous form. This is essential, of course, if the computing time is to be finite! The computer is therefore fed a set of sampled ordinates $I(n\Delta x)$, where Δx is the sampling interval, and then performs a discrete Fourier transformation on this set. For reasons of mathematical convenience and to accord with physical intuition which demands that $S(0) = 0$, i.e. that there is no radiant power at zero frequency, the computer first produces a zero-mean set,

$$I'(n\Delta x) = I(n\Delta x) - \overline{I}, \qquad (4.3.13)$$

where $\overline{I}$ is the average of the $I(n\Delta x)$ set. The machine then evaluates the sum

$$S(\tilde{\nu}) = \Delta x \sum_{-(N-1)}^{+N} I'(n\Delta x) \cos(2\pi\tilde{\nu}n\Delta x) \qquad (4.3.14)$$

which is the discrete Fourier transform. For the reasons outlined in section 1.6, however, the spectrum produced by this discrete transform will be the same as that which would have been available from a continuous transform provided the incident power spectrum contains no components lying above the "folding" or Nyquist wavenumber $\tilde{\nu} = (2\Delta x)^{-1}$. The reconstruction will then be perfect to a resolution given only by the maximum mirror travel available. The really significant differences between the various delay type spectrometers lie in the nature of the interference patterns formed in the image plane. These patterns, as discussed by Gebbie and Twiss [276], become more and more easily interpreted as the number of interfering beams rises. Thus the prism with its formally infinite number of beams gives a corresponding pattern—the spectrum—which is unambiguous. The grating with a finite number of interfering beams, i.e. the number of lines across the width of the grating, gives a spectrum which is possibly ambiguous because of the risk of overlapping orders. Grating spectroscopy is best carried out, therefore, with the help of appropriate absorptive or interference filters to do some preliminary order sorting. When one comes to consider the pattern produced by the two-beam interferometer, it is obviously hopeless to attempt the unscrambling optically—or mechanically—although Michelson [64] and others [277] have tried hard enough. Recourse must be had to the power of a computer. The analysis of Gebbie and Twiss showed that, normalised to some such parameter as resolution, this loss of information in the image plane was matched by an increase in the radiant throughput. So although it is true that in all delay-type instruments there is a trade-off between energy throughput (that is the solid angle of radiation collection) and limiting resolution this trade-off is most favourable for the interferometer. The actual form of the trade-off is surprisingly simple: either by optical arguments or else by invoking information theory, it can be shown that the limiting resolving power and the energy throughput are inversely proportional to one another. Equation (4.3.9a) is an example. A convenient measure of the energy throughput is the *étendue*, sometimes called the "throughput luminosity". It is defined by the relation

$$E_L = \Omega_L A_L \qquad (4.3.15)$$

where Ω_L is the solid angle subtended by the limiting stop at the corresponding condensing surface (lens or mirror) whose area is A_L. Good optical design calls for E_L to be an invariant of the whole chain, this being a sort of optical analogue of impedance matching in electrical engineering. However, in the design of a spectrometer it may not always be possible to achieve this invariance and there will usually be a limiting stop (or aperture). Thus in an interferometer the detector window may be smaller than the full image of the source. In a grating spectrometer the limiting stop is nearly always the entrance slit. The concept of a limiting stop gives a simple way of understanding the advantage of the interferometer mentioned above. The interferometer, which has a *circular* limiting stop, enjoys full cylindrical symmetry and therefore has the higher *étendue* for a fixed value of the resolution [278]. One would therefore expect to get better S/N results from an interferometer observing the same spectrum for the same time to the same resolution. It is worth pointing out, however, that although the concepts discussed above are modern, the actual physical content has been known almost from the first years of quantitative spectrometry. It is certainly familiar to anyone who has ever had to open up the slit of his instrument in order to get an acceptable S/N and has then watched the resolution deteriorate.

So far, the resolution of the instrument has been taken implicitly to be the width of the apparatus function. However, this is not the traditional or even the most common meaning of the term. A spectroscopist usually defines the resolution to be the closest approach in wave number or frequency terms of two monochromatic lines which can still be clearly resolved by his instrument. Fortunately, it turns out that the two definitions lead to almost identical results in practice. If one has two equal intensity monochromatic lines irradiating the instrument, then each will produce its own apparatus function (immediately in the dispersive case, after computation in the interferometric case) and the question to be asked is how far apart must two identical apparatus functions be displaced before a perceptible doublet structure develops. This question was first posed in the context of dispersive spectroscopy since this was the only type available and Rayleigh after considering it came up with his famous criterion that two apparatus functions could be said to be resolved when the grand maximum of the second lay on the first zero of the first. This criterion was developed within the framework of linearised diffraction theory and the consequent Airy functions, but it can be used just as well with the simpler sinc^2 apparatus functions. Two equal intensity sinc^2 functions separated by the specified amount will develop a minimum in the centre whose depth is 19 % of the total height. The Rayleigh criterion, if applied as it stands to the interferometer, gives the simple result [7, 12] that

$$\Delta \tilde{v} = 0 \cdot 5 / x_{max}, \tag{4.3.16}$$

but Chantry and Fleming [279] have shown that it may not be applied as it stands since two sinc functions separated by this amount do not develop a minimum in the centre. They quote instead the result

$$\Delta \tilde{v} = 0 \cdot 66 / x_{max} \tag{4.3.17}$$

as a more realistic assessment of the situation. This numerical detail apart, one again has the underlying result that the resolution interval is proportional to the reciprocal of the maximum path-difference introduced. The same result emerges from a detailed study of the prism or grating instruments. In these cases it is either prism base or grating width which decides the resolution interval. It is perhaps worth remarking that spectroscopists, in their everyday parlance, would refer to *small* values of $\Delta \tilde{v}$ as *high* resolution. This can sometimes be a confusing usage.

4.3.3 *Comparison of the performance of two-beam interferometers and dispersive instruments*

The ideas of resolution, *étendue* and their relation to the number of beams interfering have been discussed by Jacquinot [278] and by Gebbie and Twiss [276] who have shown that when normalised to some parameter such as resolution the energy throughput or *étendue* falls as the number of beams rises. This is an important matter in infrared physics, where one is often energy limited. An important consequence has been that NaCl or KBr prism spectrometers, which were so common in the fifties, are now rarities and nearly all dispersive instruments now feature gratings. To be fair, it must be admitted that modern replica gratings are often cheaper than prisms and there is not the same problem caused by the hygroscopicity of alkali halides, but the increased radiation grasp is also very important. The point about normalisation to resolution can be looked at in another way by noting that prism and grating spectrometers have narrow rectangular limiting stops (the entrance and exit slits), whilst interferometers have complete cylindrical symmetry and therefore circular limiting stops. If the source field is uniformly illuminated, i.e. if one has a circular source, the interferometer must transmit more energy since none will be intercepted by the jaws of a slit. The commercial battle between companies producing grating spectrometers, e.g. Perkin Elmer, Beckman, Pye–Unicam, etc., and those producing interferometers, e.g. Digilab, Nicolet, Bruker, Bomem, Analect etc., has naturally led to the respective instruments being optimised and grating instruments usually feature long thin sources, which match better the tall rectangular slits, whilst interferometers feature much shorter, wider sources to take full advantage of the higher *étendue*.

Nevertheless a top-flight interferometer always has better energy throughput than an equivalent grating spectrometer, and consequently gives better signal-to-noise ratios. The interferometer has another and rather more subtle advantage, which was first pointed out by Fellgett [280] this is the *multiplex advantage*. In the simplest derivation one notices that a dispersive instrument is sequential in operation, that is if one imagines the spectrum to be divided into slices each a resolution unit wide then the instrument scans the slices one after the other. If there are M such slices and the total observing time is T, then each is observed for a time T/M and the signal-to-noise ratio, from the arguments given in the previous section, will go as $M^{-1/2}$. The interferometer, on the other hand, "sees" all the frequency slices all the time and its signal-to-noise ratio will be independent of M. The interferometer will therefore be better than the grating instrument by a factor $M^{1/2}$, a number which can be very large when one is considering high-resolution instruments. This, in simple outline, is the origin of the multiplex advantage but to see how it in fact applies in practice requires a rather more rigorous analysis. Firstly, it is implied in the above argument that all the noise is additive and hence that the noise level is independent of the signal level. This is sometimes referred to as the detector noise limited condition since it is usually the detector which is the dominant source of additive noise. If the noise is not truly additive, the advantage will be reduced and may even turn into a disadvantage [7]. The two commonest examples of signal-dependent noise are shot noise, where one has the noise proportional to the square root of the signal intensity (equation 4.2.50), and signal noise where the noise is directly proportional to the signal intensity. Several authors, for example Chamberlain [7], have analysed mathematically the operation of multiplexing in the shot and signal noise limited regimes and have shown that to a first approximation the advantage just cancels in the former case and turns into a disadvantage in the latter. One says to a first approximation because in the photon noise limited condition, as Kahn [281] has shown, one can have some regions of the spectrum where there is an advantage and others where there is a disadvantage: overall there is usually cancellation. The physical reason for this is that when one has multiplicative noise, a dispersive spectrometer may be well placed since the noise it observes when recording one of its sequential channels is only the noise which belongs inescapably to that channel. A multiplex spectrometer, on the other hand, will see noise in a given channel that may be mostly generated in all the other channels. Thus if one had a spectrum made up of sharp lines of very different intensities, an interferometer might find the weaker features swamped by the noise due to the stronger features, whereas the sequential spectrometer would be able to study each feature well regardless of its intensity. In this situation, there would be a multiplex advantage for the strong lines but a multiplex disadvantage for the weaker ones. Multiplicative noise tends to be more common the higher the frequency for the reasons spelled out earlier. In practice this tends to mean the near infrared and shorter wavelengths. At these wavelengths, the combination of high signal and very sensitive detectors makes the possible loss of the multiplex advantage tolerable, but in the far infrared where neither of these applies it is the multiplex advantage which makes high resolution interferometry a practical proposition. The second major consideration in the analysis of the multiplex advantage is that it is implied that, throughout the observing time, the interferogram is sensibly oscillating. If it is not, the signal being recorded is either truly zero (with PM) or else constant and therefore essentially zero (with AM). If one spends most of one's observing time t_{max} looking at nothing, one cannot expect an advantage over an observer who, all the time, is looking at something. The point has been nicely put by Pike [282] who remarks that whereas an interferometer sees all frequencies all the time a dispersive instrument sees all delay times all the time and in this symmetric situation there is no reason to expect one approach to be superior to the other. The argument as given is correct, but in practical spectroscopy it can prove fallacious since the time t_m which enters should be that required by the interferometric practitioner to reach the point where his interferogram has stopped oscillating. For a fair comparison, the dispersive spectroscopist should be required to observe the corresponding spectrum in exactly the same time since, under this regimen, both will be observing finite signals all the time. Played according to these rules, interferometric spectroscopy enjoys a real advantage but it has to be admitted that there is often substance to Pike's criticism since it is not always carried out under ideal conditions. A case where the multiplex advantage would be very obvious would arise when the spectrum was highly structured with all the features sharp. When this is so the sequential spectroscopist will make poor use of the available observing time whereas the interferometric spectroscopist will always be recording a finite signal. An ideal example would be the emission spectrum of a low-pressure gas observed to high resolution. In cases like this multiplex advantages of remarkable magnitude have been proven. Thus the Connes have observed spectra in times of the order of hours which would have required years by conventional techniques [283].

The two-beam interferometer is the commonest form but it is certainly not the only form of multiplex spectrometer. Several other types exist [284], for example the Hadamard spectrometer which features an array of encoded slits in its exit plane. This spectrometer has been much used for pollution monitoring [285] and for high-resolution work [286]. It is even possible to multiplex conventional dispersive spectrometers especially if a particular set of measurements is to be repeated many times [287].

The advantages of interferometry discussed so far apply even in the limit of truly stationary noise but in experimental spectroscopy it is rare to have unlimited time at one's disposal and when time is restricted there is a further advantage. The reasons for the restriction of the observing and hence of the integrating time may be manifold, but some common ones are time variation of the nature or of the composition of the specimen, the onset of drifts, variations in the output of the source etc. Faced with these difficulties the spectroscopist may have to observe a large number of spectra, each individually of unacceptably low S/N, and then average them all in the hope of getting a final spectrum of acceptable quality. To do this effectively requires that each spectrum be sampled at exactly the same values of the frequency so that equivalent quantities will be averaged. In dispersive spectroscopy there may be difficulties because of "backlash" in the spectrometer drive mechanism and one may then have random abscissa shifts. In interferometry, on the other hand, not only does one have one's spectrum in the form of a set of equispaced samples, but one also knows for sure that all the samples correspond to exactly the same set of frequencies. Hence the advantage but, as in everything else, one needs to specify the conditions carefully. Thus averaging interferometric spectra rather than the original interferograms implies that one has no dynamic range problems in the digital system. If one has, then it may be necessary to average the original interferograms and backlash problems may obtrude once more. However, for an interferometer there is a neat solution in that broadband radiation from a subsidiary source propagating through the instrument to a separate detector can be used to provide a "white-light fringe" fiducial mark which can be used to line up all the individual interferograms. In this way, backlash problems can be almost eliminated. Present practice seems to be that spectra are averaged in aperiodic far-infrared spectroscopy and interferograms in rapid-scan mid-infrared spectroscopy.

The built-in computer of a rapid-scan commercial instrument is mostly there to carry out the essential Fourier transformation, but it can also be used amongst other things to store reference spectra (for example those of substrates or of solvents) and these can be used in a ratio to yield solely the spectrum of the desired component. The approach is equivalent to the double-beam operation of a commercial dispersive instrument where the incident beam is split into two beams one of which probes the specimen and the other traverses a reference channel. The interferometer suffers a slight disadvantage in that the two operations are separated in time but it more than makes up for this by the extra flexibility stemming from the computer control and from the improvement resulting from each component being observed under ideal conditions. The natural extension of this process is for the built-in computer to "talk-to" an external main frame which can have access to literally hundreds of thousands of reference spectra. The possibilities this presents for infrared spectrochemical analysis are awe inspiring. The point has not, of course, been lost on the manufacturers of dispersive instruments and modern up-market models invariably feature some computing power often in the form of a built-in microprocessor (section 4.5). The very latest models such as the Perkin–Elmer Model 983 are completely computer controlled.

The speed of operation of interferometers can be applied in areas where sequential instruments would be far too slow to give useful results. The prime examples are found in the spectroscopy of transient species such as free radicals and of gas-chromatographic fractions observed "on-the-fly"—that is as they come off the top of the column. Griffiths has given a good account of this field [288]. The speed of response is also valuable in the study of the periodic changes in the microstructure of high polymers subjected to alternating cycles of extension and compression. Clearly here the only rival technique, X-ray diffraction, would be impossibly slow.

One therefore has a largish list of the advantages of interferometry. Gebbie and Twiss [276] make the point that many of these advantages stem from removing the maximum amount of the operation of the spectrometer from the mechanical regime, where ideal behaviour is rare, into the realm of the digital computer where it is the norm. It must not be thought, however, that dispersive spectrometers lose on all counts. Thus at the moment they have a considerable advantage in price. A "state-of-the-art" "top-of-the-range" dispersive spectrometer currently costs about half the price of its interferometric equivalent and interferometric penetration into the middle and bottom ends of the spectrometer market has only just begun. How long this advantage will last is any one's guess but there are now very few people in the instrument industry who can foresee a real future for anything save the most routine dispersive infrared spectrometers.

4.3.4 *Modes of operation of infrared spectrometers*

In a dispersive instrument, the traditional and still by far the most common method of operation was to mount a single detector behind the exit slit and then to scan the spectrum over this slit by rotating the grating. The alternative to this scanning regime was only available in the very near infrared where, by using specially sensitised photographic plates, one could dispense with the exit slit and thus gain the full multiplex advantage. Very recently, attempts have been made to use either linear or two-dimensional arrays of detectors in spectroscopy. These arrays, of which the most well known is that based on the silicon photodiode, can be thought of as a low-resolution electronically scannable photographic plate. Most of the work using arrays has so far been confined to the visible and near infrared, mostly because silicon technology is so advanced, but there is no fundamental reason, unlike the photographic plate, why their use cannot be extended into the middle infrared [289]. A suitable material for the construction of such diode arrays would be mercury-cadmium-telluride.

Scanning spectrometers can operate in two distinct sub-modes. The first of these is relatively rare and involves the use of high scan rates so that the fluctuating signals from the detector will lie well into the audiofrequency range and one will not need any additional modulation. The signals from the detector can then be passed to a broadband, or video, amplifier and subsequently directly displayed, for example on a cathode ray screen. This method was originally introduced [290] in the hope that it might be useful in studying transient species, for example free radicals, but it turned out in practice to be of insufficient value to justify the considerable outlay involved and in more modern times this task has been tackled much more successfully by rapid-scan interferometers (see below). Rapid-scan grating spectrometers have thus passed into the historical curio class. The second mode, which is virtually universal, is to scan slowly but to shift the now ultra-low frequency signals into the audio band by the use of a chopper. This confers all the advantages detailed earlier especially when, as is natural, the demodulation is carried out in a PSD. Nearly all commercial infrared grating spectrometers are double beam so they give as their primary output the transmission spectrum. Further practical details are given in section 6.6.

Interferometers can operate in two distinct modes which bear more than a passing resemblance to their grating equivalents. Thus one can arrange to scan the moving mirror rapidly back and forth through the region including the zero-path-difference fringe or, alternatively, one can choose to scan the mirror slowly through this region in a single run. The former is called periodic or rapid-scan interferometry whilst the latter is called aperiodic or slow-scan interferometry. The differences between the two arise mainly, as in the dispersive case, because in rapid-scan operation, acceptably high audio-frequency signals are produced by the detector, whereas in slow-scan they are not. Thus, if the mirror is travelling at a velocity v and the interferometer is producing a cosinusoidal interferogram from input radiation of wavenumber $\tilde{v}$, then the detector will give out an audiofrequency

$$v = 2v\tilde{v} \qquad (4.3.18)$$

A mirror velocity of $1\,\text{cm s}^{-1}$ and an input mid-infrared line at $1000\,\text{cm}^{-1}$ would therefore lead to an audiofrequency of $2\,\text{kHz}$ which is quite acceptable—in fact about as high as one would want to go with many detectors. On the other hand operating in the far infrared at $100\,\text{cm}^{-1}$ a slowly scanning mirror moving at $0.01\,\text{cm s}^{-1}$ would produce the uncomfortably low frequency of $2\,\text{Hz}$. In rapid-scan operation, therefore, one uses a broad-band amplifier and no additional modulation of the beam. The output is processed without rectification. In slow-scan operation either amplitude or phase-modulation is used to shift the ultra-low frequency signals up into the audiofrequency band where they can be amplified and processed and a PSR is used to rectify the resultant back into the low-frequency region again for interpretation. The output from the final amplifier in a rapid-scan instrument can be processed in two ways. Firstly one can pass it through a tunable audiofrequency filter/rectifier combination and thereby produce, in essence, a sequential spectrometer [291]. This approach has been used only seldom because of the loss of the multiplex advantage. It was developed originally despite this handicap because filters were cheap and computers and their ancillaries were expensive but the dramatic drop in the cost of computing hardware has taken away this consideration. Nearly all modern rapid-scan instruments digitise the interferogram and Fourier transform the result. The built-in computer gives the operator great flexibility in choosing the best regimen for a particular application. The digitising signals, which trigger the analogue/digital converter, are usually provided by means of the fringes produced from the radiation of a He/Ne laser (0·6328 μm) passing simultaneously through the interferometer. The computer memory locations can be assigned one to each sampling point and a large number of interferograms, in digital form, can therefore be averaged ("co-added" is the usual term used) to give a resultant with an acceptably high (S/N) prior to the Fourier transformation which yields the final spectrum. There is also the option of averaging spectra. The operator has almost complete control over the form of printed output from the machine. Thus he or she can select a given spectral region and expand this in abscissa or in ordinate, using interpolation programmes stored in the machine, to give smooth plots. The computer will give presentation in linear wavenumber, linear wavelength or any other chosen mode. Spectra can be presented in transmission or absorption form and absorption spectra can readily be subtracted from one another to show, for example, the weak spectra due to adsorbed species. The power of these modern instruments is most impressive. They do tend, however, to be restricted in their operation to the mid-infrared region. There are several reasons for this: firstly they were designed for this part of the spectrum which, as the "fingerprint" region, is commercially and technically predominant; secondly, far infrared spectrometers are best run evacuated to suppress water vapor absorption and this presents considerable problems since the design of a periodic interferometer usually features an air-bearing to give smooth mirror motion. Far infrared interferometers are therefore usually aperiodic since this readily permits evacuation. The latest periodic interferometers, however, can be operated in the far infrared using dry nitrogen for the air bearing and filling the whole optical path with this gas. Nevertheless they are not as good there as a specially designed aperiodic instrument. In aperiodic operation the operator has two choices. He can scan continuously (albeit slowly) and use RC integration at the end of his electronic chain or else he can scan discontinuously, i.e. hopping rapidly from each sampling point to the next but spending most of the observing time with the mirror stationary. This latter approach, especially when a digital averager is used, is the ideal method. Aperiodic instruments can use mechanical devices, i.e. stepper motors coupled to micrometers, to sense the positions where the samples should be taken or else may use optical devices to achieve the same result. Moiré fringe counters have been used [292] but for the highest precision work a subsidiary laser interferometer is often employed. A mechanical system can transport the mirror to roughly its proper position and then final adjustment can be made via an electromechanical transducer (e.g. a piezoelectric crystal) controlled by the laser interferometer [293].

The real object of infrared spectroscopy is to produce the complete variation of $\hat{n}$ with frequency or wavenumber but mostly the spectroscopist is content to produce merely a record of the variation of α. This is because (a) the absorption spectrum is more important technically than the refraction spectrum and (b) it is usually very difficult to determine n as a function of frequency. The refractive index always follows at once if one knows the optical path length equivalent to a given geometrical path length so one method of determining n has been the use of channel fringes as described in section 3.5.2. In normal spectrometers one cannot do any better because of the large number of partial beams, but in a Michelson interferometric spectrometer one can do very much better indeed because one now has only two partial beams and these are well separated from one another in space. If one puts the specimen into one of the "active" (i.e. beam-splitter to mirror) arms of the interferometer (see Fig. 4.17) then the white light fringe pattern will be shifted because of the extra optical path length introduced and it will be distorted because of the dispersion in the specimen. It is readily shown [294], however, that if one takes the full complex Fourier transform about the brightest fringe as origin then this transform contains all the information necessary to calculate $n(\tilde{v})$. It is usual to write the complex transform in the form

$$\hat{S}(\tilde{v}) = P(\tilde{v}) + iQ(\tilde{v}) \tag{4.3.19}$$

where P is the cosine transform and Q is the sine transform. For a nondispersive specimen Q would be zero for AM and P would be zero for PM but with normal dispersive specimens both are finite in both cases. The spectrum, from which $\alpha(\tilde{v})$ can be calculated, is given by

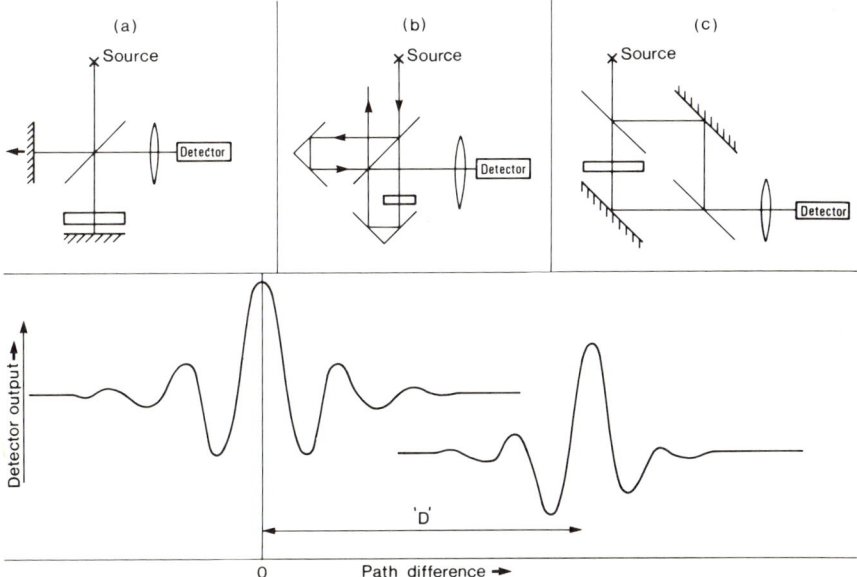

FIG. 4.17. Upper insets: three possible interferometer configurations for dispersive work, lower inset: the interferogram grand maximum plus subsidiary oscillations is both shifted and distorted when the specimen is in one of the active arms.

$$S(\tilde{v}) = |\hat{S}(\tilde{v})| = [P(\tilde{v})^2 + Q(\tilde{v})^2]^{1/2} \qquad (4.3.20)$$

and the refractive index is given by

$$n(\tilde{v}) = \bar{n} + (4\pi\tilde{v}d)^{-1}\{\arctan[Q(\tilde{v})/P(\tilde{v})] \pm m\pi\} \quad m = 0, +1 \text{ etc.} \quad (4.3.21)$$

Here d is the specimen thickness and $\bar{n}$ is the average refractive index *defined* by

$$\bar{n} = D/2d \qquad (4.3.22)$$

where D is the extra path-difference which has to be introduced in moving from the old zero path-difference position to the new one. One says *defined* above because in a dispersive medium there is, strictly speaking, no zero path-difference position; a brightest fringe in the shifted pattern is all there is. The problem is similar to that of propagation in dispersive media discussed in section (3.4.6). Equation (4.3.21) has been widely used by Chamberlain and his associates [295] to derive full complex refractive index spectra for a wide range of materials in the far infrared. This has turned out to be an important area of work with the rise in significance of near-millimetre wave technology. A comprehensive survey of the present position has been given by Birch & Parker [296].

Operation of the interferometer with the specimen in one of the active arms is usually called *Dispersive Fourier Transform Spectrometry* or DFTS (care must be taken not to confuse this with dispersive, i.e. grating spectrometry). It was originally introduced by Chamberlain and Gebbie [297], Bell [162] and others [164] just for the purpose of determining refraction spectra. However, more recently Birch, Afsar and others [298] have realised that it is *par excellence* the best way of doing *Fourier Transform Spectrometry* (or FTS). One main reason for this is that the interferogram which is produced contains *all* the information about the optical properties of the specimen and it is possible therefore to calculate absolute values of $\alpha(\tilde{v})$ without the difficulty of accounting for the interface effects [299] which plague conventional spectrometry. A second reason, and related to this, is that for specimens which are too opaque to permit measurable amounts of radiation to be transmitted, one can work in reflection and derive the full complex reflectivity $\hat{R}(\tilde{v})$ from which $\hat{n}(\tilde{v})$ is readily calculated (equation 3.5.18). Thirdly, there is no question of radiation from the specimen being modulated in the instrument thus leading to erroneous answers for $\alpha(\tilde{v})$. DFTS is usually carried out using a Michelson interferometer which when working in transmission gives the power dispersion spectrum because the beam interacts with the specimen twice (once going and once returning) and thus it is E^2 which is involved. One can, however, use either a "roof-top" Michelson interferometer (Fig. 4.17(b)) or else a Mach–Zehnder interferometer (Fig. 4.17(c)) to obtain the amplitude dispersion spectrum since the beam passes through the specimen only once [300]. In this variant approach, equations (4.3.21) and (4.3.22) require slight modifications.

Another point that is worth making about interferometric spectrometry is that the double-sided (i.e. observing from $-x_{max}$ to $+x_{max}$) operation implied in equation (4.3.14) is not strictly necessary with conventional operation. This is because the interferogram produced by a perfect interferometer is absolutely symmetrical about zero path-difference and it is therefore only necessary to observe one side of it. One has therefore single-sided operation which is much more economical of the precious limited path difference since one can arrange to have the mirror so set up that it scans from $x = 0$ to $x = 2x_{max}$. This doubles the possible resolution. The difficulty is that real interferometers do introduce distortions and, furthermore, the sampling comb will usually be misplaced so that the samples will not be taken symmetrically about $x = 0$. In double-sided operation it can readily be shown [301] that to first order these effects are cancelled but in single-sided operation they would lead to serious errors in the calculated spectrum. Several ways round this difficulty have been devised [302]. They are called "phase-correction procedures" because the incorrectly sampled interferogram can be regarded as the true one convolved with a phase error function $\phi(x)$. The phase correction procedures involve the calculation of $\phi(x)$ and the deconvolution to give the correct interferogram. It will be noted that, since the Fourier transform of a convolution is involved, one can by the convolution theorem apply phase correction before or *after* the Fourier transformation [303]. Phase correction is widely used in both rapid-scan and aperiodic interferometry and the necessary programs are usually built into the overall master program. Single-sided operation has another big advantage in that no non-linear operations like the calculation of a modulus is involved (equation 4.3.20) so there is no question of rectification of noise leading to a spurious offset in the calculated spectra. One can therefore expect that on averaging a large number of spectra one would get the correct answer. This has always been an important point in the past but recently Birch [304] has pointed out that one can considerably improve the accuracy of double-sided operation by taking the averages of the set of P's first and that of the Q's second and calculating only the modulus of the averages. This point about the effect of noise on the results of infrared spectrometry takes us on naturally to consider the effects of scanning on the output record.

4.3.5. *Spectroscopic consequences of smoothing*

With the exception of the "step-integrate-step" method just described for aperiodic interferometers, all infrared spectrometers are producing a record of something which is changing in time. Now all forms of smoothing, that is noise suppression, rely on restricting the bandwidth of the recording system and this will necessarily therefore develop a sluggishness quantified by its response time τ. It follows that the record actually produced must be different from the "true" record which would ideally be produced in noise-free circumstances by a recording system able to respond instantaneously. These remarks are general but it is helpful to analyse sequential (i.e. dispersive) instruments separately from Fourier transform instruments because the Fourier transform operation, in this connection, introduces some complications.

In sequential spectroscopy, the most obvious spectral consequence of smoothing is that if one is scanning too fast in relation to τ^{-1}, the spectral profile will be distorted. In fact the signal observed at time t will be the sum of the decaying echoes from all previous times t', that is

$$S(t) = \int_{-\infty}^{t} S(t')\exp -\left[\frac{t-t'}{\tau}\right] dt', \qquad (4.3.23)$$

This is a convolution integral and by means of the convolution theorem one may say

$$S(t) = \int_{-\infty}^{+\infty} \hat{S}'(v)\hat{T}(v)\exp(2\pi ivt)\, dt. \qquad (4.3.24)$$

In other words the transfer function modifies the spectral version of $S(t')$ (i.e. $\hat{S}'(v)$, the Fourier transform of $S(t')$) and the Fourier transform of this modified version gives the observed time-dependent signal. Obviously the distortion will only be negligible when the product of $\hat{S}'(v)$ and $\hat{T}(v)$ differs negligibly from $\hat{S}'(v)$. This will be so when τ^{-1} is greater than the maximum time frequency present in the signal. A good rule of thumb for commercial dispersive instruments [305] is that

$$[\text{Resolution (cm}^{-1})/(\text{scan speed (cm}^{-1}\,\text{s}^{-1})] = 7\tau. \qquad (4.3.25)$$

In Fourier transform spectrometry one must consider two effects, the spectral distortion introduced by the finite scanning speed and secondly the noise to be expected in the transform. The interferogram which is recorded is not the true interferogram but instead

$$I(x) = I_0(x)* R(t) \qquad (4.3.26)$$

where $I_0(x)$ is the true (ideal) interferogram and the asterisk signifies convolution. If the scan speed is too high, the interferogram will be just as severely distorted, as was the spectrum in sequential spectroscopy, but the effect on the resulting spectrum will merely be a progressive (and gentle) attenuation of the higher frequencies. Individual sharp bands may be *weaker* than one would expect but their *shape* will be hardly affected. Routine search spectroscopy can therefore be carried out with relatively fast scanning in FTS—this is another (little realised) advantage of the method.

The noise problem is at first sight quite straightforward. Fourier transformation is a process of sorting out the cosinusoids making up a given profile according to their frequency. The very act of transforming a noisy time interferogram will therefore separate all the noise, at frequencies remote from the signal, away from the signal and the signal-to-noise ratio will be determined by the strength of the signal and the noise amplitude for frequencies close to the signal frequency. Clearly this ratio does not depend on the time constant (or other smoothing) since signal and noise will be being equally attenuated. In the light of this conclusion one might ask why one uses smoothing at all in FTS!! The answer is twofold.

1. In the absence of any smoothing, the noise excursions will be very large and very expensive high-grade electronics and recording systems will be necessary to handle these with a linear response.

2. Because the interferogram has to be digitally transformed, it must be sampled at finite intervals and the resulting spectrum will be "aliassed", i.e. will be endlessly reflected and repeated in the infinite range of the abscissa variable (frequency). The high-frequency noise, which is separated from the fundamental region in which lie the signal frequencies, will promptly be returned to this region by the operation of aliassing and one will have a spectral noise power which *does* depend on the degree of high-frequency attenuation, i.e. on the time-constant. For a given time constant, the aliassing will be reduced by sampling more finely, i.e. by choosing a higher "folding" frequency. Of course if one samples a fixed length of inter-ferogram more finely one will have more samples and the improvement in S/N can be looked at in another way as a natural consequence of feeding more information into the averaging process. One can consider two extreme cases here: first where one is sampling infinitely finely (in other words one is in essence doing a continuous transform) and secondly where one is sampling very coarsely. Connes has shown [12] (see also Fleming [306]) that the ratio of the variances in the interferogram and the transformed spectrum in the low-frequency limit are respectively

fine sampling $\varepsilon_v^2 = 4Qt_m\tau\varepsilon_t^2 = 4Qt_m(P_N^e)^2,$ (4.3.27a)

coarse sampling $\varepsilon_v^2 = 2Qt_mh\varepsilon_t^2 = 2Qt_m(h/\tau)(P_N^e)^2,$ (4.3.27b)

where Q is the mean value of the weighting function, t_m the total observing time and h the sample time. In the fine sampling case the spectral noise variance does not depend on τ since $\tau\varepsilon_t^2$ is a constant (4.2.30b) and, since the signal will not be attenuated either, the S/N will likewise be independent of τ. In the coarse sampling limit ($h \gg \tau$), however, the S/N will depend on τ. This is, of course, a direct consequence of the aliassing of high-frequency noise. Normally the delay in the interferometer will be measured in terms of optical path difference x. Under these circumstances one will be interested in the spectral noise power variance in units of W^2 wave-number^{-2} rather than in $W^2 Hz^{-2}$. The conversion is readily achieved by multiplying throughout by the mirror velocity (v) squared; one then has

$$v^2\varepsilon_v^2 = \varepsilon_{\tilde{v}}^2 = 2N\Delta x^2\varepsilon_x^2$$ (4.3.28)

where N is the total number of interferogram samples.

4.3.6. *Digitisation and sampling noise*

There are two further aspects of sampling which have some bearing on the interferometric noise problem. Firstly, the digital system will necessarily have a finite lower level of discrimination—in other words the total available dynamic range is divided into a finite number of "steps". Fluctuations less than one step cannot be recorded. The spectral ordinates resulting from the Fourier transformation of these "rounded-off" ordinates will themselves not then be correct—they will contain random uncertainties which will mimic true random noise. Therefore this phenomenon is often referred to as "digitisation noise". Digitisation noise only obtrudes at high values of interferogram S/N since at low values the true random noise on the interferogram will exceed one step [307]. It is clear therefore that the desirability of using smoothing to improve the interferogram S/N comes up against an instrumental limit when the noise has fallen to a level at which the average noise fluctuation equals one step. Using further smoothing and the consequent still slower scan cannot then gain and is a waste of experimental time. The best use of observing time is secured by setting the noise level equal to one step [307] and observing several interferograms and averaging them. This can gain because the dynamic range inside the computer can be much greater than that of the primary data acquisition system [308]. Alternatively one can transform each noisy interferogram and average the spectra. Averaging interferograms is a little more difficult since, as mentioned earlier, a fiducial mark is necessary on the path-difference scale so that the set of interferograms to be averaged, each of which possibly subject to a random amount of "backlash", can be put into correct register. A subsidiary "white-light" fringe system sharing common optics can be used to uniquely locate zero path-difference and thus provide the necessary mark. When averaging spectra, there is, of course, no difficulty since the computed spectra are always, of necessity, in perfect register. With modern instruments which feature very fine ($\sim 2^{12}$, i.e. 4096) steps, digitisation noise is hardly ever encountered, unlike the pioneering instruments where, with only 256 steps, it was a constant hazard. The method then used, i.e. to put in filters to restrict the spectral band pass and thus increase the interferogram oscillations, is no longer so necessary but occasionally, even today, for the most accurate work it may be desirable.

The second type of "noise" arises from the presence of random jitter on the positions where the interferogram samples are taken. Ideally, the samples should all be equispaced, but in mechanical and some optical systems random sampling errors easily arise. When this happens, each incorrectly taken sample will contribute to the spectrum a cosine wave whose phase will not be right. The total sum of all the cosine waves will therefore contain random errors distributed throughout the spectral range. This type of spectral ordinate uncertainty is referred to as "sampling noise". The effects of sampling noise get more obvious as the resolution increases and especially so as the frequency increases [12]. Thus, whereas it may be tolerated for moderate resolution work in the far infrared, it is not tolerable at any resolution in the mid infrared. This is one of the more fundamental reasons for the use of laser-referenced sampling in all rapid-scan mid-infrared instruments.

4.3.7 *The use of detector arrays in infrared spectrometry*

Traditionally in an infrared spectrometer there is a single detector which gives out a single electrical signal. This situation is rapidly changing, however, due to the development and introduction of detector arrays. These arrays, either linear or rectangular in format, were originally made for use in infrared imaging (section 7.1) but they can be used just as well in infrared spectrometry and their availability is opening up some exciting new possibilities. A single detector with its finite aperture is limited to a low resolution account of spatial variations of intensity and can give that information only at a fixed moment in time. An array of smaller detectors of the same total area can give either much higher spatial resolution or else with *T*ime-*D*elay-and-*I*ntegrate (or TDI) circuitry can give information on how signals at different spectral frequencies are varying in time. The arrays are nearly always composed of identical semiconductor detectors either photoconductive or photovoltaic (section 6.5). The original versions were made by assembling the individual microdetectors by hand into the array format but all modern versions are made by the integrated circuitry fabrication techniques which have been evolved for the microelectronics industry. They tend, therefore, to be available only in those materials for which the advanced fabrication technology exists. Silicon photodiode arrays are thus readily available and, since they respond from 0.4 to 1.1 μm, they are widely used in spectroscopy. Examples would include, for example, automatic Raman and fluorescence spectrometers. Much work is currently being devoted to mercury-cadmium-telluride and lead-tin-telluride arrays for use in the technically important 8–16 μm atmospheric window [309]. Clearly other materials (in particular pyroelectrics such as TGS) will soon become available and one can anticipate a situation where detector arrays will be available for most of the infrared.

Some of the possible applications of arrays in infrared spectrometry being investigated at the moment would include the following.

a. In dispersive spectrometry where a linear array can be used as a sort of electronic photographic plate. Each individual detecting element feeds a corresponding channel of a multichannel analyser (or MCA) and the MCA can either just record the raw data or else process it by convolution procedures to give a high-grade continuous spectrum. In this way the need for mechanical scanning is avoided and in addition the multiplex advantage is regained.

b. Again, in dispersive spectrometry one can use scanning of the spectrum over the linear array but this time invoke TDI techniques to obtain a good quality spectrum with enhanced S/N. If the spectrum is varying in time, repeated scans with TDI enable plots of the variations of power at a set of chosen frequencies to be automatically recorded.

c. In an interferometer, larger apertures can be used without having to sacrifice resolution. The reason for this is that a two-dimensional array can detect the presence of more than one ring within the "detector" aperture and can therefore fully restore the contrast of the ring pattern. This contrast would, of course, be seriously degraded if one were obliged to use a single detector of the same equivalent aperture, since of necessity the detector would be unable to sense the variations in intensity across its aperture. This approach leads to the so-called "superthroughput" interferometer [310].

Arrays are undergoing intense development under the stimulus of large military contracts so the spectroscopist can look forward to still further applications, especially since much larger arrays will become available—those at the moment have about 100 in a two-dimensional layout or about 256 in a linear—and because the arrays plus their automatic addressing circuits go very naturally with computer control and computer data processing.

4.4 The measurement and detection of radiant infrared flux

4.4.1. *Basic definitions*

At a fundamental level there is no conceptual distinction between the measurement and the detection of radiant flux, but at the practical level this distinction is useful. Thus one might imagine three very different experimental situations.

a. In a national standards laboratory where the experimentalists are trying to determine the intensity to high precision in absolute terms, i.e. in $W\,m^{-2}$. This sort of experiment would be called *absolute radiometry*.

b. In an experimental physics or chemistry laboratory where the experimentalists are trying to determine merely the *ratio* of one radiant intensity to another—this would arise, for example, in two-beam spectro-radiometry and would be called *relative radiometry*.

c. In a research laboratory working at the frontier of the art where the experimentalists would be content if they could merely demonstrate the presence of either a very weak radiant field or else of the slight attenuation of a strong beam is traversing a weak absorber. This might be called *state-of-the-art radiometry*

All discussion of measurement leads at once to the properties of the detector, since the rest of the equipment can be assumed to be of known performance. Absolute radiometry requires a detector which is not only linear but whose output is calibrated in absolute terms, for example $V\,W^{-1}$. Relative radiometry merely requires that the detector be linear over its range of use. State-of-the-art radiometry may in extreme cases make no demands on the linearity of the detector. If that were all there were to it, everyone would choose an absolute radiometer every time, but as is implied in the above, these three types of detector tend to work in different ranges of intensity. True absolute radiometers work down to only about the milliwatt level. This can be extended by the use of secondary standard radiometers, i.e. ones calibrated against a true primary standard, down to about the microwatt level, but it is unsafe to trust them any further. Relative radiometers are available with a sensitivity that extends down to the nanowatt level and beyond, but state-of-the-art radiometers can extend this to the femtowatt level and possibly even lower. Absolute radiometers, because of their mode of construction, would be expected to be stable in time but the sensitivities of most other detectors, and especially the most sensitive ones, would be expected to be constant only over short periods.

4.4.2. *Absolute radiometry*

Making a true absolute radiometer is a very difficult task because one has to compare an unknown radiant power with a known standard power under strictly equivalent conditions. Two different approaches have emerged. The first relies essentially on calorimetry where one measures the heat rise due to the absorption of the incident radiant power. The second relies on a previous calibration of the radiometer in terms of a source of known radiant power, e.g. a black body. The first approach is sometimes thought to be more "absolute" than the second since it goes back to measurement first principles and makes no theoretical assumptions, but in actual practice there is not much difference between the two and it is best to regard both types as equally "absolute".

In the calorimetric approach, one measures the amount of heat produced when the radiation is absorbed in a thermally isolated black receiver. One could do this by having a sensitive absolute thermometer buried in the receiver and then, if one knew the thermal capacity of the receiver, one would have the magnitude of the incoming power directly. For several reasons of convenience this simple positive approach is replaced by a null technique in which the amount of electric power required to produce the same temperature rise is measured. The incoming beam is square wave modulated by a mechanical chopper and, at the same time, a heater buried in the receiving block receives a zero-based square-wave voltage exactly π out of phase with the radiation. A thermometer also buried in the block has its output connected to an AC amplifier which feeds a PSR triggered by a monitor signal from the voltage drive. When the electric power is less than the radiant power one will obtain, say, a positive DC voltage out of the PSR and when it is greater the DC output will be negative so the electrical power is adjusted until the PSR gives a zero output. One then has an exact measure of the incoming radiant power. This method is attractive because a power is being compared with a power and because absolute thermometry is not necessary, but perhaps most of all because one is going back directly to the absolute standards of voltage and current. The difficulties with the method are of two kinds, particular and general. The main particular problem is that one needs to have a receiving block, a heater and a thermometer which respond reasonably quickly so that acceptably high AC modulation frequencies can be used. In practice it becomes very difficult to work at more than a few Hz. The general problems are (a) ensuring that the receiving block is truly black, (b) ensuring that the electrical heating and the radiation heating are truly equivalent, i.e. that they act in the same place in the block, and (c) ensuring that the receiver and the radiation field are properly matched to one another. These two sets of problems are far from trivial and their combined effects limit absolute radiometry at the moment to a precision not much better than a few per cent. The main advantages of this form of absolute radiometry are that it is frequency independent so one can work from the visible down to the millimeter-wave regions, and that it can be used perfectly well with either incoherent broadband radiation or continuous-wave coherent laser radiation. A modern form of this radiometer, due to Blaney [311], suitable for measuring the power from continuous-wave far-infrared lasers is shown in Fig. 4.18(a). In this device, the radiation is absorbed in a thin (< 10 nm) nichrome metal film which also acts as the electrical heater. The absorbing metal film rests on a soda-lime-silicate glass base which is itself connected soundly to a copper disc which carries the 25 or so thermocouples. The cold ends of the thermocouples are buried in a substantial heat sink. This instrument has been shown to have a proven precision of better than 15% and to work well down to power levels of about 1 mW.

It might be thought, at first, that a metal film would be a poor absorber because, from an optical viewpoint, it will have a high reflectivity. However, provided the film is thin enough, this optical argument fails and the film can absorb quite strongly. The reason for this is best discussed from an electromagnetic viewpoint, where one replaces the interfaces—that is abrupt changes of complex refractive index—of the optical engineer by the impedance mismatches of the electrical engineer. From this point of view, the incoming beam propagating in free space with impedance 376·7 Ω (equation 1.2.15) would suffer no reflection at all provided the normal impedance of the film was also 376·7. At infrared and longer wavelengths, the impedance of the film derives mostly from the conductivity term so one can calculate it in terms of the DC resistivity. Taking a metal film square of side l and thickness d, the resistance in the film and parallel to one of the edges will be given by

$$R = \frac{\rho l}{l d} = \rho / d. \tag{4.4.1}$$

For no reflection this has to be 376·7 Ω. Taking therefore a fairly resistive metal such as nichrome with specific resistivity ρ equal to $10^{-4}\,\Omega\,\text{cm}$, it follows that the film thickness has to be about 3 nm. It is interesting to note that R is independent of l, that is of the actual size of the square. For this reason, the electromagnetic impedance of a metal film is often quoted as "so many ohms per square" or simply as $\Omega/\square$. With the matching condition satisfied, the electromagnetic energy will flow undisturbed across the interface but then suffer heavy attenuation within the metal film. The film will, not of course, be thick enough for total absorption and more detailed calculations reveal that the optimum absorption occurs not for the $376{\cdot}7\,\Omega$ condition but for exactly half of it, i.e. $188{\cdot}4\,\Omega$. An ideal film with this impedance would reflect 25 %, absorb 50 % and transmit 25 % of the incident power. Performance approaching this ideal is often achieved in practice [311].

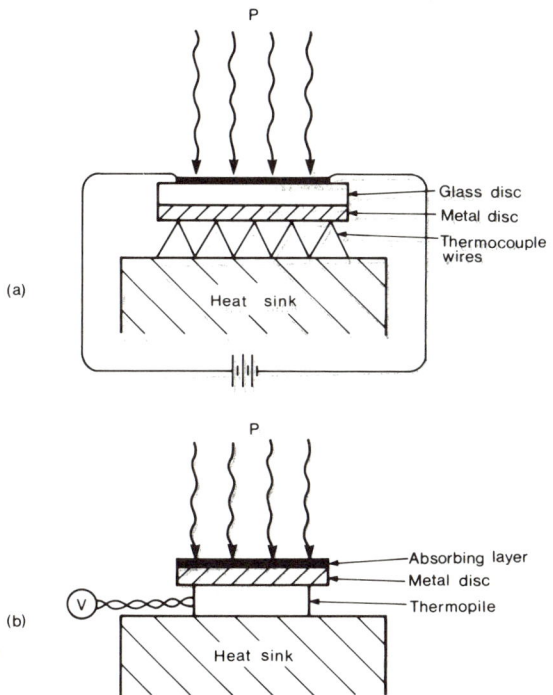

FIG. 4.18. Two forms of infrared absolute radiometer. In (a) the incoming radiant energy is determined by comparison with electrically generated energy. In (b) the voltage output of the radiometer has been previously calibrated against an absolute source.

The second type of absolute radiometer is available commercially (for example the range offered by Laser Instrumentation). A typical example is shown in Fig. 4.18(b). The incident radiation is absorbed in a 10 mm diameter flat metal disc coated with a black paint which has good absorbing properties [312] in the specified wavelength range (usually 0·3–15 μm). The disc is attached to and supported by thermocouple junctions and wires. The performance of this type of radiometer falls off as one goes to longer wavelengths because the "black" paint starts to become transparent. Blaney [311] has shown that one can get round this, to a large extent simply by using a thicker paint layer or else by incorporating microscopic glass beads into the paint layer. At these long wavelengths, the paint layer behaves like a homogeneous dielectric so one has only specular reflection to worry about and this can readily be measured. One can thus make a final correction [311] by writing

$$\mathscr{R}_{lw} = \mathscr{R}_{sw}(1 - R_{lw})/(1 - R_{sw}),\qquad(4.4.2)$$

where $\mathscr{R}_{lw}$ and $\mathscr{R}_{sw}$ are the reponsivities at long and short wavelengths respectively and R_{lw} and R_{sw} are the corresponding power specular reflectivities.

One of the more serious potential sources of systematic error in absolute radiometry is the difficulty of getting a good match of the detector receiving geometry to the actual field distribution. This is equivalent to requiring the antenna pattern, of the detector and its ancillaries, to intercept virtually all the incoming energy. This is not too difficult to achieve if (a) one is working at short wavelengths or (b) one is carrying out total power measurements. The difficulties arise when one is trying to do spectroradiometry at long wavelengths. It is rather difficult to calculate the antenna pattern of, say, a Michelson interferometer so only two possible approaches remain. A replica experiment with scaled up apparatus and microwave sources can be done or an attempt at calibration can be made. The first approach is attractive and powerful but, alas, also very costly, so it tends to be reserved only for those tasks which the Michelson interferometer alone can do—for example modelling the performance of sub-millimeter heterodyne receivers. "Absolute" spectroradiometry therefore rests on setting up a black-body source as geometrically equivalent to the source to be measured as it is possible to ensure, and then, by running it at a known temperature, one has what is hoped is an absolute source coupled to an absolute receiver and the effects of the antenna pattern can either be inferred or else (and more usually) compensated for by taking a ratio. Absolute spectroradiometry in the far infrared would usually be carried out on sources, lasers or plasmas for example, where some measure of coherence would be the norm. The virtual absence of coherence in a black-body calibration source is a great help therefore. As usual there are problems. The black-body source would have to be run hot and the black paints so useful for the receivers would burn away. The classic black body based on a cavity lined with glass or metal is therefore used but temperatures beyond 500 K present difficulties. The cavity also has a small aperture so it is not appropriate for matching a high-aperture test source. In this case a flat glass plate can be used. This is grey [313] rather than truly black, but it will give some sort of an answer. Another possibility is to use mode scramblers to smooth out the antenna pattern. This approach is very useful when dealing with a high-coherence source, a laser for example, and the mode scrambler can take the very simple form of a diffuser made from a ball of wire wool.

Secondary standard radiometers can take relatively simple forms—a pyroelectric detector (section 4.4.3) for example. The accuracy available is determined mostly by how well the calibration against the primary standard is carried out. This calibration is done by placing the detector and the absolute radiometer alternately into the beam as the power level is increased. A plot of detector reading versus radiometer reading will then be a reliable calibration curve. The main difficulty is that the secondary standards are usually used in a lower range of power than are the primary standards, so an absolute attenuator is also required if the calibration is to be trustworthy. An absolute attenuator—that is a device with a performance that can be calculated exactly from first principles—is not the easiest thing to make, but one attractive possibility is based on the concept of frustrated total reflection. The phenomena which take place when a ray, travelling in a dense medium, strikes an interface with a less dense medium at an angle greater than the critical angle were discussed in section 3.7.2. From this discussion, it will be seen that if the beam is suffering total reflection at the lower boundary of a block of transparent material (Fig. 4.19) then when a block of identical material is brought up smoothly towards the first until eventually the two blocks are in contact, then one must pass continuously from a situation where on average *no* energy flows across the interface to one where *all* the energy does so [314]. The reflection at the interface will therefore pass from 100 % to zero as the second block approaches and one will have a wide-range attenuator of calculable performance.

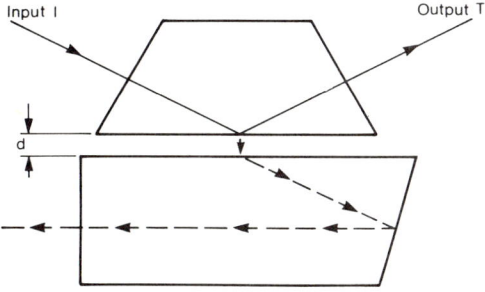

Fig. 4.19. Calculable absolute attenuator based on the principle of frustrated total reflection. The power in the beam T falls from 100 % to zero as the distance d goes from a few times λ to zero.

Attenuators based on frustrated total reflection are relatively easy to make and use in the infrared because the range of prism separation, about three times the wavelength, is a convenient sort of quantity for normal workshop techniques [314]. The disadvantage, as usual, stems from materials problems. These are particularly severe in the far infrared where non-absorbing materials are just not available. However, the losses caused by reflection at the other interfaces and by absorption in the prism material are constant, so it is still possible to use the attenuator successfully but some care is needed. Oddly enough, absorption in the prism material can sometimes be welcome. This is because it is often possible in the infrared to find two materials which hardly differ in their refractive indices but which differ markedly in their absorption coefficients. Thus if one were to chose a virtually transparent material for the first prism and a much more absorbing material for the second, the frustrated beam would be rapidly attenuated and there would be no possibility of its leaking back and either giving a false reading in the detector channel or else getting back into a laser cavity and thereby disturbing the operation of the laser.

It is important, for many applications, to determine the linearity of secondary standard radiometers not just because they are commonly used at lower power levels than the primary type but because they are commonly exposed to very wide ranges of input power. There is thus always the risk of their being driven into a non-linear region. It is also desirable to calibrate their spectral variance of sensitivity, though this may not be essential if the detector is only to be used for a specific purpose like measuring the output of a given type of laser. Secondary standard radiometers are widely used to measure the output from pulsed lasers. This is always a difficult measurement and one prone to error, especially when the duty cycle (or mark-to-space ratio) becomes small. Usually the detector is used merely as a Joulemeter since it may not be possible to follow the pulse envelope. If true power measurements are required, then two detectors may be used. A slow but reliable radiometer to measure the *energy* in the pulse and a fast if less reliable detector to measure its *duration*. It must be stressed again, however, that this is a difficult field and the optimistic claims of enthusiastic laser engineers should often be taken with a large grain of salt!

4.4.3 *Relative radiometry—infrared detectors* [315, 316, 317, 318]

In the final analysis, all infrared detectors are quantum detectors since the beam to which they react is made up of quanta, but it, nevertheless, makes sense to divide the wide range of sensitive detectors which is available, often commercially, into two broad categories. These are (a) *thermal* detectors which operate by sensing the increase in temperature brought about by the flow of radiant energy and (b) *quantum* detectors which operate by sensing the promotion of entities (usually electrons) into higher energy levels by the absorption of the incident quanta.

Thermal detectors have been used throughout the history of infrared research. The oldest type, and one still occasionally used in commercial equipment, is the thermopile which consists of a set of thermocouples arranged in series so as to multiply the individual voltage up to a usable level. Copper/constantan couples are still used but nowadays several other combinations are available. Thermopiles have the advantages of being cheap, robust and reliable but the disadvantages of being relatively insensitive and of having long response times. Bolometers, which are essentially radiation resistance thermometers, are also of respectable antiquity. They can be very sensitive and can be made to have short response times but they need also to be operated at cryogenic temperatures which tends to restrict their commercial application. For commercial work a room-temperature detector is required and the two leading candidates are the Golay cell and the pyroelectric detector. The Golay cell [319] features a sealed gas-filled cell in which the incident radiation is absorbed on a blackened surface. The heat produced is transferred to the gas which expands and distends a flexible membrane which forms part of the cell walls. The motion of the membrane is then sensed by an optical lever and a photocell. The whole arrangement is therefore rather delicate and of necessity is slow to respond but the sensitivity is good. Golay cells are therefore used in vibration-free situations where relatively low (≈ 16 Hz) modulation frequencies are acceptable. The pyroelectric detector depends on the phenomenon that certain non-centrosymmetric crystals [320], such as triglycine sulphate (or TGS), when cut in an appropriate way develop surface charges on the exposed faces and these surface charges vary as the temperature changes. This is the pyroelectric effect [321]. If electrodes are placed on these faces and the electrodes connected to an external potential circuit, then the crystal becomes a thermal detector since variations in the incident intensity will produce matching variations in the voltage developed in the external circuit. Like the Golay cell, the pyroelectric detector is used with an inbuilt preamplifier so that the voltage output will be at a reasonably high level and noise pick-up problems will be minimised. There are several possible materials for the detecting element; thus TGS [322], lead zirconate titanate [323] and even crystalline polymers such as polyvinylidene fluoride [324] (PVF_2) have all been used but TGS technology is the most advanced. These detectors have usually to be "poled", that is polarised with a DC high voltage at an elevated temperature and then cooled to ambient temperature with the field still on and this can be a nuisance if by mischance the element becomes de-poled. However, for TGS it is possible to introduce small amounts of alanine [325] and this leads to a crystal which does not spontaneously depole. The best TGS detectors approach the Golay cell in sensitivity but they are very microphonic. Their great advantage—which makes them almost irreplaceable—is that they have short response times. TGS detectors have been used with modulation frequencies up to 10^5 Hz and specially fabricated ones made from $LiTaO_3$ have been shown to have subnanosecond capabilities [326]. For rapid-scan interferometry, a high-speed detector is essential and all currently available commercial instruments feature a pyroelectric detector.

Thermal detectors have many features in common—thus they are inherently broad-band devices and all that is required to make the detector respond at a given wavelength is to ensure that its absorbing surface is "black" at that wavelength. Much work has gone into making broad-band "black" infrared paints and most thermal detectors have essentially constant response across the whole infrared. Golay cells, for example, have been used at wavenumbers less than 5 cm^{-1}. Pyroelectric crystals tend to absorb strongly due to internal modes and to reststrahlen vibrations so the crystal can be used as it is for the mid infrared. Some blackening of the receiving surface may be necessary, however, if the detector is to be used in the near and far infrared [327]. The analysis of the efficiency of thermal detectors is also very similar for the various types so we shall just briefly describe that for the bolometer [328]. The detecting system is shown schematically in Fig. 4.20.

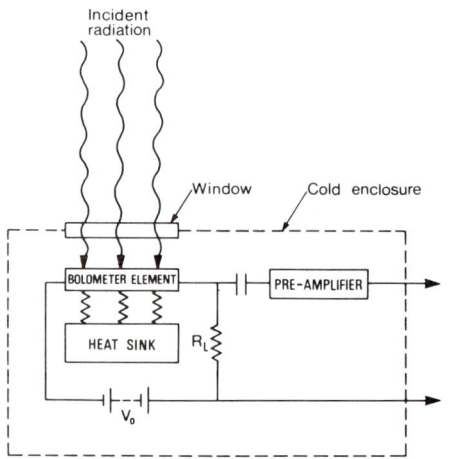

FIG. 4.20. Schematic of the arrangement of a bolometer detector.

Suppose the incident intensity I falls on a bolometer element of area A, thermal capacity C (Joules per degree) electrical resistance R_B and emissivity ε. The element is connected to a heat sink and also loses heat to its surroundings by reradiation. At a temperature differential ΔT with respect to ambient, its rate of heat loss is given by $G\Delta T$. We now suppose that with the element in thermal equilibrium with its surroundings unmodulated radiation power is suddenly switched on. The subsequent temperature change will be given by the equation

$$C\frac{d(\Delta T)}{dt} = \varepsilon I A - G\Delta T, \qquad (4.4.1)$$

the solution of which is

$$\Delta T = (\varepsilon I A/G)[1 - \exp(-t/\tau)], \qquad (4.4.2)$$

where τ, the detector time constant, is given by

$$\tau = C/G. \qquad (4.4.3)$$

Thus for high sensitivity, that is a large ΔT, we should have a high emissivity (i.e. a black detector), a large area and a low heat conductance. However, a low heat conductance will from (4.4.3) mean a sluggish response so a compromise is indicated. The heat capacity only enters into the expression for τ so the smaller is C the better. Bolometer elements are therefore usually made in the form of a film so as to maximise A in comparison with C. It is relatively easy to make a bolometer element with $C \approx 10^{-6}\,\mathrm{J\,K^{-1}}$ and $G \approx 10^{-4}\,\mathrm{W\,K^{-1}}$ so τ values of around 10^{-2} are readily available. To get shorter response times calls for great ingenuity in the construction. Clarke [329] for instance has described some very intricate bolometer elements.

The output voltage change of the bolometer when the detecting element changes its resistance by ΔR will be

$$\Delta V = \frac{V_0 R_L}{R_B^0 + R_L} - \frac{V_0 R_L}{R_B^0 + \Delta R + R_L} \approx \frac{V_0 R_L \Delta R}{(R_B^0 + R_L)^2}. \qquad (4.4.4)$$

The change in the resistance of the element will be related to the temperature change ΔT by the usual linear relationship

$$R_B = R_B^0(1 + \alpha \Delta T), \qquad (4.4.5a)$$

i.e.

$$\Delta R = R_B^0 \alpha \Delta T, \qquad (4.4.5b)$$

so, by substitution of (4.4.5b) in (4.4.4), one has

$$\Delta V = \frac{V_0 R_L R_B^0 \alpha \Delta T}{(R_B^0 + R_L)^2}. \qquad (4.4.6a)$$

To maximise ΔV, one therefore looks for a large R_B^0, that is a high resistance element and for a high α, that is for a high temperature coefficient of resistance. Film construction helps with obtaining a large R_B^0 but nevertheless metals, such as platinum, which were widely used in the past, are no longer attractive since semiconductors, e.g. germanium, are intrinsically higher resistance materials and, moreover, have much larger α values. Metals have small positive values of α, whereas semiconductors have larger negative values, but since it is only the magnitude of α which matters, the semiconductors win out. Semiconductor bolometers are sometimes called thermistors. For all types of bolometer, one will always have $R_L \gg R_B^0$, so equation (4.4.6a) is usually written in the approximate form

$$\Delta V = V_0 R_B^0 \alpha \Delta T / R_L. \qquad (4.4.6b)$$

The incoming radiation will nearly always be modulated so we need the equivalent of (4.4.2) for the case where I varies periodically with time. The mathematics is very similar to that used for analysing smoothing networks (section 4.1.4). By analogy we may write

$$\Delta T = (\varepsilon I_0 A/G)\left[1 + (2\pi v\tau)^2\right]^{-\frac{1}{2}} \tag{4.4.7}$$

Assuming, as one normally may, that $R_L \gg R_B$, the voltage responsivity, that is the voltage output per unit incident power per unit area, is then

$$\mathscr{R} = (\Delta V/I_0 A) = (\varepsilon V_0 R_0 \alpha/R_L G)\left[1 + (2\pi v\tau)^2\right]^{-\frac{1}{2}}. \tag{4.4.8}$$

The bolometer will usually only be used for frequencies much less than τ^{-1} so the second factor on the RHS of (4.4.8) can usually be set equal to unity.

With this assumption and substituting from (4.4.6b), one has that the responsivity, i.e. the voltage output per unit incident intensity per unit area, is given by

$$\mathscr{R} = \left(\frac{\Delta V}{I_0 A}\right) = \frac{\varepsilon V_0 R_B^0 \alpha}{R_L G}. \tag{4.4.9}$$

V_0 is not really an adjustable parameter since it is desired to avoid electrical heating of the bolometer element but otherwise, from (4.4.9) a high ε, a high R_B^0, a high α and a small G are sought. Putting in some appropriate values into (4.4.9) gives the result that the voltage responsivity will be of the order $10\,\text{V W}^{-1}$ for a reasonably good bolometer and $100\,\text{V W}^{-1}$ for a very good one.

However, it is not the voltage responsivity alone which determines the quality of a detector. The true measure, as mentioned several times previously, is the SNR. In the detector context one nearly always interprets this in terms of the *Noise Equivalent Power* (or NEP) which is that incident power which will lead to a SNR of unity with unit bandwidth in the electronic processing chain. It is not possible to operate bolometers in the Johnson noise limit since, of necessity, the element has to be in thermal equilibrium with its environment. What is needed, therefore, is a thermodynamic analysis of the fluctuations in the radiative exchanges. Such analyses have been given by several authors [4, 235]. The final result is

$$\text{NEP} = (16\sigma kT^5 AB/\varepsilon)^{\frac{1}{2}}, \tag{4.4.10}$$

where B is the bandwidth of the post-detector electronics. In numerical terms, for $A = 1\,\text{cm}^2$, $B = 1\,\text{Hz}$ and $\varepsilon = 1$, one has

$$\text{NEP} = 3.54 \times 10^{-17}\, T^{5/2}. \tag{4.4.11}$$

Thus for room temperature operation one would have a noise-equivalent-power of $5.5 \times 10^{-11}\,\text{W}$. From this analysis, it follows that lowering the detector element temperature is only effective provided one also lowers the temperature of its surroundings. All cooled bolometers therefore feature cold screens to prevent ambient thermal radiation from reaching the detecting element. Unfortunately, one has to have a hole in these screens so that the radiation which is to be detected can get through! Unless one is careful, "warm" radiation will also get through. Cooled filters which pass only the desired spectral pass band can be very effective here when used over the entrance aperture. With the whole detector and its screens at liquid helium temperature (4 K) the bolometer would have a limiting NEP of $1 \cdot 1 \times 10^{-15}$ W, but it is very difficult to achieve such a figure in practice because other sources of noise, for example heat conduction along the electrical connections, tend to obtrude. Cooling is nevertheless very effective and bolometers have been used at very low temperatures indeed. Thus Sievers [330] has operated one at a temperature of $1 \cdot 3$ K achieved by pumping on liquid ^{3}He and Chanin *et al.* [331] have described a bolometer operating below 1 K. They find that for $T = 0 \cdot 1$ K, the best NEP is $4 \cdot 5 \times 10^{-17}$ W which is rather worse than one might expect from (4.4.11) but which nevertheless represents an extremely sensitive detector. Britt and Richards [332] have described the use of an adiabatic demagnetisation refrigerator based on manganese ammonium sulphate which permits temperatures of $0 \cdot 3$ K to be maintained for periods of up to 60 hours. A bolometer cooled in this way would be a very attractive proposition for infrared astronomy from space stations.

Noise equivalent power is a convenient way of assessing detector performance but it has three drawbacks. These are (a) it is an inverse sort of quantity which gets smaller as the detector performance improves, (b) there is no explicit mention of the detector area in its definition so comparison of similar detectors with different areas becomes an uncertain matter, (c) there is no indication of how the detector performance will vary as its bandwidth is altered. The first point is readily dealt with by introducing the detectivity $D(\lambda)$, defined by

$$D(\lambda) = (\text{NEP}(\lambda))^{-1}. \qquad (4.4.12)$$

The second and third points require that some hypothesis be introduced which will indicate how $D(\lambda)$ will vary with area A and bandwidth B. The variation with A can be derived by means of a thought experiment in which one imagines the actual receiving surface divided up into a large number of elemental smaller area detectors all arranged either in series or in parallel. From the fundamental noise theorem (see, for example, equation 4.2.20) of incoherent addition, it readily follows that

$$(V_s/V_n) = A^{\frac{1}{2}}(\Delta V_s/\Delta V_n), \qquad (4.4.13)$$

where V_s and V_n are the signal and the noise voltages respectively and the Δs indicate these quantities for the elemental detectors. One next observes that the definition of NEP can be expressed as

$$(\text{NEP}) = IA/(V_s/V_n) \qquad (4.4.14)$$

and thus derives the result in agreement with (4.4.10) that

$$(\text{NEP}) \sim A^{\frac{1}{2}}. \qquad (4.4.15)$$

The detector noise will be expected to vary as the square root of the bandwidth (q.v. equation 4.2.31) so it follows that

$$(\text{NEP}) \sim B^{\frac{1}{2}}. \tag{4.4.16}$$

Equations (4.4.12), (4.4.15) and (4.4.16) can be combined to give the overall result

$$D(\lambda) \sim (AB)^{-\frac{1}{2}}. \tag{4.4.17}$$

This at once suggests a way of defining a more universal figure of merit for detectivity, since one can write

$$D^*(\lambda) = (AB)^{\frac{1}{2}} D(\lambda). \tag{4.4.18}$$

The quantity on the LHS of (4.4.18), which is usually called "D-star", is widely used [333] to compare practical detectors. Its use does, however, presume that the areal dependence of (NEP) is correctly given by (4.4.15). This assumption is generally true but there are exceptions. From (4.4.18) it follows that the dimensions of D-star are $W^{-1} L (Hz)^{\frac{1}{2}}$ but it is most important to notice that, rather unfortunately, most detector constructors use the centimetre for the length dimension. In general, thermal detectors have smaller ($\sim 10^9$) D-star values than do photon detectors ($\sim 10^{11}$), but this does not reflect any fundamental inferiority, only that thermal detectors must of necessity have worse noise figures because they respond to *all* spectral frequencies. Photon detectors, as will emerge later, are only sensitive in restricted spectral ranges and so they have better noise figures. A thermal detector used with effective cold screens and a matching cold filter might be expected to give the same performance as the equivalent photon detector but, if response over a narrow range is all that is required, it is usually easier and often cheaper to use the photon detector.

Most of the commonly available available photon detectors operate on the basis of the photoelectric effect in which the absorption of an incident photon frees an electron from its bound state so that it is then able to contribute to the current. There are two distinct types of photoelectric effect, *external* when the electron is freed from the bulk of the material and liberated into the surrounding vacuum and *internal* when the electron stays within the material but is transferred to a conduction band (see section 6.5). The external type does require that the energy of the incident photon be at least equal to the work function and, since this will be about 1 eV or greater, external photo-detectors tend to be restricted to the visible and ultraviolet regions and are rather rare in the infrared. Internal ionisation into the conduction band, on the other hand, commonly involves near- to mid-infrared photons for intrinsic material and even quite long wave far-infrared photons for the extrinsic type. Photoconductors therefore tend to be concentrated in the infrared and in fact provide some of the best available infrared detectors. The sensitivity of external photodetectors can be enormously increased by using them as the primary photocathode in a photomultiplier [334]. The initial photoelectrons are attracted to a so-called dynode which has positive potential relative to the photocathode. The electrons are accelerated by the dynode and on collision with it produce a larger number of secondary electrons which in their turn are attracted to a second dynode and so on. Commercial photomultipliers often have as many as ten dynodes so the initial photocurrent can be enormously enhanced. The only photocathode which has seen significant use in the infrared is the so-called S1 type which is based on a silver-oxygen-caesium formulation. This has a maximum response at 0·8 μm and is still usable as far out as 1·1 μm. The chief source of noise in a photomultiplier is thermionic emission, principally from the cathode. This leads to a so-called "dark current" and it is the fluctuations in the dark current which contribute to the noise. Because of the mode of construction of the photomultiplier this noise is close to pure shot noise and it can therefore be diminished by reducing the dark current. Cooling the photocathode is an effective way of achieving this and it is particularly important in the infrared where, because the work function is low, room temperature thermionic emission can be significant. With S1 photo-cathodes it is found that the quantum efficiency starts to fall when the temperature becomes too low, so from an S/N point of view there is an optimum temperature. This is commonly in the region of − 30°C (240 K).

Photomultipliers, like all real photon detectors, show a far from ideal absolute (i.e. Volts output per Watt incident) response. The basic reason for this is that at most one photoelectron per photon is absorbed regardless of the frequency of the photon. Thus, since for a constant beam intensity the number of photons varies inversely with the frequency, it follows that the detector sensitivity will vary in the same way. The sensitivity therefore rises linearly with the wavelength until the region of the wavelength λ_{WF} corresponding to the work function energy is reached, after which there is very rapid fall-off. There is consequently a sharp peak in response in the neighbourhood of λ_{WF}. The analysis for photoconductors is the same, except that it is the band-gap energy E_g and its equivalent wavelength λ_g which are involved. The detecting element may have to be in vacuum as in photomultipliers, so there will have to be a window and its spectral characteristics will be impressed on the overall response curve. Also, for semiconductors, there is an enormous increase in reflectivity for wavelengths shorter than λ_g and this will lead to an impaired efficiency. However, by choice of suitable windows, and, if necessary, antireflection coatings, it is always possible to tailor the detector to give the desired response. The designer of infrared photoconductors is also well placed in that a wide range of suitable intrinsic semi-conductors is available for the shorter wavelengths and an even wider range of doped (section 6.5) materials for the longer wavelengths. Some typical response curves are shown in the lower inset of Fig. 4.21. The equivalent electrical circuit is shown schematically in the upper inset. The curve A in the lower inset represents the response of a purely hypothetical Background Limited Intrinsic Photoconductor (or BLIP) at 295 K. It is ideal in the sense that it has a wavelength independent responsivity and the only source of noise is the radiation arriving from the warm (295 K) surroundings. It is interesting to observe that several practical photoconductors approach the BLIP condition at the wavelengths of their peak response.

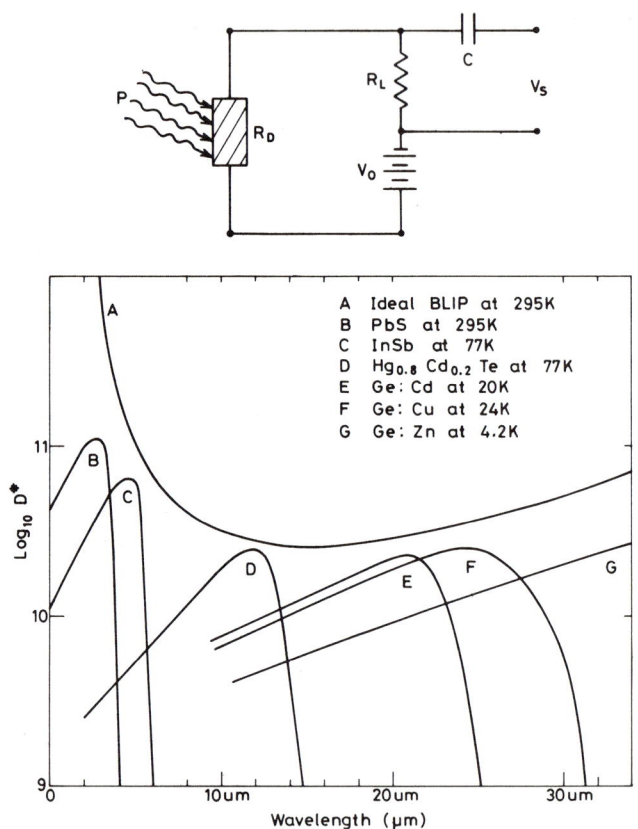

FIG. 4.21. Schematic electrical circuit (upper inset) and some typical response curves for some selected photon detectors.

4.4.4 *"State-of-the-art" infrared detectors*

4.4.4.1 *Introduction*

A state-of-the-art detector may be defined as one which does something which is either beyond the reach of a conventional detector or else so close to its limits that the resulting performance would be unacceptable. For example, if one were looking for hypersensitivity, even the best bolometers or photodetectors might not be adequate and one might have to use a superconducting transition bolometer or else a heterodyne detector. If ultra-high speed of response were necessary, then a photon-drag detector would be required and if the measurement of extremely low absorptances was being contemplated then it would be necessary to use photoacoustic detection. Obviously all of these detectors have drawbacks since otherwise they would always be preferred, but when they are used for those tasks which suit them best they can give quite outstanding performance.

4.4.4.2 Superconducting transition bolometers

The superconducting transition bolometer was first successfully operated by Martin and his colleagues in 1961 [335] but since then several groups have developed working versions [336, 329]. The detecting element is a metal such as lead or aluminium which is held exactly at its normal/superconducting transition temperature. Any small rise in temperature will thus produce a large change in the resistance of the element and one will have, at least in principle, a very high sensitivity. Practical difficulties make the construction and operation of this type of detector rather tricky so it is not used routinely, but for measurements on incoherent radiation at the very limits of detectivity it is very attractive. NEPs of about 2×10^{-15} W have been reported by Clarke et al [329].

4.4.4.3 Heterodyne detection—introduction

Heterodyne detection is fundamentally different from all the other types discussed so far since it involves coherent operation [337]. The signal field, either coherent or incoherent, of frequency v_s and power P_s is combined exactly, in space, with a coherent "local oscillator" field of frequency v_{LO} and power P_{LO}. The average power of the combined field will then (cf. equations 1.4.12a and 1.4.14a) be given by [337]

$$\langle P_{in} \rangle = \langle [(2P_{LO})^{1/2} \cos(2\pi v_{LO} t) + (2P_s)^{1/2} \cos(2\pi v_s t + \phi)]^2 \rangle.$$
(4.4.19)

If this resultant is incident on a "square-law" detector, that is one which gives an output proportional to the incident intensity and therefore to the *square* of the field strength, the detector output, say a current, will be given by

$$i_S = K \langle P_{in} \rangle$$
$$= K[P_{LO} + P_s + 2(P_{LO} P_s)^{1/2} \cos(2\pi v_{IF} t - \phi)].$$
(4.4.20)

In this equation $v_{IF} = |v_{LO} - v_s|$ is the so-called "intermediate frequency". For coherent operation, it is obviously essential that the detector be able to respond at the intermediate frequency but, since one will seldom have a detector capable of responding at the signal frequencies, it follows that $v_{IF} \ll v_s \approx v_{LO}$. Clearly it does not matter whether v_s lies above or below v_{LO}: there will still be an output at the intermediate frequency. Following radio terminology, one says that the heterodyne detector is a double sideband (or

DSB) receiver. In the radio region where RF tuning can be used, it is possible to discriminate against one or other of the sidebands and thus always guarantee to pick up only the desired signal. At millimetre and submillimeter wavelengths this is not possible, so heterodyne detection tends to be reserved for the detection of single emission or absorption lines when it is fairly certain that there is no corresponding feature on the other side of v_{LO}. At a fixed value of v_{LO}, the accessible spectral range is then given by twice the maximum intermediate frequency, that is by twice the reciprocal of the response time of the mixer. Within this range, the actual spectral interval being explored will be determined by the frequency setting of the *IF* amplifier. One thus has a simple electronically scanned spectrometer with a resolution determined by the stability and monochromaticity of the local oscillator and by the pass-band of the IF amplifier. The technique is identical to that involving spectrum analysers in microwave spectroscopy but for infrared usage there is one major difference, namely that the presently available local oscillators (i.e. gas lasers) cannot be tuned. The spectral range depends therefore on having a mixer with a very short response time but this range can obviously be extended if one has several local oscillators stategically spaced along the frequency axis.

The physical interpretation of equation (4.4.20) is that, for zero P_{LO}, the detector will merely give out a signal, either DC or at the modulation frequency, which is proportional to the incident signal power. This is, of course, just ordinary detection, but as P_{LO} increases, there will be an increasing contribution at the intermediate frequency with a magnitude which will be proportional to $(P_{LO} P_S)^{1/2}$ and for the usual experimental conditions of $P_{LO} \gg P_S$ this term will be the dominant one. There is thus "heterodyne gain". The signal at the intermediate frequency is usually passed to an intermediate frequency amplifier and then demodulated and recorded in the usual way. A heterodyne receiver of this type is thus a purely classical device since it involves the observation of fields rather than quanta. In particular, the usual visualisation of the Manley–Rowe restriction is inappropriate since one can get an arbitrarily large number of IF "photons" out for one signal photon in. Classical behaviour of this sort is inevitable if the non-linearity is merely square law but if it becomes more abrupt, quantum effects can start to be manifest. Obviously this would be of interest at very low signal levels for then the experimentalist would be glad to have a photon counter rather than an intensity meter. The main condition for quantum effects to show up is that the non-linearity in the i/V characteristic should take place within the voltage scale of the photon (i.e. hv/e); in practice this is some millivolts. Heterodyne mixers based upon superconducting tunnelling can satisfy this condition. They are discussed in some detail later on in section 4.4.4.7.

The noise sources in heterodyne receivers [337] are (1) signal noise, usually shot, $\langle i_{SN}^2 \rangle$, (2) background thermal noise $\langle i_T^2 \rangle$, (3) non-coherently detected thermal noise $\langle i_{BT}^2 \rangle$, (4) Johnson noise in the mixer $\langle i_J^2 \rangle$ and finally IF amplifier noise $\langle i_A^2 \rangle$. The important point to note is that only the first two experience the heterodyne gain. The power signal-to-noise ratio at the output of the IF amplifier will then be [337]

$$(S/N)_{IF} = \frac{\langle i^2_{IF} \rangle}{\langle i^2_{SN} \rangle + \langle i^2_T \rangle + \langle i^2_{BT} \rangle + \langle i^2_J \rangle + \langle i^2_A \rangle}. \qquad (4.4.21)$$

If the local oscillator power (assumed to be noise free) is high enough, this reduces to

$$(S/N)_{IF} = \frac{\langle i^2_{IF} \rangle}{\langle i^2_{SN} \rangle + \langle i^2_T \rangle} \qquad (4.4.22)$$

and it is possible to calculate the NEP since the two terms in the denominator can be calculated exactly from the theory of electromagnetic field fluctuations as indicated in section 4.2.5. Thus at short wavelengths where shot noise predominates, one has

$$S/N = \eta P_S/h\nu B \qquad (4.4.23a)$$

where η is the photon absorption efficiency and B the bandwidth, whilst at long wavelengths where thermal noise predominates, one has

$$S/N = \eta P_S/kTB. \qquad (4.4.23b)$$

Thus in either case one has ideal NEPs of the order 10^{-20} W per unit IF-amplifier bandwidth. It is figures of this order which make heterodyne receivers so attractive. It is important to notice that the temperature of the heterodyne mixer does not directly affect the performance: the temperature which appears in (4.4.23b) is, of course, that of the warm background. In principle, therefore, heterodyne detectors can always be used at ambient temperature unless, as is the case with the superconducting versions, the basic non-linearity depends on cryogenic operation. The availability of a super-sensitive room-temperature detector is obviously of enormous importance to the astronomers who have to work in remote high-altitude observatories. In practice, small temperature dependences are found, say a factor of three in going from room temperature down to liquid helium temperature, so some form of cooling is desirable, but one is not in the same situation as is the user of a bolometer where every little gain in detector cooling is worth striving for. Reasonably portable closed-cycle coolers of the "cryotip" type which can reach 20 K are perfectly adequate for heterodyne detection. It is worth pointing out here that although heterodyne detection is relatively new in the infrared, it has been a commonplace of broadcast reception at RF frequencies. The theoretical NEPs mentioned above can be closely approached at radio frequencies, which explains the great sensitivity of modern broadcast receivers, but in the infrared this is not yet the case and the noise figures observed are several times larger than the ideal. This is usually quoted in terms of the *noise temperature*, that is the temperature of a hypothetical black-body source which would give the same limiting long-wave NEP. For example at frequencies of around 500 GHz noise figures are observed of about 2000–10 000 K instead of the 300 K appropriate to the ambient background.

The coherent detection is inevitably associated with the need to align the two fields, both in direction of propagation and in direction of polarisation. The heterodyne receiver therefore has a well-defined antenna pattern which is polarisation sensitive. In some laboratory applications this is a nuisance but in astronomical applications where the detector is coupled to the antenna pattern of a telescope it can be an advantage since it provides additional spatial discrimination. Another aspect of the same coherence/directionality properties is that two or more coherent heterodyne detectors can be used in an aperture synthesis telescope. This is a clever way of obtaining spatial resolution appropriate to a large "dish" from combining coherently the signals obtained from a number of smaller dishes of much reduced total area. The directionality of the detection depends on the wavelength and in fact [338] if one has a detector of area A illuminated by radiation of wavelength λ, the coherent detection will be maintained within a solid angle Ω given by

$$\Omega = \lambda^2/A. \tag{4.4.23}$$

Since $A\Omega$ is the *étendue* (see equation 4.3.15), it follows that proper matching of the detector to the telescope will maximise the throughput for a given directionality. For a detector of area $1\,\text{mm}^2$, it follows from (4.4.23) that the tolerance in the matching of signal and LO beams will be 60° for $\lambda = 1000\,\mu\text{m}$, 6° for $\lambda = 100\,\mu\text{m}$, and 0·6° for $\lambda = 10\,\mu\text{m}$. The matching is therefore easy to achieve at very long wavelengths where diffraction is dominant and at very short wavelengths where one can rely on the optical approximation holding, but in the intermediate region, the far infrared, there are some difficulties. The basic problem is that one needs to concentrate the field coherently onto the non-linear element. At short wavelengths, one can use free-space, that is optical techniques, and arrange that the detector just fills the first Airy disc. At very long wavelengths the detector crystal can be mounted on a post in a waveguide and this can be followed with an adjustable short. This short can be so arranged that an antinode of the stationary (i.e. standing wave) field coincides with the crystal. The crystal can then be contacted by DC leads to provide any necessary bias and by an RF connection to take away the intermediate frequency signal. As one moves into the near millimetre and submillimetre wavelength regions, one runs into the obvious difficulty that the optical approach is invalid whilst the microwave approach is severely hampered by the very small internal dimensions required for fundamental mode waveguide. Nevertheless, Blaney and Cross [339], for example, have managed to use fundamental mode waveguide at frequencies above 400 GHz where the inner dimensions are less than a millimetre, but most workers in this frequency band prefer to use quasi-optical techniques. There are then two problems, firstly to combine the signal and the local oscillator fields and secondly to couple the combined field effectively and coherently to the detector.

The field combiners are usually called "diplexers". Several satisfactory types for millimetre and submillimetre wave operation have been devised, based on quasi-optical versions of the familiar Fabry—Perot, Michelson and Mach–Zehnder interferometers. Thus Gustincic [340] has designed Michelson forms and also an ingenious folded Fabry–Perot type. Fettermann and his colleagues [341] have devised a form of the Mach–Zehnder based on two beam dividers meeting along an edge and inclined at 90° to one another together with a pair of 90° retromirrors forming a "rooftop". This diplexer has seen widespread use: like all the others it permits phase adjustment but it has also the great merit, common to all Mach–Zehnders, of not returning any power to the source. This would be an important consideration if a laser were being used as the local oscillator. One should also remark that presently available lasers are rather noisy devices and the diplexer can provide a very welcome reduction in the local-oscillator contribution to the final noise figure. The reason for this is that when the path-difference in the diplexer is set correctly, both signal and LO inputs will encounter constructive interference and both will appear, as desired, out of the detector port. However LO noise at the signal frequency, the only kind of any consequence, will encounter destructive interference and will thus emerge out of the dump port. Some further types of diplexer have been described by Martin [342] and by Wrixon [343]. To illustrate diplexer action, one example, based on the quasi-optical equivalent of the Mach–Zehnder interferometer, is shown in Fig. 4.22. Martin [344], has pointed out that these quasi-optical diplexers, based on what at first sight look like classical interferometer designs, are the strict analogues of

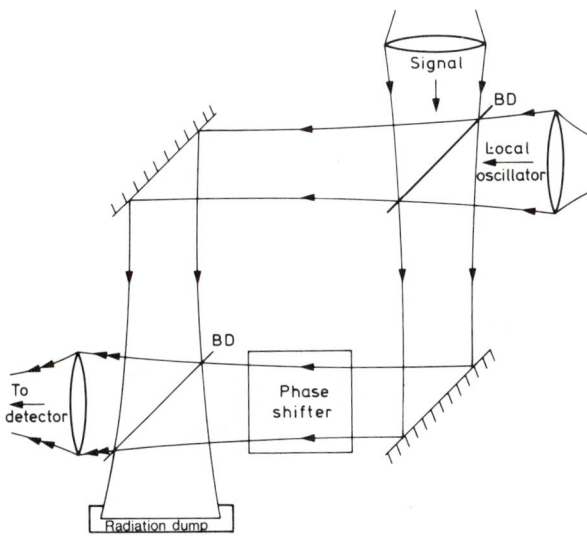

FIG. 4.22. Quasi-optical mixer based on the Mach–Zehnder interferometer.

microwave circuit elements. One is thus approaching a situation where one can think of quasi-optics in circuit terms.

The final coupling to the detector presents the same sort of problems because of the same difficulty, namely the very short wavelength. Two approaches have emerged. The earlier was the use of a "long-wire" antenna which terminated at one end in a "cats whisker" contacting the crystal and which served to couple out the IF at the other. The theory of such "open structures" has been developed by several authors [344, 345]. The coupling to the field of the long (i.e. several wavelengths) wire is not good so it is usual to use a reflector, either cylindrical or "rooftop", behind the wire in order to reinforce the field strength at the wire. The second, more modern, approach is to use monolithic techniques either on a silicon [346] or a GaAs [347] substrate. The advantage of the integrated approach is that the devices so made can be rugged, reliable and, in principle at least, cheap. Another advantage is that circuit elements can be mounted electrically close to the non-linear element so losses and parasitic elements can be controlled more easily. The coupling out of the IF in the device developed by Clifton et al. [348] is via a thin metallic stripe deposited on the substrate which acts both as a coplanar transmission line and at its sharp end as the contact to the non-linear crystal, in this case a Schottky diode. The complete assembly is very small so it can be mounted inside and parallel to the smaller dimension of a short length of WR-10 waveguide which is terminated by a waveguide horn. This approach is interesting for several reasons not least because of the possibility of scaling it down so that it may become possible to make infrared integrated circuits.

Heterodyne detectors have been mostly used so far in the near and submillimetre wavebands where they are the almost ideal detectors for the study of compact astronomical sources. Thus the supersensitivity of these detectors has permitted such *tours de force* as the observation of the $J = 6 \rightarrow J = 5$ emission line of CO at 605 GHz in the Orion Nebula from a ground-based (in fact Mauna Kea) observatory [349]. At the still higher infrared frequencies, however, they lose the sensitivity advantage since there are conventional infrared detectors which can approach the fundamental quantum limits (see Fig. 4.21). Heterodyne detectors tend, therefore, to be limited to high-resolution work, to the provision of frequency "markers" on tunable laser spectroscopic records [350] and, of course, to applications in the area of fundamental frequency metrology. In the submillimetre and near-millimetre wavebands, as remarked several times above, they are the best detectors available but only at millimetre wave frequencies and below are there tunable local oscillators (carcinotrons etc.) available so the usefulness of the heterodyne mixer is determined primarily by its response speed. The available types [351, 352, 353] are

a. rectifier diodes [354];
b. metal-oxide-metal or "MOM" detectors [355];

c. microwave-biassed reflection detectors [356];
d. photoconductive or "hot-electron" detectors [357];
e. Schottky diodes [358];
f. Josephson junctions [359];
g. "SIS" or "SIN" detectors [360].

The rectifier diodes are eminently fast enough but at the moment they are restricted to microwave frequencies and below. The MOM detectors are also fast and their frequency response extends almost up to the visible region but their sensitivities are low (noise temperatures 10^6 K) which makes them unsuitable for low-intensity signals. The microwave-biassed GaAs reflection detectors are not well developed and at the moment are mostly restricted to the 100–300 μm region which, because of atmospheric absorption, is not of major technical importance. That leaves four major candidates with reasonably advanced technologies. A brief account of each of these will now be presented.

4.4.4.4 *Heterodyne detection—photomixers*

Examples of heterodyne detection using photomixers can be found throughout the infrared. At the longest wavelengths, the "hot-electron" InSb bolometer is used as the mixing element whilst at shorter wavelengths various bulk photodetectors are available. However, at the moment, the most popular region for heterodyne work with photomixers lies between 9 and 11 μm where the combination of the sensitive HgCdTe detector and the powerful CO_2 laser acting as local oscillator makes a very attractive spectroscopic system. A good example is found in the work of Sattler and his colleagues [361]. Here the source is a tunable PbSnSe diode laser whose output is first passed through a monochromator to provide coarse spectral limitation and is then incident on a beam divider which produces two beams at right angles. One of these passes through an absorption cell containing the gas under investigation and then onto a conventional HgCdTe photodetector operating in a video mode. The output of this detector is processed in the usual way and the final analogue signal displayed on the upper trace of a two-pen recorder. The other beam is mixed with the radiation from a CO_2 laser and taken onto another HgCdTe detector which acts as a photomixer. The output from this detector goes via a spectrum analyser to the lower pen of the recorder. The output frequency of the diode laser is altered by varying the drive current usually by means of a "ramp", that is the current increases linearly from a minimum value to a maximum value and then quickly flips back to its starting position again. At the maximum and minimum frequencies which represent discontinuities, "blips" will appear in the lower trace and these can be used as frequency calibrations of the upper trace. The absolute frequencies of the blips follow at once from the known frequency of the CO_2 laser and the reading of the spectrum analyser. For rough spectroscopy, one can assume a linear scale between the two blips but for accurate frequency determinations, of a given absorption line, the amplitude of the ramp is chosen so that one of the blips exactly coincides with the line. This technique is very similar to that involved in the operation of a video microwave spectrometer where the analogue of the CO_2 laser local oscillator is provided by a phase-locked klystron connected via a chain of harmonic multiplication down to a fundamental frequency standard. The HgCdTe photomixer has an IF bandwidth in excess of 1 GHz, so, since the CO_2 laser lines are spaced at 47 GHz intervals, wide spectral coverage is available provided one has a number of diode lasers of varying stoichiometries.

Photomixers are very desirable detectors for heterodyne work but they are not absolutely essential and at the longer infrared wavelengths where these devices are not readily available one can even use a thermal detector as the mixing element. Thus hetrodyne detection of weak laser beams returning after long trips through absorbing gases has been demonstrated using merely a Golay cell [362]. A more generally useful approach is based on the use of the short response time versions of the pyroelectric detector [326]. Heterodyne detection is thus available throughout the infrared, but at the very shortest infrared wavelengths it tends not to be favoured because at the very low levels of intensity which one would be interested in detecting, photon counting is superior [363] and is readily available.

4.4.4.5 Heterodyne detection—Schottky diodes

The Schottky-barrier diode has been known for well over 100 years but the explanation of its operation was only worked out in the early years of this century. Basically it consists of a bulk semiconductor separated from a plated-on metal contact by a very thin insulating layer. It is found that the i/V characteristic of the diode is given to a good approximation by [364, 358]

$$i = i_0 [\exp(eV/kT) - 1], \qquad (4.4.24)$$

where i_0 is the saturation current. This highly non-linear relationship, implying a very non-linear resistance, is the explanation of the good performance of Schottky diodes as detectors and mixers. The operation of the diode can be understood classically in terms of "cold" thermionic emission from the semiconductor, which has a low work function, into the insulating layer or, alternatively, in a quantum mechanical formalism, in terms of electrons tunnelling across the narrow insulating barrier. The great advantage of Schottky diodes is that they can be made very small [365] so that their self-capacitances are minute and they can therefore work at very high frequencies—well up into the far infrared. Coupling to the electromagnetic field at such short wavelengths is achieved, as mentioned earlier, by means of a "long-wire" or travelling-wave antenna [366]. Schottky mixers have the great advantage that they work at room temperature, whereas the rival super-conducting mixers have to operate at cryogenic temperatures but, on the other hand, the Schottky mixers require more local oscillator power and this can be a problem at the higher frequencies. Schottky diodes can be readily fabricated by modern thin-film techniques [367] whereas the corresponding technology for, say, Josephson junctions is rather lagging behind. The original Josephson junctions were just "cats-whisker" contacts. This has now been improved on and Blaney and Cross [339], for example, have succeeded in making more stable junctions by encapsulating the structure in glass. Undoubtedly the manufacture of Josephson junctions by planar techniques cannot be far off, especially since such junctions will have valuable applications in the switching circuits of ultra-high-speed computers and the technology of SIS and SIN mixers is advancing rapidly. Nevertheless at the moment the Schottky technology has to be regarded as the established approach but the super-conducting technology is making a strong challenge and Josephson hetero-dyne receivers working at 460 Ghz [368] have already been constructed.

4.4.4.6 *Heterodyne detection—Josephson junctions*

The Josephson effect was predicted by Brian Josephson [369] in 1962 and for this brilliant piece of work he was awarded a Nobel Prize for Physics in 1973. Josephson's work rested on the accepted theory of superconductivity due to Bardeen, Cooper and Schrieffer (or BCS) [370] according to which the electrons making up the electron "gas" in a metal can couple with one another by means of exchanging virtual phonons. In simpler language, as one electron moves through the lattice it perturbs it by attracting the positive ions and the perturbation can in its turn attract another electron because of the locally increased field. The pair of electrons so connected will have zero or unit total spin and so will behave as a boson rather than as a fermion. These so-called "Cooper pairs" will tend to condense out of the electron gas as the temperature falls, in much the same way that a liquid condenses out of the corresponding vapour and there will be analogously a well-defined transition temperature. Above this temperature there are no Cooper pairs, whilst below it the number of pairs increases rapidly with further decreasing temperature. Finally at absolute zero all normal electrons will have disappeared. Cooper pairs can conduct electricity without any ohmic heating, since only virtual phonons, not real ones, are involved in their motion; therefore below the transition temperature the electrical resistance of the metal vanishes. This is the spectacular phenomenon of superconductivity discovered in 1923 by Kamerlingh–Onnes. It is assumed that the phenomenon is general—i.e. that all metals will become superconductors at sufficiently low temperature, but the main technological interest is in those which have the highest transition temperatures. Oddly enough it is not those metals which have the highest normal conductivities, viz. copper, silver and gold, which have the highest temperature superconductivity transitions. The Cooper pairs, being held together essentially by a spin-spin coupling, can become detached in a strong magnetic field so there is a subsidiary (and technically very important) interest in high-field superconductors. On both counts alloys such as niobium-tin have proved very attractive, having transition temperatures as high as 10 K. For these only conventional cryogenic techniques are required. The existence of a critical temperature implies the existence of a corresponding band gap. The two are related by the simple equation

$$E_g = 3 \cdot 5 \, k T_c \qquad (4.4.25)$$

which implies that for photons having frequencies of the order E_g/h, the superconductor will show partial transparency. The BCS theory indicates that the ratio τ_S/τ_N of the superconducting to the normal transmissivities will be zero at DC but will steadily increase with frequency, and, for a reasonably broad band centred on $v_m = E_g/h$, it will be greater than unity. At higher frequencies the ratio declines towards unity and the specimen appears to be a normal metal. The superconductors of technical interest have E_g corresponding to far-infrared frequencies, 300–750 GHz, that is 10–25 cm^{-1}. The few experiments [371] which have been carried out on the far-infrared properties of superconductors have confirmed the correctness of equation (4.4.25) and have shown a frequency dependence of τ_S/τ_N in general agreement with the theory. In particular the ratio is found to reach values of between two and three in the vicinity of the peak.

The BCS quantum theory of superconductivity has a consequence that the phase of the electronic total wave function will be coherent, that is well defined, throughout the bulk of the superconductor. Thus, although the absolute value of the phase at any point and at any time is unobservable, *differences* of phase between two points or two times are constant and observable. It was this fact which led Josephson to wonder what would happen if two pieces of bulk superconductor, made from the same material, were to be brought together from infinity to eventual contact. Initially there would be zero phase correlation but eventually there would have to be perfect phase correlation and it follows therefore that phase correlation must build up at distances which, though short, do not correspond to metallic contact. This means that a superconductor–insulator–superconductor junction has some measure of phase correlation across it. Quantum mechanically one can think of this phase correlation arising from Cooper pairs tunnelling through the insulating barrier. If a constant bias voltage V_0 is maintained across the junction, the Josephson theory shows that the phase relation across the junction becomes time dependent, being given by

$$\phi = \phi_0 + (4\pi e V_0/h)t. \qquad (4.4.26)$$

This means that, in addition to the "normal" diode current, there is an oscillatory supercurrent of frequency

$$v_0 = 4\pi e V_0/h. \qquad (4.4.27)$$

In numerical terms a voltage of 1 mV corresponds to a frequency of 483·6 GHz, that is 16·12 cm^{-1}. A relation of this type is naturally of considerable interest, not only to infrared detector workers but to people in the National Standards Laboratories who are charged with maintaining, amongst other things, traceable standards of voltage. Traditionally this has been done by the use of banks of Weston standard cells, but the possibility via equation (4.4.27) of relating voltage to the most accurately defined standard of all, namely frequency, means that progressively, in future, the Josephson effect will be used to provide the fundamental standard of voltage [372].

However, returning to the use of Josephson junctions in infrared research one sees that such a junction can act as a tunable source of far-infrared radiation [373]. The power levels available are low but when confined in a suitable cavity, for example an ellipsoid with the junction at one focus, real spectroscopy is possible [374]. However it is the converse of this observation which is of the most interest, since when radiation of frequency v is incident on a biassed Josephson junction, "steps" appear in the DC diode characteristic, that is the graph of current versus voltage. These steps occur at integral multiples of the characteristic voltage $(hv/4\pi e)$. The amplitude of these steps (that is the width of the "treads") depends on the intensity of the incident field so, by picking up the corresponding voltages, it is possible to make a very sensitive narrow-band detector [375]. The same process can be used to make broad-band video detectors [374] and these have been shown to have NEPs of the order of 5×10^{-13} W for 1 Hz bandwidth. The "risers" of the steps in the diode characteristic are almost perpendicular and exceedingly sharp which means that they contain, on Fourier decomposition, very high harmonics at appreciable amplitude (see, for example, equation 4.1.4). Josephson mixers therefore operate on very high order harmonics of the fundamental frequency. Submillimetre lasers, for example, have been directly mixed [376] with the 130th harmonic of an X-band Klystron in a Josephson junction. This is an important observation for the schemes (section 7.6) to link the microwave standard of time with the visible standard of length via a chain of laser frequency measurements spanning the infrared, since fewer steps will be needed.

Josephson heterodyne mixers have proved themselves very sensitive devices for the reception of the weak signals encountered in millimetre-wave astronomy and Thomson scattering from plasmas. The noise figures reported [377], whilst considerably above the theoretical limits given earlier (equation 4.4.22), are still quite acceptable. Thus at high microwave frequencies the values are a few hundred K whilst at several hundred GHz these figures have worsened to no more than a few thousand K. The reasons for the less than perfect performance are obscure. One suggestion [378] is that because the Josephson junction is so non-linear, one will get a very large number of sum and difference oscillatory supercurrents flowing and that these will thence span a wide frequency range. One could then get a large amount of noise, especially high-frequency noise, mixed down into the frequency range of interest. This suggestion is certainly supported by the observation of much lower noise temperatures for the very much less non-linear quasi-particle detectors (next section). The Josephson mixers have the advantage over the rival Schottky mixers in that they need much less local oscillator power—only a milliwatt or so being necessary—and this is an important consideration at frequencies above 200 GHz where powerful tunable sources hardly exist. On the other hand, Josephson junctions require cryogenic conditions and must be cyclable from room-temperature down to helium temperature may times without deterioration, whereas the Schottky diodes operate at room temperature. At the moment, the Josephson mixers are only clear winners at frequencies above 400 GHz but when stable junctions made by planar techniques become available there will be a more serious challenge in the millimetre wave band.

4.4.4.7 *Quasi-particle detectors and mixers*

At large values of the bias voltage applied to a Josephson junction, the oscillatory supercurrents may have so high a frequency that they are essentially shorted out by the junction capacitance. The Josephson effect is then suppressed. One can achieve a similar result by applying a magnetic field sufficiently intense to uncouple the Cooper pairs. The *S*uperconductor-*I*nsulator-*S*uperconductor (or SIS) structure, however, is still capable of acting as a diode and normal electrons or else "quasi-particles" can nevertheless tunnel through the insulating barrier. The i/V characteristic shows a quite sharp step or "knee" at $2 E_g/e$ and this step occurs over a voltage range which is comparable to that (i.e. $h\nu/e$) corresponding to a millimetre wave photon. In this circumstance a classical treatment of the diode non-linearity becomes inappropriate and a full quantum mechanical treatment becomes necessary. This quantum theory [378] gives some remarkable results. Thus not only is the SIS mixer an inherently low-noise device, it is, unlike all classical mixers, capable of giving gain. This is an important point since less gain will then be required in the IF amplifier and the overall system noise will be reduced.

The first experiments using quasi-particle mixers were carried out at microwave frequencies using a device called the "super-Schottky" diode. This is a piece of superconductor separated from a piece of heavily doped semiconductor by an insulating Schottky barrier [379]. Very good performance was obtained, particularly from a noise point of view, but the spreading resistance of the device set an upper frequency limit still well within the microwave band. The breakthrough, as far as millimetre and submillimetre operation was concerned, came as a spin-off from the intense development of evaporated film SIS devices to act as ultra-high-speed switches in digital computers [380, 381]. The junctions made in this programme had impedances and RC relaxation times amply low enough for them to be used as millimetre-wave mixers. Early experiments [382, 383] gave very promising results and this led to much work both theoretical and experimental on the SIS mixer. From this has emerged two key concepts. Firstly it has been shown that near millimetre-wave receivers can be buit which will approach the quantum noise limit. The form of this limit was a matter of dispute for some time since the quantum gain in the mixer seemed to lead to a paradoxical result that the product of photon-number uncertainty and phase uncertainty was less than the limit set by the uncertainty principle! This paradox was resolved by Heppner [384] who showed that zero-point fluctuations in a photon-amplifier lead to extra noise and the limiting noise temperature is not $T_N = h\nu/k$, but is instead $T_N = h\nu/k \ln 2$. For 36-GHz operation this would be 2K! Even in the millimetre and submillimetre bands the noise temperature would be less than 100 K. Secondly the theory shows that the form of the "knee" in the i/V characteristic depends on the level of the illumination, that is one has photon-assisted tunnelling [385]. It thus becomes possible to make photon counters working in the near millimetre-wave region. The experiments so far using normally Sn-SnO$_2$-Sn diodes [386] support this conclusion. With this type of detector, therefore, the era is dawning when the detectors available in the millimetre wave and submillimetre-wave region will be as good as those available in the near infrared and visible regions. Of course it will only be possible to take advantage of such low noise temperatures when the background radiation temperature is similarly low but astronomy from space, where the background temperature could be as low as 3 K, could provide one valuable application. The main problem with SIS detectors is that at higher frequencies it becomes more difficult to suppress Josephson tunnelling. However, a *Superconductor-Insulator-Normal* (metal) (or SIN) diode may work and present a solution [387].

4.4.4.8 *Photon-drag detectors*

Almost at the opposite pole to the supersensitive detectors lies the type of detector which has a very short response time since to obtain a fast response involves, at the moment, a heavy penalty in sensitivity. Conventional sensitive detectors can be engineered to give response times of about 1 μs but these clearly will not do, say, for monitoring Q-switched laser pulses where the rise time may be much less than 1 ns. Fortunately there is so much power in the brief pulses that quite insensitive detectors will be adequate provided they are fast enough. The outstanding detector of this class at present is the photon-drag detector developed by Kimmitt and his colleagues [388]. This consists simply of a slab or cylinder of a suitable semiconductor, usually germanium, moderately doped and bearing electrodes at each end. When a beam of intense radiation propagates along the axis of the detector, the carriers tend to get swept along by the radiation pressure and finish up on the far side. There is a corresponding complementary charge developed at the near face and a voltage therefore appears across the detector. This can be sensed via the two electrodes.

The sensitivity of a photon-drag detector at frequencies which satisfy the relation

$$2\pi\nu\tau_e < 1, \tag{4.4.28}$$

where τ_e is the carrier scattering time, can be derived by purely classical arguments. The result [388] is

$$V \sim I\mu[1 - \exp(-\alpha l)], \qquad (4.4.29)$$

where μ is the carrier mobility and α is the absorption coefficient (assumed to arise entirely from the free carriers). The carrier scattering times for some candidate materials are nGe, 3.7×10^{-13}, pGe, 3.2×10^{-13}, nSi 1.9×10^{-13}, pSi, 1.2×10^{-13} and nGaP, 9.5×10^{-14} s, so one cannot expect classical behaviour until well into the far infrared and perhaps even the near millimetre region. At these very long wavelengths, all these materials absorb strongly so the deciding factor in the choice of semiconductor is the mobility. The highest mobility material is n-germanium and this is preferred despite its long scattering time.

At shorter wavelengths where (4.4.28) is not satisfied the behaviour of photon-drag detectors can be very complex indeed [389]. The response is found to vary rapidly with wavelength and with temperature and may even change sign over a small wavelength or temperature interval. This phenomenon can be explained in terms of the momentum exchange between a photon and a carrier in an indirect band-gap material with a complex band structure (see section 6.5) and indeed the photon-drag response may even be used to explore the band structure of semiconductors, but from an experimental viewpoint it is a considerable nuisance since photon-drag detectors cannot be relied on as power meters for mid-infrared tunable lasers. At a *fixed* wavelength and at a fixed temperature, however, the photon-drag response has been proved to be linear over power ranges of several orders of magnitude so they are very popular for pulsed laser power monitoring. This is especially true for the CO_2 laser. The best material here is p-germanium which, despite its lower mobility, wins on the count of having a much higher absorption coefficient. Even so, the absorption coefficient is rather small in absolute terms and the monitor can actually be used "in-line", that is transmitting with negligible attenuation the beam to be monitored. For this application the monitor does have to be antireflection coated to avoid the otherwise heavy losses by reflection (for germanium n is approximately 4!) but when it is so treated it has a very high damage threshold—an attractive feature for much pulsed CO_2 work. A typical monitor would have a sensitivity of 1 μVW^{-1} and a rise time of better than a nanosecond.

4.4.4.9 *Photoacoustic detection*

One of the most difficult tasks in accurate radiometry is the determination of very low levels of absorption. In the normal approach where one measures the power in and the power out, one is in the situation, avoided whenever possible by metrologists, of having one's answer as the difference of two very large and almost equal quantities. Clearly it would be much better to find a way of measuring just this difference itself, i.e. of measuring directly the power absorbed. Perhaps the best way of doing this is to use photoacoustic detection [390].

Photoacoustic detection has been known for a long time; in fact it was first demonstrated by Alexander Grahame Bell in the 1880s but only recently has it come to the fore as an important technique for radiation detection. It can be used with solid, liquid or gaseous specimens but is mostly employed with the latter so this case will be discussed as the example. The experimental arrangement is shown schematically in Fig. 4.23. A beam of modulated radiation enters the cell where some of the power is absorbed by the gas. This causes heating and the gas as it expands produces a sound wave at the modulation frequency. This sound wave can be picked up by a microphone placed in a side-arm. The microphone signal can be amplified and phase-synchronously rectified in the usual way. The PSR output will therefore be proportional solely to the absorbed power and the first objective will have been achieved, but there is another advantage in that the signal is proportional also to the incident power. This is readily seen by writing Lambert's law in the form

$$\Delta I = (I_0 - I) = I_0 (1 - \exp(-\alpha l)). \tag{4.4.30}$$

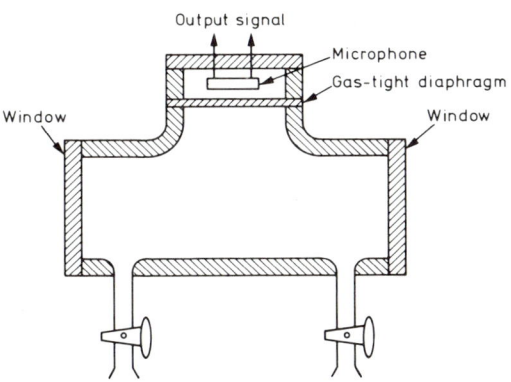

FIG. 4.23. Schematic of a photoacoustic detector.

In the classical period of infrared physics, this fact, though recognised, was only of academic interest since there were no powerful tunable sources available to take advantage of it. Now, however, with the proliferation of infrared lasers, it is becoming of major significance since it means that the only limitation to the absorptance one can measure is set by the available power in one's source. A particularly valuable application has been found [391] in the investigation of close coincidences between powerful pump laser lines and low-pressure gas phase absorption lines. The point of such investigations is the diagnosis of possible new optically pumped lasers. A similar use has been found to make simple in-line nearly transparent power monitors for optically pumped lasers [392]. Another application has been found in pollution monitoring where the line to be measured will be weak because the concentration of the pollutant will usually be low and the actual absorbance will be lower still because there will usually be a frequency mismatch between the laser frequency and the absorption line centre [393]. CO_2 laser lines have been used in this way to monitor SO_2 for example. A comprehensive account of the use of laser sources with photo-acoustic detectors has been given by Claspy [394].

The noise encountered in photo-acoustic detectors is that arising from the random or Brownian motion of the molecules making up the gas. This random motion induces random microscopic fluctuations of pressure in the gas and these are manifest as random sound waves propagating through the gas. This noise is temperature dependent and in fact it is conceptually at least, quite akin to Johnson noise. The detectors usually work at room temperature so the sensitivity though good is nowhere near as good as that of a cooled detector, but of course this is hardly the point since in a photoacoustic detector it is purely the S/N which is of interest. An analysis of noise effects in photoacoustic detection has been given by Krupnov [101]. The principle drawback to photoacoustic detection is that, almost by definition, these detectors are exceedingly microphonic. It is thus necessary to shield them from all extraneous sources of acoustic noise and from all forms of mechanical vibration. In the average research laboratory this is more easily said than done.

Photoacoustic detectors are mostly used as purely relative radiometers but there is no reason why they should not be calibrated against an electrical standard and turned into an absolute radiometer. Martin and Puplett have shown how this may be done by absorbing the beam in a metal film deposited on a polymeric septum which divides a gas-filled cell [395]. The gas is chosen to be non-absorbing, dry air for example, but it can transmit the sound waves produced when the power is absorbed in the metal. This metallic film is chosen to have a suitable surface resistance per square to maximise the absorption (ideally 50%). The film, in the usual way, also serves as the electrical heater.

The photoacoustic detector has, in distinction to other devices, the unique feature that the absorption process and the detection process are separated in space. This provides a possibility of greatly reducing observed Doppler widths since baffles can be placed in the microphone tube so that the microphone can only receive the sound waves coming from a restricted axial velocity set of molecules [101].

4.4.4.10 Frequency selective detectors

Because the absorption occurs in the filling gas, the conventional photo-acoustic detector has a highly frequency-selective response. In fact the sensitivity is confined to those narrow spectral regions defined by the line absorption spectrum of the gas. This frequency selectivity can be an advantage or a disadvantage depending on the particular application being envisaged. If one is looking for a reasonably broad-band response, it is obviously unacceptable and it was this feature which led Golay [319], when designing his celebrated detector to abandon the conventional gas-phase absorption in favour of (a) absorption in a blackened disc followed by (b) transmission of the resulting pressure wave through a non-absorbing gas. In Fourier transform spectrometry, on the other hand, it can be an advantage since it provides a neat way of side-stepping the difficulty that the dynamic range of the interferogram has to be high because it is carrying a lot of unwanted information, that is information about the source intensity in regions where the specimen does not absorb. By filling a photoacoustic detector with the gas under investigation, it is possible to obtain an interferogram which contains only the wanted information [396]. The combination of FTS and photoacoustic detection has some further promising applications in the field of surface analysis [397]. Often in industry, the conventional techniques, such as ATR, cannot be applied because the specimen is awkwardly shaped (the inside of a coated can for instance) or else is too small. In these cases, the specimen can be mounted inside a photoacoustic cell and the beam focussed onto the surface to be investigated. The output from the side-arm microphone will give, as for the gas case, an interferogram from which the spectrum can be obtained by Fourier transformation. There are two main difficulties. Firstly the high noise levels, due to microphony, are often sufficient to wash out the interferometric advantages. Secondly, in rapid-scan mid-infrared work, the only kind so far tried, each spectral component is converted into a sound wave with its own distinct frequency and the response of solid specimens to sound waves is strongly frequency dependent. This is basically because low-frequency waves penetrate much more deeply than do high-frequency waves. The net result is that, on transformation into the spectral frequency domain, there is a strong and progressive attenuation of the higher frequencies. These twin difficulties are not trivial but for many specimens it is worth combating them since otherwise one would be hard put to obtain a spectrum at all. Of course one can increase the sensitivity of the method by replacing the single interaction with the beam by multiple interactions in which one essentially sets up an integrating sphere. One then has the technique of *Diffuse Reflection Fourier Transform Spectrometry*, sometimes acronymically called DRIFTS.

The other main type of frequency-selective detector is that introduced by Luft [398] in Germany and by Veingerov [399] in the Soviet Union. A modern discussion of the physics of this instrument has been given by Hill and Powell [400]. The basic ideas can be illustrated by reference to Fig. 4.14 which, with a conventional black-body source replacing the tunable source and some suitable filters to isolate the spectral region where the gas absorbs, can be turned into a non-dispersive quantitative analyser. Thus, if an absorbing gas is introduced into the specimen channel whilst an equal pressure of a non-absorbing gas, such as N_2, H_2 or A, is introduced into the reference channel then one will get an output signal from the PSD and this signal will be correlated with the amount of absorbing gas in the cell. The instrument can be made still more exactly quantitative by the use of an alternative procedure in which, by the addition of measured amounts of the absorbing gas into the reference channel, the null reading is restored. Such an analyser will work well for pure gases, but the gases to be analysed in industry are seldom pure. More usually they consist of a mixture of many components, most of which absorb radiation somewhere within the pass band of the instrument. This difficulty can be overcome and the instrument still be used as an analyser provided *two* unconnected cells are placed in series in each arm. In the first pair, one in each arm and identical in every respect, the gas to be analysed is confined. No output signal is observed, of course, since the two channels are still identical. However, if a pure sample of one of the component gases is introduced into the second cell in the reference arm, then the symmetry will be broken just for those frequencies where this gas absorbs. The instrument therefore responds to just a single component giving a maximum signal when there is none of this component in the mixture and a minimum when there is a high concentration. In this latter case, the absorption lines of the gas will be blacked out and the second cell cannot cause any further absorption. This simple argument does presume that the lines of the other components do not overlap those of the measured gas. In general this will not be true and there will be unwanted sensitivity to the interfering gases. This sensitivity can, however, be greatly reduced by inserting a third pair of cells in the chain, each containing a suitable pressure of the interfering gas.

Such a system will work and work well but there is still the unsatisfactory feature that the large majority of the radiation arriving at the detector carries superfluous information. An elegant way of getting round this problem is to replace the broad-band detector with a frequency selective one. A photo acoustic detector could be used, but since the principal field of application of this technique is in industrial process control [401] and in pollution monitoring a rather more robust detector and one not so sensitive to vibration is usually preferred. The Luft and Veingerov detectors both have the gas confined in what is essentially a capacitor, one plate of which is flexible. The Luft is said to be balanced in that it is a two-chamber device, whereas the Veingerov is said to be single-ended because it has only a single chamber. When the gas inside the cell absorbs radiation, it warms up, expands and pushes the flexible plate outwards. The capacitance thus drops and the drop can be measured with an AC bridge. In the Luft detector, the flexible plate separates the two chambers so when equal intensities are arriving in each chamber, one from the specimen channel and one from the reference channel, there will be no output signal. The Luft detector therefore does not need a beam-splitting chopper. Otherwise, however, the experimental arrangements are very similar. The gas to be measured is placed in one of the cells in the specimen arm and a non-absorbing gas is placed in the equivalent cell in the reference arm. If there is interference, pure samples of the interfering gas can be included in a pair of filtering cells as before.

Non-dispersive analysers of this type have been widely used, particularly in industry. Because of their reliability and ruggedness, they are very suitable as automatic pollution monitors and for the automatic control of process plant. They have been shown to be capable of analysing a wide range of gases, but there is no doubt that they are at their best when the absorption spectrum of the gas in question is simple, consisting say of just one band and that made up of a set of regularly spaced non-overlapping lines. Carbon monoxide and carbon dioxide fill the bill well and these gases can be analysed with high sensitivity and good discrimination. A modern application has been in the monitoring of increases in the CO_2 content of the atmosphere due to the burning of large amounts of fossil fuels in recent years [402].

A related technique is pressure modulation spectroscopy in which the pressure in the reference gas cell is cyclically varied. This provides modulation at just the frequencies where this gas absorbs so, not only is the use of a rotating chopper avoided, one can also get high selectivity. The pressure oscillations can be provided by connecting the cell to a gas-tight cylinder carrying a piston driven by a reciprocating drive from an electric motor.

4.5 The use of computers and microprocessors in infrared instrumentation

Nowadays, almost all measurement systems have computers as essential components and infrared systems are no exceptions—in fact they are star examples. One need only cite Fourier transform spectrometry and modern infrared analytical spectrometry (section 7.2) to establish this point. Computers have developed with startling speed since their initial commercial introduction in the 1950s. The earliest forms were based on vacuum tube electronics plus crude forms of memory such as the mercury delay line. They were very bulky and very unreliable but they opened up new fields of scientific investigation since many problems that could not be solved analytically could now be readily tackled numerically. A whole branch of mathematics—numerical analysis—which had previously seen only sporadic development, for example, Adams's and Leverrier's prediction of the orbit and position of Neptune, expanded rapidly into a major art as a consequence of the availability of these new machines. By the 1960s computers were mostly transistorised and their size had shrunk dramatically, as had their power consumption. The introduction of integrated circuits (ICs) and large-scale integration (LSI) could in principle have led to a further reduction in size—at least for the most powerful machines, the so-called mainframes, but for these machines the designers have aimed for increased performance rather than for size reduction. These new forms of electronics have, however, greatly reduced the bulk of the smaller general purpose computers used about the laboratory—the so-called minicomputers. One factor which restricts the reduction in size of a computer is that the ancillaries, data inputs, data outputs and external mass memory stores such as magnetic discs and tapes cannot be shrunk down nearly so much. Here too, however, the density of information on a memory device such as a computer main store has been enormously increased for a given physical size. The cost of computer systems (the "hardware") has greatly declined since the early days and this combined with the fact that they are invaluable adjuncts to most modern research programmes has been a major factor in getting computers installed in all but the smallest laboratories. The latest stage in cost/size reduction has come about with the development of the microprocessor or MPU [403]. This is usually a single integrated circuit (or "chip") which carries all the vital functions of a central processing unit (or CPU) together with memory addressing circuits. MPUs are usually specified by the number of distinguishable "words", that is sequences of binary digits (or "bits") to which they can respond. The commonly available MPUs are eight-bit devices and can therefore accept up to 256 separate instructions. The newer generation of MPUs includes sixteen-bit and even thirty-two bit devices. These latter are much faster since operations can be specified in a single instruction rather than in many but at the moment there is a dearth of sixteen-bit software and most instruments and microcomputers continue to use merely eight-bit MPUs. Some well-known MPUs are the Intel 8080 and 8086, the Motorola 6800 and the Zilog Z80 and Z8000. The microprocessor can be used together with other chips, e.g. read–write memories (or RAMs) and read-only memories (ROMs), to assemble a true microcomputer. These can be so cheap that even with a data and instruction-entering keyboard, a suitable display and a cassette recorder for entering and storing programs they can be offered for sale at under £500. The Commodore PET is a well-known example. The availability of these chips, which are also available assembled into standard computer boards, has naturally excited instrument manufacturers who can see the sales appeal of interactive machines and nearly all scientific instruments, of any degree of sophistication, which are offered for sale now feature microprocessors.

The MPU both controls the machine once the operator has decided, in dialogue with the MPU, what it is that he wants done, and processes the resulting data. MPUs are at the moment slower than minis but the latest devices (the so-called "supermicros such as the Intel iAPX-432) promise to give us micros just as fast and just as flexible as the present generation of minis. Nevertheless, the micros available at the moment are comfortably fast enough not only to control the sequence of instrument operations but also to process the data, do their own housekeeping and perform routine checks on the general state of the instrument. The performance or speed of an MPU is often specified in terms of its driving clock (generally 3–18 MHz) but the matter is more complex than this. Fundamentally the important point is that an MPU takes only a few microseconds (say 1 to 10) to execute an elementary arithmetic or logical operation, so there is usually plenty of time to control events happening on the time scale of one every ten or so seconds. Microprocessor control removes one irksome duty from the operator, who is no longer required to act as a sort of human servo feedback element, reacting to the instrument and altering all its parameters: all this can now be done quite automatically.

Large modern mainframe computers can process enormous amounts of data extremely quickly and are generally able to work on several independent jobs at the same time. In this way they can always be kept usefully employed. They are normally accommodated separately from the rest of a laboratory and are operated by specialist staff. Most access is from remote terminals which can range from simple display units (or VDUs) to medium-sized computers which are generally connected to the mainframe by serial "communications" lines.

The actual instructions which the processor obeys are in binary format and a sequence of such instructions is known as "machine code". Machine code itself is nowadays virtually never used for the composition of programs (that is the "software"): the requirements are nearly always expressed in a more human-compatible programming language. The instructions of an "assembly-level" language, generally written one to a line, have a more or less one-to-one correspondence with those of the resulting machine code, which are, in fact, specified almost completely by the user's program. They are processed by an assembler program into a form suitable, possibly after further transformation, for loading into the machine. Assembly language is machine dependent and is nowadays considered as suitable only for those program functions critically dependent on the machine hardware and for those places in the program where speed is really important.

Since assembly code is not suitable for the lucid, machine-independent, expression of complex algorithms, a large variety of higher-level languages, such as FORTRAN, Algol, Pascal, Ada etc, has been developed. More often than not, these high-level languages are "compiled", via suitable software, to give loadable machine code. This may be done directly or else via intermediate stages, one of which may be a form of assembly-level code. This approach incurs a significant overhead in compilation but it leads to code which is of a similar order of speed and compactness to that which might have been produced by an assembly-level programmer. A compiler can apply a consistent degree of optimisation, but cannot produce those inventive tricks which are sometimes possible to the human assembly-level programmer. Frequently, however, and especially for less structured languages such as BASIC, the source program is not actually translated to machine code. Rather it is stored in main memory, possibly after some lexical compression, and then executed interpretively by a run-time machine-code system. The overhead of compilation is not avoided by this alternative approach (though it is less conspicuous), since each line of source code now has to be "understood" each time that it is executed. This manifests itself in a generally much slower program.

The main advantage of high-level languages is not that they are easier to learn, since assembly language is conceptually straightforward: it is that they are easier to use. It is not necessarily easier to write programs using a high-level language but it is certainly easier to write correct ones which will run without having to carry out extensive "debugging". It is also much easier to read programs if a high-level language has been used and especially so if it is a structured one such as Algol, Pascal etc. This word "structured" is widely used in programming circles, basically it means "systematic" with much use of conditional statements and loops rather than unconditional jumps. Also, subroutine interface specifications can be set within (or "nested" in) a part of the program to which they solely belong for structuring and access control. This means that the independent functions within a program—the subroutines—can be implemented as separate modules or software components [404]. Many older computers are still in satisfactory use in which the programming has been done by archaic techniques but for the future even the smallest microcomputers will be programmable in a high-level language, possibly with development aid from a larger machine. Any smaller device directly programmed in machine code should nowadays be regarded properly merely as an electronic component.

It is in principle possible to provide all the facilities which the machine needs for a given task within the users' program. However, since many of the operations will be common to all users, the normal arrangement is to have a so-called "operating system" which is essentially a resource management program. The operating system controls the inputting and outputting of data, the data flow within the machine and the time-sharing if the computer has that facility. For mainframe computers the operating system is particular to that family of machines, but for minis and micros it is possible to use generalized operating systems which can be employed on any machine. A widely used system is UNIX developed at the Bell Laboratories [404]. This is associated with a programming language called "C". There is a case for adopting a system such as this as a machine-independent standard so as to avoid a "Tower of Babel" situation in operating system languages.

The computer controls the instrument via a set of interfaces which in their turn operate the electromechanical transducers. A typical transducer might be, for example, a stepper motor which would give control over the moving parts of the instrument. Other types of transducer would involve straightforward electrical controls using digital/analogue converters or else data input interfaces which might be either just analogue to digital converters, or else purely digital input/outputs such as shaft encoders, synchros etc. A great deal of ingenuity has gone into the design of interfaces and those now used mainly feature self-checking devices. They are rapidly approaching almost error-free operation. The computer control via software gives enormous flexibility. Major changes to the way the instrument operates can be introduced merely by reprogramming the memory or by plugging in another externally programmed ROM or EPROM unit. In traditional instruments the same task would have required considerable engineering modifications which would have been not only costly but also very time consuming. Microcomputer controlled spectrometers for instance can now be operated by a non-perfect screw thread and do not need to have elaborate sine or cosecant bars. The MPU can calculate exactly what the wavelength or wave number is from any arbitrary position of the screw thread drive. The elimination of the need for tolerances on the working parts reduces manufacturing costs dramatically since time-consuming and expensive adjustments prior to despatch are no longer needed. Also, since drift and wear are no longer problems, expensive visits by service engineers can be dispensed with. Another major advantage of MPU controlled systems is that the digital calculations give exact results with no danger of accumulating noise. When the calculations and processing take the form of numerous sequential or cascaded operations, this can be a major consideration. In this context, there are several analogies with PCM communication systems (Section 3.6.4.2). Not least of the advantages of MPU systems is the low cost of the hardware compared with the equivalent mechanical or electrically operated servo systems. Also, since the MPU hardware is very compact, the instruments can be of convenient size yet still have a remarkable amount of interactive intelligence. The microprocessor printed circuit boards are much more reliable than the older types of electronics and down-time is much reduced. When errors or faults do occur, it is now normal practice for the MPU to provide immediate details of the fault! Nearly all modern commercial infrared spectrometers now feature microprocessor control and because of this can be self calibrating. A microprocessor controlled spectrometer is beyond doubt an enormous advance over the old-fashioned inflexible and somewhat unreliable types, but the advantages do not stop there. The microprocessor can be connected to external floppy disc storage units to give a constant assessment of how well the instrument is performing *vis-à-vis* its own historical records, the disc memory can be used to store reference spectra and to encode sections of spectra for scale expansion, background subtraction etc. The MPU can also "talk" directly to an external main-frame or mini computer to extend still further the computing power. It is clearly no exaggeration to say that the evolution of totally flexible, fully interactive, self-calibrating and self-maintaining microprocessor-controlled infrared spectrometers is an advance just as dramatic as those which gave us our first commercial instruments in the late 1940s. Equally clearly we are still only on the threshold of the computer-controlled automation revolution. What is yet to come can only be guessed at.

4.6 High, very high and ultra-high resolution techniques

In the literature, various terms are used to describe resolution in excess of that available with a good-quality grating or interferometric spectrometer, that is $\sim < 0.5$ cm^{-1}. Naturally since a large number of authors is involved, there is very little agreement on the terms used but a good consensus to adopt might be

High resolution	$\Delta\tilde{v} = 0.5{-}0.1$ cm^1	i.e. $\Delta v = 15{-}3$ GHz
Very high resolution	$\Delta\tilde{v} = 0.1{-}0.001$ cm^{-1}	i.e. $\Delta v = 3$ GHz-30 MHz
Super high resolution	$\Delta\tilde{v} = 0.001{-}0.0001$ cm^{-1}	i.e. $\Delta v = 30{-}3$ MHz
Ultra high resolution	$\Delta\tilde{v} < 10^{-4}$ cm^{-1}	i.e. $\Delta v < 3$ MHz

The lower limit of ultra-high resolution spectroscopy keeps dropping and there are schemes afoot [405] to observe line widths as small as a few tens of Hz but as these would be of more interest to the fundamental standards workers (see Section 7.6) than to the spectroscopist we will arbitrarily adopt a definition of 1 KHz as a working limit. This definition, plus the divisions given above, are purely arbitrary, but the division scheme is to some extent natural in that the first two categories define resolution worse than the Doppler line width (equation 2.6.13a) for many molecules in the middle infrared at room temperature whereas the second two define sub-Doppler resolution. High and very high resolution work therefore merely requires conventional, if albeit high grade, instruments whereas super and ultra-high resolution work requires laser sources.

High resolution spectroscopy in the mid- and near-infrared regions can be carried out with either dispersive [406] or interferometric [407] instruments. In a dispersive instrument the grating or prism (nearly always in fact a grating) has to be of high quality and of large area and is therefore expensive; the slit assemblies have to be very precise and the mechanical structure of the spectrometer very solid and well engineered. The instruments are therefore very costly but several have been constructed and have given excellent results [408]. Nowadays, however, with the availability of very high-quality interferometric spectrometers, this kind of spectroscopy is nearly always done interferometrically so the days of "home-made" high resolution grating mid-infrared spectroscopy are probably past. In the far infrared, because of the energy limitation, they never really arrived and all work of any quality has been done with interferometers. This is not, however, to cast any aspersions on the work of men like Czerny [409] and Rubens [410] whose heroic pioneer work using grating spectrometers on the pure rotation spectra of gases laid some of the foundations of quantum theory [411]. An example of a modern far-infrared high resolution spectrum taken from the work of Partridge [412], is shown in Fig. 4.24.

At higher resolutions, one finds very few examples done by grating instruments even in the energy-rich mid infrared. Nearly all the work has been done with interferometers and the experimentalists have therefore had to come to grips with the problems of moving mirrors over large spans with negligible wow and yaw and, moreover, doing so with near perfect sampling as well. These problems are easiest in the far infrared and a good example of this type of work is shown in Fig. 4.25 which is a stratospheric spectrum obtained from a high altitude balloon by Bangham et al. [413]. This spectrum shows lines due not only to the majority constituents of the stratosphere such as O_2, O_3, H_2O etc. but also some due to surprising minority components such as HF! In the middle infrared the problems are naturally more difficult but engineering advances have overcome them to a large extent and commercial instruments (for example the BOMEM [414]) are now available which offer resolution better than 0.01 cm^{-1} throughout most of the infrared. Research instruments, constructed to the limits of present technology, can reach resolutions of 0.001 cm^{-1} and the very best of these have reached a proven 27 MHz [415], but this does seem to be the limit. In the near infrared the problems become formidable and, for this reason, high resolution work at wavelengths shorter than 3 μm has tended to be neglected. There is also at these short wavelengths a major computing problem. Thus, from what has gone before, it will be realised that the encoding of the sampled interferogram corresponding to radiation of maximum wavenumber $\tilde{\nu}_m$ and to resolution $\Delta\tilde{\nu}$ requires the taking of a number of samples N given by

$$N = 1.4\,(\tilde{\nu}_m/\Delta\tilde{\nu}).\tag{4.6.1}$$

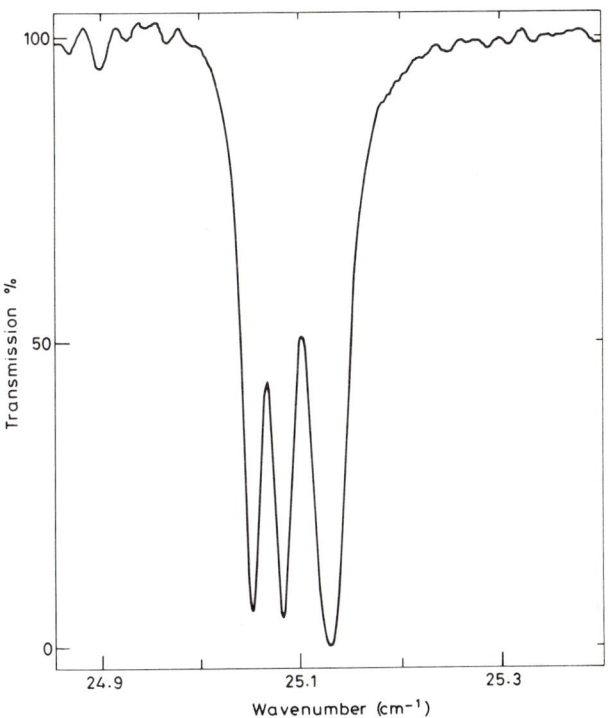

Fig. 4.24. Transmission of an $H_2O/D_2O/HDO$ mixture near 25 cm^{-1} for a path-difference of 850 cm, a gas pressure of 5 torr and at room temperature. The observations, carried out with a novel inclined mirror interferometer [412], reached a resolution of better than 0.02 cm^{-1}. The observed wave numbers of the three lines are D_2O, 25.0548, H_2O, 25.0851 and HDO, 25.1311 cm^{-1}.

Working in the near infrared, say $\tilde{\nu}_m = 10^4$ cm^{-1}, to merely high resolution, say $\Delta\tilde{\nu} = 10^{-2}$ cm^{-1}, therefore requires the observation and Fourier trans-formation of at least one million data points [415]. If the transformation is to be done by the FFT, the operation (see Appendix 5) requires array processing of N^2 points and one has a problem which can strain the resources of quite large mainframe computers. Fortunately, numbers of this order are no longer so formidable to modern "number crunching" machines, but there still remain the optical problems. These too, however, can be overcome as the classic work of the Connes [416], shows. The method that they used, a good example of experimental physics at its best, was, in brief, to replace the plane mirrors of the classic Michelson interferometer with "cat's eye" [417] or "cube-corner" [418] retroreflectors which are insensitive to tilts, to use servo control, locked to the laser reference interferometer, to ensure absolute control of the sampling, to use very high grade optics and beam splitters and to follow the interferometer with the very best analogue and digital electronics. The Connes did most of their work in emission which, of course, minimises the dynamic range problem. Some of their spectra of heavy atoms are virtual forests of lines, many of which have not so far been assigned but their use of the throughput and multiplex advantages [283, 280] of the interferometer to do planetary spectroscopy [419] from the Earth's surface was a veritable *tour de force*.

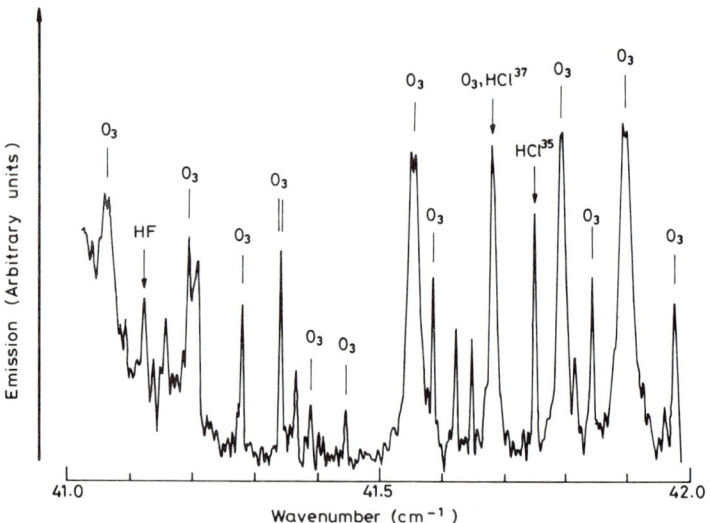

FIG. 4.25. Statospheric emission spectrum obtained from a balloon-borne platform by Bangham *et al.* [413] at an effective tangent height of 37 km. The interferometer used was of the Martin–Puplett polarising type and reached an unapodised resolution of 0·0033 cm^{-1}.

Resolution better than 10^{-3} cm^{-1} requires the use of a coherent source since any delay-type instrument would be impossibly bulky, far too costly, a designer's nightmare and would furthermore present almost insuperable computing problems. The most widely used coherent source, at the moment, is the diode laser [420] the mode of operation of which is discussed in more detail in section 6.5. The centre frequency of the laser is chosen by adjusting the stoichiometry and fine tuning is then ensured by either varying the drive current (that is temperature tuning) or else (and less commonly) by subjecting the laser to hydrostatic pressure [421]. The tuning range available with these techniques is very restricted (1 cm^{-1}) but within this range the instantaneous line width can be as small as 10^{-4} cm^{-1}. Complex many line spectra are therefore readily investigated and since such spectra usually arise from moderately heavy molecules for which the Doppler width is small, laser diode spectroscopy has proved itself to be a rather useful tool. It is particularly so for pollution monitoring [422] and for investigating the spectra of possible fuel gases for optically pumped lasers [423]. Worchesky and his colleagues, for example, have used a diode laser to work out the mechanism of the important CH_3OH optically pumped laser [424]. The sensitivity of this technique and its resolving power can both be improved by using the laser as a pump for a heterodyne system [361]. The other types of infrared tunable laser (section 6.4) are not really contenders in the super-high resolution field since they usually have a not negligible line width and their tuning characteristics are often far from smooth, but Smith and his colleagues at Heriot—Watt [425], Whiffen and his co-workers at Newcastle [426] and Patel at Bell Laboratories [427] have reported some good examples observed with a spin-flip Raman laser.

Diode lasers are not powerful enough to offer any real chance of overcoming the Doppler limitation so for super-high resolution work and better on unfavourable systems other approaches have to be sought. The Doppler broadening, as explained in Chapter 2, is inhomogeneous so there is no difficulty of principle, just of practice. The classical method was to use a molecular beam apparatus [428] in which the molecules having been evaporated from an oven are sorted out by a series of diaphragms such that when they finally emerge into the beam region they are all travelling in a direction virtually perpendicular to the beam. There is therefore negligible Doppler broadening since the beam will have a very small divergence. Also, since there is almost no component along the probing beam direction, the peak absorption frequency is a very close match to the true molecular frequency. The main difficulties with this molecular beam approach are that firstly it is very costly and secondly that the number of molecules actually interacting with the beam is very small, since UHV techniques are required, to avoid molecular scattering by collison. This solution to the Doppler problem has therefore tended to be reserved for fundamental standards work where the absence of a Doppler shift of the peak frequency is an important consideration. It is important, however, to note that molecular beam spectroscopy is linear and high power is not required for the probe beam so, although for the reasons outlined above it is unlikely to find general use, some close and still linear relatives, such as the opto-acoustic version discussed by Krupnov [101] may well find practical applications.

By far the most popular way of side-stepping the Doppler difficulty is the use of the Lamb-dip or saturated absorption technique which, as described in section 2.6.3, depends on the existence of two counter-propagating beams. Because of its non-linearity, Lamb-dip spectroscopy requires high-power lasers but this is not a real difficulty since there is no shortage of high-power infrared lasers. The real difficulty lies in the absence of broadly tunable high-power infrared lasers and because of this problem one is restricted, at the moment, to investigating lines which happen to be accidentally more or less in coincidence with an available laser line. At first sight such a restriction might be thought to be fatal but in fact is not nearly so serious for the following reasons.

a. There are several available infrared lasing systems, each usually made up of a large number of lines and this number can usually be extended by the use of isotopic or isoelectronic substitutions. Thus the CO_2/N_2O system alone can provide over one hundred possible lines.

b. Molecules of interest tend to absorb richly in just those regions where the best laser systems emit and each band is made up of very many lines so the chance of coincidence becomes very good.

c. The positions of the lines in a band are determined by just a small number of parameters (section 5.2) so if one can determine the absolute frequencies of only a few lines, to very high precision, one may nevertheless be able to derive the total information.

d. The near coincidences can often be improved and in fact made exact by the use of either the Zeeman or the Stark effect—this leads to the new topics of laser magnetic resonance and laser electric resonance spectroscopy respectively.

Given that one then has a near coincidence to within the tuning range of the laser (i.e. about 50 MHz) then, as the cavity length is varied or else as an intracavity frequency sensitive element, for example an etalon, is rotated, the cavity fringe pattern will show strong Lamb dips at the centre frequencies of all the absorption lines which lie within the laser's tuning range. The gas whose spectrum is being investigated can be contained within either an intracavity or else in an external cell. The intracavity cell gives much more sensitivity but is a little less convenient since it has to be fitted with Brewster windows. A typical example of a Lamb-dip spectrum is shown in Fig. 4.26. Naturally much work in this field has concentrated on the good coincidences such as SF_6 and OsO_4 with CO_2 [429] and of $^{15}NH_3$ with N_2O [430], but these studies have served to highlight several new spectroscopic phenomena, in particular the mechanisms of rotational, vibrational and translational relaxation. But for the reasons given above, most molecules will be expected to have at least one useful coincidence with a powerful laser line [431] so there remains very much more further work to do.

A second, and related, way of achieving Doppler-free spectroscopy is the use of two-photon absorption [432]. This is a highly non-linear effect which depends on the second and higher terms in the expansion of the molecular susceptibility in terms of the electric field (see equation 7.6.31). Powerful laser beams and/or microwave sources are therefore required but the effect does not depend on saturation in the ordinary sense of the word since the system under study is not usually resonant at the fundamental laser frequency. Again one uses counterpropagating beams and notes that a molecule with velocity component v along the tube axis will in the non-relativistic limit see one photon with an apparent frequency

$$v_+ = v_0 \, (1 + v/c), \tag{4.6.2a}$$

whereas it will see the counterpropagating photon with a frequency

$$v_- = v_0 \, (1 - v/c). \tag{4.6.2b}$$

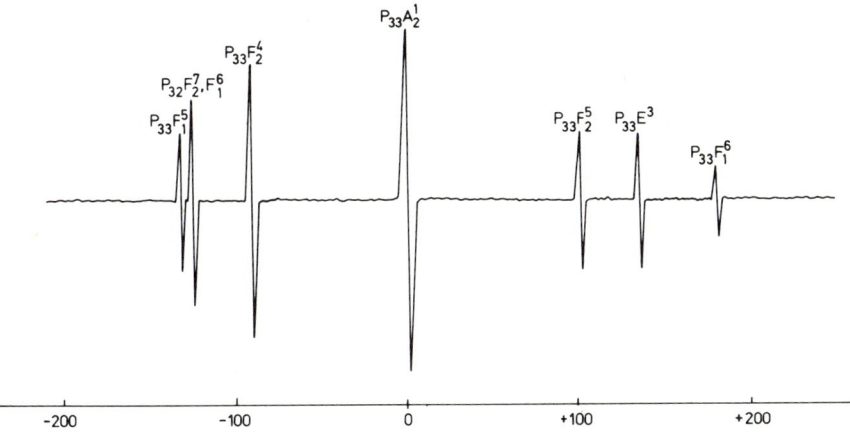

FIG. 4.26. Some Lamb dips due to the fine structure of the P(32) and P(33) lines of the v_3 band of SF_6 which happen to lie within the tuning range of the 10P(18) line of the CO_2 laser After Borde *et al.* [429]

If both photons are absorbed simultaneously the absorber will be transferred to an excited state of frequency $2v_0$. This must be a stationary state so the molecule or atom will absorb at *exactly half* the true resonant frequency and the absorption line will show no Doppler broadening since all the molecules or atoms will absorb together at the *same* frequency. Double-photon spectroscopy has the interesting property that it connects levels of the same parity so one can by this technique probe levels which are not normally accessible by single-photon electric dipole spectroscopy. A famous example is provided by the study of the 2S metastable level of hydrogen (see section 5.1) whose displacement from the nearly coincident 2P level, which *is* accessible, via Lyman α, has been shown to be strictly in accord with the Lamb shift prediction of quantum electrodynamics [433]. Most of the examples shown so far have involved visible region photons because the sharply tunable dye-lasers which are available in the visible region provide superb sources; but there are many possibilities for spectroscopy near 5 μm using two CO_2 laser photons. In the infrared though there are several competing approaches involving various types of coherent Raman scattering [434] which also connect states of the same parity.

In the middle infrared, however, the most promising form of two-photon spectroscopy is that in which one photon is provided by a CO_2 laser or one of its relatives and the other is provided by a tunable microwave source [138]. In essence one can think of this as a means of providing a tunability range of tens of GHz to the otherwise fixed frequency laser line. The Doppler broadening is not now cancelled but, since it is the microwave Doppler width which is involved, it is usually negligible. The topic has been developed by several workers such as Oka [435], Shimizu [436], Jones [138] and others [437]. The sensitivity depends just on the power levels available and since these can be large one can have very high sensitivity and thus, despite the limited number of coincidences which can be investigated, a truly remarkable amount of information has emerged [138]. One example would be the elucidation of the mechanism of molecular "collision" by Oka [132] which has already been mentioned in Chapter 2. Another example which is of particular interest to the molecular spectroscopist is the demonstration of the extreme complexity of what one might call the infrastructure of an infrared absorption band. The familiar strong P, Q and R lines (see section 5.3) are surrounded by a virtual continuum of much weaker lines arising from hot-bands, Coriolis induced bands, Fermi resonances, "forbidden" transitions etc., many of which, it has to be admitted, are at the moment of unknown origin [431]. Clearly two-photon infrared spectroscopy has unearthed a fertile field for the next generation of infrared spectroscopists.

To progress beyond super-high into the realms of ultra-high spectroscopy requires techniques in which coherent interaction is maintained for considerable times. In the microwave region, this is done by means of the elegant separated interaction method introduced by Ramsey [428]. In this a beam of molecules is excited by interaction with a microwave field in one cavity or waveguide; they then travel on for a distance in free space without any microwave field and then enter a second cavity or waveguide where they encounter the same field again. The idea is that for a slight detuning, that is for a point on the homogeneous line shape away from the peak, dephasing will set in as soon as the beam leaves the first interaction zone and by the time it enters the second the coherence with the field will be lost and the interaction will become random. Only at line centre will the coherence be rigidly held so the width of the observed absorption in the second chamber will be determined by the transit time and by the natural line width. In the microwave region this latter is very small indeed and with cavity separations of the order of a few centimetres it is possible to get resolutions of the order of hundreds of hertz. Because coherent interference is involved the observed phenomena are interference fringes which are almost universally called Ramsey fringes. In the infrared there are two important differences [438]; firstly the wavelength is now much smaller than the dimensions of the equipment and secondly the Doppler broadening is no longer negligible. However the same sort of techniques mentioned above, that is interaction with counter propagating waves (in this context that means a standing wave) can be used to overcome Doppler broadening without having to go over to molecular beams. Resolutions obtained so far have been really remarkable. Thus Chebotayev routinely achieves resolutions of 1 part in 10^{11}, that is a few hundred hertz and the superbly equipped group [439] at NBS Boulder are aiming for resolutions of the order of 100 Hz! At this level several new phenomena manifest themselves. Thus to conserve momentum the absorption of a photon necessarily involves a molecular recoil and this gives a systematic Doppler offset to the observed frequency. This leads to values of the order of 20 kHz for a medium-sized ($N \sim 40$) atom or molecule. The ordinary or classical Doppler effect can be completely cancelled by the use of counterpropagating beams but the full relativistic expression indicates that there will be a second-order or transverse Doppler effect whose magnitude can be about 100 kHz at room temperature [440]. This can only be reduced by cooling the gas but at liquid He temperature it would only be 1·5 kHz. This transverse Doppler effect is mostly just a systematic offset but it also leads to some broadening of spectral lines. Other causes of broadening are the AC or dynamic Stark effect [441] which is due to the electric field of the light itself and new forms of collisional perturbation due to correlation effects between atoms or molecules passing close to one another. The first of these is nearly always negligible but the second can be large (> 1 MHz) for ordinary temperatures and moderately low pressures. At cryogenic temperatures and very low ($< 0·01$ torr) pressures, however, these two also become negligible. A major experimental cause of line-broadening is now realised [442] to come from transit time effects. Early experiments used very narrow laser beams and the time for a lightish atom or molecule to cross the beam could be very short. It follows that the line width will be appreciable—of the order of MHz. The transit time broadening can be reduced by using beam-expanding optics or else by using molecular beams in which the molecules travel *axially* down the tube. With suitable large aperture optics, low-temperature operation and the use of the minimum allowable gas pressure, line widths of the order of 17 kHz are now being observed [443]. Such line widths are well below the natural line width of normal dipole-allowed near-infrared visible and ultraviolet transitions and this has suggested the investigation of either electric dipole-forbidden transitions or else of dipole allowed but in fact very weak transitions such as those between high Rydberg states of atoms [439]. The electric-dipole forbidden states reached by two-photon spectroscopy are attractive candidates for frequency standards whilst the Rydberg states give information about subtle electronic interactions both within and between these highly excited atoms. Perhaps the most remarkable advance in this area has been the use of resonant laser pumping to "cool" atoms or ions down to the state where their translational velocities are very low indeed [444]. Such "cold" atoms or ions can then be held for longish periods in so-called "Penning Traps" which are magnetic or electrostatic confinement devices [445]. This technique provides the possibility of studying single atoms under conditions where *all* Doppler and interactive effects are negligible.

These radically new techniques are thus not only giving us a new insight into intra- and intermolecular interactions and providing us with possibilities for a fundamental standards system based entirely on molecular or atomic properties, they are also wiping out the old and deplorable division between microwave and infrared spectroscopy and incidentally providing still another deadly riposte to the old jibe of the laser being "a solution in search of a problem"!

APPENDIX 1

Derivation of the Kramers–Kronig Relations via Fourier Integrals

In Chapter 3, the Kramers–Kronig results were derived by making explicit use of the analyticity of $\hat{X}(\hat{v})$ and then applying Cauchy's integral formula in the form of the theory of residues. The same results can be derived via more elementary methods and, since these give some physical feeling for the meaning of the Kramers–Kronig relations, it is worth going through the derivation in some detail.

We start by noting that the principle of causality obliges the response function to be "one-sided"—that is it is zero for all negative times and finite, at least in principle, for all positive times. The dispersion relations then emerge as a consequence of this "one-sidedness". A one-sided function can always be written as the sum of an even function and an odd function. This is shown pictorially in Fig. A1.1.

In more formal terms the one-sided function $F(t)$ can be written

$$F(t) = F_s(t) + F_a(t), \tag{A1.1}$$

where $F_s(t)$ is a symmetric function and $F_a(t)$ is an antisymmetric function and the range of t is now $-\infty$ to $+\infty$. Since $F(t)$ is perfectly one-sided, and since

$$F_s(+t) = F_s(-t) \quad \text{and} \quad F_a(+t) = -F_a(-t), \tag{A1.2}$$

it follows that

$$F_s(+t) = F_a(+t) \quad \text{and} \quad F_s(-t) = -F_a(-t). \tag{A1.3}$$

The same symmetry relations can be summarised in terms of the signum function ($\text{sgn}(x) = +1$ for $x > 0$ and -1 for $x < 0$), thus:

$$F_a(t) = F_e(t)\,\text{sgn}(t). \tag{A1.4}$$

The next step is to introduce a complex Fourier transform of $F(t)$, namely

$$\hat{\phi}(v) = \int_{-\infty}^{+\infty} F(t)\exp 2\pi ivt, \tag{A1.5}$$

399

which, by the symmetry properties listed above, can be written

$$\hat{\phi}(v) = \int_{-\infty}^{+\infty} F_s(t)\cos 2\pi v t\, dt + i \int_{-\infty}^{+\infty} F_a(t)\sin 2\pi v t\, dt. \qquad (A1.6)$$

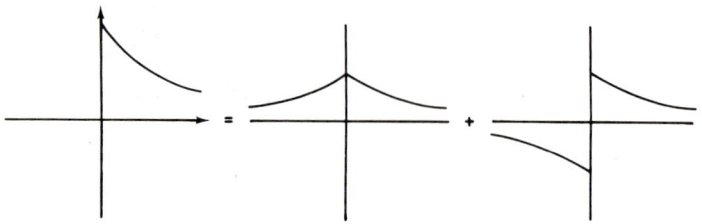

Fig. A1.1

It then follows, since $F(t) = 2F_s(t) = 2F_a(t)$ for $t > 0$, that

$$\phi'(v) = \int_0^{+\infty} F(t)\cos 2\pi v t\, dt \quad \text{and} \quad \phi''(v) = \int_0^{+\infty} F(t)\sin 2\pi v t\, dt \qquad (A1.7)$$

Thus the real and imaginary components of the Fourier transform of a one-sided function are simply the cosine and sine transforms of the function. It follows by inversion that

$$F_s(t) = \int_{-\infty}^{+\infty} \phi'(v)\cos 2\pi v t\, dv \quad \text{and} \quad F_a(t) = \int_{-\infty}^{+\infty} \phi''(v)\sin 2\pi v t\, dv. \qquad (A1.8)$$

Writing now (A1.7) and (A1.4) in the form

$$\phi''(v) = 2\int_0^{\infty} F_a(t)\sin 2\pi v t\, dt = \int_{-\infty}^{+\infty} F_s(t)\,\mathrm{sgn}\,(t)\sin 2\pi v t\, dt$$

$$= -i \int_{-\infty}^{+\infty} F_s(t)\,\mathrm{sgn}\,(t)\exp 2\pi i v t\, dt, \qquad (A1.9)$$

followed by the application of the convolution theorem, then yields

$$\phi''(v) = -\frac{1}{\pi} \int_{-\infty}^{+\infty} \frac{\phi'(v)}{v' - v}\, dv, \qquad (A1.10)$$

since the Fourier transform of the signum function is $i/\pi v$. By entirely similar arguments, one derives,

$$\phi'(v) = \frac{1}{\pi} \int_{-\infty}^{+\infty} \frac{\phi''(v)}{v' - v} \, dv. \tag{A1.11}$$

Equation (A1.10) and (A1.11) are the Kramers–Kronig relations.

APPENDIX 2

Derivation of the Fresnel Equations

The Fresnel equations give the amplitudes of the reflected and transmitted beams which are produced when an incident beam encounters a discontinuity of refractive index. The situation is illustrated in Fig. 3.10. A monochromatic transverse wave is incident, at an angle θ, on the infinitely sharp interface between two perfectly transparent media. At the interface the refractive index changes discontinuously from a value n_1 to a value n_2. Both media are assumed to be non-magnetic dielectrics.

   The boundary conditions can be derived by strict methods but they also follow at once from the intuitive ideas that (1) the tangential components of the electric field are conserved (that is they are continuous) across the boundary, and (2) the tangential components of the magnetic field are likewise conserved. Using these assumptions, the boundary conditions are found to be

$$E_i + E_r = E_t \quad \text{and} \quad H_i + H_r = H_t \tag{A2.1}$$

where the subscripts i, r and t refer to the incident, reflected and transmitted beams respectively. These boundary conditions completely specify the propagation at the interface. Thus the geometric aspects follow from a consideration of the perpendicular components of E. If one writes these as

$$E_i^y = E_i^\perp \sin\left[2\pi v_i t - n_1 k_0 \left(l_i x + n_i z\right)\right],$$

$$E_r^y = E_r^\perp \sin\left[2\pi v_r t - n_1 k_0 \left(l_r x + n_r z\right)\right], \tag{A2.2}$$

and
$$E_t^y = E_t^\perp \sin\left[2\pi v_t t - n_2 k_0 \left(l_t x + n_t z\right)\right].$$

Where the l and n are the direction cosines of the wave vector $\mathbf{k}$, it follows from the boundary conditions that the arguments of the trignometrical functions must be identical for all t, x and z. By identifying terms, it follows at once that the three waves have the same frequency, that their wave vectors lie in a plane which contains the normal, that the reflected ray makes an equal angle to the normal as does the incident ray and that the angles of incidence and refraction are related by the expression

$$n_1 \sin \theta = n_2 \sin \phi. \tag{A2.3}$$

These results are all so well known that, with the possible exception of (A2.3) (Snell's law), they are often taken to be axiomatic, but it is intellectually satisfying that they can be shown to follow from such reasonable boundary conditions.

The derivation of the reflected and transmitted amplitudes requires the use of both boundary conditions and we need therefore to be able to calculate the amplitudes of the magnetic vectors given those of the electric vectors. For transverse propagation, the magnetic vector is perpendicular to both the electric vector and the wave vector and the direction cosines of the magnetic vector follow at once when those of the other two vectors are known. The amplitude of the magnetic vector is given by Maxwell's relation.

$$\sqrt{(\mu_0/\varepsilon_0)}H = \pm\sqrt{\varepsilon_r}E = \pm nE. \tag{A2.4}$$

It will be convenient, in what follows, to use the notation $H^{(x)}$ to mean the magnetic amplitude associated with E^x though it must be stressed that $H^{(x)}$ is *not* a vector parallel to the x axis.

It can readily be shown that the treatment of the behaviour of a wave having an arbitrary polarization can be divided into two separate problems, that where the electric vector lies in the plane of incidence and that where it is perpendicular. Consider first the case where E lies in the plane, that is parallel polarization. Because the boundary conditions involve only the tangential components one need only specify these and one has therefore

$$E_i^x = -E_i'' \cos\theta \sin(2\pi vt - kx),$$
$$E_r^x = E_r'' \cos\theta \sin(2\pi vt + kx), \tag{A2.5}$$
$$E_t^x = -E_t'' \cos\phi \sin(2\pi vt - nkx),$$

where the E_i'' etc. are the amplitudes of the three waves. Using (A2.4), the magnetic amplitudes are found to be

$$\sqrt{(\mu_0/\varepsilon_0)}H_i^{(x)} = -n_1 E_i'',$$
$$\sqrt{(\mu_0/\varepsilon_0)}H_r^{(x)} = -n_1 E_r'', \tag{A2.6}$$
$$\sqrt{(\mu_0/\varepsilon_0)}H_t^{(x)} = -n_2 E_t''.$$

Applying now the boundary conditions (A2.1) at the point of intersection, i.e. where $x = 0$, one has immediately

$$\cos\theta\left[E_i'' - E_r''\right] = E_t'' \cos\phi,$$
$$n_1\left[E_i'' + E_r''\right] = n_2 E_t'', \tag{A2.7}$$

the solutions of which are

$$E_r''/E_i'' = \frac{n_2 \cos\theta - n_1 \cos\phi}{n_2 \cos\theta + n_1 \cos\phi}, \tag{A2.8a}$$

and

$$E_t''/E_i'' = \frac{2n_1 \cos\theta}{n_2 \cos\theta + n_1 \cos\phi}. \tag{A2.8b}$$

The second case, perpendicular polarization, is treated in an entirely similar fashion; one has for the tangential components

$$E_i^y = E_i^\perp \sin\left[2\pi vt - n_1 k_0 \sin\theta x - n_1 k_0 \cos\theta z\right],$$
$$E_r^y = E_r^\perp \sin\left[2\pi vt - n_1 k_0 \sin\theta x + n_1 k_0 \cos\theta z\right], \qquad \text{(A2.9a)}$$
$$E_t^y = E_t^\perp \sin\left[2\pi vt - n_2 k_0 \sin\phi x - n_2 k_0 \cos\phi z\right],$$

and

$$\sqrt{(\mu_0/\varepsilon_0)}H_i^{(y)} = n_1 E_i^\perp \cos\theta,$$
$$\sqrt{(\mu_0/\varepsilon_0)}H_r^{(y)} = -n_1 E_r^\perp \cos\theta, \qquad \text{(A2.9b)}$$
$$\sqrt{(\mu_0/\varepsilon_0)}H^{(y)} = n_2 E_t^\perp \cos\phi,$$

the solutions of which, under the boundary conditions, are

$$E_r^\perp / E_i^\perp = \frac{n_1 \cos\theta - n_2 \cos\phi}{n_1 \cos\theta + n_2 \cos\phi}, \qquad \text{(A2.10a)}$$

$$E_t^\perp / E_i^\perp = \frac{2n_1 \cos\theta}{n_1 \cos\theta + n_2 \cos\phi}. \qquad \text{(A2.10b)}$$

It is interesting to observe that these Fresnel relations give amplitudes which satisfy the other obvious boundary condition, the conservation of energy. The rate of flow of energy is given by the Poynting vector (section 1.2.1)

$$\mathbf{S} = (c/4\pi)\mathbf{E} \times \mathbf{H}, \qquad \text{(A2.11)}$$

so that, for say the parallel case, the incident rate of flow of energy density is proportional to $n_1 (E_i'')^2$. Now, by Lambert's cosine law, the rate of flow of energy density into a small area surrounding the point of intersection will be

$$\frac{d\rho_i}{dt} = \tfrac{1}{2}c\varepsilon_0 n_1 (E_i'')^2 \cos\theta, \qquad \text{(A2.12)}$$

and the rate of departure of energy density will be

$$\frac{d\rho_r}{dt} = \tfrac{1}{2}c\varepsilon_0 n_1 (E_i'')^2 \cos\theta \left[\frac{n_1 \cos\theta - n_2 \cos\phi}{n_1 \cos\theta + n_2 \cos\phi}\right]^2, \qquad \text{(A2.13)}$$

and

$$\frac{d\rho_t}{dt} = \tfrac{1}{2}c\varepsilon_0 n_2 (E_i'')^2 \cos\phi \left[\frac{2n_1 \cos\theta}{n_1 \cos\theta + n_2 \cos\phi}\right]^2, \qquad \text{(A2.14)}$$

and hence

$$\frac{d\rho_i}{dt} = \frac{d\rho_r}{dt} + \frac{d\rho_t}{dt}. \qquad \text{(A2.15)}$$

The Fresnel equations are derived by the application of field continuity conditions whereas Snell's law is derived only from geometrical reasoning and therefore the two are independent. They can, however, readily be combined to give the alternative forms

$$E_r''/E_i'' = \frac{\tan(\theta - \phi)}{\tan(\theta + \phi)},$$

$$E_r^\perp/E_i^\perp = -\frac{\sin(\theta - \phi)}{\sin(\theta + \phi)},$$

$$E_t''/E_i'' = \frac{2\sin\phi\cos\theta}{\sin(\theta + \phi)\cos(\theta - \phi)}, \qquad (A2.16)$$

$$E_t^\perp/E_i^\perp = \frac{2\sin\phi\cos\theta}{\sin(\theta + \phi)}.$$

These forms, especially the reflection formulae, are much simpler than the corresponding formulae (equations A2.8 and A2.10) and for this reason the Fresnel relations are often quoted in this way. However, it must be borne in mind that since the law of Snell becomes indeterminate at normal incidence so do the above relations which have been derived using it. With this sole proviso the purely trignometrical form of the Fresnel equations can be safely used. In particular the existence of a polarizing (or Brewster) angle follows at once since for $\theta + \phi = \pi/2$, $\tan(\theta + \phi) = \infty$ and hence $E_r^\perp$ is zero.

The Propagation of Electromagnetic Radiation in Circular Section Pipes

The theory of the propagation of electromagnetic waves in pipes has been thoroughly developed by Stratton [445a]. This general theory covers all kinds of pipes from those made of transparent dielectric up to those made of superconducting metal, but Stratton was mostly interested in pipes made of metals with high, but not infinite, conductivity—for example copper. It is found in general that two types of mode exist, *T*ransverse *E*lectric (TE or H) and *T*ransverse *M*agnetic (TM or E), but pure forms only occur for infinite conductivity or for the special case which arises when the first mode number $n = 0$. For $n > 0$ and for finite conductivity, mixtures of the two mode types result which, in the extreme, become hybrid modes which are designated EH.

Stratton shows that application of the boundary conditions (continuity of the tangential components of **E** and **H**) leads to the four equations:

$$\frac{ni\hat{\gamma}_p}{u^2} J_n(u)a_n^i + \frac{2\pi iv}{\varepsilon_0 c^2 u} J_n'(u)b_n^i = \frac{ni\hat{\gamma}_p}{v^2} H_n^{(1)}(v)a_n^e + \frac{2\pi iv}{\varepsilon_0 c^2 v} H_n^{(1)\prime}(v)b_n^e, \quad (A3.1)$$

$$J_n(u)a_n^i = H_n^{(1)}(v)a_n^e, \quad (A3.2)$$

$$\frac{-i\hat{\gamma}_1{}^2 \varepsilon_0 c^2}{2\pi vu} J_n'(u)\,a_n^i - \frac{ni\hat{\gamma}_p}{u^2} J_n(u)b_n^i = -\frac{i\hat{\gamma}_2{}^2 \varepsilon_0 c^2}{2\pi vv} H_n^{(1)\prime}(v)a_n^e$$

$$-\frac{ni\hat{\gamma}_p}{v^2} H_n^{(1)}(v)b_n^e, \quad (A3.3)$$

$$J_n(u)b_n^i = H_n^{(1)}(v)b_n^e, \quad (A3.4)$$

Where the superscripts i and e mean interior of the pipe and exterior of the pipe respectively, $\hat{\gamma}$ is the complex propagation factor of the pipe, $\hat{\gamma}_1$ that of the interior material and $\hat{\gamma}_2$ that of the external material, n is the principal mode number and the parameters u and v are defined by

$$u = a\left|\sqrt{(\hat{\gamma}_p{}^2 - \hat{\gamma}_1{}^2)}\right|, \quad v = a\left|\sqrt{(\hat{\gamma}_p{}^2 - \gamma_2{}^2)}\right|, \quad (A3.5)$$

where a is the radius of the pipe. The Bessel functions of the first kind, $J_n(u)$, ensure proper behaviour of the field inside the pipe and the Hankel functions $H_n^{(1)}(v)$ ensure that the field radiates properly away to zero amplitude at infinity. The prime superscripts on these functions indicate differentiation with respect to the argument. It is assumed throughout that both the interior and the exterior material are non-magnetic with susceptibilities identically equal to that of free space.

The quantities a_n^i, a_n^e, b_n^i and b_n^e are field coefficients which are determined by the initial conditions. Clearly one solution of equations (A3.1–A3.4) is that all four field coefficients are zero. A more interesting set of solutions is arrived at by using (A3.2) and (A3.4) to eliminate the external coefficients from (A3.1) and (A3.3) and then writing the resulting two equations in matrix vector form:

$$
\begin{bmatrix}
n\hat{\gamma}_p\left(\dfrac{1}{u^2}-\dfrac{1}{v^2}\right), & \left(\dfrac{2\pi v}{\varepsilon_0 c^2}\dfrac{1}{u}\dfrac{J_n'(u)}{J_n(u)}-\dfrac{1}{v}\dfrac{H_n^{(1)'}(v)}{H_n^{(1)}(v)}\right) \\[3mm]
\dfrac{\varepsilon_0 c^2}{2\pi v}\left(\dfrac{\hat{\gamma}_1^{\,2}}{u}\dfrac{J_n'(u)}{J_n(u)}-\dfrac{\hat{\gamma}_2^{\,2}}{v}\dfrac{H_n^{(1)'}(v)}{H_n^{(1)}(v)}\right), & n\hat{\gamma}_p\left(\dfrac{1}{u^2}-\dfrac{1}{v^2}\right)
\end{bmatrix}
\begin{bmatrix} a_n^i \\[3mm] b_n^i \end{bmatrix} = 0 \qquad \text{(A3.6)}
$$

The alternative solutions then arise from demanding that the matrix be singular, that is that its determinant be zero; this gives

$$
n^2\hat{\gamma}_p^{\,2}\left(\dfrac{1}{u^2}-\dfrac{1}{v^2}\right)=\left(\dfrac{1}{u}\dfrac{J_n'(u)}{J_n(u)}-\dfrac{1}{v}\dfrac{H_n^{(1)'}(v)}{H_n^{(1)}(v)}\right)\left(\dfrac{\hat{\gamma}_1^{\,2}}{u}\dfrac{J_n'(u)}{J_n(u)}-\dfrac{\hat{\gamma}_2^{\,2}}{v}\dfrac{H_n^{(1)'}(v)}{H_n^{(1)}(v)}\right). \qquad \text{(A3.7)}
$$

Equation (A3.7) is highly transcendental and no general solution may be derived. However, in many practical situations one may derive approximate solutions which can often be used perfectly well. One of the most important of these arises when the conductivity of either medium is very high. In this case one can assume that the modes are predominantly either transverse electric or transverse magnetic with only small mixing due to the finite conductivity. If the mode number n is zero—and the field therefore is circularly symmetric—the waves are necessarily of the pure types *even if* the conductivity is less than infinite. For microwave and infrared work one is usually concerned with the case where the external medium is a metal and the interior medium is a dielectric—usually air or vacuum. The converse case of a solid metal cylinder (that is a wire) embedded in a dielectric is of enormous significance and interest to electrical engineers. A major difference between these two cases is that for the latter the modes are determined by the ratio of Hankel functions and since, say, $H_n^{(1)}(0) = \infty$, it is possible for this ratio to be zero when $v = 0$ and therefore waves of *any* frequency can propagate down the wire. For the dielectric tube with a metal outer, on the other hand, the modes are determined by the ratio of first kind Bessel functions and since $J_0(0)$ is *not* zero it follows that propagation for any arbitrary frequency may not be possible. This is sometimes put succinctly by saying that there is no principal wave in a hollow metallic pipe. The equations (A3.7) have for each value of n an infinite (denumerable) number of solutions each of which is designated by a second mode label m. Since the solutions are discrete, only discrete values of u occur and these are designated u_{nm}. From (A3.5) rearranged with proper care for the complex quantities one has

$$\hat{\gamma}_p = \sqrt{\left[\left(\frac{u_{nm}}{a}\right)^2 + \hat{\gamma}_1^2\right]}.$$

(A3.8)

This equation has some interesting consequences. If $\hat{\gamma}_1$ is pure imaginary as would occur if the dielectric were transparent, then $\hat{\gamma}_p$ is either pure real or pure imaginary depending on the relative magnitude of $\hat{\gamma}_1$ and (u_{nm}/a). The cross-over condition occurs when $\hat{\gamma}_1 = i(u_{nm}/a)$ and therefore

$$2\pi n v_c/c = u_{nm}/a,$$

(A3.9a)

that is

$$v_c = \frac{u_{nm} c}{2\pi n a},$$

(A3.9b)

where v_c is the cut-off frequency. It is important to note that unlike the resonant cavity, the infinite pipe can propagate any frequency, in a given mode, provided the frequency is *greater* than the corresponding cut-off frequency. For this reason the various cut-off frequencies are also usually designated by the mode labels v_{nm}.

The modes of the hollow metallic pipe are, as mentioned above, determined by the first kind Bessel functions. Explicitly we have the following conditions

$$\text{TE modes } J'_n(u) = 0,$$

(A3.10a)

$$\text{TM modes } J_n(u) = 0.$$

(A3.10b)

The symmetrical case $n = 0$ is interesting because of the coincidence that

$$J_0'(x) = J_1(x) = -J_{-1}(x).$$

(A3.11)

The mode conditions for TE_0 modes are therefore often quoted in terms of $J_1(x)$ instead of $J_0'(x)$ which can sometimes cause confusion. The lowest possible value of u_{nm} is therefore u'_{11}, that is the first root of $J'_1(u)$, namely 1·84. The longest wavelength, therefore, that can be propagated by the pipe is (via A3.9b)

$$\lambda'_{11} = 2\pi a/u'_{11} = 3·415\, a,$$

(A3.12)

where λ is the wavelength *in the dielectric*. This mode is, however, rather strongly attenuated compared with the TE_{01} mode which, since $u_{01} = 2·405$, has the second longest critical wavelength,

$$\lambda_{01} = 2\pi a/u_{01} = 2·612\, a.$$

(A3.13)

λ_{01} is therefore the longest practical wavelength which the pipe will propagate and as a very rough rule of thumb one can say that the critical wavelength is equal to the tube diameter. Hollow metal pipes are therefore only of use for transmitting microwaves and shorter wavelength radiation. However, it must be borne in mind that the theory applies to a *perfect* cylinder and as a decreases it becomes more and more difficult to make perfect cylinders. If to overcome this practical difficulty one makes $a \gg \lambda$ then the pipe is overmoded and the desired mode may be able to decay into other more heavily attenuated modes. Clearly in any situation a compromise has to be struck.

The phase and group velocities in the pipe are found by writing (A3.8) in the form

$$\gamma_p = \gamma_1 \left[1 - \left(\frac{v_{nm}}{v} \right)^2 \right]^{\frac{1}{2}}, \tag{A3.14}$$

which follows because γ_1 is pure imaginary. Then if $v_1 = c/n$ is the phase velocity in an unbounded specimen of the inner dielectric it follows that

$$v_{nm}^p = v_1 \left[1 - \left(\frac{v_{nm}}{v} \right)^2 \right]^{-\frac{1}{2}}, \tag{A3.15a}$$

$$v_{nm}^g = v_1 \left[1 - \left(\frac{v_{nm}}{v} \right)^2 \right]^{\frac{1}{2}}, \tag{A3.15b}$$

and $v_{nm}^p \times v_{nm}^g = v_1^2$ as one would expect for normal dispersion. This dispersion can be troublesome if the pipe is being used to propagate a modulated wave for, since different frequencies travel at different speeds, a message can become garbled if it is being sent over a long distance. To reduce the effects of dispersion v_{nm}/v must be small, that is $\lambda_{nm} \ll a$ and one is back with the mode break-up problem again. The solution that has been found is to wind the pipe as a helix from insulated wire. This suppresses longitudinal currents and helps to preserve a pure TE mode. Long-distance communication links operating on TE_{01} at 70 GHz (4·3 mm) would use helical waveguides with $a \approx 70$ mm.

The most practically interesting aspect of hollow-pipe propagation is the attenuation per unit length because only if this is low enough is such a mode of propagation likely to be chosen in the first case. If the outer metal is a perfect conductor, the dissipation can be found from (A3.8) where $\hat{\gamma}_1$ is now complex. In this it is assumed that the absorption is not too strong; this is necessary since without this proviso the basic assumption that u_{nm} is a root of the appropriate Bessel function fails. In this case one would have to solve equation (A3.7) directly. The usual case, however, is where the inner material is a perfect dielectric and the outer is an imperfect metal. The metallic conductivity, although not infinite, will still be very high and equation (A3.7) can be solved by a perturbation method in which one expands the Bessel functions about the appropriate value of u_{nm}. Stratton calculates the attenuation by a different method and arrives at the approximate result

$$\alpha_{nm}' = a^{-1} \left(\frac{\pi \varepsilon_0 \varepsilon_1' v}{\sigma_2} \right)^{\frac{1}{2}} \left(1 - \left(\frac{v_{nm}'}{v} \right)^2 \right)^{-\frac{1}{2}} \left[\left(\frac{v_{nm}'}{v} \right)^2 \right.$$

$$\left. + \left(\frac{n}{u_{nm}'} \right)^2 \left(1 - \left(\frac{n}{u_{nm}'} \right)^2 \right)^{-1} \right], \tag{A3.16}$$

where σ_2, the conductivity of the outer material, is defined by

$$\sigma_2 = 2\pi\varepsilon_0\varepsilon_2'' v = 4\pi\varepsilon_0 n_2 k_2 v. \tag{A3.17}$$

It is most interesting to observe that α_{01}'—the attenuation of the TE_{01} mode—falls to zero as $v \to \infty$ even in the lossy case. When the inner dielectric is air or vacuum we have $\varepsilon_1' = 1$ and $\varepsilon_1'' = 0$ and therefore

$$\alpha_{01} = \frac{1}{2a} \frac{1}{\sqrt{(n_2 k_2)}} \left(\frac{v_{01}'}{v}\right)^2, \tag{A3.18}$$

with the assumption that $v \gg v_{01}'$. Marcatili and Schmeltzer [185] give a different approximate solution,

$$\alpha_{nm} = \left(\frac{u_{nm}}{2\pi}\right)^2 \frac{\lambda^2}{a^3} R_e \begin{cases} \dfrac{1}{\sqrt{(\hat{n}^2 - 1)}} & \text{for } TE_{0m} \text{ modes,} \\[2mm] \dfrac{\hat{n}^2}{\sqrt{(\hat{n}^2 - 1)}} & \text{for } TM_{0m} \text{ modes,} \\[2mm] \dfrac{\frac{1}{2}(\hat{n}^2 + 1)}{\sqrt{(\hat{n}^2 - 1)}} & \text{for } EH_{nm} \text{ modes,} \end{cases} \tag{A3.19}$$

which is more useful in the case of relatively low conductivity. However, their solution reduces to that of Stratton in the high-conductivity limit, thus one has

$$\begin{aligned} \alpha_{01} &= \left(\frac{u_{01}}{2\pi}\right)^2 \frac{\lambda^2}{a^3} R_e\left(\frac{1}{\sqrt{(\hat{n}^2 - 1)}}\right), \\ &= \left(\frac{v_{01}'}{v}\right)^2 \frac{1}{a} R_e\left(\frac{1}{\sqrt{(n_2^2 - k_2^2 - 1 - 2ink)}}\right). \end{aligned} \tag{A3.20a}$$

Assuming now, as is reasonable for a metal, that $n \approx k \gg 1$ one has

$$\begin{aligned} \alpha_{01} &= \left(\frac{v_{01}'}{v}\right)^2 \frac{1}{a} \frac{1}{\sqrt{(n_2 k_2)}} R_e\left(\frac{1}{\sqrt{(-2i)}}\right), \\ &= \left(\frac{v_{01}'}{v}\right)^2 \frac{1}{2a} \frac{1}{\sqrt{(n_2 k_2)}}, \end{aligned} \tag{A3.20b}$$

in agreement with (A3.18).

References

[1] SI: the International System of Units (translation of the International Bureau of Weights and Measures publication "Le Système international d'unites") (Eds C. H. Page and P. Vigoureux), HMSO, 3rd Edition (1977), ISBN 0.11.480056.

[2] D. J. E. Knight and W. R. C. Rowley, Survey Review XXIV, 185 (1977).

[3] E. R. Cohen and B. N. Taylor, CODATA task group on Fundamental Constants, CODATA Bulletin No. 11 (1973), *J. Phys. Chem. Ref. Data* **2** (4), 663 (1973).

[4] R. A. Smith, F. E. Jones and R. P. Chasmar, "The Detection and Measurement of Infrared Radiation", First Edition (1957) Revised Second Edition (1968), Oxford University Press.

[5] L. J. Bellamy "The Infrared Spectra of Complex Molecules", Methuen, London (1958).

[6] C. G. Granquist, *Appl. Opt.* **20**, 2606 (1981).

[7] John Chamberlain, "The Principles of Interferometric Spectroscopy" (completed, collated and edited by G. W. Chantry and N. W. B. Stone), John Wiley, Chichester, UK (1979).

[8] G. W. Chantry, "Submillimetre Spectroscopy", Academic Press, London and New York (1971).

[9] For an account see Reference 15 pp. 413–415.

[10] R. F. Hoskins, "Generalised Functions", Ellis Horwood, Chichester, UK (1979).

[11] R. Bracewell, "The Fourier Transform and Its Applications", McGraw Hill, New York (1965).

[12] J. Connes, Thesis, University of Paris, 7 October 1960, subsequently published as a special issue of *Rev. Opt. Théor. Instrum.* serie, A No. 3579, No. d'Ordre 4451 and in *Rev. Opt. Théor. Instrum.* **40**, 45, 116, 171 and 231 (1961). Mme Connes' thesis was also issued as a US Air Force Special Report, AFCRL Rep No. 11471–0019.

[13] A. H. Filler, *J. Opt. Soc. Am.* **54**, 762 (1964).

[14] R. H. Norton and R. Beer, *J. Opt. Soc. Am.* **66**, 259 (1976).

[15] G. Arfken, "Mathematical Methods for Physicists", Academic Press, 2nd Edition, New York and London (1970).

[16] R. B. Blackman and J. W. Tukey, "The Measurement of Power Spectra from the Point of View of Communication Engineering". Dover, New York (1959).

[17] For good introductions to information theory see L. Brillouin, "Science and Information Theory", 2nd Edition, Academic Press, New York (1962), and Shannon and Weaver "The Mathematical Theory of Information", University

of Illinois Press, Urbana, Illinois (1949). A good modern article is by A. D. Wyner, *Proc. IEEE*, **69**(2), 239 (1981).

[18] M. S. Bartlett, "An Introduction to Stochastic Processes", 2nd Edition, Cambridge University Press (1966).

[19] J. K. Kauppinen, D. J. Moffatt, H. H. Hantsch and D. G. Cameron, *Appl. Spectrosc.* **35**, 211 (1980); J. K. Kauppinen, D. J. Moffatt, D. G. Cameron and H. H. Hantsch, *Appl. Opt.* **20**, 1866 (1981).

[20] For good general accounts of maximum entropy methods see D. G. Childers, "Modern Spectrum Analysis", IEEE Press, New York (1978); T. E. Barnard, Rep. TR–75–01, Texas Instruments, June (1975); L. G. Griffiths and R. Prieto–Diaz, *Trans. GeoSci. Electron.* **GE-15**, 13 (1977); T. E. Landers and R. T. Lacoss, *IEEE Trans. GeoSci. Electron.* **GE-15**, 26 (1977).

[21] "Information Theory and Statistical Mechanics" E. T. Jaynes, Papers I and II, *Phys. Rev.* **106**, 620 (1957) and **108**, 171 (1957); see also E. T. Jaynes New engineering applications of information theory, *in* "Proceedings of the First Symposium on Engineering Applications of Random Function Theory and Probability" (Eds J. R. Bogdanoff and F. Kozin), pp. 163–203, John Wiley, New York (1963).

[22] R. T. Lacoss, *Geophysics*, **36** (1971).

[23] G. White, M.Sc. Thesis, Jodrell Bank, University of Manchester (1973); an account of the non-linear theory can be found in C. H. Chen, *Proc. IEEE*, **69**, 839 (1981); for a criticism of the MEM see M. A. Fiddy and T. J. Hall, *J. Opt. Soc. Am.* **71**, 1406 (1981).

[24] V. Cappellini, A. G. Constantides and P. Emiliani, "Digital Filters and their Applications", Academic Press, New York and London (1978), ISBN 0–12–159250–2.

[25] G. Winnewisser, E. Churchwell and C. M. Walmsley, Astrophysics of interstellar molecules, *published in* "Modern Aspects of Microwave Spectroscopy" (Ed. G. W. Chantry), Academic Press, New York and London (1979).

[26] H. P. Baltes, *Infrared Phys.* **16**, 1 (1976).

[27] J. R. Benoit, C. Fabry and A. Perot, *Trav. Mém. Bur. Int. Poids Mes.* **15**, 1 (1913); see also, H. Barrell *Proc. R. Soc.* **A186**, 164 (1946).

[28] R. Y. Chiao and B. P. Stoicheff, *J. Opt. Soc. Am.* **54**, 1286 (1964).

[29] E. A. Baker and B. Walker, *J. Phys. E.* (*Sci. Instrum.*) (1982).

[30] A. G. Fox and T. Li, *Bell. Syst. Tech. J.* **40**, 453 (1961); G. D. Boyd and J. P. Gordon, *Bell. Syst. Tech. J.* **40**, 489 (1961).

[31] H. Kogelnik and T. Li. *Proc. IEEE*, **54**, 1312 (1966); see also P. W. Smith, Single-frequency lasers *in* "Lasers" (Eds A. K. Levine and A. J. De Maria), Vol. 4, Marcel Dekker, New York (1976).

[32] A. F. Harvey, "Coherent Light", Wiley Interscience, London and New York (1970).

[33] L. A. Weinstein, "Open Resonators and Open Waveguides", Golan Press Boulder Colorado (1969).

[34] G. Goubau and F. Schwering, *IRE Trans.* **AP-9**, 248 (1961).

[35] D. H. Martin and J. Le Surf, *Infrared Phys.* **18**, 405 (1978).

[36] A. Abraham and S. D. Smith, *J. Phys. E.* (*Sci. Instrum.*), **15**, (1982).

[37] A. L. Cullen and P. Nagenthiram, Millimetre open resonators, *in* "High Frequency Dielectric Measurement" (Eds J. Chamberlain and G. W. Chantry), IPC Science and Technology Press, Guildford, England (1973).

[38] R. N. Clarke and C. B. Rosenberg, *J. Phys. E.* (*Sci. Instrum.*), **15**, (1982).

[39] A. Crocker, H. A. Gebbie, M. F. Kimmitt and L. E. S. Mathias, *Nature* (*London*) **201**, 250 (1964); H. A. Gebbie, N. W. Stone and F. D. Findlay, *Nature* (*London*) **202**, 685 (1964).

[40] See for example F. K. Kneubuhl and E. Affolter, Infrared and submillimetre-wave waveguides, *in* "Infrared and Millimeter Waves", Vol. 1. "Sources of Radiation" (Ed. K. J. Button), Chapter 6, Academic Press, New York and London (1979).

[41] P. Belland, D. Veron and L. B. Whitbourn, *J. Phys. D: Appl. Phys.* **8**, 2113–2122 (1975).

[42] D. T. Hodges and T. S. Hartwick, *Appl. Phys. Lett.* **23**, 252–253 (1973).

[43] R. H. Dicke, P. J. E. Peebles, P. G. Roll and D. T. Wilkinson, *Astrophys. J.* **142**, 414 (1965).

[44] S. Weinberg "The First Three Minutes", Andre Deutch, London (1977).

[45] A. A. Penzias and R. W. Wilson, *Astrophys. J.* **142**, 419 (1965).

[46] D. P. Woody, J. C. Mather, N. S. Nishioka and P. L. Richards *Phys. Rev. Lett.* **34**, 1036 (1975); see also E. I. Robson, D. G. Vickers, J. S. Huizinger, J. E. Beckman and P. E. Clegg, *Nature* **251**, 591 (1974); also article by P. E. Clegg *in* "Infrared Astronomy" (Eds G. Selti and G. G. Fazio), D. Reidel, North Holland (1978).

[47] The phenomenon was apparently first reported by C. H. Townes and A. C. Cheung for cosmic formaldehyde in *Astrophys. J. Lett.* **157**, L103 (1969).

[48] This work is lucidly reviewed by P. C. W. Davies *in Rep. Prog. Phys.* **41** (8), 1313 (1978).

[49] S. W. Hawking, *Nature* (London) **248**, 30 (1975); *Commun. Math. Phys.* **43**, 199 (1975); but also see reference [48] for an illuminating discussion.

[50] J. D. Barrow and F. J. Tipler, *Nature* (*London*) **276**, 453 (1978).

[51] J. E. Beckman and A. F. M. Moorwood, Infrared astronomy *published in* Rep. Prog. Phys. **42**, 87 (1979).

[52] J. W. Fleming and G. W. Chantry, *IEEE Trans. Instrum. Meas.* **IM-23**, 473 (1974).

[53] H. A. Willis and M. E. A. Cudby, private communication.

[54] J. E. Harries *J. Opt. Soc. Am.* **67**, 880 (1977).

[55] E. B. Wilson Jr. and A. J. Wells, *J. Chem. Phys.* **14**, 578 (1946).

[56] Th. Encrenaz, *Infrared Phys.* **19**, 353–373 (1980).

[57] J. Hansen, D. Johnson, A. Lacis, S. Lebedeff, P. Lee, D. Rind and G. Russell, *Science* **213**, 957 (1981).

[58] A. Shkolnik, C. R. Taylor, V. Finch and A. Borat, *Nature* (*London*) **283**, 373 (1980); R. Dmiel, A. Prevulotzky and A. Shkolnik, *Nature* (*London*) **283**, 761 (1980).

[59] H. M. Nuzzenzveig, "Introduction to Quantum Optics", *in* the series "Documents on Modern Physics", Gordon and Breach, London and New York (1975); G. J. Troup and R. G. Turner, Optical coherence theory, *Rep. Prog. Phys.* **37**, 771 (1974).

[60] M. Born and E. Wolf, "Principles of Optics", Pergamon Press, Oxford (1959).

[61] M. J. Halmos and J. Shamir, *Appl. Opt.* **21**, 265 (1982).

[62] G. W. Chantry and John Chamberlain, Far infrared spectra of polymers, *in* "Polymer Science", a Materials Science Handbook (Ed. A. D. Jenkins), Vol. 2, North-Holland, Amsterdam (1972).

[63] L. Jannossy, *Nuovo Cim.* **6**, 111 (1957); ibid. **12**, 369 (1959).

[64] A. A. Michelson, "Light Waves and their Uses", University of Chicago Press (1902); "Studies in Optics", Phoenix Edition, University of Chicago Press (1962).

[65] L. Mandel and E. Wolf, *J. Opt. Soc. Am.* **51**, 815 (1961).

[66] R. H. Brown and R. Q. Twiss, *Proc. R. Soc.* A242, 300 (1957); ibid. A243, 291 (1957); ibid. A248, 199 (1958).

[67] P. W. Smith, *IEEE J. Quant. Elect.* **QE-1** (8), 343 (1965).

[68] M. Yamanaka, H. Yoshinaga and S. Kon. *Jap. J. Appl. Phys.* **7**, 827 (1967).
[69] W. E. Lamb, *Phys. Rev.* **134**, A1429 (1964); the original "hole-burning" concept is due to W. R. Bennett, *Phys. Rev.* **126**, 580 (1962).
[70] C. K. N. Patel and E. D. Shaw, *Phys. Rev. Lett.* **24**, 451 (1970).
[71] O. R. Wood and S. E. Schwarz, *Appl. Phys. Lett.* **11**, 88 (1967).
[72] R. G. Jones, C. C. Bradley, J. Chamberlain, H. A. Gebbie, N. W. B. Stone and H. Sixsmith, *Appl. Opt.* **8**, 701 (1969).
[73] C. K. N. Patel, Gaseous optical masers, *in* "Lasers and Applications", Chapter 2, Ohio State University (1963); Gas Lasers *in* "Lasers" (Eds A. K. Levine and A. J. De Maria), Vol. 2, Marcel Dekker, New York (1971).
[74] T. H. Maiman, *Phys. Rev. Lett.* **4**, 564 (1960).
[75] J. C. Polanyi, *J. Chem. Phys.* **34**, 347 (1961); J. V. V. Kasper and G. C. Pimentel, *Phys. Rev. Lett.* **14**, 352 (1965); P. J. Kuntz, E. M. Nemeth and J. C. Polanyi, *J. Chem. Phys.* **50**, 4607 (1969); P. J. Kuntz, M. H. Mok and J. C. Polanyi, *J. Chem. Phys.* **50**, 4623 (1969).
[76] K. L. Kompa and G. C. Pimentel, *J. Chem. Phys.* **47**, 857 (1967); see also G. C. Pimentel, Infrared study of transient molecules in chemical lasers, *in* "Molecular Spectroscopy IX", General Lectures presented at the Ninth European Molecular Spectroscopy Congress, Butterworth's (London) (1969).
[77] A. J. De Maria and C. J. Ultee, *Appl. Phys. Lett.* **9**, 67 (1966).
[78] B. Lax, Progress in semi conductor lasers, *in* IEEE Spectrum, p. 62 (1965); see also J. F. Butler, Semiconductor diode lasers *in* "Applied Optics and Optical Engineering" (Eds R. Kingslake and B. J. Thompson), Vol. 6, Academic Press, New York and London (1980); E. D. Hinkley and P. L. Kelley, *Science* **171**, 635 (1971); E. D. Hinkley, *Appl. Phys. Lett.* **13**, 49 (1968).
[79] G. W. Chantry and G. Duxbury, Molecular lasing systems, *in* "Methods of Experimental Physics", Vol. III, "Molecular Spectroscopy" (Ed. D. Williams), Academic Press, New York and London (1974).
[80] D. R. Lide and A. G. Maki, *Appl. Phys. Lett.* **11**, 2 (1967).
[81] L. O. Hocker and A. Javan, *Phys. Lett.* **25A**, 489 (1967.
[82] B. Hartman and B. Kleman, *Appl. Phys. Lett.* **12**, 168 (1968); W. S. Benedict, ibid. p. 170; M. A. Pollack and W. J. Tomlinson, ibid. p. 173; W. S. Benedict, M. A. Pollack and W. J. Tomlinson, *IEEE J. Quant. Electron.* **QE5**, 108 (1969).
[83] M. A. Pollack, *IEEE. J. Quant. Electron.* **QE5**, 558 (1969).
[84] J. Reid and K. Siemsen, *Appl. Phys. Lett.* **29**, 250 (1976); K. J. Siemsen and J. Reid, *Opt. Commun.* **20**, 284 (1977).
[85] M. Yamanaka, *Rev. Laser Engng* **3**, 253 (1976); M. Rosenbluh, R. J. Temkin and K. J. Button, *Appl. Opt.* **15**, 2635 (1976); J. J. Gallagher, M. D. Blue, B. Bean and S. Perkowitz, *Infrared Phys.* **17**, 43 (1977); D. T. Hodges and D. J. E. Knight, "Laser Handbook", to be published.
[86] T. Y. Chang and T. J. Bridges, *Opt. Commun.* **1**, 423 (1970); T. Y. Chang, T. J. Bridges and E. G. Burkhardt, *Appl. Phys. Lett.* **17**, 249 (1970).
[87] T. Oka, *Adv. Atom. Mol. Phys.* **9**, 127 (1974).
[88] T. Y. Chang and J. D. McGee, *Appl. Phys. Lett.* **29**, 725 (1976); D. G. Biron, R. J. Temkin, B. Lax and B. G. Danly, *Opt. Lett.* **4**, 381 (1979).
[89] J. C. Slater, "Microwave Electronics", Van Nostrand, New York (1950).
[90] J. M. Manley and H. E. Rowe, *Proc. IRE* **47**, 2115–2116 (1959).
[91] J. R. Birch, *Electron. Lett.* **16**, 799 (1980).
[92] C. H. Townes and A. L. Schawlow, "Microwave Spectroscopy", McGraw Hill, New York (1955); W. Gordy, Microwave spectroscopy, *in* "Handbook of Physics", 2nd Edition (Eds E. U. Condon and H. Odishaw), Chapter 6, McGraw Hill, New York (1967); W. Gordy and R. L. Cook, "Microwave Molecular Spectra", Wiley Interscience, New York (1970); for further re-

ferences see G. Roussy and G. W. Chantry, Microwave spectrometers, *in* "Modern Aspects of Microwave Spectroscopy" (Ed. G. W. Chantry), Academic Press, London and New York (1979).

[93] See for example T. Oka, Infrared and radio frequency spectroscopy in the laser cavity, *in* "Frontiers in Laser Spectroscopy" (Eds Balian *et al.*). North Holland, Amsterdam (1977).

[94] H. Hellwig, "Frequency Standards and Clocks", NBS Technical Note 616 (2nd Revision); Atomic Frequency Standards Survey, *Proc. IEEE* **63**, 212 (1975); C. Audoin and J. Vanier, Atomic frequency standards and clocks. *J. Phys. E.* (*Sci. Instrum.*) **9**, 697 (1976).

[95] A. F. Pearce and D. J. Wootton, Reflex Klystrons, *in* "Millimetre and Submillimetre Waves" (Ed. F. A. Benson), Iliffe, London (1969).

[96] W. Gordy, *Chim. Pure Appliquee* **11**, 403 (1965); see also as an example, F. C. De Lucia, D. Helminger and W. Gordy, *Phys. Rev.* **A3**, 1849 (1971).

[97] H. Kilp, *J. Phys. E* (*Sci. Instrum.*), **10**, 985 (1977).

[98] Ph. Helminger, F. C. De Lucia and W. Gordy, *Bull. Am. Phys. Soc.* **16**, 531 (1971).

[99] G. Kantorowicz and P. Palluel, Backward-wave oscillators, *in* "Infrared and Millimetre Waves" (Ed. K. J. Button), Vol. 1, Academic Press, New York and London (1979).

[100] A. F. Krupnov and A. V. Burenin, *in* "Molecular Spectroscopy: Modern Research" (Ed. K. N. Rao), Vol. II, pp. 93–126, Academic Press, New York and London (1976).

[101] A. F. Krupnov, Modern submillimetre microwave-scanning spectroscopy, *in* "Modern Aspects of Microwave Spectroscopy" (Ed. G. W. Chantry), Academic Press, London, New York (1979).

[102] C. E. Cleeton and N. H. Williams, *Phys. Rev.* **45**, 234 (1934).

[103] R. Q. Twiss, *Aust. J. Phys.* **11**, 424 and 564 (1958).

[104] J. Schneider, *Phys. Rev. Lett.* **2**, 504 (1959).

[105] A. V. Gaponov, Addendum, *Izv. Vyssh. Uchebn. Zaved. Radiofiz.* **2**, 450, 836 (1959).

[106] See for example A. A. Andronov, V. A. Flyagin, A. V. Gapanov, A. L. Goldenberg, M. I. Petelin, V. G. Usov and V. K. Yulpatov, *Infrared Phys.* **18**, 385 (1978).

[107] D. V. Kisel, G. S. Korablev, V. G. Navel'yev, M. I. Petelin and Sh. Ye Tsimring, *Radiotekh. Elektron.* **19**, 782 (1974). (English translation, *Radio Eng. Electron. Phys.* **19**, 781 (1974).

[108] R. M. Gilgenbach, M. E. Read, K. E. Hackett, R. Lucey, B. Hui, V. L. Granatstein, K. R. Chu, A. C. England, C. M. Loring, O. C. Eldridge, H. C. Howe, A. G. Kulchar, E. Lazarus, M. Marakami and J. B. Wilgen, *Phys. Rev. Lett.* **44**, 647 (1980).

[109] V. A. Flyagin, A. V. Gaponov, M. I. Petelin and V. K. Yulpatov, *IEEE Trans. Microwave Theory Tech*, **MTT-25**, 514 (1977).

[110] J. L. Hirshfield and V. L. Granatstein, The electron cyclotron maser—an historical survey, *IEEE Trans. Microwave Theory Tech.* **MTT-25**, 522 (1977).

[111] J. L. Hirshfield, Gyrotrons, *in* "Infrared and Millimeter Waves" (Ed. K. J. Button), vol. I, pp. 1–54, Academic Press, New York and London (1979).

[112] K. R. Chu, Y. Y. Lau, L. R. Barnett and V. L. Granatstein, *IEEE, Trans. Electron Devices*, **ED-28**, 866 (1981); L. R. Barnett, Y. Y. Lau, K. R. Chu and V. L. Granatstein, *ibid.* p. 872.

[113] P. Lagarde, *Infrared Phys.* **18**, 395 (1978): see also "Handbook on Synchrotron Radiation", in four volumes (series editors D. E. Eastman and Y. Farge), North-Holland, Amsterdam and New York (1982).

[114] F. E. Close, The quark-parton model, *Rep. Prog. Phys.* **42**, 1285 (1979).
[115] J. H. Poole, Daresbury Technical Memorandum DL/SRF/TM4 (Revised). Science Research Council, Daresbury Laboratory, Warrington, UK (1978); P. Meyer and P. Lagarde, *J. Phys.* **37**, 1387 (1976).
[116] P. Sprangle, R. A. Smith and V. L. Granatstein, Free electron lasers and stimulated scattering from relativistic electron beams, *in* "Infrared and Millimeter Waves", (Ed. K. J. Button), Vol. 1, Academic Press, New York and London (1979).
[117] P. L. Kapitza and P. A. M. Dirac, *Proc. Camb. Phil. Soc.* **29**, 297 (1933).
[118] L. R. Elias, W. M. Fairbank, J. M. J. Madey, H. A. Schwettman and T. I. Smith, *Phys. Rev. Lett.* **36**, 717 (1976).
[119] D. A. G. Deacon, L. R. Elias, J. M. J. Madey, G. J. Ramian, H. A. Schwettman and T. I. Smith, *Phys. Rev. Lett.* **38**, 892 (1977).
[120] V. L. Granatstein, S. P. Schlesinger, M. Herndon, R. K. Parker and J. A. Pasour, *Appl. Phys. Lett.* **30**, 384 (1977).
[121] S. J. Smith and E. M. Purcell, *Phys. Rev.* **92**, 1069 (1953).
[122] R. S. Rusin and G. D. Bogomolov, *JETP Lett*, **4**, 160 (1966).
[123] R. P. Leavitt, D. E. Wortman and C. A. Morrison, *Appl. Phys. Lett.* **35**, 363 (1979).
[124] K. Mizuno, S. Ono and Y. Shibata, *IEEE Trans. Electron Devices* **ED-20**, 749 (1973).
[125] R. P. Leavitt, D. E. Wortman and H. Dropkin, *IEEE J. Quant. Electron.* **QE-17**, 1333 (1981); D. E. Wortman, H. Dropkin and R. P. Leavitt, ibid. 1341 (1981).
[126] J. M. Wachtel, *J. Appl. Phys.* **50**, 49 (1979).
[127] D. H. Martin and K. Mizuno, *Adv. Phys.* **25**, 211, (1976): this review also covers many other "electronic" sources of infrared and submillimetre-wave radiation.
[128] M. P. Shaw, H. L. Grubin and P. R. Solomon, "The Gunn-Hilsum Effect", Academic Press, London and New York (1979).
[129] H. J. Kuno, IMPATT devices for the generation of millimeter waves *in* "Infrared and Millimeter Waves", (Ed. K. J. Button), Vol. 1, Academic Press, New York and London (1979).
[130] J. Nishizawa, 5th International Conference on Infrared and MM-Waves, FRG, Digest, p. 201 (1980).
[131] S. Ahmad and J. Freyer, *IEEE Trans. Electron Devices* **ED-26**, 1370 (1979).
[132] T. Oka, "Molecular spectroscopy using infrared lasers; a study of radiative and collisional processes", published in Horizons of Quantum Chemistry (Eds K. Fukui and B. Pullman), pp. 151–167, D. Reidel (1980).
[133] J. M. Stone, "Radiation and Optics", pp. 252ff, McGraw-Hill Book Company, New York (1963).
[134] J. H. Van Vleck and V. F. Weisskopf, *Rev. Mod. Phys.* **17**, 227 (1945).
[135] C. Brot, *Phys. Lett.* **30A**, 101 (1969).
[136] J. Cuthbert and E. J. Denney, "Application of microwave spectroscopy to chemical analysis", published in Molecular Spectroscopy, Institute of Petroleum (London), (1971).
[137] R. Karplus and J. Schwinger, *Phys. Rev.* **73**, 1020 (1948); see also H. S. Snyder and P. I. Richards, *Phys. Rev.* **73**, 1178 (1948).
[138] H. Jones, Infrared-microwave double-resonance techniques, *in* "Modern Aspects of Microwave Spectroscopy" (Ed. G. W. Chantry), Chapter 3, Academic Press London and New York (1979).
[139] L. Allen and J. H. Eberley, "Optical Resonance and Two-level Atoms", Wiley Interscience (1975).
[140] I. I. Rabi, *Phys. Rev.* **51**, 652 (1937).
[141] A. Javan, *Phys. Rev.* **107**, 1579 (1957).

[142] T. Shimizu and T. Oka, *Phys. Rev.* **A2**, 1177 (1970): T. Yajima, *J. Phys. Soc. Japan*, **16**, 1594 (1961); M. Takami, *Jap. J. Appl. Phys.* **15**, 1063, 1889 (1976); ibid. **17**, 125 (1977).

[143] A. Eyer and H. Jones, *J. Mol. Spectrosc.* **52**, 420 (1974).

[144] J. G. Baker, Microwave-microwave double resonance, *published in* "Modern Aspects of Microwave Spectroscopy" (Ed. G. W. Chantry), Academic Press, New York and London (1979).

[145] P. Glorieux, J. Legrand, B. Macke and J. Messelyn, *J. Quant. Spectrosc. Radiat. Transfer* **12**, 731 (1972).

[146] F. Bloch, *Phys. Rev.* **70**, 460 (1946).

[147] S. L. McCall and E. L. Hahn, *Phys. Rev. Lett.* **18**, 908 (1967); *Phys. Rev.* **183**, 457 (1969); ibid. **2A**, 861 (1970).

[148] R. G. Brewer, A. Z. Genack and S. B. Grossman, Coherent transients and pulse Fourier transform spectroscopy, *published in* "Laser Spectroscopy III" (Eds J. L. Hall and J. L. Carlsten), Springer Verlag Berlin, Heidelberg, New York (1977); R. G. Brewer, Coherent Optical Transients, *published in* "Coherence in Spectroscopy and Modern Physics" (Eds F. T. Arecchi, R. Bonifacio and M. O. Scully), Plenum Press, New York and London (1978).

[149] H. M. Gibbs and R. E. Slusher, *Appl. Phys. Lett.* **18**, 505 (1971).

[150] G. L. Lamb, Jr., *Phys. Lett.* **25A**, 181 (1967); *Rev. Mod. Phys.* **43**, 99 (1971); see also T. W. Barnard, *Phys. Rev.* **A7**, 373 (1973).

[151] F. H. Read, "Electromagnetic Radiation", John Wiley, Chichester, UK (1980).

[152] R. W. Ditchburn, "Light" (3rd Edition), Appendix 6D.1, Academic Press, London and New York (1976).

[153] F. Grum and R. J. Becherer, "Optical Radiation Measurements": Vol. 1 "Radiometry", Academic Press, New York and London (1979).

[154] T. Pearcey, "Tables of Fresnel Integrals to Six Decimal Places", Cambridge University Press (1956)

[155] H. E. Bennett and J. O. Porteus, *J. Opt. Soc. Am.* **51**, 123 (1961); J. O. Porteus, ibid. **53**, 1394, (1963).

[156] R. Kompfner, *Appl. Opt.* **11**, 2412 (1972).

[157] P. K. Yu and A. L. Cullen, *Proc. R. Soc. Lond.* **A380**, 49 (1982).

[158] J. A. Arnaud, "Beam and Fiber Optics", Academic Press, New York and London (1976).

[159] J. S. Toll, *Phys. Rev.* **104**, 1760 (1956); F. Stern, *Solid State Phys.* **15**, 299 (1963).

[160] E. C. Titchmarsh, "Introduction to the Theory of Fourier Integrals", Oxford University Press, Oxford (1937).

[161] J. Chamberlain, F. D. Findlay and H. A. Gebbie, *Nature (London)* **206**, 886 (1965).

[162] E. E. Bell, *Jap. J. Appl. Phys.* **4**, 412 (1965); *Infrared Phys.* **6**, 57 (1966); *Handbuch Phys.* **25**, 1 (1967); *J. Phys.* (Suppl. 3–4), **28**, C2-18, C2-25 (1967).

[163] E. E. Russell and E. E. Bell, *Infrared Phys.* **6**, 75 (1966).

[164] J. Gast and L. Genzel, *Opt. Commun.* **8**, 26 (1973).

[165] T. J. Parker and W. G. Chambers, *IEEE Trans. Microwave Theory Techn.* **MTT-22**, 1032 (1974); T. J. Parker, W. G. Chambers and J. F. Angress, *Infrared Phys.* **14**, 207 (1974); D. A. Ledsham, W. G. Chambers and T. J. Parker, ibid. **16**, 515 (1976); D. A. Ledsham, W. G. Chambers and T. J. Parker, ibid. **17**, 165 (1977).

[166] J. R. Birch, G. D. Price and J. Chamberlain, *Infrared Phys.* **16**, 311 (1976); J. R. Birch and D. K. Murray, *Infrared Phys.* **18**, 283 (1978); see also J. Chamberlain, M. N. Afsar, D. K. Murray, G. D. Price and M. S. Zafar, *IEEE Trans. Instrum. Meas.* **IM-23**, 483 (1974).

[167] P. R. Staal and J. E. Eldridge, *Infrared Phys.* **17**, 299 (1977).

[168] M. S. Zafar, J. B. Hasted and J. Chamberlain, *Nature Phys. Sci.* **243**, 106 (1973).

[169] J. Chamberlain, M. N. Afsar, J. B. Hasted, M. S. Zafar and G. J. Davies, *Nature Phys. Sci.* **255**, 319 (1975); M. N. Afsar and J. B. Hasted, *J. Opt. Soc. Am.* **67**, 902 (1977).

[170] T. S. Robinson and W. C. Price, *Proc. Phys. Soc.* **B65**, 910 (1952); T. S. Robinson and W. C. Price, ibid. **B66**, 969 (1953); see also D. M. Roessler, *Brit. J. Appl. Phys.* **16**, 1119 (1965); ibid. **17**, 1313 (1966).

[171] J. W. Fleming, D. Siapkis, J. Lewis and G. R. Wilkinson, "High Frequency Dielectric Measurement" (Eds J. Chamberlain and G. W. Chantry), pp. 122–126, IPC Press, Guildford, UK (1973).

[172] H. M. Gibbs, S. L. McCall, T. N. C. Venkatesan, A. C. Gossard, A. Passner and W. Wiegmann, *Appl. Phys. Lett.* **35**, 451 (1979).

[173] D. A. B. Miller, S. D. Smith and A. Johnston, *Appl. Phys. Lett.* **35**, 658 (1979).

[174] D. A. B. Miller, S. D. Smith and C. T. Seaton, *IEEE J. Quantum. Electron.* **QE-17**, 312 (1981); Optical Bistability (Eds C. M. Bowden, M. Ciftan and H. R. Robl), Plenum Press, New York and London (1981); see also A. Zardecki, *Phys. Rev.* **A22**, 1664 (1980); and also P. D. Drummond, K. J. McNeil and D. F. Walls, *Phys. Rev.* **A22**, 1672 (1980).

[175] D. A. B. Miller and S. D. Smith, *Opt. Commun.* **31**, 101 (1979).

[176] D. A. B. Miller, R. G. Harrison, A. M. Johnston, C. T. Seaton and S. D. Smith, *Opt. Commun.* **32**, 478 (1980).

[177] J. Chamberlain and H. A. Gebbie, *Nature (London)* **206**, 602 (1965); J. Chamberlain, E. G. C. Werner, H. A. Gebbie and W. Slough, *Trans. Faraday Soc.* **63**, 2605 (1967).

[178] E. V. Loewenstein and D. R. Smith, *Appl. Optics* **10**, 577 (1971); **12**, 398 (1973).

[179] P. J. Severin, *Appl. Opt.* **9**, 2381 (1970); P. F. Cox and A. F. Stalder, *J. Electrochem. Soc. Solid-State Sci. Technol.* **120**, 287 (1973).

[180] H. Kilp *J. Phys. E. (Scientific Instruments)* **10**, 985 (1977); H. Kilp, D. C. Barnes, F. W. J. Clutterbuck, M. N. Afsar and G. W. Chantry, *Infrared Phys.* **18**, 11 (1978).

[181] D. E. Williamson, *J. Opt. Soc. Am.* **42**, 712 (1952).

[182] H. Witte, *Infrared Phys.* **5**, 179 (1965); E. V. Loewenstein and G. Newell, *J. Opt. Soc. Am.* **59**, 407 (1969); D. H. Martin, *Infrared Physics*, **15**, 67 (1975).

[183] K. D. Moller and W. G. Rothschild, "Far Infrared Spectroscopy", pp. 68–72, Wiley Interscience, New York (1971).

[184] A. F. Harvey, "Microwave engineering", pp. 1040ff, Academic Press, New York and London (1963).

[185] E. A. J. Marcatili and R. A. Schmeltzer, *Bell System Tech. J.* **43**, 1783 (1964).

[186] H. Steffen, J. Steffen, J. F. Moser and F. K. Kneubuhl, *Phys. Lett.* **20**, 20 (1966); ibid. **21**, 425 (1966); P. Schwaller, H. Steffen, J. F. Moser and F. K. Kneubuhl, *Appl. Opt.* **6**, 827 (1967).

[187] H. Steffen and F. K. Kneubuhl, *IEEE J. Quant. Electron.* **QE-4**, 992 (1968).

[188] F. K. Kneubuhl and E. Affolter, Infrared and submillimetre-wave waveguides, *in* "Infrared and Millimetre Waves" (Ed. K. J. Button), Vol. 1, pp. 235–278, Academic Press, New York (1979).

[189] P. W. Smith, *Appl. Phys. Lett.* **19**, 132 (1971).

[190] T. J. Bridges, E. G. Burkhardt and P. W. Smith, *Appl. Phys. Lett.* **20**, 403 (1972); R. E. Jensen and M. S. Tobin, *Appl. Phys. Lett.* **20**, 508 (1972).

[191] G. Lockhard III and R. Yusek, *IEEE Electron Devices Meeting*, Washington DC, December 4–6 (1972).

[192] P. Belland, D. Veron and L. B. Whitbourn, *J. Phys. D. (Appl. Phys.)* **8**, 2113 (1975).

[193] J. J. Degnan and D. R. Hall, *IEEE J. Quant. Electron.* **QE-9**, 901 (1973); J. J. Degnan, *Appl. Phys.* **11**, 1 (1976); R. L. Abrams, Waveguide gas lasers, Chapter A2 of the "Laser Handbook" (Ed. M. L. Stitch), North Holland, Amsterdam (1979)

[194] E. Loh, *Phys. Rev.* **166**, 673 (1968).

[195] D. T. Hodges, F. B. Foote and R. E. Reel, *IEEE J. Quant. Electron.* **QE-13**, 491 (1977).

[196] F. K. Kneubuhl and E. Affolter, Distributed-feedback gas lasers, *in* "Infrared and Millimetre Waves" (Ed. K. J. Button), Vol. 6, Academic Press, New York and London (1982).

[197] F. K. Kneubuhl and E. Affolter, *J. Opt.* **11**, 449 (1980).

[198] C. V. Shank, J. E. Borkholm and H. Kogelnik, *Appl. Phys. Lett.* **18**, 395 (1971).

[199] P. Czerski and S. Baranski, "Biological Effects of Microwaves", Dowden, Hutchinson and Ross, Stroudsberg, PA (1976).

[200] P. J. B. Clarricoats, *Proc. IEE*, **108C**, 170 (1961); E. Snitzer, *J. Opt. Soc. Am.* **51**, 491 (1961); W. Schlosser and H. G. Unger *in* "Advances in Microwaves" (Ed. L. Young), Vol. 1, p. 319, Academic Press, New York and London (1966) N. S. Kapany and J. J. Burke, "Optical Waveguides", Academic Press, New York (1972); P. J. B. Clarricoats and K. B. Chan, *Proc. IEE*, **120**, 1371 (1973).

[201] D. Marcuse, "Theory of Dielectric Optical Waveguides", Academic Press, New York and London (1974).

[202] D. Marcuse, "Light Transmission Optics", Van Nostrand Reinhold, Princeton, New Jersey (1972).

[203] J. B. Keller and S. J. Rubinow, *Ann. Phys.* **9**, 24 (1960).

[204] E. G. Neumann and H. D. Rudolph, *IEEE J. Microwave Theory Tech.* **MTT-23**, 142 (1975).

[205] K. C. Kao and G. A. Hockman, *Proc. IEE* **113**, 1151 (1966); D. Gloge, *Rep. Prog. Phys.* **42**, 1777 (1979).

[206] E. A. J. Marcatili, *Bell Syst. Tech. J.* **48**, 2103 (1969); D. Gloge, *Appl. Opt.* **10**, 2252 (1971).

[207] Texas Instruments Inc. P.O. Box 225012, M/S 308, Dallas Texas 75265, USA.

[208] D. B. Keck and R. E. Love, Fiber optics for communications, *in* "Applied Optics and Optical Engineering" (Ed. R. Kingslake and B. J. Thompson), Vol. 6, Academic Press, London and New York (1980).

[209] J. E. Midwinter, "Optical Fibers for Transmission", John Wiley, (1979); see also C. P. Sandbank, Optical Fibre Communication Systems, John Wiley (1980), and Optical Fibre Communications, Vol. 4, Ed. M. J. Howes and D. V. Morgan, John Wiley, Chichester and New York (1980).

[210] S. E. Miller and A. G. Chynoweth, Optical Fiber Telecommunications, Academic Press, London and New York (1979).

[211] P. C. Schultz, Progress in Optical Waveguide processes and materials. Proceeding of the Optical Society of America Topical Meetings on Optical Fiber Communication, Washington DC, 6–8 March (1979); D. Charlton and P. C. Schultz, Electro Optical Systems Design, December 1980, pp. 23–29; M. R. Montierth, *J. Electron. Mat.* **6**, 271 (1977).

[212] J. LeSergent, Brit. Pat. Spec. 1554978, 31 Oct. (1979).

[213] T. Moriyama, O. Fukuda, K. Sanada and S. Tanaka, Fabrication of ultra-low OH⁻ optical fibers with the VAD method. Proceedings of the Fifth European Conference on Optical Communications, York, England, 16–19 September (1980).

[214] K. J. Beals and C. R. Doug, Physics and Chemistry of Glasses, **21** (1980); R. Olshansky, *Rev. Mod. Phys.* **51**, 308 (1979).

[215] C. P. Sandbank, *Electl Commun.* **50**, 10 (1975).

[216] C. H. L. Goodman, *Solid-St. Electron Devices* **2**, 129 (1978).

[217] A. G. Steventon, R. E. Spillett, R. E. Hobbs, M. G. Burt, P. J. Fiddyment and J. V. Collins, *IEEE J. Quant. Electron.* **QE-17**, 602 (1981).

[218] D. R. Smith, R. C. Hooper and I. Garrett, *Opt. Quant. Electron.* **10**, 293 (1978).

[219] D. R. Smith, A. K. Chatterjee, M. A. Z. Rejman, D. Wake and B. R. White, *Electronics Lett.* **16**, 750 (1980).

[220] N. Susa, H. Nakagome, O. Mikami, H. Ando and H. Kanbe, *IEEE, J. Quant. Electron.* **QE-16**, 864 (1980).

[221] J. P. Gordon, Optics of general guiding media, *Bell Syst. Techn.* **15**, 321 (1966); R. Olshansky, *Appl. Opt.* **15**, 782 (1976).

[222] P. C. Hensel, *Electronics Lett.* **13**, 734 (1977); see also M. J. Adams, "An Introduction to Optical Waveguides", John Wiley, Chichester and New York (1981), Chapter 8.

[223] P. R. Cooper, J. S. Leach, A. B. Harding and M. A. Mathews, *Laser Technol.* **14**(2), 87, April (1982).

[224] British Telecom Research Laboratories, Optical Fibre Systems and Components, publicity leaflet, BPO Tel Consult, Room 202, Lutyens House, Finsbury Circus, London EC2M 7LY.

[225] M. Eve, P. C. Hensel, D. J. Malyon, B. P. Nelson, J. R. Stern, J. V. Wright and J. E. Midwinter, *Opt. Quant. Elec.* **10**, 253 (1978); J. E. Midwinter and J. R. Stern, *IEEE Trans. Commun.* **COM-26**, 1015 (1978).

[226] S. E. Miller, *IEEE J. Quant. Electron.* **QE-8**, 199 (1972).

[227] T. Tamir (Ed.) Integrated Optics, Topics in *Appl. Phys.* **7**, Springer Verlag, Berlin, Heidelberg, New York (1975); see also P. K. Tien and J. A. Giordmaine, Bell Laboratories Record, December (1980), p. 371, The Proceedings of the European Conference on Integrated Optics, IEE Conference Series (1981); and H. Kogelnik, *Proc. IEEE* **69**, 232 (1981).

[228] N. J. Harrick, "Internal Reflection Spectroscopy", John Wiley, New York (1967).

[229] J. Fahrenfort, *Spectrochim. Acta* **17**, 698 (1961).

[230] P. A. Wilks, A practical approach to internal reflection spectroscopy, *in* Laboratory Methods in Infrared Spectroscopy" 2nd edition (Eds R. G. J. Miller and B. C. Stace), Chapter 14, Heyden and Son, London (1972). H. A. Willis and V. J. I. Zichy, The examination of polymer surfaces by infrared spectroscopy, *in* "Polymer Surfaces" (Ed. D. T. Clark and W. J. Feast), Chapter 15, Wiley Interscience, Chichester and New York (1978).

[231] P. A. Wilks, *Appl. Spectrosc.* **22**, 782 (1968).

[232] B. L. Crawford, T. G. Goplen and D. Swansen, The measurement of optical constants in the infrared by attenuated total reflection, in "Advances in Infrared and Raman Spectroscopy" (Eds R. J. H. Clark and R. E. Hester), Vol. 4, Heyden and Son, London (1978).

[233] R. W. Ditchburn, Light, 3rd Edition, Vol. 1, Section 5.22, pp. 122–124, Academic Press, London and New York (1976).

[234] C. S. Evans, R. Hunneman and J. S. Seeley, Proceedings of the Conference on Infrared Techniques, University of Reading, Sept (1971), IERE Conference Proceedings, No. 22, p. 125; *J. Phys. D (Appl. Phys.)* **9**, 309 (1976).

[235] J. T. Houghton and S. D. Smith, "Infrared Physics", Oxford University Press (1966).

[236] J. S. Seeley, R. Hunneman and A. Whatley, *Infrared Phys.* **19**, 429 (1979).

[237] G. Mie, *Ann. d. Physik.* **25**(4), 377 (1908), but see Born and Wolf [ref. 60], pp. 633 onwards; for a modern application see W. J. Glantschnig and S. H. Chen, *Appl. Opt.* **20**, 2499 (1981).

[238] P. C. T. Stillwell, private communication, but see the article on Thermal Imaging by this author in *J. Phys. E. (Sci. Instrum.)* **14**, 1113 (1981).

[239] G. Duyckaerts, *Analyst* **84**, 201 (1959).

[240] A. E. Costley, K. H. Hursey, G. F. Neill and J. M. Ward, *J. Opt. Soc. Am.* **67**, 979 (1977).

[241] J. P. Auton, *Appl. Optics*, **6**, 1023 (1967).

[242] T. Larsen, *IRE Trans. Microwave Theory Tech.* **MTT-10**, 191 (1962); see also R. Petit, *Nouv. Rev. Optique* **6**, 129 (1975).

[243] J. A. Beunen, A. E. Costley, C. L. Mok, G. F. Neill, T. J. Parker and G. Tait, *J. Opt. Soc. Am.* **71**, 184 (1981).

[244] F. S. Ham and B. Segall, *Phys. Rev.* **124**, 1786 (1961); see also W. G. Chambers, C. L. Mok and T. J. Parker, *J. Phys. A. Math. Gen.* **13**, 1433 (1980).

[245] C. L. Mok, W. G. Chambers, T. J. Parker and A. E. Costley, *Infrared Phys.* **19**, 437 (1979).

[246] See for example M. N. Afsar, J. B. Hasted and J. Chamberlain, *Infrared Phys.* **16**, 301 (1976).

[247] D. H. Martin and E. Puplett, *Infrared Phys.* **10**, 105 (1970).

[248] P. A. R. Ade, A. E. Costley, C. T. Cunningham, C. L. Mok, G. F. Neill and T. J. Parker, *Infrared Phys.* **19**, 599 (1979); see also ref. 240.

[249] B. Walker, E. A. M. Baker and A. E. Costley, *J. Phys. E. (Sci. Instrum)* **14**, 832 (1981).

[250] R. Ulrich, K. F. Renk and L. Genzel, *IEEE Trans* **MTT-11**, 363 (1963).

[251] R. Ulrich, *Infrared Phys.* **7**, 37 (1967); ibid. p. 65; *Appl. Opt.* **7**, 1981 (1968).

[252] G. D. Holah and N. Morrison, *J. Opt. Soc. Am.* **67**, 971 (1977).

[253] M. Francon, "Optical Image Formation and Processing", Chapter 7, Academic Press, London and New York (1979); for specifically infrared applications see T. L. Williams, *Proc. Soc. Opt. Inst. Eng.* **46**, 305 (1974), and also C. J. Hutchinson, J. P. Jennings, C. Lewis and G. N. Turner, *J. Phys. E. (Sci. Instrum)*, **14**, 846 (1981).

[254] J. R. Birch, J. D. Dromey and E. A. Nicol, *Infrared Phys.* **21**, 17 (1981).

[255] R. L. Petritz, *Phys. Rev.* **104**, 1508 (1956); A. V. MacRae and H. Levinstein, *Phys. Rev.* **119**, 62 (1960); V. Radeka, Proc. ISPRA Nuclear Electronic Symposium, p. 1 (1969).

[256] F. N. Hooge, T. G. M. Kleinpenning and L. K. J. Vandamme, *Rep. Prog. Phys.* **44**, 479 (1981).

[257] J. Chamberlain, *Infrared Phys.* **11**, 25 (1971); J. Chamberlain and H. A. Gebbie, *Infrared Physics*, **11**, 56 (1971).

[258] D. P. Blair and P. H. Sydenham, *J. Phys. E. (Sci. Instrum)*, **8**, 621 (1975).

[259] P. C. G. Danby, *Electron. Engng*, January (1970).

[260] J. D. W. Abernethy, *Phys. Bull.* **24**, 591 (1973).

[261] R. W. Harris and T. J. Ledwige, "Introduction to Noise Analysis", Pion Press, Applied Physics Series No. 7 (1974).

[262] C. E. Shannon, Communication in the presence of noise, *Proc. IRE*, **37**, 10–21 (1949).

[263] G. Shorter, *Wireless World* 200–205 May (1975).

[264] B. Austin Barry, "Errors in Practical Measurement in Science, Engineering and Technology", Wiley Interscience, New York and Chichester (1978).

[265] F. Oberhettinger, Fourier transforms of distributions and their inverses; A collection of tables, "Probability and Mathematical Statistics Series", Academic Press, New York and London (1973).

[266] See for example "Handbook of Chemistry and Physics" published by the Chemical Rubber Publishing Company, Cleveland, Ohio.

[267] J. W. Goodman, Introduction to Fourier Optics, McGraw Hill New York (1968); H. J. Caulfield (Ed.), "Handbook of Holography", Academic Press, New York and London (1979); A Van der Lugt, *Proc. IEEE* **62**, 1300 (1974); J. W. Goodman, *Proc. IEEE*, **65**, 29 (1977).

[268] D. Marcuse, "Principles of Quantum Electronics", Academic Press, New York (1980).

[269] R. Loudon, "The Quantum Theory of Light", Clarendon Press, Oxford (1973).

[270] See L. C. Robinson, "Methods of Experimental Physics", Vol. 10, "Physical Principles of Far-Infrared Radiation" (Ed. L. Marton) Academic Press, New York and London (1973); 219ff, for a detailed description of this point.

[271] R. H. Hanbury-Brown, "The Intensity of Interferometer", Taylor and Francis, London (1974).

[272] E. R. Pike and E. Jakeman, Photon-statistics and photon-correlation spectroscopy, *in* Advances in Quantum Electronics, Vol. 2 (Ed. D. W. Goodwin), Academic Press, London and New York (1974).

[273] M. J. Colles and C. R. Pidgeon, Tunable lasers, *Rep. Prog. Phys.* **38**, 329, (1975).

[274] J. W. Fleming and J. Chamberlain, *Infrared Phys.* **14**, 277 (1974).

[275] J. W. Fleming, *IEEE Trans. Microwave Theory Tech.* **MTT-22**, 1023 (1974).

[276] H. A. Gebbie and R. Q. Twiss, *Rep. Prog. Phys.* **29**, 729 (1966).

[277] L. Genzel and A. Hadni, Private communications.

[278] P. Jaquinot, *J. Opt. Soc. Am.* **44**, 761 (1954): *Rep. Prog. Phys.* **23**, 267 (1960).

[279] G. W. Chantry and J. W. Fleming, *Infrared Phys.* **16**, 655 (1976).

[280] P. Fellgett, Thesis, University of Cambridge (1951); *J. Phys. Radium, Paris* **19**, 187, 236 (1958).

[281] F. Kahn, *Astrophys. J.* **129**, 518 (1959).

[282] E. R. Pike, Review of Reference (7), *Nature (London)* **283**, 700 (1980).

[283] P. Jaquinot, *Appl. Opt.* **8** (3), March (1969).

[284] J. Strong, Multiplex spectrometry *in* "Essays in Physics" (Eds G. K. T. Conn and G. N. Fowler), Vol. 5, Academic Press, New York and London (1973).

[285] J. A. Decker, *Appl. Opt.* **10**, 24 (1971).

[286] P. Hansen and J. Strong, *Appl. Opt.* **11**, 502 (1972).

[287] M. Harwit, P. G. Phillips, T. Fine and N. J. A. Sloane, *Appl. Opt.* **9**, 1149 (1970).

[288] P. R. Griffiths, Infrared fourier transform spectrometry: applications to analytical chemistry, *in* "Transform Techniques in Chemistry" (Ed. P. R. Griffiths), Chapter 6, Heyden, London and New York (1978).

[289] C. Corsi, G. Cappucio, A. D'Amico, G. Petrocco and G. Vitali, *Infrared Phys.* **16**, 37 (1976).

[290] H. W. Thompson, *Nature (London)* (1946).

[291] L. Genzel, *J. Molec. Spectrosc.* **4**, 241 (1960); H. Happ and L. Genzel, *Infrared Phys.* **1**, 39 (1961).

[292] R. C. Milward, *in* "Molecular Spectroscopy", p. 81, Institute of Petroleum (1969).

[293] J. Connes and P. Connes, *J. Opt. Soc. Am.* **56**, 896 (1966).

[294] Reference (7), pp. 197ff; see also J. R. Birch and C. E. Bulleid, *Infrared Phys.* **17**, 279 (1977).

[295] J. Chamberlain, Submillimetre-wave techniques, *in* "High Frequency Dielectric Measurement" (Eds J. Chamberlain and G. W. Chantry), IPC Press, Guildford, UK (1972).

[296] J. R. Birch and T. J. Parker, "Dispersive Fourier Transform Spectroscopy in Infrared and Millimetre Waves" (Ed. K. J. Button), Vol. 2, Academic Press, New York and London (1979).

[297] J. Chamberlain, J. E. Gibbs and H. A. Gebbie, *Infrared Phys.* **9**, 185 (1969).

[298] M. N. Afsar, J. Chamberlain, G. W. Chantry, *IEEE Trans. Instrum. Meas.* **IM-25**, 290 (1976).

[299] J. Chamberlain, *Infrared Phys.* **12**, 145 (1972).

[300] J. E. Allnutt and J. A. Staniforth, *J. Phys. E. (Sci. Instrum.)* **4**, 730 (1971); J. Chamberlain, J. Haigh and M. J. Hine, *Infrared Phys.* **11**, 75 (1971).

[301] J. Connes, *Rev. Opt. Theor. Instrum.* **440**, 45, 116, 171, 231 (1961); P. L. Richards in "Spectroscopic Techniques" (Ed. D. H. Martin), pp. 58ff, North Holland, Amsterdam (1967).

[302] The best known is that of M. L. Forman, W. H. Steel and G. Vanasse, *J. Opt. Soc. Am.* **56**, 59 (1966), but the subject is also discussed very informatively by L. Mertz, *Appl. Opt.* **2**, 1331 (1963); *Infrared Phys.* **7**, 17 (1967).

[303] J. W. Fleming, NPL Report DES 49, September (1978).

[304] J. R. Birch, *Infrared Phys.* **20**, 349 (1980), but see also R. P. Lowe, R. J. Niciejewski and D. N. Turnbull, *Infrared Phys.* **21**, 189 (1981).

[305] D. P. C. Thackeray, the infrared spectrometer, *in* Laboratory Methods in Infrared Spectroscopy" (Eds R. G. J. Miller and B. C. Stace), Chapter 1, Heyden, New York and London (1972).

[306] J. W. Fleming, *Infrared Phys.* **17**, 263 (1977).

[307] G. A. Vanasse and H. Sakai, "Prog. Opt.", Vol. 6, North Holland, Amsterdam (1967).

[308] A similar point is discussed by J. Butterworth, D. E. MacLaughlin and B. C. Moss, *J. Sci. Instrum.* **44**, 1029 (1967).

[309] T. S. Moss, Infrared Detectors, Pergamon Press (1976): this is a hardback version of *Infrared Phys.* **15** (1975); see also various papers in "Advanced Infrared Detectors and Systems" Proceedings of the International Conference, held at the Institution of Electrical Engineers, 29–30 Oct. (1981), IEE Conference Publication No. 204.

[310] N. E. Johnson, Spectral imaging with the Michelson interferometer, *published in* "Infrared Imaging Systems Technology", *Proc. Soc. Photo-Optical Instrum. Engineers (SPIE)*, **226**, 2 (1980).

[311] T. G. Blaney, Proceedings of the SPIE Conference on New Developments and Applications in Optical Radiation Measurement, NPL, Teddington UK, May 7–8 (1980), *SPIE* **234**, 22–26.

[312] Details of "black" paints suitable for infrared work have been given by J. L. Pipher and J. R. Houck, *Appl. Opt.* **10**, 567 (1971) and by S. Takahashi, *Infrared Phys.* **13**, 1 (1973).

[313] M. F. Kimmitt, A. C. Prior and V. Roberts, Far-infrared techniques, *in* Plasma Diagnostic Techniques (Eds R. H. Huddlestone and S. L. Leonard), Chapter 9, Academic Press, New York and London (1965).

[314] R. R. Selleck, E. W. McDonald and J. C. Wiltse, Digest 2nd International Conference on Submillimetre Waves and their Applications, San Juan Puerto Rico (1976), IEEE, 76 CH1152-8 MTT, pp. 49–50.

[315] E. H. Putley, Solid-State Devices for Infrared Detectors, *J. Sci. Instrum.* **43**, 857 (1965); Detectors, *in* "Spectroscopic Techniques" (Ed. D. H. Martin), North Holland, Amsterdam (1967); Modern infrared detectors, *Phys. Technol.* **4**, 202 (1973).

[316] T. G. Blaney, *J. Phys. E. (Sci. Instrum.)*, **11**, 856 (1978).

[317] D. E. Bode, Infrared detectors, *in* "Applied Optics and Optical Engineering" (Eds R. Kingslake and B. J. Thompson) Vol. 6, Chapter 8, Academic Press, New York and London (1980).

[318] T. G. Blaney, "Detection Techniques at Short Millimetre and Submillimetre

Wavelengths: An Overview in Infrared and Millimeter Waves", Vol. 3, "Submillimeter Techniques" (Ed. K. J. Button), Academic Press, New York and London (1980).

[319] M. Golay, *Rev. Sci. Instrum.* **18**, 347, 357 (1947).

[320] E. H. Putley, "Semiconductors and Semi-metals" (Ed. R. K. Williardson and A. C. Beer), Vol. 5, pp. 259–285, Academic Press, London (1970).

[321] A. Hadni, Pyroelectricity and pyroelectric detectors, *in* "Infrared and Millimetre Waves", Vol. 3, "Submillimetre Techniques" (Ed. K. J. Button), Academic Press, New York and London (1980); *J. Phys. E (Sci. Instrum.)* **14** (1981).

[322] A. Hadni, *IEEE Trans. Microwave Theory Tech.* **MTT-22**, 1016 (1974); E. L. Dereniak and F. G. Brown, *Infrared Phys.* **15**. 39 (1975).

[323] R. J. Mahler, R. J. Phelan and A. R. Cook, *Infrared Phys.* **12**, 57 (1972).

[324] R. J. Phelan, R. J. Mahler and A. R. Cook, *Appl. Phys. Lett.* **19**, 337 (1971); A. M. Glass, J. H. McFee and J. G. Bergman, *J. Appl. Phys.* **42**, 5219 (1971).

[325] P. J. Lock, *Appl. Phys. Lett.* **19**, 390 (1971).

[326] C. B. Roundy and R. L. Byer, *Appl. Phys. Lett.* **21**, 512 (1972).

[327] W. R. Blevin and J. Geist, *Appl. Opt.* **13**, 1171 (1974)

[328] M. R. Holter, S. Nudelman, G. H. Suits, W. L. Wolfe and G. J. Zissis, "Fundamentals of Infrared Technology", MacMillan, New York and London (1962).

[329] J. Clarke, P. L. Richards and N. H. Yeh, *Appl. Phys. Lett.* **39**, 664 (1977).

[330] H. D. Drew and A. J. Sievers, *Appl. Opt.* **8**, 2067 (1969).

[331] G. Chanin, J. P. Torre and L. Peccoud, *Infrared Phys.* **18**, 657 (1978).

[332] R. D. Britt and P. L. Richards, *Int. J. Infrared Millimetre Waves* **2**, 1083 (1981).

[333] The concept came originally from R. C. Jones, *Proc. Inst. Radio Eng.* **47**, 1481 (1959).

[334] A. H. Sommer, "Photo-emissive Materials", John Wiley, New York (1968).

[335] D. H. Martin and D. Bloor, *Cryogenics* **1**, 159 (1961).

[336] C. L. Bertin and K. Rose, *J. Appl. Phys.* **42**, 163 (1971); M. K. Maul and M. W. P. Strandberg, *J. Appl. Phys.* **40**, 2822 (1969); G. Gallinaro and R. Varone, *Cryogenics*, **15**, 292 (1975).

[337] T. G. Blaney, *Space Sci. Rev.* **17**, 691 (1975).

[338] A. E. Siegman, *Proc. IEEE*, **54**, 1350 (1966).

[339] T. G. Blaney and N. R. Cross, *J. Phys. E.* (Sci. Instrum.) **10**, 146 (1977).

[340] J. J. Gustincic, *Proc. Soc. Photo-Optical Instrum. Engineers (SPIE)*, **105**, 40 (1977).

[341] H. R. Fettermann, P. E. Tannenwald, B. J. Clifton, C. D. Parker, W. D. Fitzgerald and N. R. Erickson, to be published.

[342] D. H. Martin, *in* Infrared Detection Techniques for Space Research, Reidel (1972).

[343] G. T. Wrixon and W. M. Kelly, *Infrared Phys.* **18**, 413 (1978).

[344] D. H. Martin, Invited lectures to the Fifth International Conference on Infrared and Submm Waves, Wurzburg FRG (1980), and to the XXth Congress of URSI, Washington DC (1981).

[345] H. Krautle, E. Sauter and G. V. Schultz, *Infrared Phys.* **18**, 705 (1978).

[346] D. B. Rutledge, S. E. Schwarz and A. T. Adams, *Infrared Phys.* **18**, 713 (1978).

[347] D. B. Rutledge, S. E. Schwarz, T. L. Hwang, D. J. Angelakos, K. K. Mei and S. Yokota, *IEEE J. Quant. Electron.* **QE-16**, 508 (1980).

[348] B. J. Clifton, G. D. Alley, R. A. Murphy and I. H. Mroczkowski, *IEEE Trans. Electron. Devices*, **ED-28** 155 (1981).

[349] D. Buhl, G. Chin, G. A. Koepf, N. McAvoy, H. R. Fettermann, B. J. Clifton, D. D. Peck, P. E. Tannenwald, P. F. Goldsmith and N. R. Erickson, to be published.

[350] J. P. Sattler, T. L. Worchesky, K. J. Ritter and W. J. Lafferty, *Opt. Lett.* **5**, 21 (1980).

[351] J. R. Tucker, *IEEE Trans. Quant. Electron.* **QE-15**, 1234 (1979).

[352] W. G. Chambers, *J. Phys. A.* **14**, 138 (1981).

[353] M. McColl, A review of submillimeter-wave mixers, *Proc. Soc. Photo-Optical Engineers (SPIE)*, **105**, 24 (1977); P. E. Tannenwald, *Int. J. Infrared Millimetre Waves* **1** 159 (1980).

[354] C. D. Payne and B. E. Prewer, *Radio Electron. Eng.* **39**, 167 (1970); A. A. M. Saleh, "Theory of Resistive Mixers", MIT Press, Cambridge, Massachusetts (1971).

[355] J. W. Dees, *Microwave J.* **9**, 48 (1966); V. Daneu, D. Sokoloff, A. Sanchez and A. Javan, *Appl. Phys. Lett.* **15**, 398 (1969); S. I. Green, P. D. Coleman and J. R. Baird, The MOM electric tunnelling detector, *in* Submillimeter Waves, Proceedings of the Symposium, Polytechnic Press, Polytechnic Institute of Brooklyn, Volume XX, (1970); H. D. Riccius, *Appl. Phys.* **17**, 49 (1978); P. J. Epton, W. L. Wilson, F. T. Tittel and T. A. Rabson, *Infrared Phys.* **19**, 335 (1979).

[356] J. N. Crouch, *Infrared Phys.* **18**, 89 (1978); J. E. Muller and C. Hanke, *Infrared Phys.* **19**, 533 (1978).

[357] T. G. Phillips, *Astrophys. J. Lett.* **186**, L19 (1973); see also A. Arams, *Proc. IEEE.* **54**, 612 (1966).

[358] W. M. Kelly and G. T. Wrixon, Optimisation of Schottky barrier diodes for low conversion-loss operation at near mm wavelengths, *in* "Infrared and Millimeter Waves", Vol. 3, "Submillimeter Techniques" (Ed. K. J. Button), Academic Press, New York and London, (1980); B. J. Clifton, *IEEE Trans. Microwave Theory* Tech. **MTT-25**, 457 (1977); *Radio Electron. Engng* **49**, 333 (1979).

[359] P. L. Richards, *in* "Semiconductors and Semimetals", Vol. 12, "Infrared Detectors II" (Eds R. K. Willardson and A. C. Beer), p. 395, Academic Press, New York and London (1977); Y. Taur, J. H. Claasen and P. L. Richards, *IEEE Trans. Microwave Theory Tech.* **MTT-22**, 1005 (1974); J. Edrich, Proceedings of the Fourth International Conference on Infrared and Millimetre Waves, Conference Digest IEEE Cat. No. 79 CH 1384–7 MTT, p. 152 (1979).

[360] P. L. Richards, T. M. Shen, R. E. Harris and F. L. Lloyd, *Appl. Phys. Lett.* **34**, 345 (1979).

[361] J. P. Sattler, T. L. Worchesky, M. S. Tobin, K. J. Ritter and T. W. Daley, *Int. J. Infrared Millimetre Waves* **1**, 127 (1980); W. J. Lafferty, J. P. Sattler, T. L. Worchesky and K. J. Ritter, *J. Molec. Spectrosc.* **87**, 416, (1981).

[362] H. A. Gebbie, N. W. B. Stone, E. H. Putley and N. Shaw, *Nature (London)* **214**, 165 (1967).

[363] S. D. Personick, *Proc. IEEE*, **69**(2), 262 (1981).

[364] S. M. Sze, "Physics of Semiconductor Devices", John Wiley, New York (1969).

[365] M. McColl, D. T. Hodges and W. A. Garber, *IEEE Trans. Microwave Theory Tech.* **MTT-25**, 463 (1977).

[366] L. M. Matarrese and K. M. Evenson, *Appl. Phys. Lett.* **17**, 8 (1970).

[367] R. A. Murphy, C. O. Bozler, C. D. Parker, H. R. Fetterman, P. E. Tannenwald, B. J. Clifton, J. P. Donnelly and W. T. Lindley, *IEEE Trans. Microwave Theory Tech.* **MTT-25**, 494 (1977).

[368] T. G. Blaney, N. R. Cross and Th. de Graauw, to be published.

[369] B. D. Josephson, *Phys. Lett.* **1**, 251 (1962); *Rev. Mod. Phys.* **36**, 216 (1964); *Adv. Phys.* **56**, 14 (1965); see also J. R. Waldram, The Josephson effects in weakly-coupled superconductors. *Reps. Prog. Phys.* **39**, 751 (1976).

[370] J. Bardeen, L. N. Cooper and J. R. Schrieffer, *Phys. Rev.* **108**, 1175 (1957).

[371] P. L. Richards and M. Tinkham, *Phys. Rev.* **119**, 575 (1960); P. Wyder, *Infrared Phys.* **16**, 243 (1976).

[372] I. K. Harvey, J. C. Macfarlane and R. B. Frenkel, *Metrologia* **8**, 114, (1972); B. F. Field, T. F. Finnegan and J. Toots, *Metrologia*, **9**, 155 (1973); A. Hartland, T. J. Witt, D. Reymann and T. F. Finnegan, *IEEE Trans. Instrum. Meas.* **IM-27**, 470 (1978).

[373] T. G. Blaney, *Radio Electron. Engineer*, **42**, 303 (1972).

[374] C. C. Grimes, P. L. Richards and S. Shapiro, *J. Appl. Phys.* **39**, 3905 (1968); T. Poorter and H. Tolner, *Infrared Phys.* **19**, 317 (1979).

[375] P. L. Richards and S. A. Sterling, *Appl. Phys. Lett.* **14**, 394 (1969).

[376] D. G. McDonald, V. E. Kose, K. M. Evenson, J. S. Wells and J. D. Cupp, *Appl. Phys. Lett.* **15**, 121 (1969); D. G. McDonald, A. S. Risley, J. D. Cupp and K. M. Evenson, *Appl. Phys. Lett.* **18**, 162 (1971); T. G. Blaney, Josephson mixers at submillimetre wavelengths: Present experimental status and future developments, *in* "Future Trends in Superconductive Electronics" (Eds B. S. Deaver, C. M. Falco, J. H. Harris and S. A. Wolf), pp. 230–238, AIP Conference Proceedings, No. 44 (1978).

[377] T. G. Blaney, N. R. Cross and R. G. Jones, to be published, but see also J. Edrich, D. B. Sullivan and D. G. McDonald, *IEEE Trans. Microwave Theory Tech.* **MTT-25**, 476 (1977).

[378] J. R. Tucker, *Appl. Phys. Lett.* **36**, 477 (1980).

[379] F. L. Vernon, M. F. Millea, M. F. Bottjer, A. H. Silver, R. J. Pederson and M. McColl, *IEEE Trans. Microwave Theory Tech.* **MTT-25**, 286 (1977).

[380] J. H. Greiner, C. J. Kircher and I. Ames, *IBM J. Res. Devel.* **24**, 195 (1980); T. Gheewala, *Proc. IEEE*, **70**, 26 (1982).

[381] A. Matsuda, *J. Appl. Phys.* **51**, 4310 (1980).

[382] G. J. Dolan, T. G. Phillips and D. P. Woody, *Appl. Phys. Lett.* **34**, 347 (1979).

[383] P. L. Richards and T. M. Shen, *IEEE Trans. Electron Devices, ED-77*, **ED-77**, 1909 (1980).

[384] Work attributed to Heppner in private communication from P. L. Richards.

[385] G. J. Dolan, R. A. Linke, T. C. L. G. Sollner, D. P. Woody and T. G. Phillips, *IEEE Trans. Microwave Theory Tech.* **MTT-29**, 87 (1981); T. M. Shen and P. L. Richards, *IEEE Trans. Magn.* **MAG-17**, 677 (1981).

[386] D. W. Peterson, A. R. Kerr, M. J. Feldman, P. H. Siegel and S. K. Pan, to be published.

[387] T. M. Shen, *IEEE J. Quant. Electron.* **QE-17**, 1151 (1981).

[388] A. F. Gibson and M. F. Kimmitt, Photon-drag detection, *in* "Infrared and Millimeter Waves", Vol 3. "Submillimeter Techniques", (Ed. K. J. Button), Academic Press, New York and London (1980).

[389] A. A. Grinberg, *Sov. Phys. JEPT*, **31**, 531 (1970); K. Cameron, A. F. Gibson, J. Giles, C. B. Hatch, M. F. Kimmitt and S. Shafik, *J. Phys. C; Solid-State Phys.* **8**, 3137 (1975); A. F. Gibson and S. Montasser, *J. Phys. C; Solid-State Phys.* **8**, 3147 (1975).

[390] A. Rosencwaig, *Adv. Electronics Electron Phys.* **46**, 207 (1979); "Photoacoustics and Photoacoustic Spectroscopy", John Wiley, New York (1980); G. Busse, *J. Opt. (Paris)* **11**, 454 (1980); J. Badoz, D. Fournier and A. C. Boccara, Ibid. 399; see also "Optoacoustic Spectroscopy and Detection" (Ed. Y. H. Pao), Academic Press, New York and London (1977).

[391] G. Busse and K. F. Renk, *Infrared Phys.* **18**, 517 (1978).

[392] G. Busse and H. Schultz, *Infrared Phys.* **19**, 313 (1979).

[393] C. F. Dewey, R. D. Kamm and C. E. Hackett, *Appl. Phys. Lett.* **23**, 633 (1973).

[394] P. C. Claspy, Infrared optoacoustic spectroscopy and detection, *in* "Optoacoustic Spectroscopy and Detection" (Ed. Y. H. Pao), Chapter 6, Academic Press, New York and London (1977).

[395] D. H. Martin and E. Puplett, to be published.

[396] G. Busse and B. Bullemer, *Infrared Phys.* **18**, 631 (1978).

[397] D. W. Vidrine, *Appl. Spectrosc.* **34**, 314 (1980).

[398] K. F. Luft, *Z. Tech. Phys.* **24**, 97 (1943); *Angew. Chem.* **B19**, 2 (1947); *Compt. Rend.* **238**, 1651 (1954).

[399] M. L. Veingerov, *Dokl. Akad. Nauk. S.S.R.* **19**, 687 (1938); J. C. Waters and N. W. Hartz, *Instruments* **25**, 622 (1952).

[400] D. W. Hill and T. Powell, "Non-dispersive Infrared Gas Analysis in Science, Medicine and Industry", Plenum, New York (1968).

[401] C. O. Peterson, W. V. Dailey and W. G. Amthein, *Instrum. Technol.* **14**, 8, 45 (1967); D. J. Troy, *Control Eng.* **4**, 11, 116 (1957).

[402] T. Takahashi, M. O. Weaver and L. A. Prince, *J. Geophys. Res.* **81**, 3736 (1976).

[403] H. M. J. M. Dortmans, *J. Phys. E.* (Sci. Instrum.) **14**, 777 (1981); see also various articles in *Electronics and Power*, **27(3)**, March (1981).

[404] B. W. Kernighan and P. J. Plauger, Software Tools.

[405] W. Neuhauser, M. Hohenstatt and P. E. Toschek, *in* Laser Spectroscopy IV (Eds H. Walther and K. W. Rothe), Springer Verlag, London, Heidelberg and New York (1979) pages 73–78.

[406] E. K. Plyler and L. R. Blaine, *J. Res. Nat. Bur. Stand.* **62**, 7 (1959).

[407] G. Guelachvili, *Appl. Opt.* **17**, 1322 (1978); L. Genzel and K. Sakai, *J. Opt. Soc. Am.* **67**, 871 (1977); see also K. Sakai, High resolving power fourier spectroscopy *in* "Spectrometric Techniques" (Ed. G. Vanasse), Vol. 1, Academic Press, New York and London (1977).

[408] See for example A. G. Maki and J. Sams, *J. Mol. Struct.* **26**, 107 (1975).

[409] M. Czerny, *Z. Phys.* **34**, 227 (1925); ibid. **45**, 476 (1927).

[410] H. Rubens and E. Aschkinass, *Astrophys J.* **8**, 176 (1898); H. Rubens and G. Hettner, *Ver. Deut. Phys. Ges.* **18**, 154 (1916); H. Rubens, *Berl. Ber.* **8** (1921).

[411] N. Bjerrum, *Nernst Festschrift*. 90 (1912).

[412] R. H. Partridge, *Infrared Phys.* **19**, 571 (1979).

[413] M. J. Bangham, A. Bonetti, R. H. Bradsell, B. Carli, J. E. Harries, F. Mencaraglia, D. G. Moss, S. Pollitt, E. Rossi and N. R. Swann, to be published.

[414] D. J. W. Kendall, H. L. Buijs and J. W. C. Johns, to be published.

[415] G. Guelachvili, *Appl. Opt.* **17**, 1322 (1978).

[416] J. Connes, H. De Louis, P. Connes, G. Guelachvili, J. P. Mailland and G. Michel, *Nouv. Revue d'Optique appliquee* **1**, 3 (1970).

[417] M. Cuisenier and J. Pinard, *J. Phys. (Suppl.)* **28**, 97 (1967); R. Beer and D. Marjamiemi, *Appl. Opt.* **5**, 1191 (1966); see also R. B. Sanderson and H. E. Scott, *Appl. Opt.* **10**, 1097 (1971).

[418] W. H. Steel, Interferometers for Fourier Spectroscopy, Aspen International Conference on Fourier Spectroscopy (1970) (Eds G. A. Vanasse, A. T. Stair Jr. and D. J. Baker), p. 43, AFCRL-71-0019, 5 Jan 1971, Spec. Rep. No 114.

[419] J. Connes and P. Connes, *J. Opt. Soc. Am.* **56**, 876 (1966); P. Connes, Astronomical fourier spectroscopy, *Ann. Rev. Astron. Astrophys.* **8**, 209 (1970).

[420] H. R. Schlossberg and P. L. Kelley, Infrared spectroscopy using tunable lasers, *in* Spectrometric Techniques (Ed. G. Vanasse) Vol II, chapter 4, Academic Press, London and New York (1981).

[421] J. M. Besson, W. Paul and A. R. Calawa, *Phys. Rev.* **173**, 699 (1968); A. S. Pine, C. J. Glassbrenner, and J. A. Kafalas, *IEEE J. Quant. Electron.* **QE-9**, 800 (1973).

[422] J. C. Hill, W. Lo and J. A. Sell, *Laser Focus*, **16**, 86 (1980); R. T. Ku, E. D. Hinkley and J. O. Sample, *Appl. Opt.* **14**, 854 (1975); E. D. Hinkley, *Opt. Quant. Electron.* **8**, 155 (1976); E. D. Hinkley, R. T. Ku, K. W. Nill and J. F. Butler, *Appl. Opt.* **15**, 1653 (1976); L. W. Chaney, D. G. Rickel, G. M. Russwurm and W. A. McCleany, *Appl. Opt.* **18**, 3004 (1979).

[423] J. P. Sattler and G. J. Simonis, *IEEE J. Quant. Electron.* **QE-13**, 461 (1977); T. L. Worchesky, K. J. Ritter, J. P. Sattler and W. A. Riesser, *Opt. Lett.* **2**, 70 (1978).

[424] J. P. Sattler, T. L. Worchesky and W. A. Riessler, *Infrared Phys.* **18**, 521 (1978).

[425] S. D. Smith, "Very High Resolution Spectroscopy" (Ed. R. A. Smith), Chapter 2, Academic Press, London and New York (1977).

[426] P. G. Buckley, J. H. Carpenter, A. McNeish, J. D. Muse, J. J. Turner and D. H. Whiffen, *J. Chem. Soc. (Faraday Trans. II)*, **74**, 129 (1978).

[427] C. K. N. Patel, *Appl. Phys. Lett.* **25**, 112 (1974).

[428] N. F. Ramsey, "Molecular Beams", Oxford University Press, New York and London (1956).

[429] Ch. J. Borde, M. Ouhayan, A. van Lerberghe, C. Salmon, S. Avrilliev, C. D. Cantrell and J. Borde, High resolution saturation spectroscopy with CO_2 lasers, application to the v_3 bands of SF_6 and OsO_4 *published in* "Laser Spectroscopy IV" (Eds H. Walther and K. W. Rothe), Springer-Verlag, Berlin, Heidelberg and New York (1979).

[430] T. Oka and T. Shimizu, *Appl. Phys. Lett.* **19**, 88 (1971).

[431] H. Jones and F. Kohler, *J. Molec. Spectrosc.* **58**, 125 (1975). H. Jones, ibid. **78**, 452 (1979).

[432] G. Grynberg and B. Cagnac, *Rep. Prog. Phys.* **40**, 791 (1977); see also S. M. Freund and T. Oka, *Phys. Rev.* **A13**, 2178 (1976); for an example see H. Jones, *Infrared Phys.* **18**, 449 (1978).

[433] T. W. Hansch, S. A. Lee, R. Wallenstein and C. Wieman, *Phys. Rev. Lett.* **34**, 307 (1975). C. Wieman and T. W. Hansch, *in* "Laser Spectroscopy III" (Eds J. L. Hall and J. L. Carlsten), Springer Series in Optical Sciences, Vol. 7, p. 39, Springer-Verlag, Berlin, Heidelberg and New York, (1977).

[434] A. Owyoung, High resolution coherent Raman spectroscopy of gases, *published in* "Laser Spectroscopy IV" (Eds H. Walther and K. W. Rothe), Springer-Verlag, Berlin and New York (1979).

[435] T. Oka, "Laser Spectroscopy" (Eds H. G. Brewer and A. Mooradian), pp. 413–431, Plenum, New York (1974).

[436] K. Shimoda and T. Shimizu, "Non-Linear Spectroscopy of Molecules", Pergamon, Oxford (1972).

[437] J. Lemaire, J. Thibault, F. Herlemont and J. Houriez, *Mol. Phys.* **27**, 611 (1974).

[438] V. P. Chebotayev, Multiple coherent interaction in optically separated fields, *in* "Laser Spectroscopy IV" (Eds H. Walther and K. W. Rothe), pp. 106-119, Springer-Verlag, Berlin, Heidelberg and New York (1979); see also the article by J. C. Berquist, R. L. Berger and D. J. Glaze in the same volume, pp. 120–129, and also J. C. Berquist, S. A. Lee and J. L. Hall, *Phys. Rev. Lett.* **38**, 159 (1972).

[439] S. A. Lee, J. Helmcke and J. L. Hall, High-resolution two-photon spectroscopy of Rb Rydberg levels, *in* "Laser Spectroscopy IV" (Eds H. Walther and K. W. Rothe), pp. 130–141, Springer-Verlag, Berlin, Heidelberg and New York (1979); see also article by J. L. Hall in *Science*, October (1978).

[440] J. L. Hall, Bureau Int. Poids Mesures, CCDM 78–9, (1978).

[441] P. Glorieux, J. Legrand, B. Macke and J. Messelyn, *J. Quant. Spectrosc. Radiat. Transfer* **12**, 731 (1972).

[442] C. Borde, *C. R Acad. Sci. (Paris)* **282B**, 341 (1976); Ye V. Baklanov and B. Ya Dubetskii, *Sov. J. Quant. Electr.* **8**, 51 (1978).

[443] J. L. Hall, C. J. Borde and K. Uehara, *Phys. Rev. Lett.* **37**, 1339 (1976).

[444] R. E. Drullinger and D. J. Wineland, Laser cooling of ions bound in a Penning trap, *in* "Laser Spectroscopy IV" (Eds H Walther and K. W. Rothe), Springer-Verlag, Berlin, Heidelberg and New York (1979).

[445] D. J. Wineland, R. E. Drullinger and F. L. Walls, *Phys. Rev. Lett.* **40**, 1639 (1978); W. Neuhauser, M. Hohenstatt, P. Toschek and H. Dehmett, *Phys. Rev. Lett.* **41**, 233 (1978).

[445a] J. A. Stratton, "Electromagnetic Theory", McGraw Hill, New York (1941).